PUBLICATIONS OF THE ISRAEL ACADEMY
OF SCIENCES AND HUMANITIES

SECTION OF SCIENCES

THE GENUS TRIFOLIUM

Trifolium resupinatum L. – Israel, Sharon Plain near Binyamina
(Photo Dr. D. Darom)

Trifolium stellatum L. – Israel, Judean Mts. near Har Gillo
(Photo Dr. D. Darom)

THE GENUS TRIFOLIUM

BY

M. ZOHARY and D. HELLER

JERUSALEM 1984
THE ISRAEL ACADEMY OF SCIENCES AND HUMANITIES

ISBN 965-208-056-X

Printed in Israel
Typesetting and makeup by S.T.I., Jerusalem
Plates by Printiv, Jerusalem
Printed at Ahva Printing Press, Jerusalem

PREFACE

THE IDEA OF PREPARING a revision of the genus *Trifolium* did not come to the authors as a sudden inspiration. The task of tackling 240 species which so closely resemble one another is a serious undertaking. Things developed slowly, starting from the revision of 47 local species of cloves for the *Flora Palaestina* by the senior author, who later expanded it into a revision of the 100 species included in the *Flora of Turkey*

During the treatment of the above species, three basic traits were revealed. Firstly, the bulk of the species are rather smooth taxa, well differentiated, not very variable and with very few hybrids between them. Secondly, the 8–10 sections into which the genus has been subdivided by various authors, sometimes under different names and assigned to various taxonomic levels (genera or subgenera), are fairly discrete and tenable. Thirdly, Section *Lotoidea* is an assemblage of primitive species or species groups which may have served as a source taxon for the evolution of the other sections. These points have made the subject more meaningful and attractive. One of the side-issues was to establish which particular groups within the *Lotoidea* should be accredited with the ancestry of the other sections.

Another goal, though only secondary to the main descriptive part, was the tracing of the carpological and carpobiological evolution in the various sections of the genus. Bearing in mind that *Trifolium* is carpologically the most extremely reduced but highly elaborated genus in the tribe *Trifolieae*, it was indeed very interesting to follow the trend of retrogressive evolution here.

These are some of the points which triggered the present authors to accomplish this revision in full.

The main aim, however, was to offer to all those individuals and bodies interested in this genus a reference book providing full descriptions and illustrations of all the clover species. As a matter of fact, it is surprising how few species of this huge genus have actually been cultivated to date. It is hoped that this monograph will help both the taxonomist and the breeder to become better acquainted with this useful genus and make more use of it.

Emotionally, it was a pleasure to work with such handsome and often beautiful plants, most of which do not lose their looks even when dry and mounted on herbarium sheets. In the dried state all of them also retain all the parts necessary for taxonomic examination.

Nevertheless, the present work serves only as a basis for further investigation of the genus. Much remains to be done in elucidating its morphology, ecology and cytogenetics. Unfortunately, even the question of the origin of the genus has not been answered definitively in this monograph. Its Northwest American centre of origin, as assumed here, has yet to be documented. One is amazed by the fact that there is a rich centre of clovers in the Northwestern Pacific while the Eastern Pacific is virtually devoid of clovers. This and a few other phytogeographic problems are still open to speculation.

With the hope that this revision will serve everyone who is in need of it, we can only express our gratitude to all the bodies and individuals who placed the thousands of specimens at our disposal, and also to those bodies that have burdened us with the identification of unnamed specimens. All of them share a part in this work.

Regretfully, the senior author of this volume, Prof. Michael Zohary, passed away on April 15, 1983. He had read the proofs and approved them for printing.

ACKNOWLEDGEMENTS

The authors are greatly indebted to the directors and curators of the following herbaria for putting their collections at our disposal by sending us material on loan, including type specimens (the abbreviations used for herbaria in this revision are those standardized in the *Index Herbariorum 1974*, and listed later on): AAR, ANK, B, BAB, BM, COI, DAO, E, EA, FI, G, GRA, H, HUJ, ILL, K, L, LE, LINN, M, MEXU, MPU, NA, NY, OSC, P, POM, PRE, R, RB, RSA, S, SGO, SI, SP, SRGH, TUR, UC, US, UT, W, Z.

We are especially obliged to the directors and keepers of the following herbaria, who enabled us to study at their institutions and libraries : B, BAB, BM, E, FI, G, HUJ, K, L, M, MEXU, P, SGO, SP, Z.

The senior author wishes to thank fullheartedly the authorities of the United States Department of Agriculture not only for the financial aid granted by them but also for the sponsorship and interest in the accomplishment of this work.

We are deeply grateful to Dr Irene Gruenberg-Fertig for her critical reading of the text and valuable advice, and to Mrs Stephanie Grizi for typing the manuscript in its various stages.

The plates were drawn by Mrs Esther Huber, Mrs Katty Torn, Mrs Dulic Amsler and Ms Heather Wood. Compliments and appreciation are due to these artists for their skillful work.

The authors wish to thank Mr. R. Amoils for editing and preparing the manuscript.

The junior author wishes to express his sincere thanks to Prof. Chaia C. Heyn for her help and devoted supervision in preparing Section *Lotoidea* which was submitted by him, as a Ph. D. thesis, to the Senate of the Hebrew University.

And, finally, we are grateful to the Israel Academy of Sciences and Humanities and to its editor Mr. S. Reem for the effort invested in having this work published.

M.Z. and D.H.

LIST OF HERBARIUM ABBREVIATIONS

AAR	Herbarium Aaron Aaronsohn, Zikhron-Ya'aqov, Israel.
ANK	Ankara Üniversitesi, Fen Fakültasi, Botanik Kürsüsü, Ankara, Turkey.
ARIZ	The University of Arizona Herbarium, Tucson, Arizona, U. S. A.
B	Botanisches Museum Berlin-Dahlem, Federal Republic of Germany.
BAB	Unidad Botánica Agrícola del I.N.T.A., Castelar, Prov. de Buenos Aires, Argentina.
BM	British Museum (Natural History), London, Great Britain.
BR	Jardin Botanique National de Belgique, Meise, Belgium.
C	Botanical Museum and Herbarium, Copenhagen K., Denmark.
CAN	National Herbarium of Canada, Vascular Plant Section, Ottawa, Ontario, Canada.
COI	Botanical Institute of the University of Coimbra, Portugal.
DAO	Vascular Plant Herbarium, Biosystematics Research Institute, Research Branch, Department of Agriculture, Ottawa, Canada.
DS	Dudley Herbarium, Department of Biological Sciences, Stanford University, California, U. S. A.
E	Royal Botanic Garden, Edinburgh, Great Britain.
EA	The East African Herbarium, Nairobi, Kenya.
ETH	Haile Selassie University, Addis Ababa, Ethiopia.
F	John G. Searl Herbarium, Field Museum of Natural History, Chicago, Illinois, U. S. A.
FI	Herbarium Universitatis Florentinae, Istituto Botanico, Florence, Italy.
FR	Forschungsinstitut und Naturmuseum Senckenberg, Frankfort o/M, Federal Republic of Germany.
G	Conservatoire et Jardin Botaniques, Geneva, Switzerland.
GE	Istituto ed Orto Botanico "Hanbury" dell'Universita, Genoa, Italy.
GH	Gray Herbarium of Harvard University, Cambridge, Massachusetts, U. S. A.
GJO	Botan.-Abt. des Steirrm. Landesmuseums Joanneum, Graz, Austria.
GRA	Herbarium of the Albany Museum, Grahamstown, South Africa.
H	Botanical Museum, University of Helsinki, Finland.
HUJ	Department of Botany, Hebrew University of Jerusalem, Israel.
IDS	Ray J. Davis Herbarium, Idaho State University, Pocatello, Idaho, U. S. A.
ILL	Herbarium of the Department of Botany, University of Illinois, Urbana, Illinois, U. S. A.
JEPS	Jepson Herbarium and Library, Department of Botany, University of California, Berkeley, U. S. A.
K	Herbarium and Library, Royal Botanic Gardens, Kew, Great Britain.
L	Rijksherbarium, Leiden, Netherlands.
LE	Herbarium of the Department of Higher Plants, V. L. Komarov Botanical Institute of the Academy of Sciences of the U. S. S. R., Leningrad, U. S. S. R.
LINN	The Linnean Society of London, Great Britain.
LISU	Museu Laboratorio e Jardim Botânico, Faculty of Sciences, Lisboa, Portugal.
M	Botanische Staatssammlung, Munich, Federal Republic of Germany.
MEXU	Instituto de Biologia, Universidad Nacional Autónoma de Mexico, Mexico.
MO	Herbarium of Missouri Botanical Garden, Saint Louis, Missouri, U. S. A.
MONT	Herbarium, Department of Biology, Montana State University, Bozeman, Montana, U. S. A.
MPU	Institut de Botanique, Université de Montpellier, France.
MVFA	Laboratorio de Botánica, Facultad de Agronomia, Montevideo, Uruguay.

NA	Herbarium, United States National Arboretum, Washington D. C., U. S. A.
NAP	Istituto Botanico della Università di Napoli, Italy.
ND	The J. A. Nieuwland Herbarium, University of Notre Dame, Indiana, U. S. A.
NDG	The E. L. Greene Herbarium, University of Notre Dame, Indiana, U. S. A.
NY	Herbarium, The New York Botanical Garden, Bronx, New York, U. S. A.
ORE	Herbarium of the Museum of Natural History, University of Oregon, Eugene, Oregon, U. S. A.
OSC	Herbarium, Department of Botany and Plant Pathology, Oregon State University, Corvallis, Oregon, U. S. A.
P	Museum National d'Histoire Naturelle, Laboratoire de Phanérogamie, Paris, France.
PH	Department of Botany, Academy of Natural Sciences, Philadelphia, Pennsylvania, U. S. A.
PI	Istituto Botanico dell'universita, Pisa, Italy.
POM	Herbarium of Pomona College, Department of Botany, Claremont, California, U. S. A.
PR	Botanické odděleni Přirodoved muzea Národniho muzea v Praze, Prague, Czechoslovakia.
PRE	Botanical Research Institute, National Herbarium Pretoria, South Africa.
R	Divisão de Botânica do Museu Nacional, Rio de Janeiro, Brazil.
RB	Jardim Botânico do Rio de Janeiro, Brazil.
RM	Rocky Mountain Herbarium, University of Wyoming, Laramie, Wyoming, U. S. A.
RSA	Rancho Santa Ana Botanic Garden, Claremont, California, U. S. A.
S	Section for Botany, Swedish Museum of Natural History (Naturhistoriska Riksmuseem), Stockholm, Sweden.
SGO	Museo Nacional de Historia Natural, Santiago, Chile.
SI	Instituto de Botánica Darwinion, San Isidro, Prov. Buenos Aires, Argentina.
SIU	Southern Illinois University Herbarium, Department of Botany, Carbondale, Illinois, U. S. A.
SP	Instituto de Botânica, São Paulo, Brazil.
SRGH	National Herbarium, Department of Research and Specialist Services, Salisbury, Rhodesia.
TUR	Herbarium of the Department of Botany, University of Turku, Finland.
UC	Herbarium of the University of California, Berkeley, California, U. S. A.
UNM	Herbarium, University of New Mexico, Albuquerque, New Mexico, U. S. A.
UPS	The Herbarium, Institute of Systematic Botany, University of Uppsala, Sweden.
US	U. S. National Herbarium, Department of Botany, Smithsonian Institution, Washington D. C., U. S. A.
UT	Garrett Herbarium University of Utah, Salt Lake City, Utah, U. S. A.
UTC	Intermountain Herbarium, Botany Department, Utah State University, Utah, U. S. A.
W	Naturhistorisches Museum, Vienna, Austria.
WAG	Laboratory for Plant Taxonomy and Plant Geography, Wageningen, Netherlands.
WS	Washington State University Herbarium, Department of Botany, Pullman, Washington, U. S. A.
WTU	Herbarium, Department of Botany, University of Washington, Seattle, Washington, U. S. A.
Z	Botanischer Garten und Institut für systematische Botanik der Universität Zürich, Switzerland.

CONTENTS

THE GENUS TRIFOLIUM

GENERAL PART

Introduction

The genus *Trifolium* is one of the most important genera of the Fabaceae family, both in its agricultural value and in the number of its species which amounts to 237. It is closely related to the genera *Trigonella*, *Medicago* and *Melilotus*. Its range of distribution extends throughout the temperate and subtropical regions of the globe. It also occurs, though not very abundantly, in the tropics of W. Africa and S. America, where it is mainly restricted to the montane and alpine zones.

One of the main centres of distribution is the Mediterranean region and the countries adjacent to it. Here, no less than 110 species are encountered, representing seven of the eight sections. Another centre of distribution is the Californian region and its adjacent areas. Though harbouring a smaller number of species, it is considered a primary centre of speciation of the genus.

The Mediterranean region was obviously also the centre of domestication and breeding of most of the cultivated species or cultivars of the clovers. The earliest available data on cultivation date from the beginning of the Christian era. The Hebrew name for clover, "tiltan", is already mentioned in the Mishna, but it is not certain whether this name referred exclusively to *Trifolium*. In any event, the cultivation of the red clover (*T. pratense*) is known in Europe since the 4th century A.D. This and a few other species of clover were bred mainly for cold resistance. It is amazing that out of such a large number of species so few [about 15, Hermann, 1953] have been subject to cultivation on a commercial scale. In addition to serving as forage, clovers are widely used in bee-keeping and rarely also as pot herbs and medicinal plants (e.g., *T. pratense* and *T. repens*).

While in tropical Africa and America and partly also in the northern hemisphere (N. America and Europe), *Trifolium* is confined to alpine meadows and similar natural habitats and many of the species occur as rhizomatous perennials, in the Mediterranean countries a number of species grow also in secondary habitats and very few as weedy annuals.

The Mediterranean country richest in clover species is Turkey (over 100 sp.). The poorest is Egypt. The only species which has penetrated the margin of the Sahara is a variety of *T. tomentosum*, notwithstanding the fact that several species are quite aggressive and occupative. Since the phytogeographical relations of the genus are closely linked with its evolution, a special section will be devoted to this subject.

On the Taxonomic History of the Genus

The genus *Trifolium* was well known to the classical naturalists, e.g., Theophrastus, who included the clover within *Lotus*, Dioscorides, who referred his *Trifolium* to *Psoralea bituminosa*, and Plinius, who mentioned it several times in his accounts.
Mediaeval herbalists mentioned *Trifolium* chiefly as a forage plant, as did L. Fuchs (*New Kreüterbuch*, 1543), who described and illustrated seven species of *Trifolium*, of which only four are true clovers. W. Turner (*The names of the herbs*, 1548) lists five species under Trifoly, of which only one is a clover; the others are *Anagyris*, *Medicago*, *Astragalus* and even *Oxalis*. In J. Gerard's *The Herball* (1597), 21 species of *Trifolium* are treated, but only 10 belong to this genus. C. Clusius (*Rariorum plantarum historia*, 1601) does not record more than seven species, while C. Bauhin (*Prodromus*, 1620) mentions 24 species, of which only nine belong to the genus. In his second book (*Pinax theatri botanici*, 1623), Bauhin groups the species of *Trifolium* into 11 units. Of the 55 species known to him, 22 belonging to six groups can be identified as clovers, while the others are various plants of other genera or even of other families, such as *Anemone hepatica*. Almost the same conception of *Trifolium* is encountered in the work of J. Bauhin & J. H. Cherler (*Historia plantarum universalis*, 1650–51). Although they clearly distinguished *Medicago* from *Trifolium*, they still included species of *Lotus* and *Melilotus* within the latter. Only 23 of the 47 species described and illustrated by them belong to *Trifolium*. R. Morison (*Plantarum historiae oxoniensis*, 1680) also delineated the genus and listed 45 species, but only 29 are clovers. Although A. Q. Rivinus (*Ordo plantarum*, 1690–99) and L. Plukent (*Almagestum botanicum*, 1696) misidentified many of their clovers, almost all of the 44 species recorded by J. P. Tournefort (*Institutiones rei herbariae*, 1700) are true clovers. In P. A. Micheli's book (*Nova plantarum genera*, 1729), which was greatly admired by Linnaeus, the clovers are cited under the name *Trifoliastrum* and grouped in seven orders. The orders used by Micheli closely resemble the presently accepted sections of the genus. J. F. Gronovius (*Flora virginica*, 1739), recorded 10 species of *Trifolium* from Virginia, but only one of them is indigenous, all the others having been introduced. A. van Royen & A. von Haller, cited by Linnaeus, also grouped the species of *Trifolium*, but less appropriately.
One of the problems encountered in the above-mentioned and following treatises is the separation of *Melilotus* from *Trifolium* proper. Even Linnaeus himself (*Species Plantarum*, 1753), who was already acquainted with 40 species, included *Melilotus* as one of his five *Trifolium* groups under *Trifolium-Melilotus*.
After Linnaeus, study of the genus advanced fairly rapidly, both in regard to critical corroboration of the genus whithin local floras and in the preparation of monographs and revisions.
Only a few of the taxonomic treatments in local Floras will be mentioned here. Thus, for instance, Savi (1808–10) was among the first to revise the Italian clovers, dividing the genus into two groups – bracteate and ebracteate. This distinction, which is essential for the sectional subdivision, was adopted by Lojacono (1883b), Taubert (1896), Bobrov (1947), Vicioso (1952–3), and others.

Boissier (1872) grouped the 113 species in his *Flora Orientalis* into fairly clear-cut sections, corresponding almost entirely to those of the present day. He described all of them and added some 15 species new to science.

Gibelli & Belli (1887–1893) monographed the Italian species of clovers very critically. Their delimitation of the sections and species and division of the sections are admirable and testify to their profound knowledge of the species. The dichotomic keys and illustrations provided by these authors are also praiseworthy.

Ascherson & Graebner (1907–8) followed Gibelli & Belli in most of the species they treated, but contributed much to the knowledge of the infraspecific taxa of many species. However, this meticulous treatment of the lower units, often based on literature data, has made it difficult to gain a general view of the species of this genus.

Bobrov (1947), who has reviewed the clovers of the Soviet Union, also provides an excellent historical review and a fruitful discussion on the taxonomic treatment of the genus by various authors. His critical approach to the origin and phyletic position of the various sections constitutes a considerable advance in the study of this genus.

The revision of Gillett (1952) and his later notes (1970) have added greatly to our knowledge of the tropical clovers of Africa, south of the Sahara.

Vicioso (1952–3), in his revisions of the Spanish clovers, has provided solutions to some intricate taxonomic problems relating to the genus.

Hossain (1961), for the first time since Boissier, revised the clovers of the Near East and supplied new data on the distribution of many species. He also described some new species and varieties. Despite his critical approach and exactitude in presenting most of the species, his treatise is faulted by omissions and misinterpretations, especially concerning species already recorded by Boissier.

Only few works covering the New World will be mentioned: Lojacono (1883a) made a valuable contribution by revising the 54 species of American clovers known to him; he also ranked them in sections and described seven new species.

For many years, McDermott's (1910) key to the species was a most useful tool for determining N. American species.

The very comprehensive doctoral thesis of Martin (1943) offers an excellent revision of the N. American species, but has never been published.

Quite recently, J. M. Gillett (1965–1980) devoted a series of studies to the N. American clovers, using modern taxonomic methods. His contributions to the knowledge of New World clovers are enormous.

There are very few monographic revisions or summaries dealing with the entire genus, though the first steps in this direction were already taken by Linnaeus (1753).

Seringe (1825) greatly advanced the taxonomy of *Trifolium*. His revision embraced some 150 species, including several American ones, reasonably divided and accepted by many botanists. However, some of the species of the various sections were later transferred to others and even to other genera.

Presl (1832) attempted to divide the genus into nine genera, but virtually no one (except Bobrov, 1967) has followed him in this regard, though his genera, with few exceptions, were later rightly accepted as sections.

Koch (1835), who was probably unaware of Presl's work, divided the genus into sections corresponding to Presl's genera. Though the names of his sections were not

accepted, he was correct in assigning sectional rank to the groups; this approach has been followed by almost all botanists up to the present day.
Čelakovsky (1874) considerably extended our knowledge of the genus by his critical approach to the delimitation of the sections, and by attempting to elucidate the taxonomic relations of the 11 sections known in his time. Although retaining most of the sections within their previous limits, he proposed some transfers and tried, rather unsuccessfully, to outline the phyletic relations of the genus and its sections.
Lojacono (1883b) provided a key to the identification of the 211 then known species of clovers. He divided the genus into two subgenera, *Trifoliastrum* and *Lagopus*, which were also accepted by certain other botanists. These subgenera are, however, very unnatural and include remotely related sections. One needs only mention the hyatic differences between Section *Trifolium* and Section *Trichocephalum*, both of which were included within his subgenus *Lagopus*. The same holds true for the other subgenus, which is even more artificial, encompassing such different sections as *Lupinaster* and *Vesicaria*.
Bobrov (1967) advanced a theory according to which the genus *Trifolium* should not only be split up into 11 genera (including one new genus), but that some of these genera should also be excluded from the tribe *Trifolieae* and transferred to a new tribe, *Lupineae*. This view is, in our opinion, not adequately reasoned.
Individual sections have been treated by various authors : Section *Trichocephalum* by Katznelson & Morley (1965) and Katznelson (1965, 1974); Section *Vesicaria* by Zohary & Heller (1970); Section *Trifolium* by Zohary (1971–72); Section *Lotoidea* by Heller (1978); parts of Sections *Lotoidea* and *Involucrarium* by J. M. Gillett (1965, 1969, 1971, 1972, 1976, 1980), who treated single groups or individual species.
In summarizing the above overview, it is most surprising that a key genus of such an outstanding family as the Fabaceae has not as yet been the subject of a complete and comprehensive taxonomic monograph, notwithstanding, the considerable efforts invested in its study over the centuries. It was perhaps the fortunate opportunity offered to one of the authors to become closely acquainted with over a hundred species of clovers in the field that led us to undertake the immense task of preparing a monograph of this sort.

The Position of Trifolium *within the Tribes of the Fabaceae*

Since early times the pea family has been variously divided into tribes. In the present treatise we follow the classification of Melchior (1964), who includes the following genera within the tribe *Trifolieae* : *Parochetus*, *Trigonella*, *Medicago*, *Melilotus* and *Trifolium*. The genus *Factorovskya* was obviously omitted by him because he had included it within *Trigonella*. The exclusion of *Ononis* from this tribe is, in our opinion, fully justified. The following is a brief description of the tribe :
Annuals or perennials, rarely dwarf shrubs. Leaves stipulate, trifoliate (with the exception of few *Trifolium* species which are 5–9-foliolate); nerves of leaflets

generally reaching teeth of margin. Inflorescences usually arranged in (1-) few- or many-flowered axillary heads or spikes. Petals free or sometimes adnate to staminal tube. Stamens 10, diadelphous. Ovary 1- to many-ovuled. Pod included in calyx or exserted, variously shaped, mostly indehiscent, sometimes dehiscent by 2 valves. Seeds 1 or few or many.

The delineation of this tribe is largely disputed in the literature and has a long history. Bronn (*Diss. Legum.* 132, 1822) was the first to describe it, including within it also the genera *Ononis* and *Hymenocarpus*.

Seringe (1825) considered *Trifolium* as a member of his subtribe *Loteae*, which also includes *Lotus*, *Pocockia*, *Dorycnium*, *Tetragonolobus* and *Cyanopsis*. Endlicher (1841) followed Seringe, but added to the subtribe two more genera : *Hosackia* and *Melinospermum*. Lindley (1846) went so far as to divide the tribe *Loteae* into eight subtribes, one of which accommodates *Trifolieae* together with several other genera.

Bentham & Hooker (1865) were perhaps the first to delimit the tribe *Trifolieae* from the other tribes of the family; they included in it the genus *Ononis*, a view prevailing in the literature for a long time.

Schulz (1901) divided this tribe into two subtribes : the *Trifolieae* comprising *Ononis*, *Parochetus* and *Trifolium*; and the *Trigonelleae* with *Trigonella*, *Melilotus* and *Medicago*.

Ascherson & Graebner (1907–8) also divided the tribe into two subtribes : the *Ononidinae* with the stamens all united, comprising only *Ononis*; and the *Trifoliinae* with diadelphous stamens, comprising the genera *Trifolium*, *Melilotus*, *Medicago* and *Trigonella*.

Wettstein (1935) raised the above two subtribes to the rank of tribes. He was followed by Hutchinson (1964), who included *Ononis* and *Passaea* within the tribe *Ononideae*, while the rest of the genera were attributed to the tribe *Trifolieae*. In all the above groupings, even the earliest ones, the genus *Trifolium* was regarded as the most advanced one.

Bobrov (1967) adopted a very extreme attitude towards the division of *Trifolium*. He considered the 11 sections of the genus as independent genera which he grouped into two tribes. The genera *Trifolium* (Section *Trifolium*) and *Calycomorphum* (Section *Trichocephalum*), together with the genera *Ononis*, *Parochetus*, *Medicago* and *Trigonella*, were included in the tribe *Trifolieae*. His other nine genera, together with the genus *Lupinus*, were grouped in his new tribe *Lupineae*.

Conceptually, this procedure is fairly attractive and undoubtedly merits much reflection. However, in view of the lack of morphological argumentation, it is unlikely to be accepted by traditional botanists.

The Position of Trifolium *within the Tribe* Trifolieae

The tribe *Trifolieae* seems to be a natural unit. Its five genera do not constitute a phylletic group, though there are a few morphological links between some genera, e.g., between *Medicago* and *Trigonella*. The markers characterizing the genera of this tribe are as follows :

1. *Parochetus* Buch.-Ham. ex Don – a monotypic genus occurring in the mountains of tropical Africa and tropical Asia. Perennial, creeping herb, rooting at nodes. Flowers blue-purple, axillary, solitary or 2–3 borne on common peduncle. Bracts similar to stipules at base of pedicels. Pods many-seeded, linear, dehiscent.
Parochetus is distinguished from the other five genera of this tribe by its caducous petals, digitate leaves, etc.
2. *Trigonella* L. – comprises some 135 species, distributed mainly in the Mediterranean and Irano-Turanian regions; a few occur also in the Saharo-Arabian and S. African regions. Annuals or perennial herbs. Flowers variously coloured, solitary or capitate. Bracts inconspicuous. Pods many-seeded, linear, dehiscent or indehiscent.
3. *Factorovskya* Eig – one E. Mediterranean annual. Stems creeping. Flowers yellow, cleistogamous, geocarpous, penetrating into the ground by means of elongating peduncles and gynophores. Fruit subterranean, ovate to globular, densely hairy, 1–2-seeded, indehiscent.
4. *Medicago* L. – about 100 species, mostly Mediterranean and Irano-Turanian, some Euro-Siberian and N. African. Annuals or perennials, rarely shrubs. Flowers small, yellow or violet, solitary or in few-flowered racemes or heads. Bracts minute or 0. Pods mostly coiled, many-seeded, indehiscent.
5. *Melilotus* Mill. – about 25 species, mostly Mediterranean and Irano-Turanian, a few Euro-Siberian. Annuals or perennials. Flowers small, yellow or white, arranged in spike-like racemes. Bracts inconspicuous or 0. Pods globular or ovoid, as long as or somewhat longer than calyx but not enclosed within it, indehiscent. Seeds solitary or few.
6. *Trifolium* L. – about 240 species, mainly Mediterranean, also Irano-Turanian, Afro-Alpine and Neotropic-Alpine, some Euro-Siberian and N.W. American. Annuals or perennials. Leaves 3-, rarely 5–9-foliolate. Flowers variously coloured, small to medium-sized, mostly arranged in heads or spike-like racemes, umbels or very rarely solitary. Bracts conspicuous to minute or 0; sometimes inflorescences involucrate by entire, dentate or lobed stipules of upper leaves or by connate lower bracts. Pod oblong or obovoid, usually enclosed within calyx but sometimes longer than calyx, 1–2-, rarely many-seeded, mostly indehiscent.

Three of the genera, *Trigonella*, *Medicago* and *Melilotus*, are phyletically close to one another and were rightly included by Schulz in the tribe *Trigonelleae*. In addition to their morphological similarity, they are also geographically and cytologically akin (Heyn, 1968). However, the limits between them are generally quite distinct. *Factorovskya* is closely related to *Trigonella* and was formerly included in the latter. The most isolated genus is *Trifolium*, although a few of its species (such as *T. ornithopodioides*, *T. multinerve* and *T. schimperi*) have long, exserting, many-seeded, dehiscing pods quite similar to those of *Trigonella* and for this reason were previously included in the latter. Similar pods are also found in a few other species of Subsection *Lotoidea*. They are no doubt the only vestiges pointing to the origin of the pod in the genus *Trifolium*. Although other features of this genus are also far from homogeneous, it can readily be divided into well-delineated sections and series. Generally speaking, it is the most advanced genus not only within its tribe but also in the family as a whole. This is best seen in the trend towards reduction of the pod size and the number of seeds per pod, as well as the loss of the separation tissue

leading to indehiscence, and the exploitation of some floral organs as dispersal accessories. In this direction *Trifolium* has reached its highest evolutionary point.
In all the genera of the tribe the basic number of chromosomes is x = 8. However, in *Medicago* the number is reduced to x = 7, and in *Trifolium* an entire aneuploid series of x = 8, 7, 6, 5 has been encountered, though 8 remains the prevalent number.
Geographically, five of the genera have not extended their distribution areas to the western hemisphere, while *Trifolium* is richly represented in the Old World as well as in the New World. This fact is largely connected with the question of the origin of this genus, which will be discussed in detail further on.

Main Morphological Features of the Genus and Their Organophyletic Trends

The Leaves

The leaves in all the sections of the genus are trifoliolate. Only species of Subsection *Lupinaster* of Section *Lotoidea* have a peculiar digitate leaf consisting of 5–9 leaflets, a feature indicating the origin of *Trifolium* from an extinct ancestral lupinoid stock. Another deviant form of leaf is that of African Subsection *Ochreata*, in which the petiole is adnate to the stipule for its entire length, the internodes thus being sheathed. Such adnation is, to a certain extent, encountered in the upper leaves of many species, but in *T. patulum* of Section *Trifolium* it closely resembles that of the *Ochreata* species.
The leaves are usually alternate; in some species the uppermost leaves appear to be constantly opposite because of the approximation of the uppermost buds, which sometimes gives rise to a false dichotomy. Usually, however, one branch of this fork is undeveloped and the stem terminates in a single pedunculate or sessile head subtended by a pair of opposite leaves.
The lower leaves often differ markedly in shape and size from the upper ones, their leaflets being usually ovate, obovate or broadly elliptical, and always broader than the upper leaflets. The shape of the leaflets, varying but to a limited extent, is generally a reliable feature. The venation of the leaflets is constant for each species; the lateral nerves are more or less parallel, they and their branches reaching the margin.
The configuration of the stipules is also a very reliable feature in certain groups, ranging from narrow and abruptly cuspidate in the upper part to almost leaf-like; they are mostly entire, except in Sections *Involucrarium* and *Paramesus* where they are dentate or even deeply laciniate or lobed.

Ramification

Almost all the species are branched from the base or along the stem so that branching is of minor taxonomic significance. In a few species, however, the stem is constantly scapose due to the crowding of the leaves near its base. The position of the stem – erect, decumbent or creeping – is a most reliable marker.

Indumentum

The hair cover of the plants displays considerable variation, but the nature of the indumentum is definitely constant. Modification may occur in the density of the hair cover, ranging from densely hairy to glabrescent or glabrous. Highly constant is also the direction of the hairs, which may be appressed or patulous, antrorse or retrorse. There are some closely related species which can be distinguished, at a first glance, by the character of the indumentum (especially in Section *Trifolium*).

Inflorescence

The inflorescences are commonly pedunculate and axillary, and less often terminal; sometimes they may be sessile. The trend of evolution from many-flowered heads to few-flowered ones is not marked in the genus; many-flowered heads are standard and few-flowered ones are an exception.
Zohary (1972c) presented examples of the main types of flowering heads taken from Section *Lotoidea* in order to illustrate the plasticity and diversity of forms encountered within this primitive section. However, these types do not exhaust the entire gamut of head polymorphism in *Trifolium*. The sequence of the types of inflorescences recorded does not necessarily present a successive trend in head evolution, but rather one of the possibilities of such a trend. Organophyletically, the heads of all other sections of the genus could readily be derived from this rich source of patterns.
From the types given by Zohary, we can summarize that the racemose inflorescence is the primary state in the genus. The trends here are towards a spicate inflorescence which is met with in the more elaborate sections. The stages of evolution of the clover head are linked with the length of the pedicels and their direction at and after anthesis.

Bracts

Bracteate flowers are found in the more primitive sections of the genus, revealing a primary state of a racemose inflorescence. The two trends manifested in the regressive evolution of the bracts are rudimentation and disappearance of the bracts, or aggregation of the bracts at base of the head, leading to involucration.
There is a trend here towards debracteation of the spike or the raceme, thus separating the latter from the vegetative parts of the plant. This is fully accomplished in Section *Trifolium*. In Section *Lotoidea*, some species have well-developed bracts (Fig. 1A–B), while others have inconspicuous or rudimentary ones. In this section, the successive stages of this trend can be followed, and in the elaborate American Subsection *Neolagopus* one notes the rudimentation or total disappearance of the bracts. Well-developed bracts can be found in *Mistyllus* which is regarded, according to its other features, as one of the most primitive sections (Fig. 1C).
That the bractless state constitutes the most advanced stage in the evolution of the clover head is evident from the fact that the most elaborate sections of *Trifolium* are

either bractless or only rudimentarily bracteate. In some of the ebracteate sections, such as Section *Trifolium*, bracts may occur rarely, as in *T. noricum*. In *T. alexandrinum* and *T. cinctum* bracts form an involucre at the base of the head. This is highly indicative of the evolutionary process leading to the abortion of this organ. Both distinct or connate bracts forming an involucre are encountered in Section *Lotoidea*. While in Series *Macrochlamis* of this section the bracts of the involucre are not fully connate (Fig. 1B), in some species of Subsection *Lupinaster* they are united into a small ring. Similar involucres of bracteal origin are characteristic of certain species of Section *Vesicaria* and Section *Paramesus*. In some cases involucral structures are possibly derived from abortive lower flowers of the head, and could perhaps be considered an ancestral point of departure of Section *Involucrarium* (Fig. 1E–F). The origin of the involucre is not always the same. In Section *Involucrarium* the involucre is sometimes of stipular and not bracteal origin, as evidenced by the identical structure of the involucre and stipules. Of similar origin is the involucre of *T. cherleri* (Fig. 1D), *T. hirtum, T. congestum* and *T. andinum*. While these structures are reliable markers in identification, their functional significance is not understood.

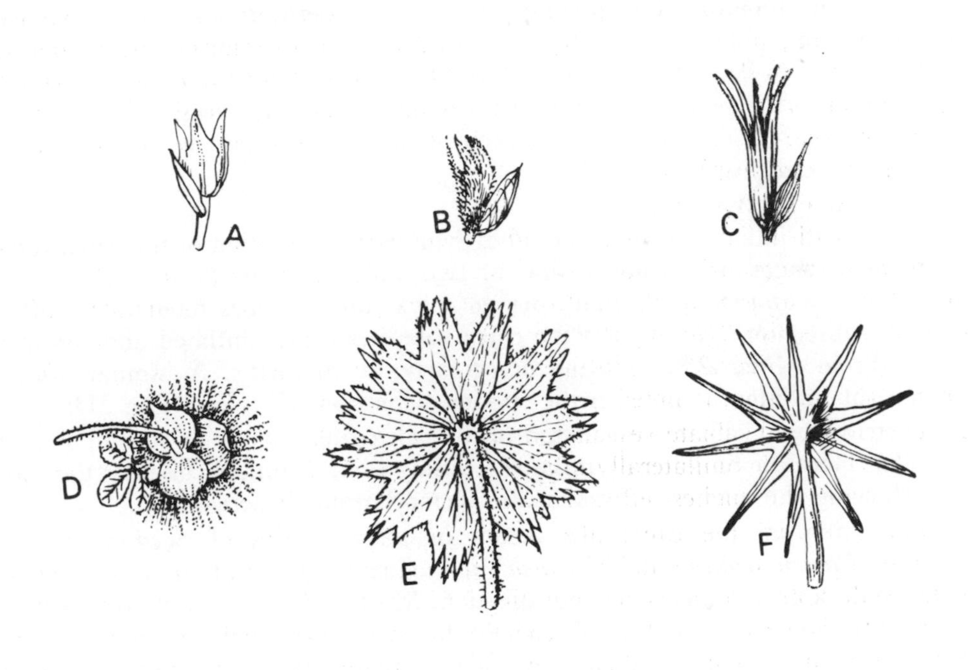

Fig. 1. Bracts and involucres
A–C : Calyx subtended by a bract
A – *T. ambiguum* (× 4); B – *T. riograndense* (× 3); C – *T. setiferum* (× 3)
D–F : Involucres subtending inflorescences
D – *T. cherleri* (natural size); E – *T. microdon* (× 3); F – *T. pinetorum* (× 3)

Pedicel

The presence, nature and size of the pedicel are of diagnostic value. There is a well-marked evolutionary trend from umbellate inflorescences with long pedicellate flowers towards shortening of pedicels, leading to spike-like inflorescences. This trend can be followed in the primitive Section *Lotoidea*, whereas in Subsection *Lotoidea* the pedicels are long and well-developed, reflexing soon after anthesis; in the species of Subsection *Platystylium* the pedicels are short and remain erect or only those of the lower flowers become reflexed (e.g., *T. ambiguum* or *T. montanum*) due to the pressure of the upper flowers of the inflorescences. The length and growth direction of the pedicel have given rise to the numerous intermediate forms between the umbellate and the spicate types of inflorescences. This is especially notable in species of Section *Lotoidea* (Zohary, 1972c). In the advanced Subsection *Neolagopus* (of Section *Lotoidea*), as in other advanced sections of the genus (e.g., Section *Trifolium*), pedicels are totally absent and the flowers of the spicate inflorescences remain erect in fruit. The reflexing of the pedicels after anthesis is associated with seed dispersal, the seeds thus being easily released from the pods and dropping to the ground in close proximity to the mother plant.

Calyx

The prevailing type of calyx in the genus, prior to permutation, has a symmetric, tubular or campanulate form with an open throat and five more or less equal lobes or teeth (Fig. 2A–B). This type has undergone many modifications in the course of the evolution of the genus, both in structure and in function. Originally serving as a protective organ of the flowers, it became a protective organ of the seed and an accessory in seed dispersal.

The three main calyx types are:

1. Calyx with inflated or vesicular tube. Symmetrical inflation of the calyx tube in its incipient stages is already visible in two subsections of Section *Lotoidea*. In Subsection *Loxospermum* the multi-nerved calyx tube becomes moderately inflated, while in Subsection *Calycospatha* the calyx tube is strongly inflated and constricted at the throat (Fig. 2C), splitting irregularly at maturity. A similar form of symmetrical inflation is noted in the species of Section *Mistyllus* (Fig. 2D).
Asymmetrical or bilabiate vesiculation of the calyx tube occurs in Section *Vesicaria* (Fig. 2E). Here the unilaterally inflated fruiting calyx is almost closed at the throat, thus achieving the highest efficacy as an anemochorous dispersal unit.
2. Bilabiation of the calyx also occurs in some species of Sections *Lotoidea*, *Trifolium*, *Chronosemium* and *Vesicaria*. Its incipient manifestation is the inequality of the teeth, both in dimensions and direction. Most striking is the two-lipped calyx in Section *Chronosemium* (Fig. 2F), where the two upper teeth are shorter than the three lower ones. The bilabiate, vesicular, inflated calyx of species of Section *Vesicaria* has already been mentioned; in this case it is the upper (adaxial) side of the calyx that swells after anthesis, the entire calyx becoming a vesicular body crowned with two teeth much longer than the others. In this section bilabiation appears to be more perfect in the annual species than in the perennial ones.
3. Calyx closure. In Section *Trifolium* there is a clear tendency towards closure of

the calyx tube by outgrowth of the throat. There are various morphological devices for closing the throat of the calyx, thus retaining the single seed within the tube. They range from a hairy or callous ring around the throat (Fig. 2G–H) to a bilabiate, callous outgrowth shutting the calyx tube very tightly (Fig. 2I). In some other sections of the genus the monospermous pods are also retained within the calyx, but without special structural devices (e.g., Section *Vesicaria* and Section *Chronosemium*). Koller (1964) inferred that retention of seeds in the calyx is a means of delaying their germination in the field until the calyx becomes thoroughly moistened by an amount of rain sufficient not only to remove the inhibiting substances contained in the calyx and the testa of the seed, but also to ensure further development of the seedlings. In arid and semi-arid regions with sporadic rainfall, this is probably an efficient means of ensuring the survival of the species by avoiding precocious germination. However, the enclosure of the seed within the calyx may also be viewed as a means of dispersal or, in some cases, as a means of seed protection.

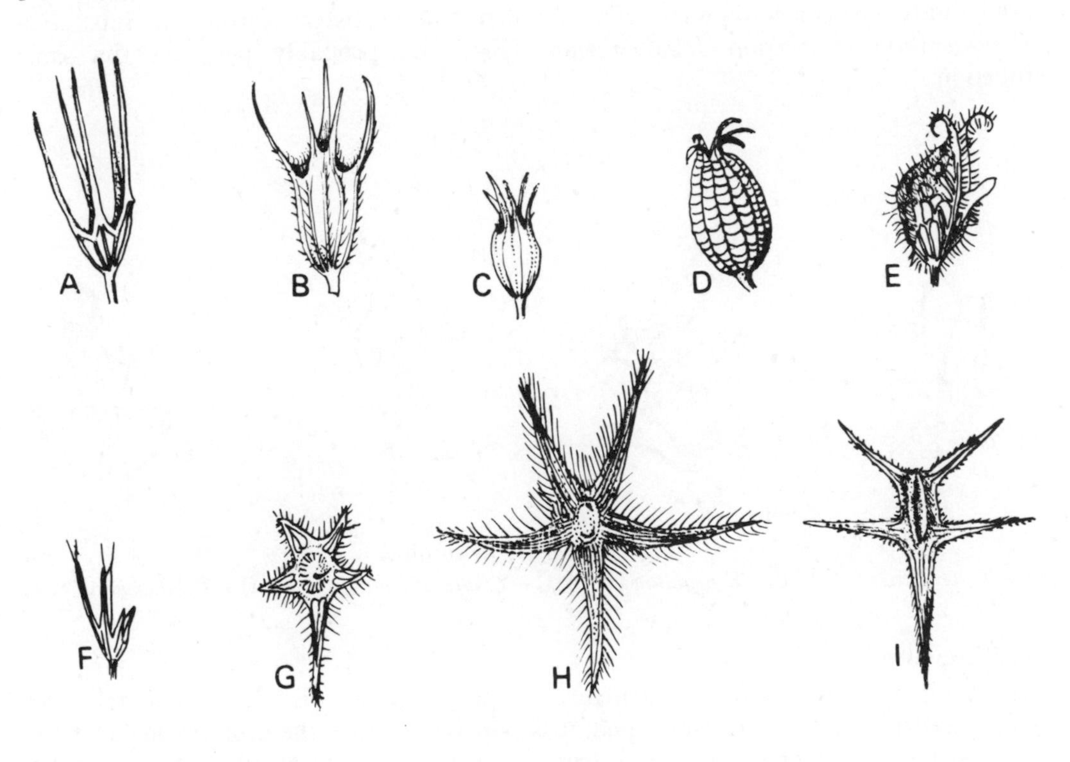

Fig. 2. Calyces
A–C : In flower
A – *T. michelianum* (× 4); B – *T. montanum* ssp. *humboldtianum* (× 4)
C – *T. mattirolianum* (× 4)
D–I : Fruiting calyces
D – *T. spumosum* (× 2.5); E – *T. tomentosum* var. *curvisepalum* (× 4); F – *T. brutium* (× 4)
G – *T. apertum* var. *kilaeum* (× 3); H – *T. stellatum* (× 2); I – *T. canescens* (× 2)

Corolla

The corolla is papilionaceous and is made up of five petals (Fig. 3A); the four lower ones are loosely connate and adnate to the stamens, while the fifth, the standard, is larger and free. The wings are usually much longer than the keel. The small nectary is located at the base of the upper free stamen. The corolla is variously coloured : from pure white to dark purple. Bicoloured corollas occur in many species. Quite often flowers change their colour after anthesis so that inflorescences appear bicoloured.
The corolla is generally caducous, but in some species it is persistent. Early shedding of the corolla is prevalent in Section *Trifolium*. Persistent corollas are characteristic of Sections *Lotoidea*, *Mistyllus*, *Involucrarium* and *Chronosemium*. In Section *Mistyllus* the persistent corolla also becomes scarious (Fig. 3B), and only in Section *Chronosemium* does it serve as an effective anemochorous means of seed dispersal (Fig. 3D). In species of the last-mentioned section, both the boat- or spoon-shaped standard and the spreading wings together with calyx (in which the seed is enclosed) are readily dispersed by wind. The inflated and persistent corolla in Subsection *Physosemium* of Section *Involucrarium* (Fig. 3C) probably performs the same function.

Fig. 3. Persistent corollas
A – *T. somalense* (× 3); B – *T. argutum* (× 5); C – *T. minutissimum* (× 6); D – *T. billardieri* (× 4)

The Legumen

Assuming that the primary structure of the legumen is that of a folicular, many-seeded, suturally dehiscent pod, it is possible to trace the evolutionary trend of the pod within the genus *Trifolium* from its very archaic form, through gradual reduction, up to the most extreme derivative stage. In Section *Lotoidea*, where the most primitive features of the genus have been preserved, we found quite a number of species with typical, many-seeded legumes dehiscent by two sutures at maturity (Fig. 4A–B). From this point of departure, one can follow various stages of quantitative and qualitative reduction, leading to that type representing a one-seeded utricle or nutlet with a very tiny membranous pericarp sometimes consisting of the epidermal layers only. By loss of its separation tissue, the pod is no more dehiscent but splits transversely or irregularly. This extremely reduced type is characteristic of the advanced groups of Section *Trifolium* (e.g., Subsection *Alexandrina*), but is also a terminal stage in other sections, even in Section *Lotoidea*. In the latter 2–9-seeded pods with well-developed sutures protruding from the calyx and dehiscing by two sutures are not rare [e.g., *T. tembense*, *T. ornithopodioides* (Fig. 4A), *T. multinerve*, *T. gracilentum*, *T. bejariense*]. Some species of Sections *Mistyllus* and *Vesicaria* also have two-seeded pods with well-developed sutures. A further stage of reduction is no doubt the one-seeded pod in which both sutures still exist; it is encountered in several species of Section *Lotoidea* [e.g., *T. burchellianum* (Fig. 4D), *T. usambarense* and many others]. This type also occurs in some species of Section *Chronosemium* and even in a few species of Section *Trifolium*. A penultimate stage in the course of reduction is the one-seeded membranous pod that still retains a single suture (found in various sections but rather rare), while the ultimate stage of the trend is the membranous, sutureless, one-seeded pod that ruptures irregularly or transversely (Fig. 4E, this type is prevalent in Section *Trifolium*).

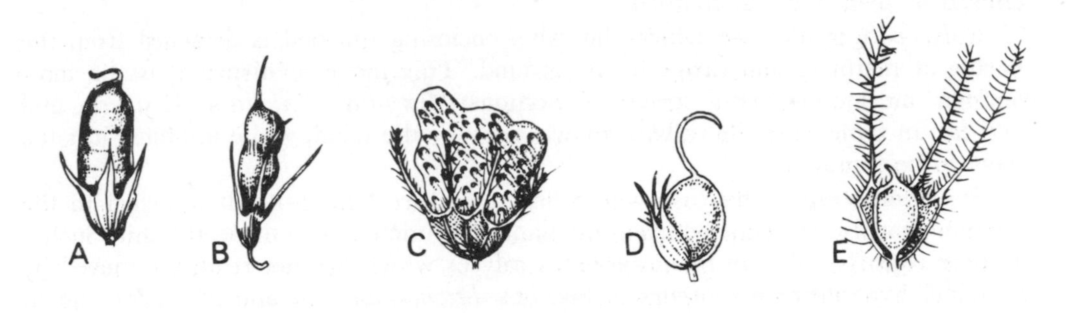

Fig. 4. Fruiting calyces with legumen
A – *T. ornithopodioides* (× 3); B – *T. alpinum* (× 3); C – *T. subterraneum* (× 4); D – *T. burchellianum* (× 3); E – Opened calyx of *T. diffusum* (× 3)

Seed

The seeds of the genus vary in their dimensions, weight and shape. They are usually ovoid, elliptical or oblong-elliptical and laterally compressed.

Some measurements of clover seeds were recorded by Isley (1948), Heinish (1955) and Peinado et al. (1971).

The shape of the seeds varies as a result of the position of the hilar notch which separates the radical lobe from the lobe of the cotyledons. In species belonging to three subsections of Section *Lotoidea* (*Lotoidea*, *Ochreata* and *Oxalioidea*), the notch is terminal; in this case the radical lobe is nearly equal in length to that of the cotyledons (Fig. 11A). In the other species of Section *Lotoidea* and in almost all other sections of the genus, the hilar notch of the seeds has a marginal or intermediate position; the radicle here is shorter than the cotyledons (Fig. 11B).

The surface of the seed coat may be smooth, roughened, tuberculate, wrinkled or pitted. However, our observations with the scanning electron microscope did not reveal any correlation between the morphological structure of the seed coat and the taxonomy of the genus (see p. 47).

Dispersal

The diversity of the means and accessories connected with dispersal in this genus has been treated by Zohary (1972c) from an evolutionary point of view.
Here we wish to exemplify the two main categories of dispersal : topochory and telechory.

(a) *Topochory* (Zohary, 1962; atelechory, Van der Pijl, 1972) is the phenomenon of the seeds or fruit remaining in the vicinity of the mother plant due to structural properties of the dispersal units. Some of the cases of this category may be mentioned here :

1. *Barychory* (Van der Pijl, 1972; baryspermy, Zohary, 1962) is the dispersal of solitary, naked, heavy seeds detached from the protruding polyspermous pod which is suturally dehiscent. Such dispersal is common in species of the subsections of Section *Lotoidea* (*Falcatula*, *Loxospermum* and *Lotoidea*), where the many-seeded pod is also reflexed or bent down at maturity.
2. *Calycoboly* is the case where the calyx enclosing the pod is detached from the rhachis at maturity and drops to the ground. This mode of dispersal is the most common one among many species of Sections *Mistyllus*, *Trifolium*, *Vesicaria* and *Lotoidea*; in some cases the calyx is blown away by the wind, while in others it is too heavy to be removed.
3. *Synaptospermy* is the case where heavy, mature, entire heads attached to the common rhachis are detached from the plant and function as a dispersal unit. Such a diaspore consists of many monospermous calyces which are not readily removed by the wind. Synaptospermy occurs in Sections *Trichocephalum* and *Mistyllus* and in species of Section *Trifolium* (e.g., *T. cherleri* and *T. scutatum*).
4. *Aestatiphory* (Zohary, 1937) is the phenomenon of the calyces being attached to one another and to the rhachis of the head which does not separate from the stem until the latter is broken or decays. In this case the pods remain closed in their calyces (e.g., *T. echinatum* and *T. latinum*).
5. *Basicarpy* is the case where the fruiting heads are congested at the base of the plants and remain pressed to the ground until germination, e.g., in *T. suffocatum*, *T. congestum* and *T. tomentosum* var. *chtonocephalum*, etc.
6. *Geocarpy* is noted in two species of the subterranean clovers (Section *Trichocephalum*). Here, the manv upper (inner) sterile flowers are converted into a drilling apparatus that encloses the few fertile flowers and enables the heads to penetrate into the ground, where they ripen their fruits.

Katznelson & D. Zohary (1970) have shown that there is a tendency towards geocarpy in two other species of the above section (*T. batmanicum* and *T. chlorotrichum*). In these species the heads reach the ground by means of the long, deflexed peduncle, but do not penetrate into it.

(b) *Telechory* is long-distance dispersal by wind and animals. It is the common mode of dispersal in this genus.

1. *Anemochory* is facilitated when the plant produces light calyces with or without accessories. Such diaspores can readily be dispersed by the wind with the aid of light and feathery (e.g., *T. arvense*), winged (*T. campestre*) or vesiculate (*T. resupinatum*) calyces.

Sometimes entire fruiting heads are easily dispersed by the wind due to their anemochorous accessories, such as those found in five species of Section *Trichocephalum* where the wooly or feathery head is made up of many sterile calyces surrounding a few fertile ones. In addition, many globular, entire heads of species of Section *Vesicaria*, such as *T. tomentosum* and *T. bullatum*, comprise a large number of inflated calyces easily borne or moved by the wind.

Other cases of anemochory are those in which mature seed-bearing flowers are detached from the axis, and the persistent corolla serves as a wind-dispersal accessory (e.g., some species of Sections *Lotoidea* and *Mistyllus*), while a more elaborate wind-borne apparatus is found in the species of Section *Chronosemium*. Here, the persistent corolla, especially the standard (sometimes also the wings), is converted into a spoon- or boat-shaped aerodynamic apparatus readily transported by the wind.

2. *Epi- or exo-zoochory* occurs in a few species where single calyces with recurved teeth furnished with spreading hairs or also entire heads provided with recurved calyx teeth give the head a burr-like appearance (e.g., *T. retusum* and *T. scabrum*). The wooly heads of species of Section *Trichocephalum* are also readily carried away by animals.

3. *Amphicarpy* is a phenomenon of the plant producing fruit of two kinds. The aerial fruit, borne on long peduncles, is detached from the rhachis at maturity and spread. The second type of fruit develops from small basal inflorescences produced by flowers borne on the prostrate stolons from which the elongated peduncle bends down, burying the flowers in the ground. The only case of amphicarpy known to us is *T. polymorphum* from S. America and the southern part of N. America.

The Subgeneric Division of the Genus Trifolium

a. *Historical Survey*

The first attempts to divide the genus into natural groups were made by pre-Linnean botanists, though the number of clovers known to them was very small. Classification into clear-cut groups commenced with Linnaeus (1753), who divided his 41 species into the following five units, some of which were later accepted as sections :

1. *Meliloti* – with eight species belonging partly to *Trigonella* and partly to *Melilotus*;
2. *Lotoidea* – with six species, all presently included in Section *Lotoidea*;
3. *Lagopoda* – with 18 species, of which 14 belong to Section *Trifolium*, two to Section *Trichocephalum*, and two to Section *Lotoidea*;

4. *Vesicaria* – with four species, all, with the exception of *T. spumosum*, belonging to Section *Vesicaria*;
5. *Lupulina* – with five species, all of Section *Chronosemium*.

In the second edition of *Species Plantarum* (1762–63), Linnaeus proposed another division into five groups: 1) *Oppositifolia*, 2) *Racemosa*, 3) *Aggregata*, 4) *Capitata*, and 5) *Spicata*. This division was subsequently rejected.

Moench (1794) followed Linneaus' first classification and divided his 17 species of *Trifolium* into four groups.

Savi (1808–10) divided the genus into two groups: 1) *Ebracteata* which is very heterogeneous in comparison with the present-day sectional division; 2) *Bracteata* with several subdivisions that are in closer conformity with the present-day sectional division. This division based on the bract is still partly applied as a diagnostic feature distinguishing Section *Lotoidea* from Section *Trifolium*.

In his comprehensive revision of the genus, Seringe (1825) distinguished seven sections, not all conforming to the present-day division. He regarded the following as differential markers: the form of the head, the structure and form of the calyx, inflation of the calyx and the corolla, and the venation of the leaflets.

The division of the genus into the following nine genera by Presl (1830) constituted a landmark in its treatment: *Paramesus*, *Amarenus*, *Lupinaster*, *Amoria*, *Micranthemum*, *Trifolium*, *Mistyllus*, *Galearia*, and *Calycomorphum*. Virtually all of these groups are retained today as sections, but some of them have been renamed, e.g., *Amarenus* = *Chronosemium*, *Amoria* = *Lotoidea*, *Galearia* = *Vesicaria* and *Calycomorphum* = *Trichocephalum*. Except for the ranking, his division is excellent and his comments are most instructive.

Koch (1835) reduced the number of sections to seven, as did Boissier (1872). These sections (sometimes under different names) were accepted by most botanists and applied as such in various Floras.

Most of the above treatments related to Old World species of *Trifolium*, and little work has been done on the subgeneric division of the American species.

In his treatise on American clovers, Lojacono (1883a) attempted to extend his Section *Lupinaster* by including within it many species of Section *Lotoidea*, thus blurring the limits between the two sections. This has led to some difficulties in the grouping of the American clovers. Lojacono ranked the American species in six additional sections, four of which were described by him for the first time (Sections *Cyathiferum*, *Physosemium*, *Micranthoidea* and *Neolagopus*).

In another paper consisting mainly of a key to the 211 species of *Trifolium* known to him, Lojacono (1883b) divided the genus into two subgenera: *Trifoliastrum* incorporating the bracteate species, 2) *Lagopus* encompassing the bractless species.

The first subgenus is divided into 11 sections, one of which, Section *Ochreata* comprising three African clovers, is described for the first time. The second subgenus is divided into only two sections.

Another approach was adopted by Hossain (1961), who raised all eight sections known to him to the rank of subgenera.

Coombe (1968) divided the European clovers into three subgenera: *Falcatula*, with one species; *Lotoidea*, with seven sections, and *Trifolium*, with two sections.

b. The Taxonomic Position of the Sections and Their Subdivision

The present study has led the authors to the conviction that the genus *Trifolium* is morphologically a natural unit which should neither be split up into independent genera nor be assigned to any tribe other than the *Trifolieae*.
The genus includes 237 species (1,000 or more binomials), and should be divided into eight sections as follows :

1. Section *Lotoidea* – America, Africa, Eurasia
2. Section *Paramesus* – Eurasia
3. Section *Mistyllus* – Eurasia, Africa
4. Section *Vesicaria* – Eurasia
5. Section *Chronosemium* – Eurasia
6. Section *Trifolium* – Eurasia, S. Africa
7. Section *Trichocephalum* – Eurasia
8. Section *Involucrarium* – America

1. Section *Lotoidea* is the largest section of the genus and taxonomically the most difficult. Its heterogeneity led the authors to divide it tentatively into nine subsections and 13 series. This division is tentative because many of the species are not adequately known. Formerly, the group of species equivalent to Section *Lotoidea* was very irrationally divided on the basis of obscure and unreliable features. This applies particularly to American members of this section, which form the large majority. The identification of the species and their grouping into natural units has proved troublesome for many authors, including the present ones. After much hesitation, we decided not to regard ourselves as being bound by the conventional treatment of the group and to reassess the grouping and ranking of the members of this section.
The following are the main traits that characterize this section : the umbellate shape of the inflorescences, the pedicellate and bracteate flowers, and the usually two- to many-seeded legume often protruding from the calyx and its sutural dehiscence. The above features, though not occurring in all the species, have led us to the conclusion that this section should be considered the most primitive group of the genus. It is the heterogeneity that accredits it to feature as the issuing point for the origin of the other sections.
The two main subsections, *Lotoidea* and *Platystylium*, subdivided into a number of series, differ from one another in the nature of the pedicel and its post-floral movements. They include species from both hemispheres. The other subsections are restricted geographically. This applies both to the African subsections (*Loxospermum*, *Ochreata* and *Calycospatha*) and American subsections (*Oxalioidea* and *Neolagopus*), the sole exception being Subsection *Lupinaster* which includes five American and two Eurasian species.

2. Section *Paramesus* with its two species is presently considered as a separate section, but in future it should perhaps be incorporated in Section *Lotoidea* as a special series. The only features distinguishing it from Section *Lotoidea* are the structure of the pod and the occurrence of gland-bearing teeth on the stipules and calyx teeth.

3. Section *Mistyllus* comprises nine species. The unique structure of the symmetrically vesiculate calyx and the persistent corolla, the manifestly bracteolate flowers and the 2–4-seeded pod dehiscing suturally, sharply delimits this section from the others. It is also unique in its distribution: six species appear in the Mediterranean region and three on the Afro-Alpine sites of the tropical Eritreo-Arabian province.
4. Section *Vesicaria*, characterized by the unique morphological and ontogenetical development of the asymmetrical vesicular calyx which serves most efficiently in dispersal, has no parallel within the genus. The occurrence of resupinate flowers here is also quite exclusive.
5. Section *Chronosemium* is a very clear-cut group possessing the following distinctive features : a peculiar bilabiate calyx; a persistent corolla with a spoon- or boat-shaped standard, serving as a very efficient anemochorous device; a small, generally stipitate, one-seeded pod.
6. Section *Trifolium* ranks second in the number of species, after *Lotoidea*. It is heterogeneous in appearance but has several distinctive markers. It appears to comprise a large number (17) of small but natural clusters which were regarded as subsections (Zohary, 1971,1972a,b). The diagnostic features of this section are : flowers sessile and ebracteate; throat of calyx tube usually provided with a hairy or callous ring or entirely closed by a bilabiate protrusion; calyx limb often with unequal teeth; the one-seeded pod enclosed in the calyx tube has a membranous pericarp with no sutures and ruptures transversely at maturity.
7. Section *Trichocephalum* sharply differs from all other sections in the fact that most of the flowers of the head are sterile and converted into a mass of hairs or bristles serving for the dispersal of the seeds produced in the few fertile flowers. Dispersal is here anemochorous, zoochorous or geocarpic. True geocarpy is encountered in only two species; it is achieved by the drilling action of the peduncle and the sterile flowers, which push the fruit of the few fertile flowers into the ground.
8. The exclusively American Section *Involucrarium* is well differentiated from all other sections of the genus by three features : the dentation of the calyx teeth, the sharply dentate or incised stipules, and the presence of an involucre at the base of the head displaying dentation similar to that of the stipules.This section is divided into two subsections; in one of them, Subsection *Physosemium*, the fruiting corolla develops into a vesicular body which appears to be quite effective in seed dispersal.

Evolutionary and Phylogenetic Trends within the Genus

Section *Lotoidea* is not only the largest among the sections of *Trifolium* but also the oldest and most primitive. As will be seen later, it is this section that contains the most ancestral forms of the genus. It also has the widest range of distribution and is probably the only stock from which all other American and Eurasian sections could have been derived directly. Accordingly, one should regard various parts of this section as the points of issue of the evolutionary lines of other sections in this genus. Moreover, this is the only section with apparent links to such genera as *Lupinus* or *Trigonella*. Its wide distribution also suggests its antiquity and its existence at a time when the two hemispheres were connected, thus facilitating the spreading of these plants.

On the assumption that the other sections were derived from certain parts of the *Lotoidea*, we can outline the following scheme, based on morphological and biological evidence :

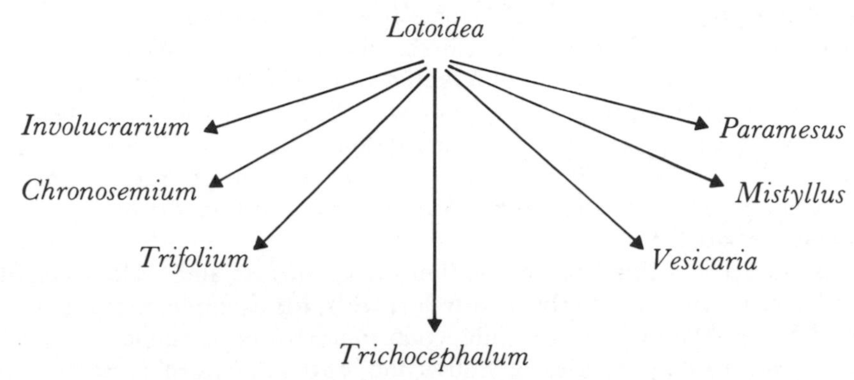

The loose racemose inflorescence, the bracteolate, long-pedicelled and non-elaborate flowers, and the typical leguminous pod can be regarded as primitive markers when compared with the derived floral and carpological features predominant in the other sections of this genus. Even if these features do not appear in their entirety in all species of the *Lotoidea* section, one is justified in viewing this section as the original, not necessarily monophyletic stock from which the other sections of the genus have been derived.

The bracteoles, dispermy and usually regular calyx of *Paramesus* indicate its relationship to Section *Lotoidea*. However, it differs from the latter in a few other markers. The occurrence of an involucre and the dentation of the stipules suggest, at a first glance, some connection with Section *Involucrarium* but this may be considered a parallel and convergent rather than phyletic evolution.

Section *Mistyllus* with its well-developed bracteoles and many-nerved, symmetrically vesicular calyx can be linked with one of the two subsections of Section *Lotoidea* : *Loxospermum* and *Calycospatha*. These subsections as well as three species of the Section *Mistyllus* are homopatric and have the same distribution range in the tropical Eritreo-Arabian province, but the latter differ from the former especially in their many-nerved bracteoles.

The apparent similarity of the calyx structure in Section *Vesicaria* and Subsection *Calycospatha* (of Section *Lotoidea*) may suggest a phyletic link between the two.

Section *Chronosemium* has attained the highest level of organization and functional efficacy of the fruiting and floral organs. This section is, no doubt, a terminal link in the anemochorous trend of seed dispersal, the corolla forming a wind-dispersed body. Only few features of this section are comparable to those of Subsection *Platystylium* (of Section *Lotoidea*), such as the pedicellate and bracteolate flowers or the persistent corolla.

Certain features of Section *Trifolium* may have been derived directly from Section *Lotoidea* via loss of the bracts and pedicels and the elaboration of the fuiting calyx as an efficient dispersal unit. The presence of vestiges of bracts and pedicels in some species, especially the perennial ones, strongly supports this suggestion. Some of the features of Section *Trifolium* appear in the American Subsection *Neolagopus* of Section *Lotoidea*. In Section *Trifolium* seed dispersal is highly diversified and the fruiting calyx usually separates from the rhachis of the head at maturity; in some cases, the fruiting calyx persists on the rhachis and on the plant; and in yet other cases, the fruiting head separates as a whole from the peduncle.

Section *Trichocephalum* should be considered the most advanced one, if only on account of its very elaborate and differentiated flower heads, forming a highly specialized dispersal apparatus. Faint links with Section *Lotoidea* can perhaps be discerned among those species in which sterilization of the upper flowers of the head is a constant feature (Series *Producta*). It is also possible to trace a link to the amphicarpic species of Subsection *Oxalioidea*. However, Section *Trichocephalum* will always remain isolated.

It is not easy to detect a link between Section *Involucrarium* and Section *Lotoidea* in its present form. However, a thorough search reveals, for example, that the involucre in species of Series *Macrochlamis* of Subsection *Platystylium* is similar to that found in Section *Involucrarium*. Species of Subsection *Lupinaster* also possess a similar small involucre consisting of connected bracts.

Plant Geography

a. *Geographical Distribution and Centres of Diversity*

The wide range of distribution of the genus encompasses virtually all temperate regions of both hemispheres, where three disconnected centres of distribution can be discerned (Map 1).

The Eurasian continent harbours 59% of the species of the genus. They are centred in the Mediterranean region, while the Euro-Siberian and the Irano-Turanian regions are very poor in species – some 7%. Almost every country bordering on the Mediterranean possesses some endemic species, the greatest number (nine) being found in Turkey. While the northern distribution limit in Europe passes through Sweden and Norway, a few species are encountered in the eastern part of the USSR, including Soviet Central Asia, and only one (*T. lupinaster*) extends as far as East Asia.

Africa serves as a habitat for 15% of the species, some of which occur in that continent's are Mediterranean. The North African species are Mediterranean ones and only two of them are known to be endemic to the Atlas Mountains. About half of the North African species are also found in the Macaronesian region. Most of the non-Mediterranean species occur in the Eritreo-Arabian province, chiefly in the subalpine and alpine highlands of Ethiopia, Sudan, Eritrea, Kenya, Tanganyika and

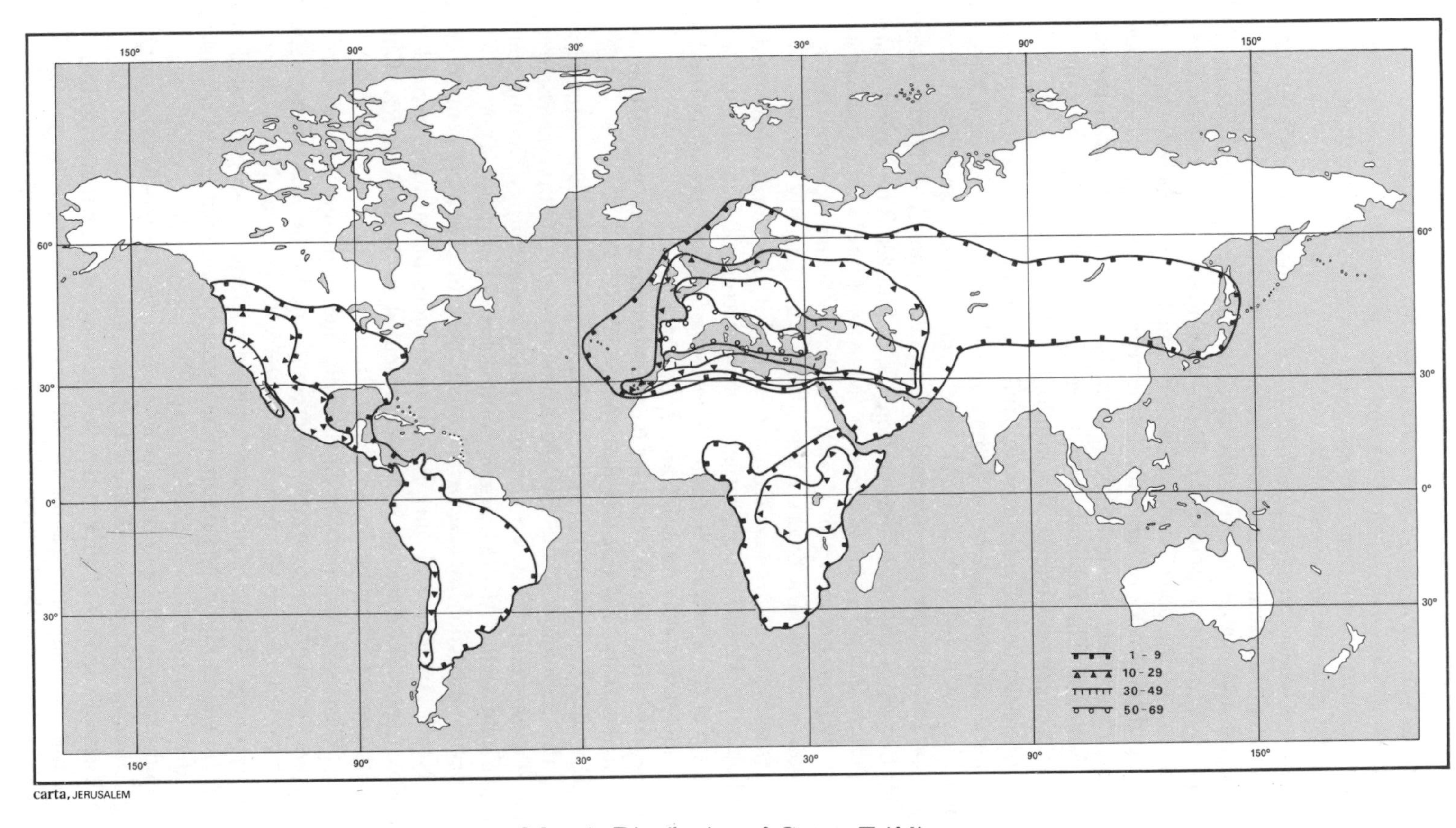

Map 1. Distribution of Genus *Trifolium*

Uganda. Only a few are found in West and Central Africa (Congo, Cameroon, Nigeria). Three species, one of which is endemic, penetrate as far south as the Cape. Ethiopia is richest in endemic species – 10, Tanganyika has three, while Kenya, Uganda and Cameroon have one each.

In the western hemisphere, the American continent harbours 26% of the species of the genus. Their distribution in North America starts in British Columbia and Vancouver Island and extends southwards to California and Central Mexico. From the west coast and the areas around the Rocky Mountains some species have spread to the east as far as the delta of the Missouri and the Appalachian Mountains. The west coast and the Rocky Mountains are the regions richest in species (about 65% of the continent's species), while no more than 8% are encountered in the eastern part of North America. The endemic species are distributed as follows: nine in California, two in Oregon and one in Washington.

Only 13 species occur in South America and their range of distribution extends from southern Peru along the Andes to Bolivia and Chile. A few of them have succeeded in penetrating eastwards into Central Argentina, Uruguay and Brazil. Four are endemic to Chile while the rest are mainly annuals common in North America.

The distribution of the individual sections is as follows. The largest centre of Sections *Trifolium* (Map 7) and *Paramesus* (Map 3) is the Mediterranean region, the number of their species decreasing fairly suddenly towards Central Europe, and abruptly towards the Sahara desert where this genus is totally absent.

The species of Section *Chronosemium* are centred in the North Mediterranean countries and show a marked concentration in the direction of Central Europe (Map 6).

The majority of the species of Section *Trichocephalum* are centred in the East Mediterranean and in Central Anatolia and Iran (Map 8).

Section *Vesicaria* has two centres: the North Mediterranean region and in the eastern part of the Irano-Turanian region (Map 5).

Section *Mistyllus* is encountered in two isolated centres: the East Mediterranean region and the highlands of the Eritreo-Arabian region (Ethiopia, Eritrea, Kenya, Uganda, Tanganyika). This is a very interesting range pattern, having some bearing on the problem of the Afro-Montane floral distribution (Map 4).

All the species of Section *Involucrarium* are American. They are centred on the west coast of North America and only few have outliers penetrating beyond the Rocky Mountains region. The section is also represented in South America (Chile) where seven of its species are to be found (Map 9).

Section *Lotoidea* is widespread, being the only one in the genus with three distribution centres: two in the eastern hemisphere and one in the western hemisphere. The largest centre, comprising 40 species, is on the west coast of North America and in the vicinity of the Rocky Mountains. Only few species have penetrated eastwards, reaching the slopes of the Appalachian Mountains. Another group consisting of five species extends southwards to Chile (Map 2).

The second centre, with 33 species, is located in the Eritreo-Arabian highlands, including Ethiopia (25 species), Kenya, Uganda, etc.

The 26 Eurasian species of this section are centred in the Mediterranean territory of southern Europe; they are equally divided between Central Europe, the Pontic and Caucasus regions, and the Central Asian region.

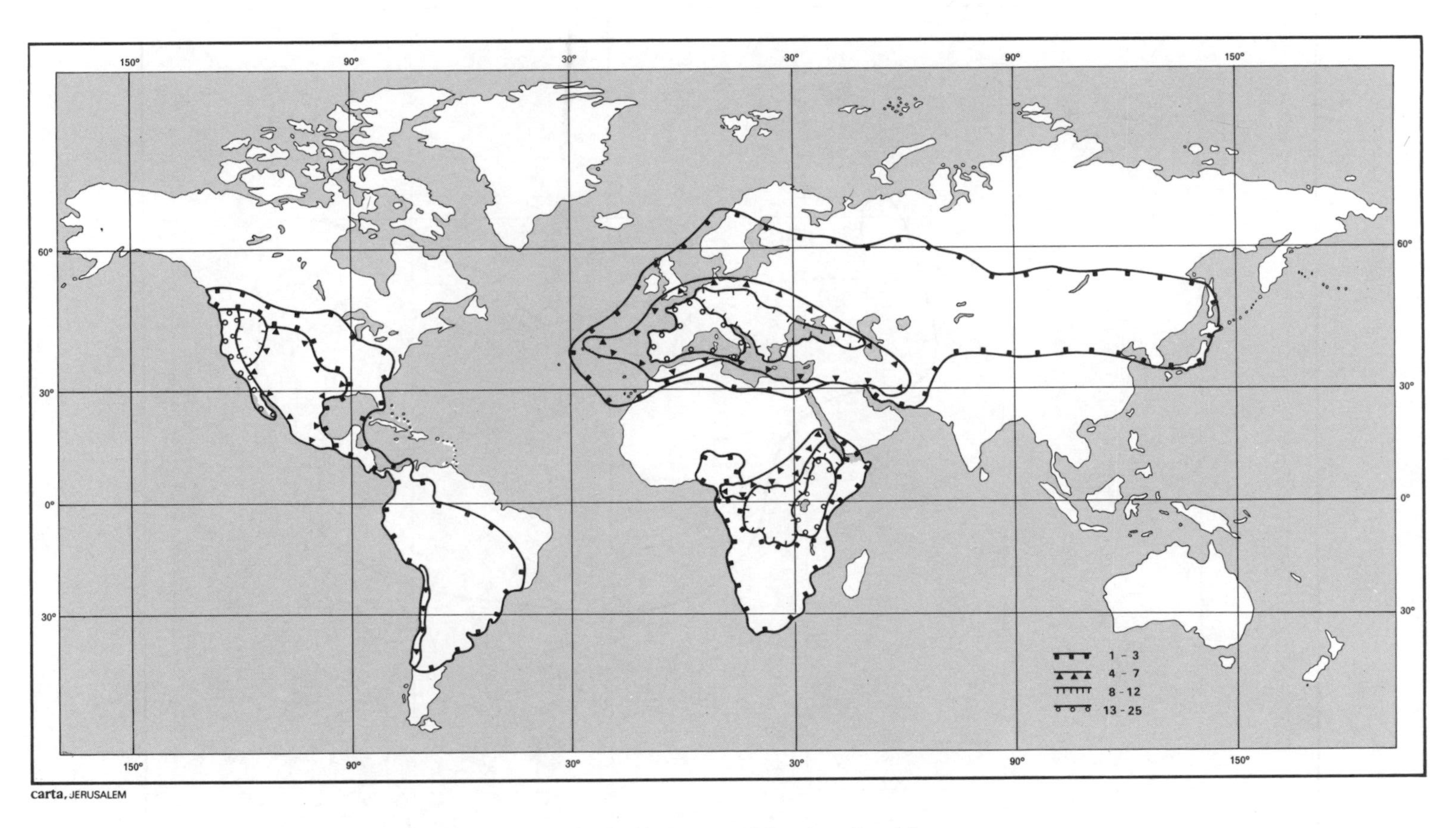

Map 2. Distribution of Section *Lotoidea*

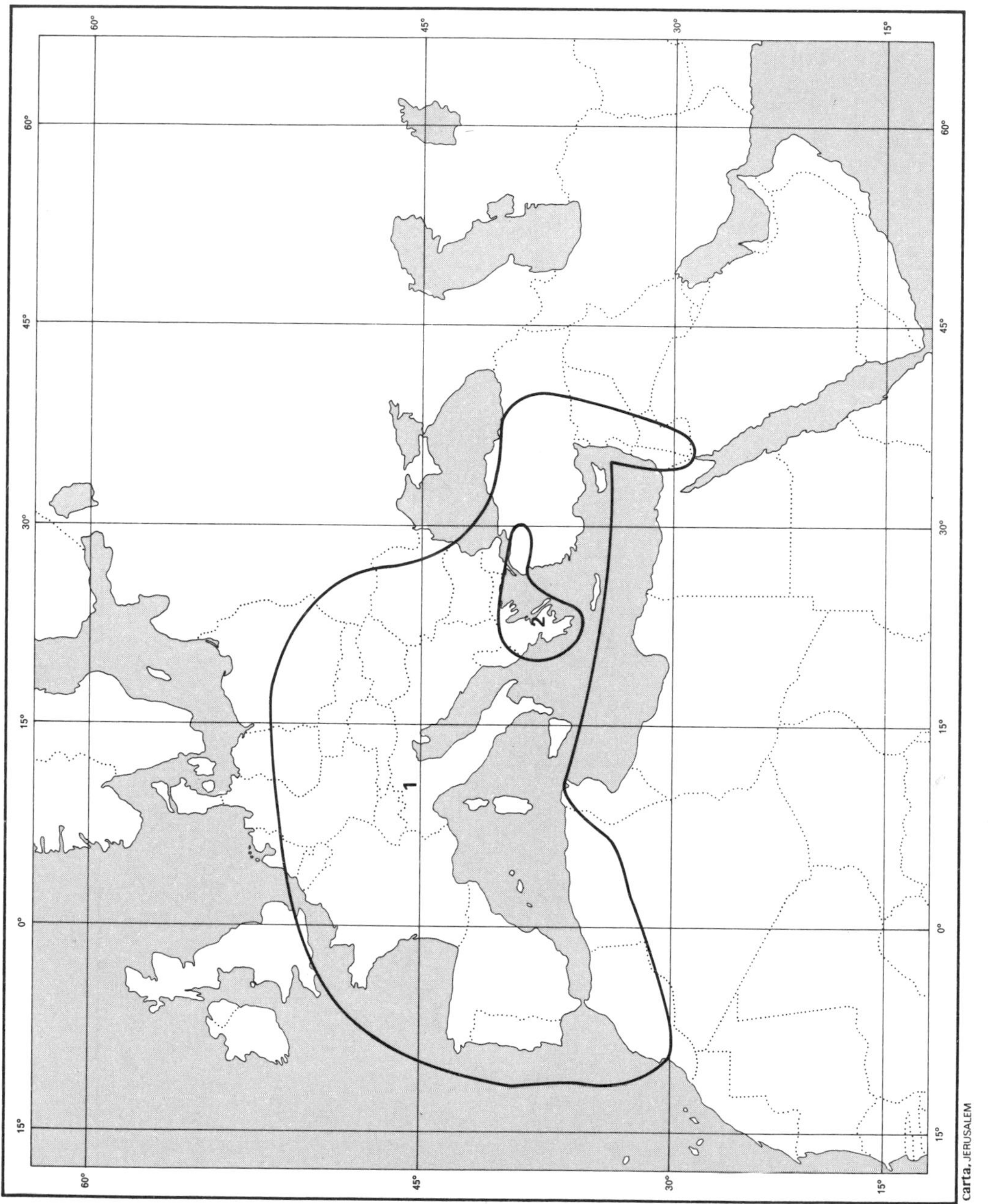

Map 3. Distribution of Section *Paramesus*

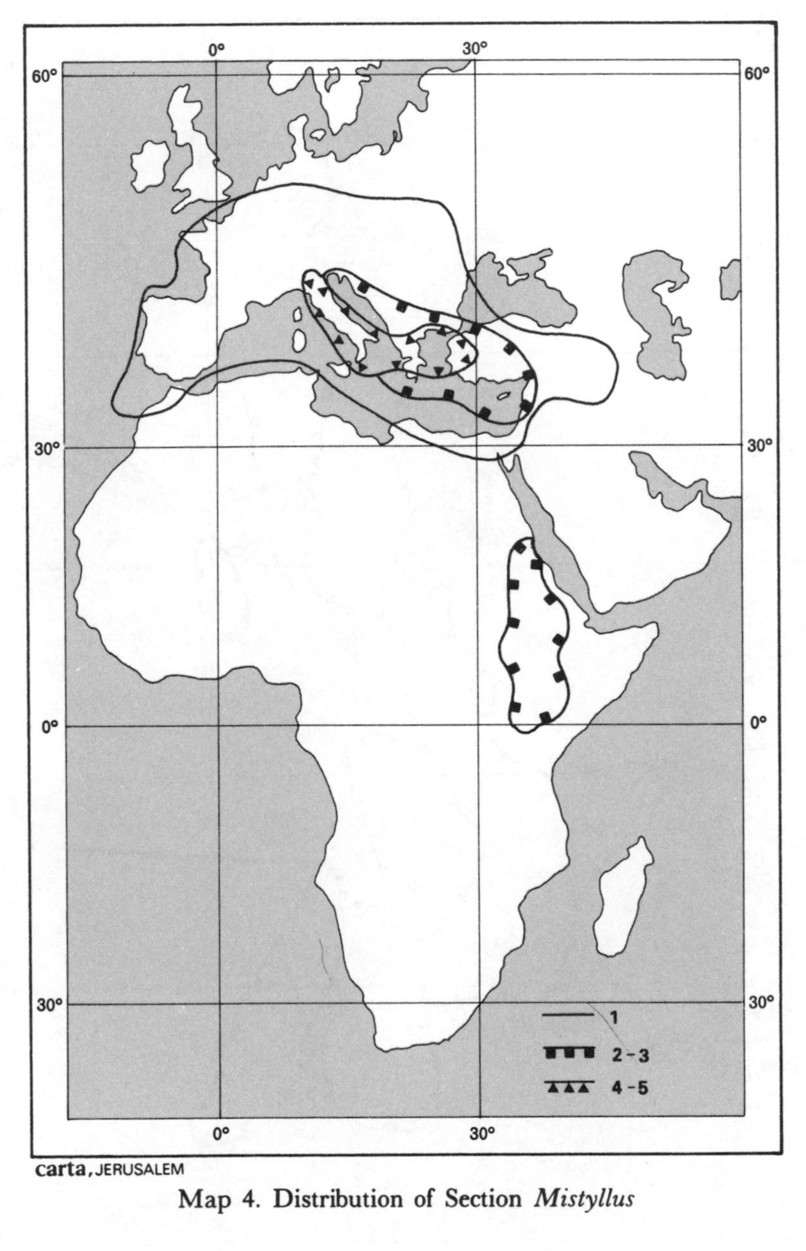

Map 4. Distribution of Section *Mistyllus*

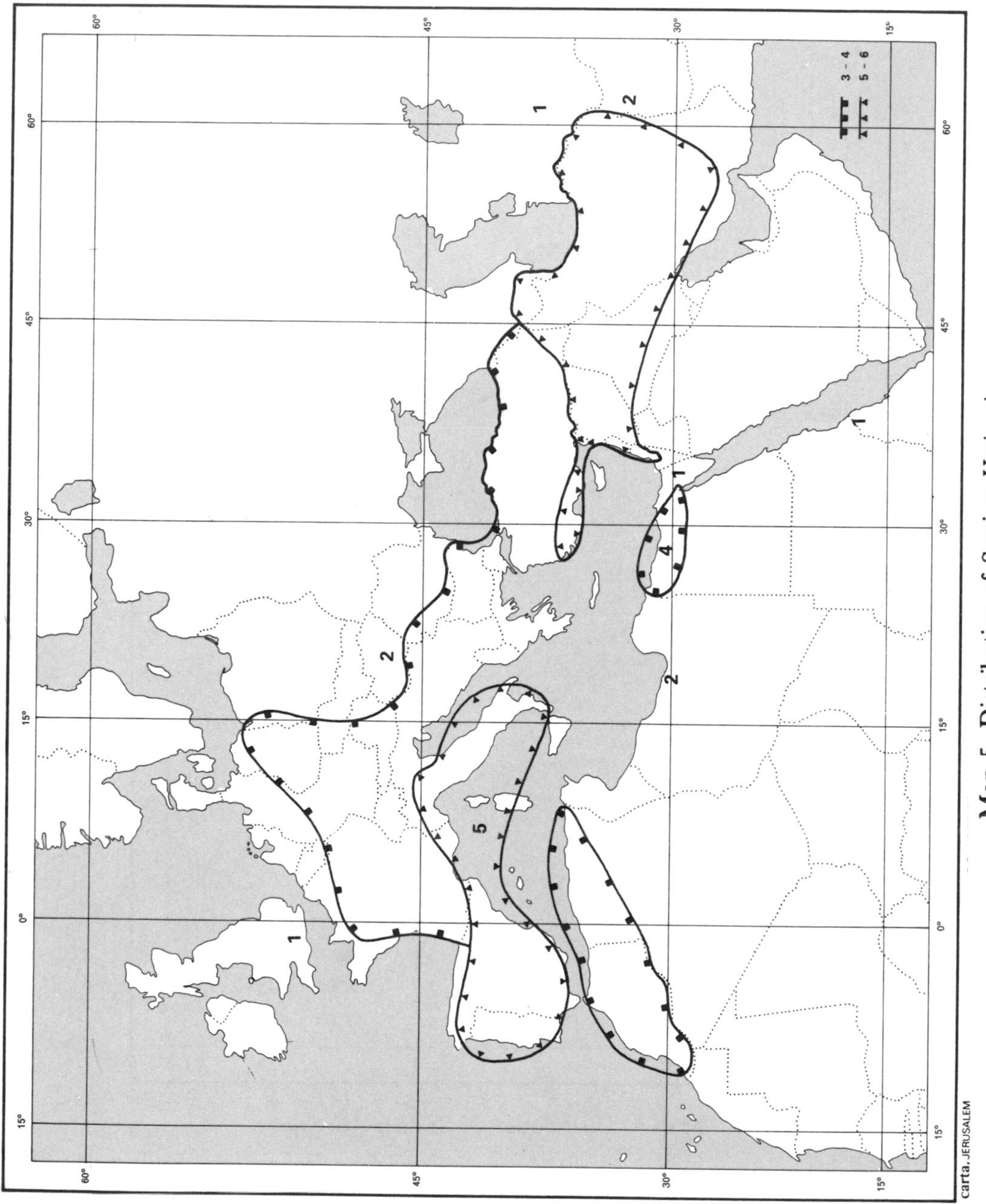

Map 5. Distribution of Section *Vesicaria*

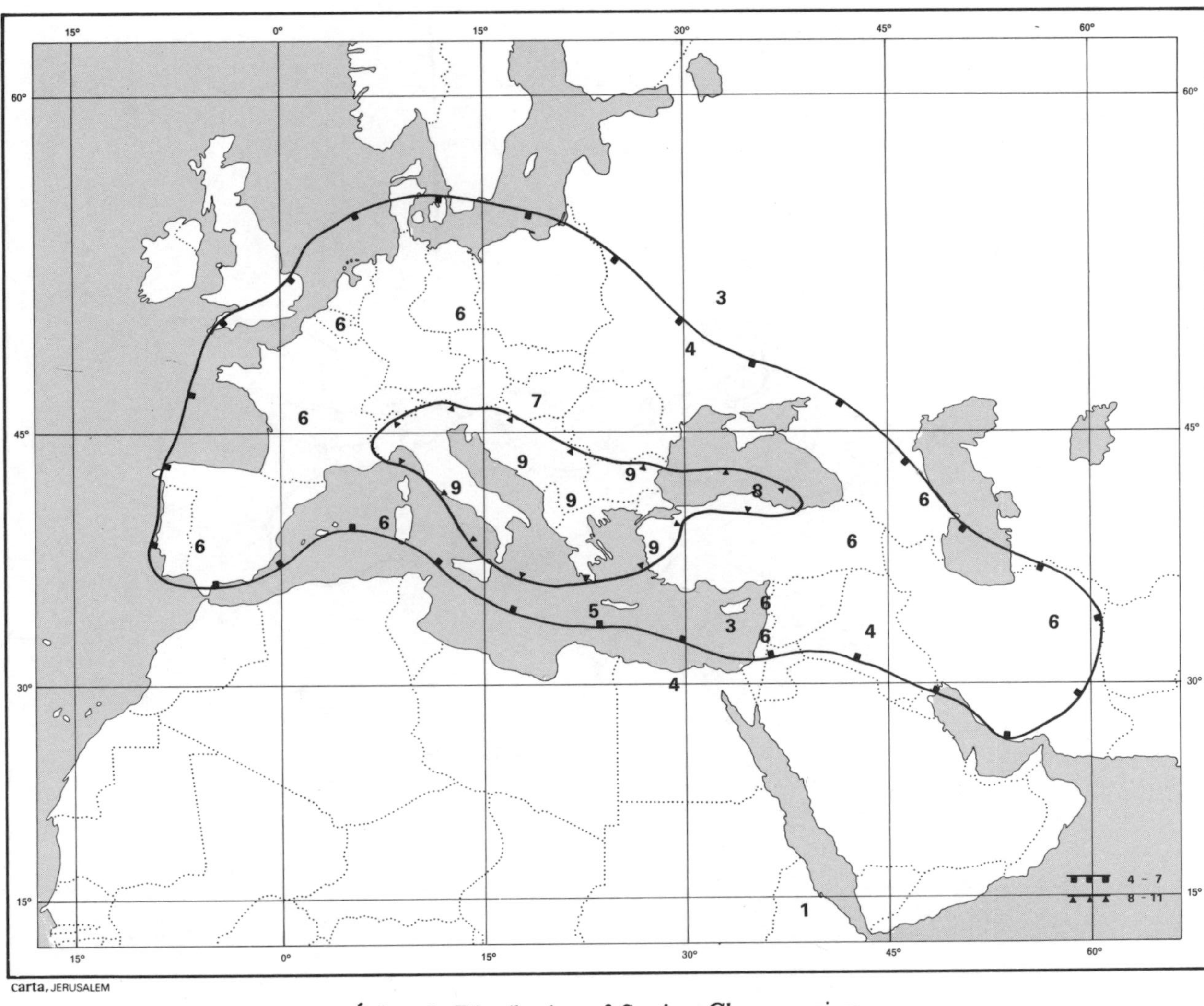

Map 6. Distribution of Section *Chronosemium*

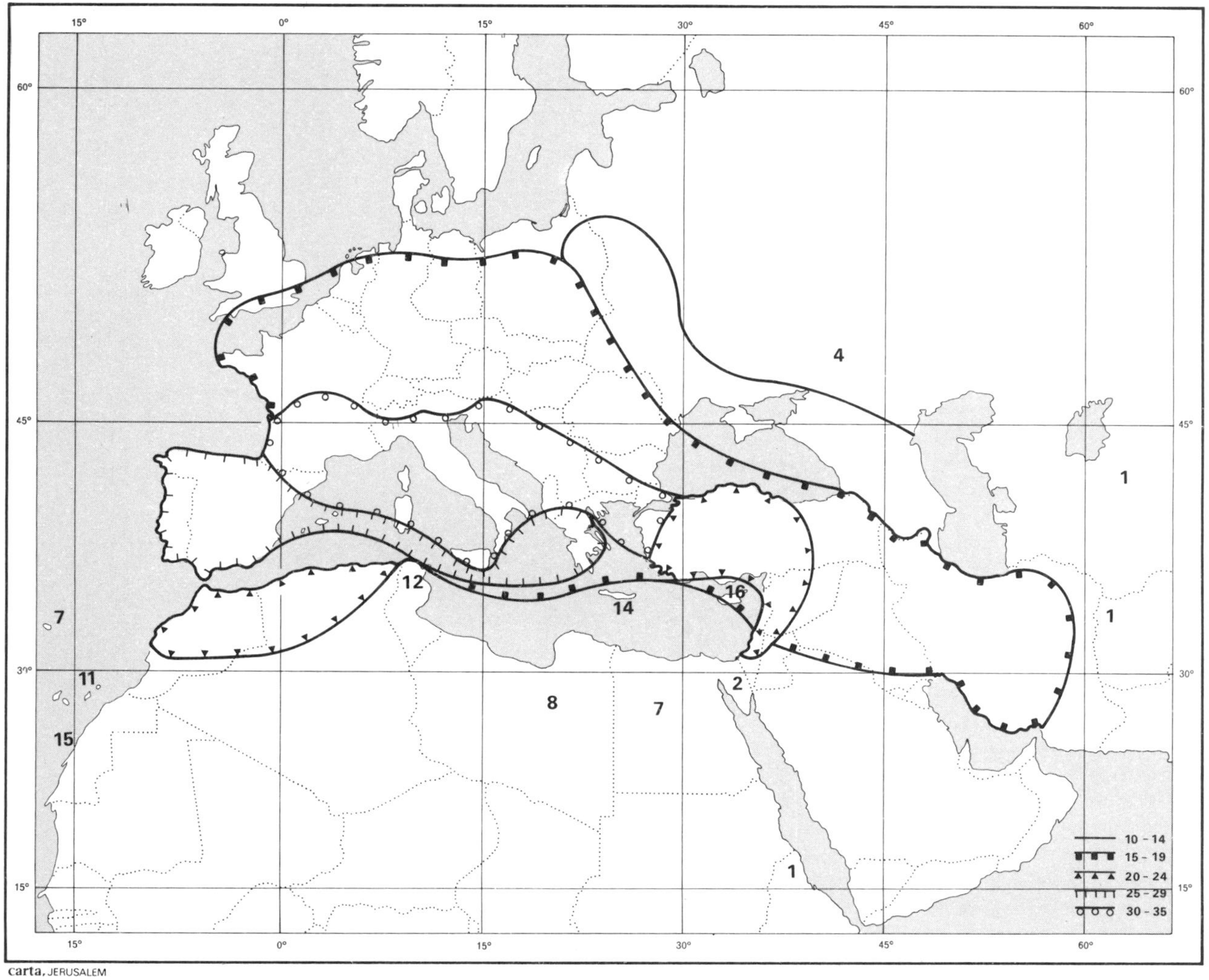

Map 7. Distribution of Section *Trifolium*

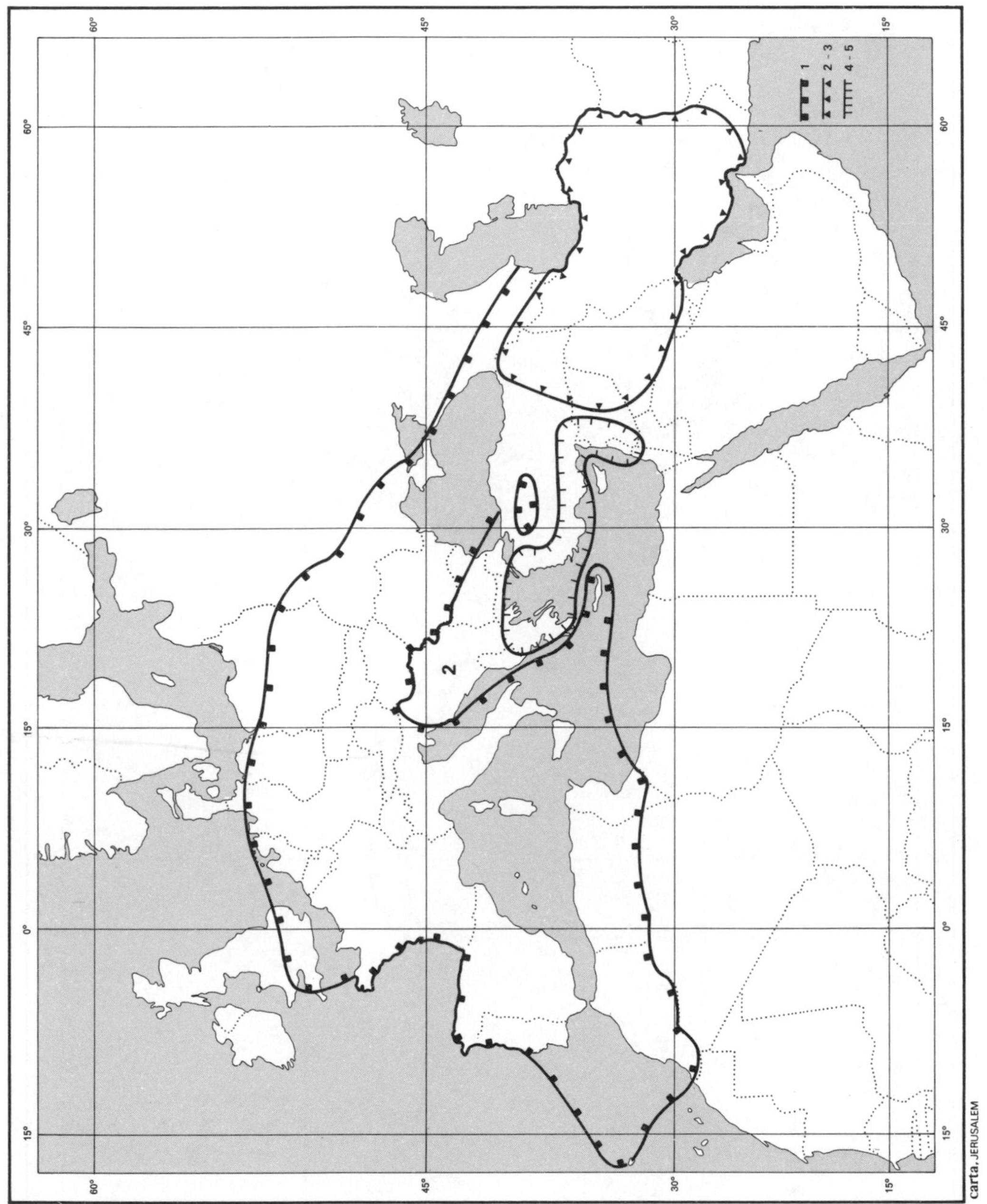

Map 8. Distribution of Section *Trichocephalum*

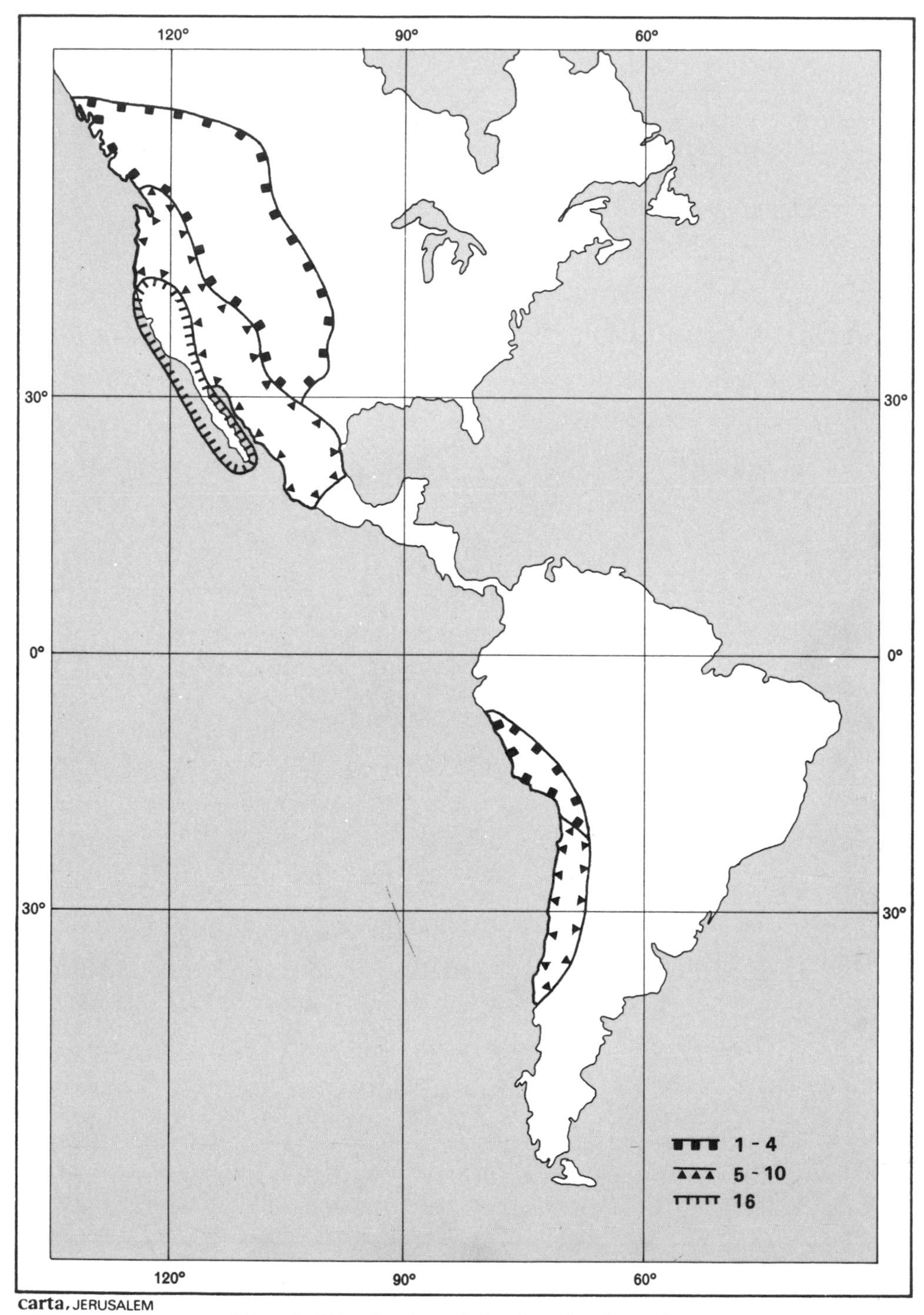

Map 9. Distribution of Section *Involucrarium*

b. *Origin, Speciation and Migration in the Genus*

A few theories have been advanced on the relationship between the three differentiation centres on the migration routes of the species and the possible centre of origin or primary site of the evolution of Section *Lotoidea* and the entire genus.

Let us first survey the relations between the Mediterranean region and continental Africa. According to Gillett (1952), the genus presumably originated in the Mediterranean region, where the largest number and greatest variety of species are now concentrated, together with all the other genera of the tribe *Trifolieae*. Thus, the genus is considered to have migrated from the Mediterranean region into equatorial Africa via Ethiopia and Eritrea (the sites of greatest concentration of the species), and not via the Tibesti Mountains of central Sahara. The species of the genus then spread across the tropics and also reached the Cape, although they are absent from several regions where they might be expected to occur.

On the other hand, Norris (1956) suggests that the highlands of East Africa could be regarded as the centre of origin of the genus, which later migrated to the north, west and south, attaining its fullest extent in the Mediterranean region, where the other sections evolved. Pritchard (1962) endorses this view and claims that the genus later migrated northwards from East Africa to the Mediterranean region and not eastwards into Asia.

Zohary (1972c) is of the opinion that the tropical African clover species of the two subsections of *Lotoidea* (*Loxospermum* and *Ochreata*) originated independently in this centre from an older Lotoidean stock which entered Africa along the migratory routes known as the Mediterraneo-Afro-Alpine Line. This assumption is also supported by the fact that Section *Mistyllus* has three species endemic to the East African highlands, the other three being endemic to the Mediterranean region. This centre of differentiation suggests that the migration of clovers to Africa must have taken place in the remote geological past, probably during the Early Miocene when Euro-African migration routes became available as a result of the shrinking of the Tethys.

The following comment should be made about the centre of the genus in the western hemisphere: the concentration of species of Sections *Lotoidea* and *Involucrarium* on the west coast of North America is considerably richer than that of Section *Lotoidea* in the eastern hemisphere.

According to Zohary (1972c), some of these species must have migrated across the Pacific from the northwestern part of North America to East Asia, probably via the Bering Strait which repeatedly served as a migration route during the Tertiary. This assumption is supported by the following evidence: 1) the great diversity of forms concentrated in the northwestern part of North America, which alone could serve as a source of further differentiation and expansion of the genus; 2) the total lack of indigenous representatives of other Eurasian *Trifolium* sections in America and the occurrence of five species of the highly complex *Lupinaster* subsection, which, from the morphological viewpoint, represents the most primitive group of the genus. This group might have originated from a lupinoid stock which is widespread in the above-mentioned region of America. Moreover, of the two additional species of this subsection, one is mainly East Asian and the other is Euro-Siberian; 3) certain other species of this North American stock might have migrated into Europe via East Asia, subsequently disappearing from Asia but surviving in the Mediterranean region where they formed an exceedingly rich and highly diversified speciation centre. Since such massive speciation in and probably also around the Mediterranean region calls for a protracted geological span, in all likelihood it started prior to the Neogene period; 4) this Mediterranean diversification into the modern sectional patterns of the genus could not have taken place without an appropriate ancestral source of forms for further evolution. This source is almost exclusively comprised of the North American representatives of Section *Lotoidea*, as mentioned above.

The first step was probably the migration to Eurasia from the northwestern part of North America of several species of Section *Lotoidea*, which differentiated here into a multitude of species of the white clover group and its relatives. These in their turn gave rise to the evolution of the various subsections of *Lotoidea* as well as to the six other sections of *Trifolium* on Eurasian territory.

The next step was the partial migration of the *Lotoidea* and *Mistyllus* species into Africa, which probably took place during the Neogene.

The few species in South America are the result of a later migration from North America after the Americas had become adjoined to one another. Obviously the migration route here was from north to south, and many similar groups followed the same direction.

c. *Ecology*

Many species of the genus *Trifolium* are able to flourish in markedly varying habitats, ranging from sub-Arctic regions through mesic to subtropical, tropical and equatorial areas of Africa and South America. The majority are restricted to specific habitats, a detailed discussion of which is beyond the scope of the present work. A few examples, however, may suffice to demonstrate this property of ecological restriction. Several species are limited to coastal plains (e.g., *T. palaestinum*); inundated soils are required by *T. berytheum*; marshy habitats are tolerated by *T. fragiferum*. Many perennial clovers are confined to acid soils of alpine heights. Not a single species is encountered in the flora of the Sahara desert. Quite a few annuals are widespread but not a single species occurs as an obligatory weed among field crops.

At least two-thirds of the species of the genus are limited to natural habitats, such as mountain peaks, rocky slopes, mountain forests, forest clearings, meadows and alpine zones with elevations even exceeding 4,000 m.

More than 75% of the American species of Section *Lotoidea* and some 87% of its African species grow in mountain regions, while only about 50% of its Eurasian species are encountered in such habitats.

Cytotaxonomy

To date chromosome counts (including 54 by the present authors) have been carried out on 70% of the species in the genus *Trifolium*.

Table 1: Chromosome numbers of certain *Trifolium* species studied

T. gracilentum	2n = 16	*T. latinum*	2n = 28
T. multinerve	2n = 16	*T. miegeanum*	2n = 16
T. pichisermollii	2n = 16	*T. purpureum*	
T. polymorphum	2n = 16, 32	var. *desvauxii*	2n = 14
T. glanduliferum	2n = 16	*T. mutabile*	2n = 16
T. affine	2n = 16	*T. bullatum*	2n = 16
T. dichroanthum	2n = 16	*T. clusii*	2n = 16
T. haussknechtii	2n = 16	*T. meduseum*	2n = 14

From these counts and a summary of the literature data the following conclusions can be drawn :

1. The basic number in 80% of the examined species is x = 8. This number also predominates in the other genera of the tribe *Trifolieae* (see Bolkhovskikh et al., 1969). Other basic numbers are : x = 7 (in 15% of the counts), x = 6 (in 2%) and x = 5 (in 3%).
2. While x = 8 has been recorded for 46% of the annuals, x = 7 plants are mostly annuals and counts of x = 6 and x = 5 were noted only for the Mediterranean annuals.
3. All the species of Sections *Lotoidea, Paramesus, Mistyllus, Vesicaria* and *Involucrarium*, as well as all the African and American species, have x = 8.

The following data appearing in the literature merit re-examination : *T. glomeratum* 2n = 14, 28 (Bleier, 1925); *T. eximium* 2n = 36 and *T. polyphyllum* 2n = 18 (Sokolovskaja & Strelkova, 1948); *T. resupinatum* 2n = 14 (Wipf, 1939).
The basic number x = 8 is characteristic of 86% of the species of Section *Chronosemium*, the remainder having x = 7. For two species of this section both of the above numbers have been recorded.
All four basic numbers of the genus have been noted in species of Section *Trichocephalum*. The number x = 6 was recorded for *T. israeliticum* and x = 5 for *T. globosum* (Kozuharov et al., 1975), previously reported as having x = 8.
Section *Trifolium* is also characterized by all four basic numbers. In two species (*T. bocconei* and *T. ligusticum*) the basic numbers were found to be 7 and 6. Five annual Mediterranean species (*T. scabrum*, *T. dalmaticum*, *T. hirtum*, *T. cherleri* and *T. purpureum* var. *desvauxii*) have x = 5; in the first species the number x = 8 has also been recorded and in our counts we found that the last species has x = 7.

The occurrence of reduced chromosome numbers and a descending aneuploid series among the species of the two last-mentioned Mediterranean sections (Section *Trichocephalum* and Section *Trifolium*) supports the assumption that the Mediterranean species are at a more advanced stage of evolution than others.

4. In the genus as a whole only 20% of the examined species are either polyploid or display polyploid and diploid strains. According to Bandel (1974), somewhat less than 30% of the entire family are polyploid.

Table 2: Summary of distribution of different basic chromosome numbers and polyploidy in *Trifolium*

Section	No. of species with basic chromosome no.				Number of species		No. of polyploid species
	x = 8	x = 7	x = 6	x = 5	studied	total in the section	
Lotoidea	71	2*	2*		71	99	22
Paramesus	2				2	2	
Mistyllus	5				5	9	
Chronosemium	8	6			12**	17	2
Trifolium	29	16	2	5	52	72	8
Involucrarium	12				12	22	1
Vesicaria	7	1*			7	7	1
Trichocephalum	4	3	1	1	8**	9	
Total	138	28	5	6	169	237	34

* Doubtful records.

**For some species more than one number is recorded.

Of the 60 perennial species investigated, 27 (45%) were found to be polyploids, while only seven annuals out of 107 species proved to be polyploid. There are no polyploid species in the following three sections: *Paramesus*, *Mistyllus* and *Trichocephalum*.

The three annual species (*T. resupinatum*, *T. dubium*, *T. micranthum*) of Sections *Vesicaria* and *Chronosemium* have both tetraploid and diploid races (Fig. 5).

Only one species, *T. wormskioldii* (Section *Involucrarium*), is recorded as being diploid, tetraploid and hexaploid.

In the two large sections, *Trifolium* and *Lotoidea*, 12 of the 30 known polyploid species are exclusively polyploid. In these sections there are species with a polyploidy level exceeding that of the hexaploids, the highest level known in the other sections of the genus.

The majority of polyploid species (65%) are concentrated in Section *Lotoidea*; they constitute about one-third of the species for which chromosome counts have been

carried out to date. Half of the polyploid species are exclusively polyploid. Only two of the 22 polyploids in this section are annuals. There is a marked trend towards polyploidization among the American species of this section; 12 of the 33 species for which there are counts are polyploid. The same section in the Old World features only 10 polyploid species out of the 38 counted, while the corresponding figure for African species is even lower: three of the 17 species counted to date.
Another interesting feature of the genus is that only 15% of the polyploid species have a basic chromosome number of less than x = 8. Britten (1963) explains this small number of teraploid species with 2n = 28 by assuming that species with reduced chromosome numbers are derived and therefore younger forms.

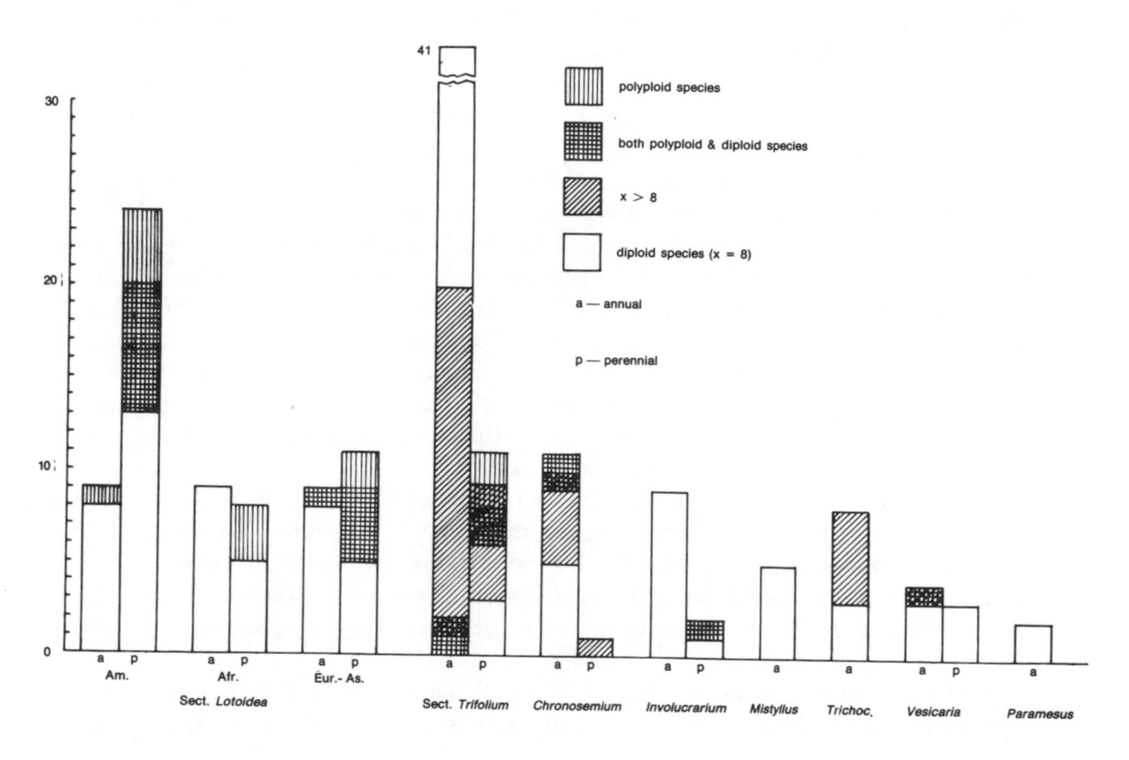

Fig. 5. The number of chromosomes and the ploidy level in the sections and species of *Trifolium*

Evans (1976) suggests that induced polyploids, such as those known in *T. hybridum* and *T. pratense*, enhance resistance to diseases. Future breeding practice should thus tend to develop polyploid varieties.
5. There is very little available information on the chromosome morphology or the karyotypes of the species of the genus. This is probably due to technical difficulties arising from the small size of the chromosomes in most of the species.
Pritchard (1962) studied the chromosome morphology of 14 African species, all (except one) of Section *Lotoidea*. He showed that the chromosome complement consists of 1–3 pairs of metacentric, 4–6 pairs of submetacentric and 1 pair of acrocentric

chromosomes with large satellites on their short arms. According to Pritchard, the pair of satellite chromosomes are the largest. In most of the species he noted a gradual decrease in size from the largest to the smallest chromosome.

Chen & Gibson (1971a) found that the 15 European species of Section *Lotoidea* which they studied differ in their karyotypes from the African species. They observed that the acrocentric pairs are usually the smallest and bear small satellites on their short arms. In two species, *T. hybridum* and *T. michelianum*, the satellites are somewhat larger. As in the African species of Section *Lotoidea*, the asymmetry between the chromosomes is generally small, being greater only in one species, *T. isthmocarpum*.

Similar are the findings of González et al. (1973) for three European species that were also studied by the last-mentioned authors.

According to our findings, both in the West European species *T. ornithopodioides* and in the African species *T. multinerve* the pair of satellite chromosomes are the largest.

In two American species of Section *Lotoidea* (*T. gracilentum* and *T. polymorphum*), we observed that the satellites are located on the small pair of acrocentric chromosomes, a phenomenon similar to that noted for the European species of this section.

One African species, *T. steudneri* of Section *Mistyllus*, studied by Pritchard (1962), shows large satellites on the short arms of the acrocentric chromosome pairs, being thus similar to the African *Lotoidea* species. However, most of the other pairs of chromosomes differ from those of *Lotoidea* by being acrocentric. Another difference from the *Lotoidea* species is the marked gradient in size between the chromosomes. González et al. (1973) observed approximately the same karyotype for a Mediterranean species, another member of Section *Mistyllus*.

Angulo et al. (1969, 1972b), who studied certain species of Section *Trifolium*, have shown that large satellites are located on the metacentric chromosome pair or on the submetacentric ones, which are also the largest. In one species, *T. squarrosum*, they detected a pair of chromosomes with two so-called secondary constrictions. Their study revealed the marked asymmetry between the chromosomes in species of Section *Trifolium*.

The results of our studies of two additional species (*T. ligusticum* and *T. miegeanum*) of the last-mentioned section are similar to those obtained by Angulo et al.

In one species of Section *Vesicaria*, *T. physodes*, we noted that the large satellites are located on the short arms of the submetacentric pair of chromosomes.

On the other hand, in two species (*T. campestre* and *T. dubium*) of Section *Chronosemium*, Angulo et al. (1971) found that the satellite chromosome pair is the largest within the complement and that the next in length are the metacentric chromosomes.

The largest satellite chromosome found in all the species of the genus was recorded by Angulo et al. (1971, 1975) for two species of Section *Trichocephalum* (*T. subterraneum* and *T. israeliticum*). This large satellite is located on the short arm of the submetacentric chromosome pair. In one of them, *T. subterraneum*, they also observed a pair of chromosomes with two constrictions. The gradient between the chromosomes of the above species is readily discernible.

According to Stebbins (1950), the symmetric or homogeneous karyotype should be regarded as being characteristic of comparatively primitive species, while the asymmetric or heterogeneous karyotype is to be found in more developed species. Stebbins (1971) states that Angiosperms show a tendency towards increased asymmetry of the karyotype, and he emphasizes the familiar hypothesis that the symmetric karyotype is probably primitive.

The above hypothesis supports the available data on the karyotypes in the genus *Trifolium*. Accordingly, in the sections (*Trifolium* and *Trichocephalum*) known to be more developed, the karyotype reported for the species is asymmetric, while in the primitive Section *Lotoidea* the karyotype of the species studied is symmetric or more homogeneous.

6. Natural hybrids are very rare among species of the genus. Evans (1976) attributes the rarity of hybridization in clovers (or in other forage legumes) to their predominant adaptation to insect pollination. Another possible reason for the scarcity of hybrids is the predominance of self-fertile pollination, especially among the annuals.

Several attempts to produce interspecific hybrids in *Trifolium*, with the aim of introducing new genetic variations, have met with little success. Evans (1962) has suggested methods of overcoming interspecific incompatability by means of grafting and embryo culture techniques.

In a series of papers, Chen & Gibson (1970a,1970b,1971b,1972a,1972b,) have shown that in some species of Section *Lotoidea (T. repens, T. uniflorum, T. occidentale, T. nigrescens)* successful crosses can be made and that some chromosome pairing does occur between the various genomes. They suggested that such species should be considered as sharing a common basic structural genome. In another paper (1971a), they point out the karyotypic similarity of those species, indicating their close phylogenetic relationship. Their close relationship in interspecific crossings and karyotypic similarity are more obvious than their morphological affinities.

Kazimierski & Kazimierska (1973) state that since the diploids *T. nigrescens* and *T. isthmocarpum* cross with the tetraploid *T. repens*, the former should be considered the probable ancestors of white clover which is believed to be a natural tetraploid.

In a series of studies, Putiyevsky & Katznelson (1970, 1972, 1973) and Katznelson & Putiyevsky (1974) have attempted, by means of interspecific crossings among 12 species related to *T. alexandrinum*, to elucidate the cytogenetics of the group and the biosystematic relationships between the species. As a result, the species were divided into five crossability groups, seed-set or crossability being generally high within each group and low or very low in inter-group combinations.

Kazimierski & Kazimierska (1972) also discovered that *T. alexandrinum*, *T. apertum* and *T. vavilovii* readily intercross and produce viable hybrids. These three species differ from one another in their genetic systems which govern seed setting.

Interspecific hybridization of *T. subterraneum* with *T. eriosphaerum* and *T. pilulare* by Katznelson (1967) yielded one new cultivar line. Katznelson (1974) suggests that the production of these hybrids is accomplished by chromosomal reduction via translocation and inversions from 2n = 16 to 2n = 12, involving some anemochorous species which are obviously morphologically similar.

As is evident from the above examples, success in interspecific hybridization has been achieved only between species of the same section. Attempts at intersectional hybridization, such as those made by Evans (1962), have failed. However, in nature even intrasectional hybrids are very rare, and thus no problems are encountered in the delimitation of the species.

Breeding System

Trifolium has a typical papilionaceous flower suitable for insect pollination. Most of the insects visiting the flowers have fairly long proboscises, e.g., the honeybee, bumble bee and some *Lepidoptera*, and come to collect nectar and pollen. When the insect visitor applies pressure, especially with its head, on the standard and wings, the stigma and anthers protrude from the keel, reverting to their former position inside the keel when the pressure is released. Thus, the anthers and stigma are pressed against the underside of the insect's head and pollination occurs. The pistil is usually longer than the stamens and slightly curved. The anthers are dehiscent even in bud. The nectar is secreted at the base of the stamens and collects at the base of the corolla tube. Weaver (1965) and other investigators recorded that in cultivated species the nectar has an average sugar concentration of 29–66%. Weaver also found that a flower of *T. repens* contained 0.02–0.08 μl of nectar with a sugar content of 42–65%.

In *T. pratense* fertilization takes place between 18 to 50 hours after pollination, depending on the temperature. Pritchard & t'Mannetje (1967) have shown that a relatively high humidity of 93–98% is essential for good pollen germination and growth.

Most species are allogamous and the best seed-set results from cross-pollination with another plant of the same species.

Certain species are autogamous and although the flowers do open normally, pollination occurs spontaneously prior to this event. However, some autogamous species need to be tripped before setting seeds.

The flowers of some species, such as *T. baccarinii* and *T. lanceolatum*, do not open during blossoming (see Bogdan, 1966). No bees or other pollen-bearing insects were observed on such cleistogamous flowers. The pollen of their dehiscing anthers reaches the stigma. In some cases, the pollen may germinate while still in the anther, the pollen tubes piercing through the wall of the anther on their way to the stigma.

Of the 20 species of Section *Lotoidea* grown for examination of their breeding system, 14 were annuals and proved to be autogamous; of these autogamous species four (*T. cernuum*, *T. retusum*, *T. suffocatum* and *T. tembense*) with minute flowers can be regarded as cleistogamous. Pritchard & t'Mannetje (1967), experimenting with *T. pseudostriatum*, *T. rueppellianum*, *T. lanceolatum*, *T. steudneri* and *T. baccarinii*, noted their high fertility due to autogamous pollination. According to Piergentili (1970), the subterranean flowers of the South American amphicarpic species *T. polymorphum* are cleistogamous. According to Taylor et al. (1979), six annual Eurasian species with 10 and 12 diploid chromosomes are autogamous.

Self-pollination resulting in approximately 60% fertility was observed by us in *T.*

bejariense, *T. michelianum* var. *balansae*, *T. multinerve* and *T. ornithopodioides*.
In the following five annual self-pollinated species an average of 22% fertility was obtained : *T. glomeratum*, *T. gracilentum*, *T. isthmocarpum*, *T. michelianum* var. *michelianum* and *T. nigrescens*.

Weaver & Weihing (1960) have shown that the Mediterranean species, *T. michelianum*, *T. isthmocarpum* and *T. nigrescens*, set only few seeds without honeybee pollination. The first two are very attractive to bees in the field as they produce 0.12–0.14 μl nectar per floret, while *T. nigrescens* was found to be only fairly attractive since it provides less nectar (0.09 μl).

Two annual African species (*T. masaiënse* and *T. mattirolianum*) were found by us to be allogamous, setting seed only in cross-pollination. Bogdan (1966) pollinated the flowers of *T. mattirolianum* with their own pollen and found that only 42% formed seed, while the corresponding figure for pollination with the pollen of *T. rueppellianum* was 70%. Bogdan also tested the hybrids of the two species and noted that self-pollination occurs in F_1. He thus concluded that self-pollination in *T. rueppellianum* is inherited as a dominant trait.

We found that three perennial species (*T. repens*, *T. semipilosum* and *T. uniflorum*) are allogamous, and the same applies to *T. africanum* and *T. burchellianum* (from Africa) tested by Pritchard & t'Mannetje (1967). Cross-pollination is also necessary in the cultivated clovers of *T. alexandrinum*, *T. fragiferum*, *T. hybridum* and *T. pratense*.

Chen & Gibson (1973) revealed that self-incompatibility in *T. repens* is due to abnormal growth of the pollen tubes; however, once a pollen tube enters the ovule, there appears to be no failure of fertilization and early seed development. Higher temperatures enhance pollen-tube growth, thus leading to the breakdown or circumvention of the incompatibility mechanisms. Relatively high temperatures were reported to increase self-seed production in several self-incompatible species, such as *T. pratense* and *T. hybridum*; some authors have termed this phenomenon pseudo-self-compatibility.

Cross- and self-incompatibility in *Trifolium* species such as *T. pratense* are reportedly controlled by a series of alleles at a single locus, which are conventionally designated S-genes.

In summary, autogamy and, rarely, also cleistogamy are fairly common in all sections of *Trifolium*, but some species need to be tripped before setting seeds. Obligatory allogamy is not rare, especially among perennials of Section *Lotoidea*. However, a large number of species are probably self-compatible, though they yield more seeds through allogamy than through autogamy.

Palynology

a. *Pollen Shape and Dimensions*

The most valuable contribution to the knowledge of pollen grain morphology in indigenous N. American *Trifolium* species was made by Gillett et al. (1973). They

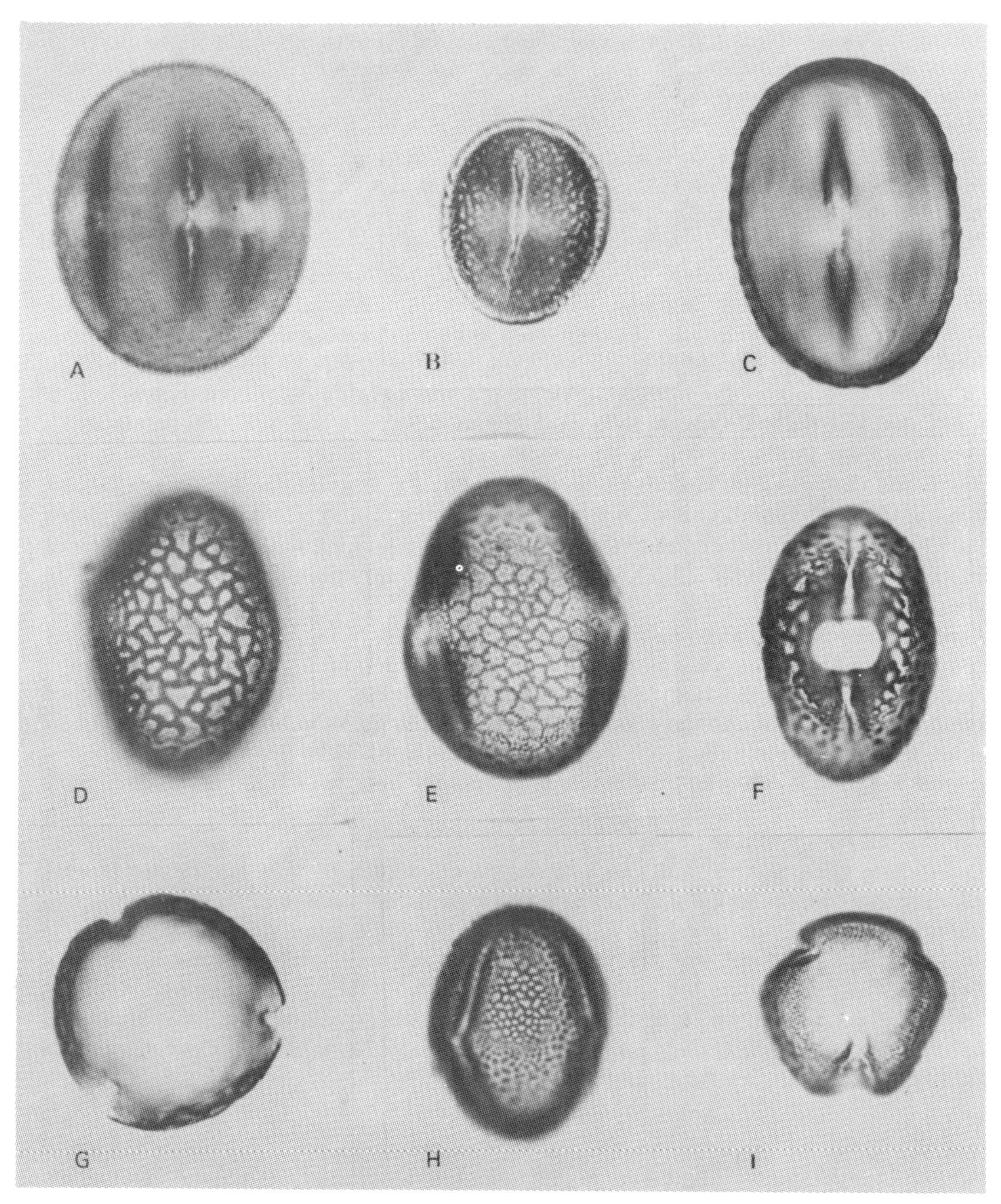

Fig. 6. Acetolized pollen (× 800)
A–F and H : Equatorial view
A – *T. mattirolianum;* B – *T. suffocatum;* C – *T. alexandrinum;* D – *T. gracilentum;* E – *T. eriosphaerum*; F – *T. glanduliferum*; H – *T. nanum*
G, I : Polar view
G – *T. multinerve*; I – *T. glomeratum*

described the pollen grains of 45 American species with the aid of light and scanning electron microscopes and provided a key for their identification. In this work an attempt was made to correlate pollen grain morphology with the taxonomy.

Prior to the publication of the above work, the scanty information on pollen grains available in the literature referred only to light microscope descriptions. For example, Erdtman et al. (1963) produced a key to the pollen of 11 Scandinavian clovers based on the size and shape of the grains.

Our studies on pollen grain morphology with the aid of light and scanning electron microscopes cover 55 species : 45 of Section *Lotoidea* and 10 belonging to other sections.

Only a summary and a brief discussion of the results can be provided here.

The dimensions of pollen grains in *Trifolium* are as follows: 14–35 μ in diameter; 21–49 μ in length; length/diameter ratio 1.19–1.64. Their prolate, tricolporate exine is 1–2.4 μ thick.

The smallest pollen grains have been observed in the two, probably cleistogamous species, *T. retusum* and *T. suffocatum* (Fig. 6B); some of their pollen was found to be abortive.

The cross-pollinating species of Section *Lotoidea* have fairly large pollen grains (Fig. 6A). In the two cultivated species of Section *Trifolium*, *T. alexandrinum* (Fig. 6C) and *T. pratense*, studied by us and by Erdtman et al.(1963), the pollen grains are the largest among the species measured to date.

Almost spherical pollen grains were found in the American species *T. albopurpureum*, while *T. uniflorum* has markedly elongated pollen grains.

Large (36–38 μ) pollen grains were observed in the African species of the *Lotoidea* subsections and also in species of Subsection *Neolagopus* (38 μ) and Subsection *Falcatula* (37 μ). The species of Subsection *Lotoidea* have smaller grains (31 μ), while a further decrease in size (29 μ) is observed in those of Subsection *Platystylium*.

Generally, the average grain size of annual species is approximately the same as that of the perennial ones, and no significant differences have been noted between the species of the three continents.

b. *Pollen Surface as Viewed under the Scanning Electron Microscope*

The sculpturing of the pollen grains either extends over the entire grain surface or only over part of it. The type of sculpturing at the polar ends and along the furrow edges differs from that on the rest of the grain surface. Sometimes the surface is smooth or foveolate, with small, mainly rounded pits.

The surface morphology of the grains shows considerable variation. According to the sculpturing, the species were grouped into seven distinct pollen types. Four of these types were found to occur each in one species only, while the remaining three are common in Section *Lotoidea* and also appear in two other sections of the genus. These three common pollen types have also been observed by Gillett et al. (1973). They are not specific to *Trifolium* and are also noted in representative species of the three other genera of the tribe (*Medicago*, *Trigonella* and *Melilotus*) and even in *Ononis*.

The following seven types of pollen grains have been observed:

1. The grain is wrinkled, giving the appearance of reticulate venation, and also shows scrobiculate sculpturing, but not at the furrow edges which are smooth or only slightly wrinkled; the polar ends are foveolate and wrinkled; the luminae are irregular in shape, wrinkled along their margin and sometimes V-shaped. This type has been observed only in *T. cernuum* (Fig 7A–B).
2. The entire surface of the grain is rugulate to ruminate with the exception of the polar ends which are reticulately sculptured; the luminae are irregular in shape and ruminate within. This type has been observed only in *T. bivonae* (Fig. 7C–D), but Gillett et al. (1973) observed faint ruminate sculpturing at the polar ends of three American species (*T. latifolium*, *T. productum* and *T. parryi*).
3. The sculpturing of the grain is reticulate-foveolate over part of its surface, areolate-reticulate or smooth at the polar ends and along the furrow edges; the luminae are irregular in shape. This type has been observed in seven species [*T. andersonii*, *T. bilineatum*, *T. dasyphyllum*, *T. hybridum*, *T. thompsonii*, *T. uniflorum* (Fig. 8A–B), *T. virginicum*] and has also been described by Gillett et al.(1973) in many other American species. It has also been encountered in grains of *Trigonella arabica* and *Medicago radiata* and on part of the surface of the grains of *Melilotus italicus* (the remaining part shows sculpturing of type 5).
4. The sculpturing of the grain is coarsely rugulate over part of the surface, areolate to foveolate or nearly smooth at the polar ends and along the furrow edges; the luminae are deep, their shape and interspaces being irregular. This type is common in 20 species of Section *Lotoidea* [*T. albopurpureum*, *T. alpinum*, *T. ambiguum*, *T. andinum*, *T. bejariense*, *T. glomeratum*, *T. isthmocarpum*, *T. macraei*, *T. montanum*, *T. nanum*, *T. nigrescens*, *T. ornithopodioides* (Fig. 8C–D), *T. pichisermolii*, *T. polymorphum*, *T. repens*, *T. retusum*, *T. semipilosum*, *T. stolzii*, *T. thalii*, *T. usambarense*]. Some of the American species of Section *Involucrarium* described by Gillett et al. (1973) show the same type of sculpturing, and this also applies to the grains of *Ononis ornithopodioides* (Fig. 10D).
5. The grain shows irregular reticulate-scrobiculate sculpturing over part of its surface; sparsely foveate or quite smooth at the polar ends and along the furrow edges; micro-verrucate within the irregularly shaped luminae, the interspaces of which are very unequal. This type is common in 14 species of Section *Lotoidea* [*T. africanum*, *T. burchellianum* ssp. *johnstonii* (Fig. 9C), *T. crytopodium*, *T. decorum*, *T. gracilentum*, *T. lanceolatum*, *T. masaiënse*, *T. michelianum* var. *michelianum* & var. *balansae*, *T. multinerve*, *T. rueppellianum*, *T. somalense*, *T. stoloniferum*,*T. suffocatum*, *T. tembense* (Fig. 9A–B)] and in two species of Section *Trifolium* [*T. stellatum* (Fig.10C), *T. alexandrinum*]. Two species of Section *Involucrarium* described by Gillett et al. (1973) show the same type of sculpturing, as does one half of the grain surface of *Melilotus italicus*.
6. The surface of the grain of *T. baccarinii* is almost smooth, with an irregular net of nerves and few rugae; this sculpturing appears at the polar ends and along the furrow edges (Fig. 9D).
7. The surface of the grain of *T. mattirolianum* features two types of sculpturing which are intermixed: foveate and scrobiculate, the latter appearing only on part of

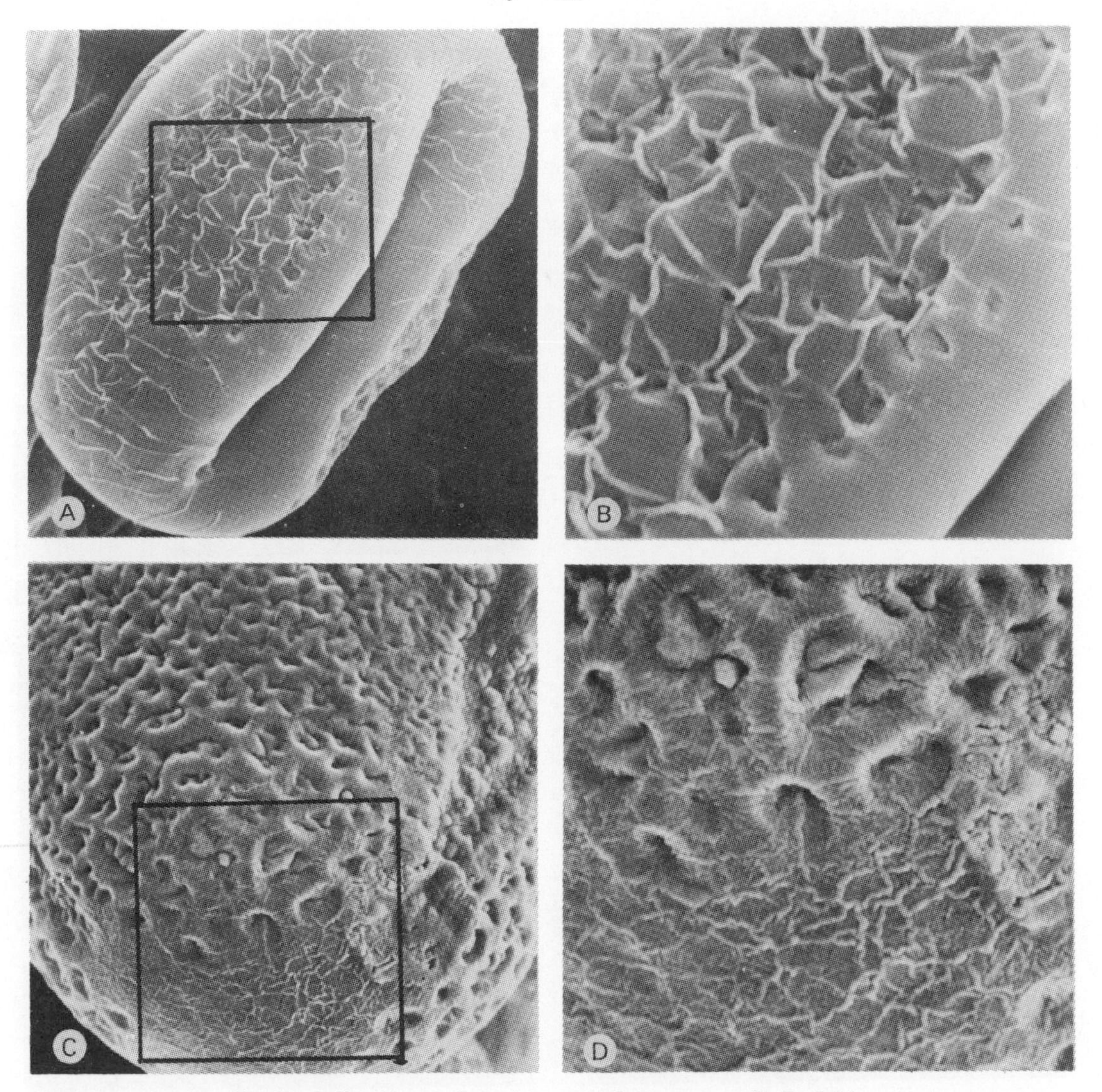

Fig 7. Pollen (scanning electron microscope – S. E. M.)
A – *T. cernuum,* side view (× 3,200); B – same, enlarged (× 8,000); C – T. ***bivonae,*** near polar end (× 4,000); D – same, enlarged (× 8,000)

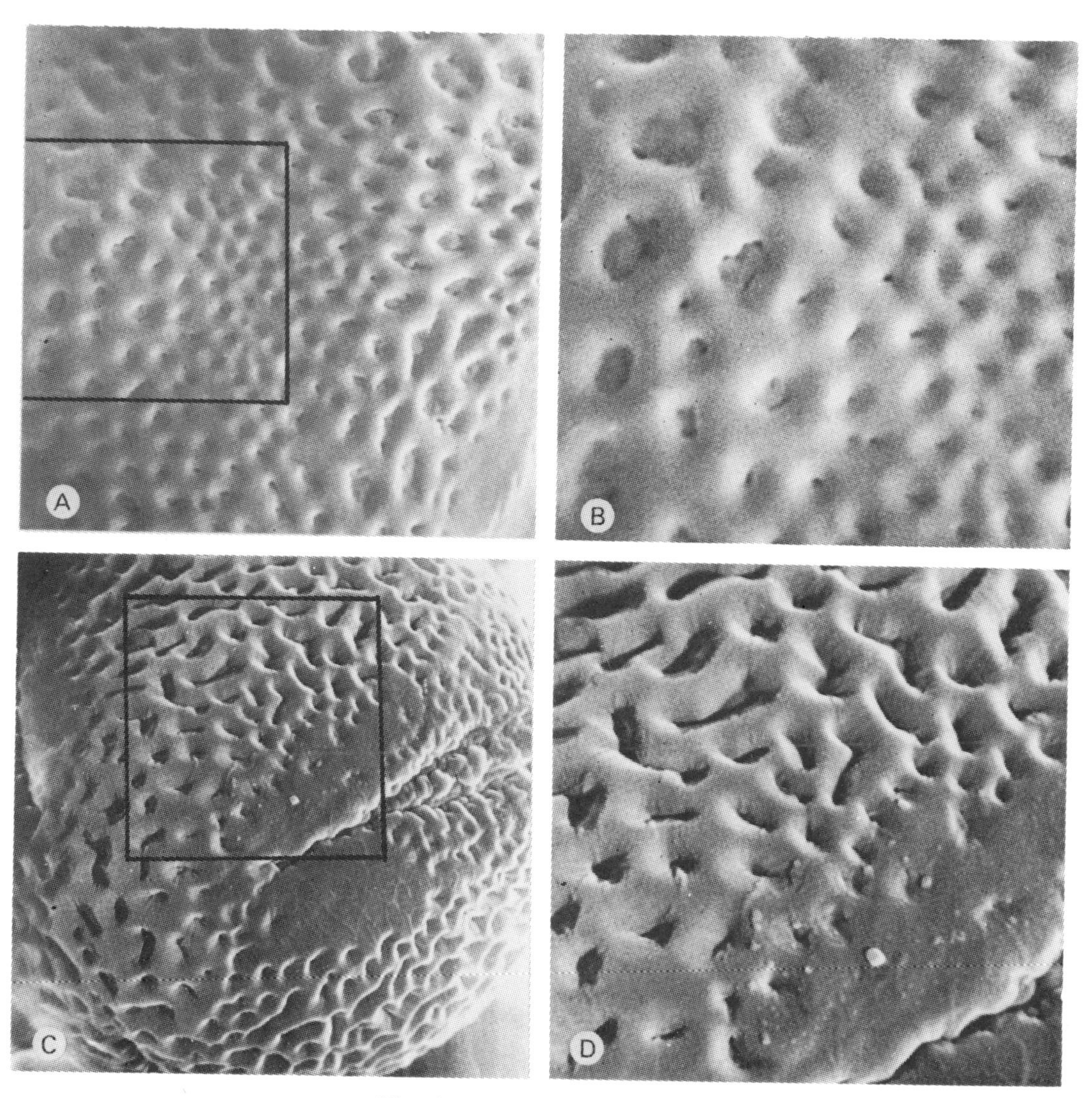

Fig. 8. Pollen (S. E. M.)

A – *T. uniflorum*, near polar end (× 4,400); B – same, enlarged (× 8,000); C – *T. ornithopodioides*, near polar end (× 4,000); D – same, enlarged (× 8,000)

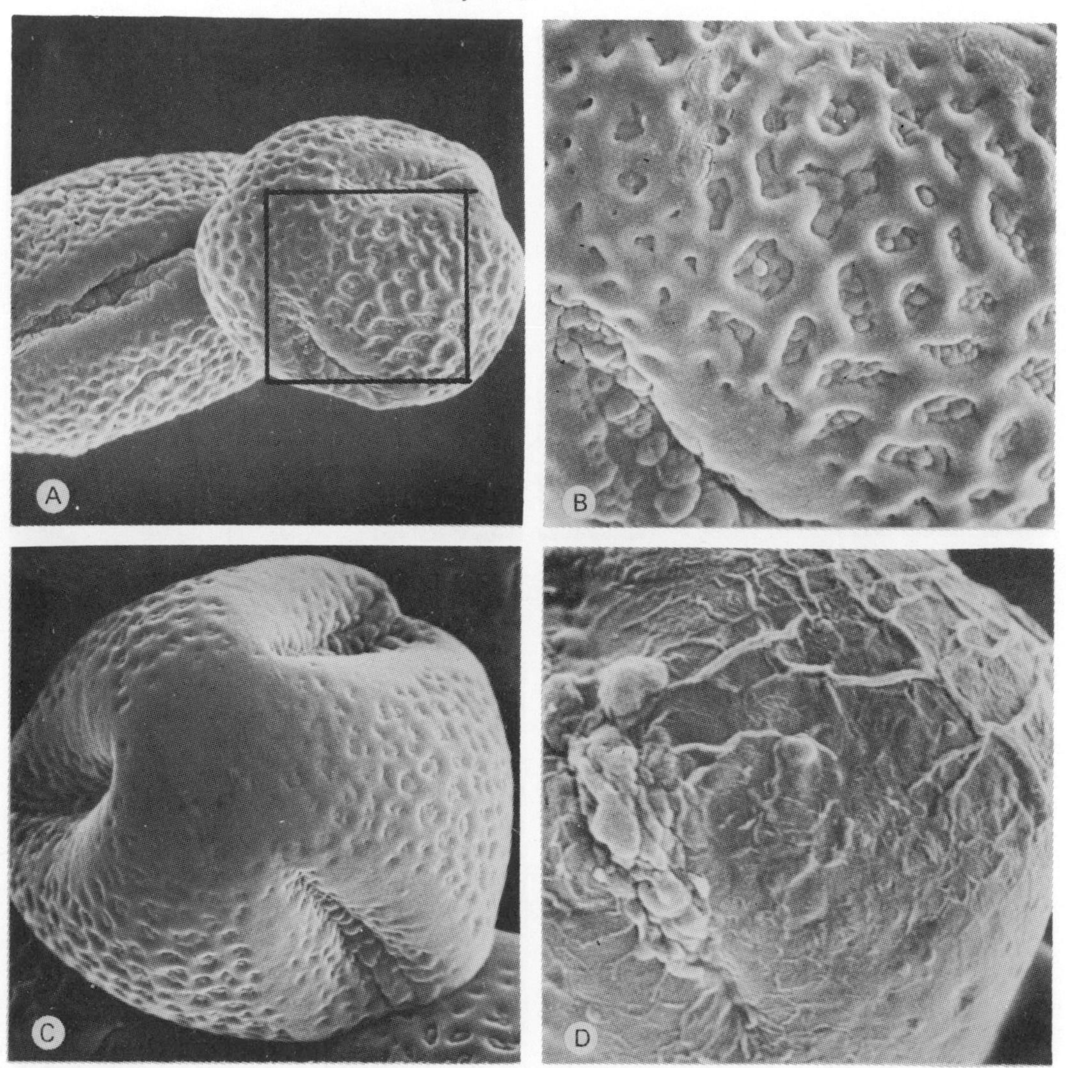

Fig. 9. Pollen (S. E. M.)
A – *T. tembense,* near polar end (× 1,500); B – same, enlarged (× 4,000); C – *T. burchellianum* ssp. *johnstonii,* near polar end (× 2,000); D – *T. baccarinii,* near furrow (× 4,000)

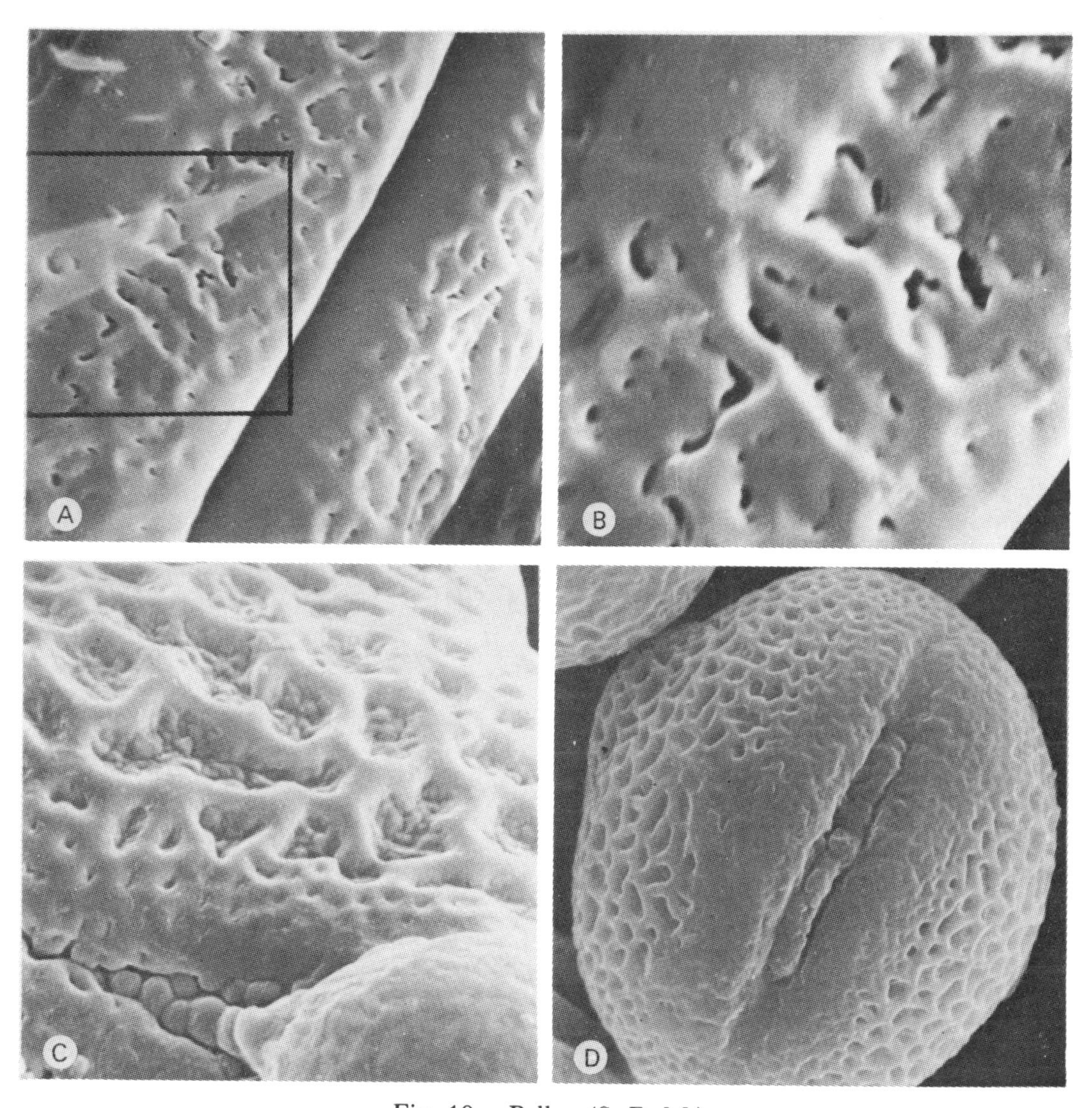

Fig. 10. Pollen (S. E. M.)
A – *T. mattirolianum*, near furrow (× 4,000); B – same, enlarged (× 8,000); C – *T. stellatum*, near furrow (× 4,000); D – *Ononis ornithopodioides*, near furrow (× 4,000)

the grain surface; the polar ends are foveolate and the furrow edges are smooth; the luminae have very irregular shapes and interspaces and are alveolate within (Fig. 10A–B).

Notwithstanding the meticulous labour invested in these examinations, no correlation has yet been found between the pollen grain morphology and the taxonomic divisions of Section *Lotoidea* as proposed here. Several pollen types may occur within each subsection, and the same pollen type is encountered in several subsections. Furthermore, morphologically related species differ in their pollen grains, while other unrelated species may have the same pollen type.

Gillett et al. (1973) suggest that, with few exceptions, pollen ornamentation in Section *Lupinaster* (sensu Gillett) is of a simpler type, and this lends some support to Zohary's (1972a) theory that this section is more primitive than the others. Additional evidence in favour of this opinion is based on morphological features as well as the linkage of this section with an extinct lupinoid stock. However, our observations on pollen types do not corroborate this view, since species of Subsection *Lupinaster* do not show grain ornamentation of a simpler type; they are included in our type 3 together with species of two other subsections, as well as those of other genera of the tribe *Trifolieae*.

Graham & Tomb (1977), in their studies of the pollen of *Erythrina*, found that pollen types do not consistently parallel the taxonomic categories; in this genus, too, the same type of pollen may occur in several different and unrelated sections and subgenera.

The Seed Coat as Viewed under the Scanning Electron Microscope

Preliminary studies were made of the seed coat surface of 24 species : 17 of Section *Lotoidea*, the others being representatives of the various sections of the genus. The surface of the seed may be smooth (Fig. 11C), roughened, tuberculate (Fig. 11A), wrinkled or pitted. In our observations we were able to distinguish five different types of seed coat patterns according to the nature of the uppermost layer and the various substances deposited on the coat surface.

The following types of seed coat morphology have been observed :

1. The surface of the seed coat is almost smooth, showing only very slight granulation. This type of seed is found in only one African species : *T. lanceolatum* (Fig. 11C).
2. The surface of the seed coat is covered with shallow scale-like elevations and scattered small granules. This type was found in four species of Section *Lotoidea* [*T. baccarinii*, *T. burchellianum* ssp. *johnstonii* (Fig. 11D), *T. michelianum* var. *michelianum* and *T. repens*] and in *T. meduseum* of Section *Trichocephalum*. Heyn & Herrnstadt (1977) reported that *Lupinus albicaulis* (Fig. 5F) has seed of approximately the same type.
3. The surface of the seed coat is densely covered by tubercles of two kinds: smooth [as noted in *T. glomeratum*(Fig. 11A and Fig. 12A–B) of Section *Lotoidea* and in *T.*

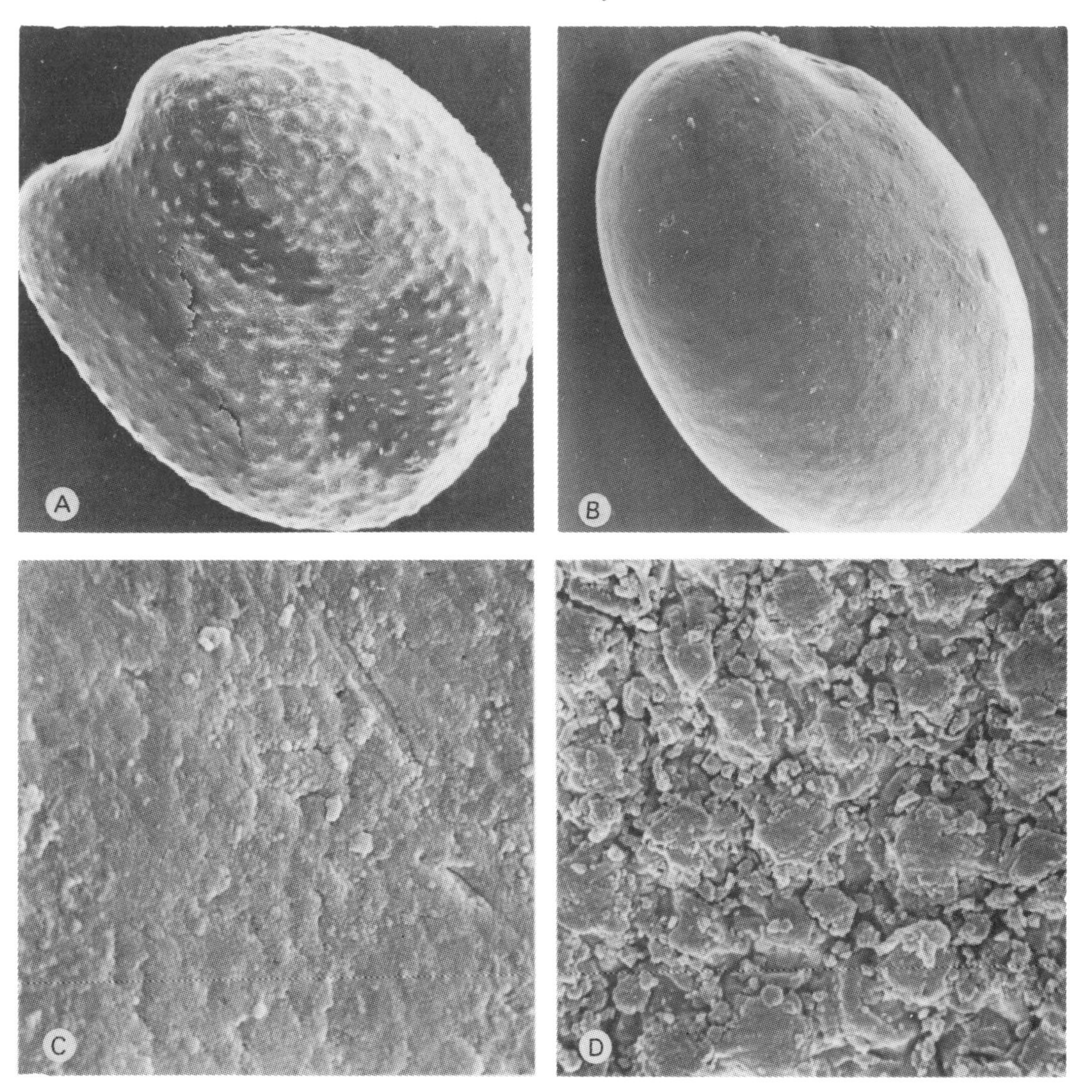

Fig. 11. **Seeds** (S. E. M.)
A–B : Seed shape from side view
A – *T. glomeratum* (× 60); B – *T. microcephalum* (× 40)
C–D : Seed coat
C – *T. lanceolatum* (× 1,600); **D** – *T. burchellianum* ssp. *johnstonii* (× 1,600)

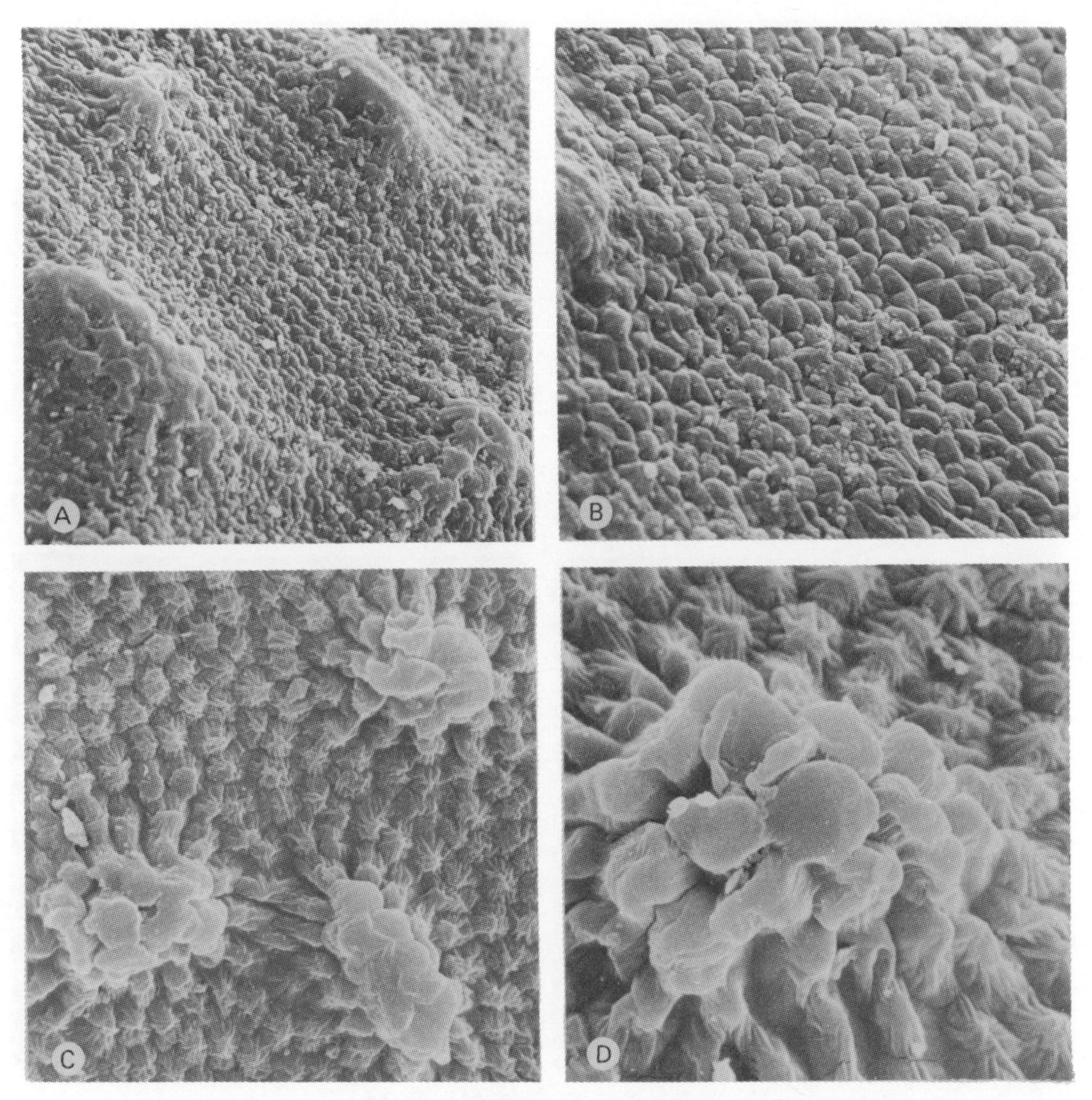

Fig. 12. Seed coat (S. E. M.)
A – *T. glomeratum* (× 800); B – same, enlarged (× 1,600); C – *T. polymorphum* (× 800); D – same, enlarged (× 1,600)

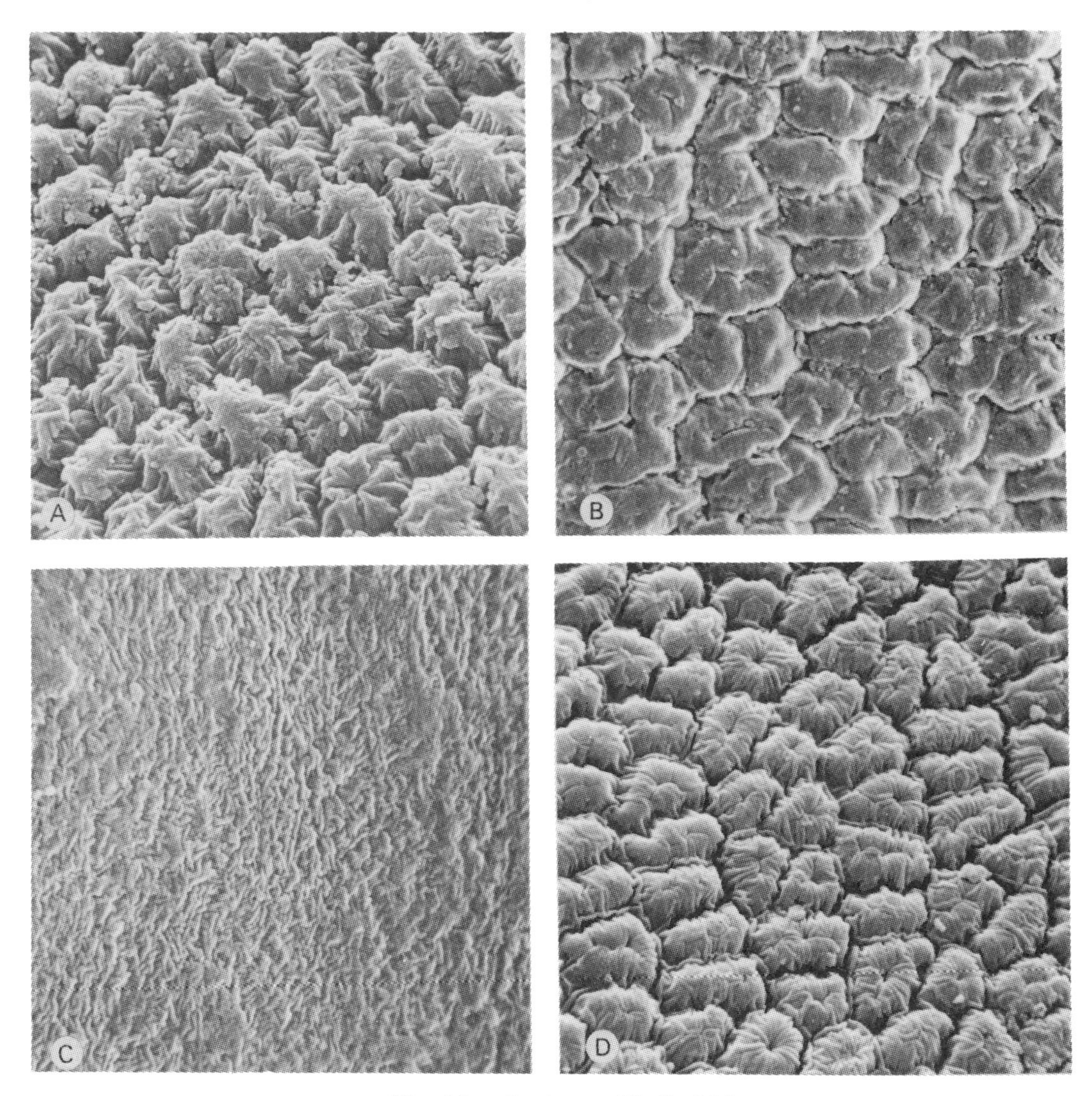

Fig. 13. Seed coat (S. E. M.)
A – *T. hybridum* (× 1,600); B – *T. ambiguum* (× 1,600); C – *T. masaiënse* (× 1,600); D – *T. microcephalum* (× 1,600)

glanduliferum of Section *Paramesus*] and wrinkled or folded [as found in five species of Section *Lotoidea* (*T. africanum*, *T. bejariense*, *T. hybridum* (Fig. 13A), *T. isthmocarpum*, *T. polymorphum* (Fig. 12C–D) and also in *T. lugardii* of Section *Mistyllus*].

In most of these species, with the exception of *T. africanum* and *T. hybridum*, the top of the tubercles is beset with a few larger protuberances. The latter have the appearance of a group of tubercles and can be seen with the aid of a magnifying glass.

4. The surface of the seed coat is densely covered with shallow protuberances, the tops of which resemble a volcanic crater. These protuberances are wrinkled throughout, including the mouth of the crater. This type was observed in four species of Section *Lotoidea* [*T. ambiguum* (Fig. 13B), *T. gracilentum*, *T. mattirolianum*, *T. nigrescens*] and in *T. microcephalum* (Fig. 13D) of Section *Involucrarium*.

These protuberances differ somewhat in *Lupinus luteus* (Heyn & Herrnstadt, 1977). Such protuberances seem to be derived from the tuberculate type 3, where a few tubercles of this form appear on the surface of the seed coat of *T. hybridum*.

5. The surface of the seed coat is covered by a very dense network of delicate, irregular but variously oriented wrinkles. This type was observed in two species of Section *Lotoidea* [*T. masaiënse* (Fig. 13C) and *T. ornithopodioides*] and in three other species : *T. purpureum* var. *desvauxii* (Section *Trifolium*), *T. dubium* (Section *Chronosemium*) and *T. batmanicum* (Section *Trichocephalum*). The seeds of *Lupinus mutabilis* (Fig. 5A), recorded by Heyn & Herrnstadt (1977), closely resemble this type.

As in the case of the pollen grains, the seed coat morphology does not appear to correlate with the proposed taxonomic divisions. This serves as additional support for the conclusion reached by Gunn & Barnes (1977), in their study of *Erythrina*, that there is no apparent correlation between seed features and current phylogenic hypotheses.

TAXONOMIC PART

Description of the Genus

TRIFOLIUM L.

Sp. Pl. 764 (1753)

Lectotypus: *T. pratense* L. (Britten & Brown, Ill. Fl. N. U. S. ed. 2, 2:353, 1913).

Annual or perennial herbs with erect, ascending or procumbent stems: sometimes with creeping rootstocks. Leaves with 3, rarely with 5 (or up to 9) dentate or entire leaflets; stipules partly connate and adnate to petioles, often sheathing, entire or sometimes dentate. Inflorescences many, axillary or terminal, often head-like, spicate or umbellate, pendunculate or sessile, many-, rarely few-flowered. Bracts evident or absent, free or connate ones sometimes forming involucre. Flowers pedicellate or sessile, all or rarely only outer (lower) ones fertile. Calyx generally tubular or campanulate, 5–10 or –20 (rarely 30–36)-nerved, with 5 equal or unequal, entire or dentate teeth; sometimes calyx bilabiate, with upper teeth often shorter or longer and partly connate at base; throat of calyx open or closed by callosity or ring of hairs; tube of calyx sometimes inflated in fruit. Corolla persistent, marcescent or caducous, sometimes inflated in fruit, white, yellow, purple, pink, flesh-coloured, lilac, violet or 2-coloured; standard free or connate at base with wings and keel; the latter two sometimes connate with one another and mostly adnate to stamens; wings often longer than keel. Stamens diadelphous; all filaments, or 5 of them, dilated at apex; anthers uniform. Ovary sessile or stipitate, with 1–12 ovules. Pod indehiscent or sometimes opening by sutures, usually 1–2 (rarely 4–9)-seeded, enclosed in persistent calyx and sometimes also in persistent corolla, rarely exserted, usually membranous, rarely leathery , ovoid to oblong or linear. Seeds globular to ovoid and oblong, sometimes reniform or lenticular. Dispersal units: seed, fruiting calyx or entire head.

Distr.: Eurasia, Africa, N. and S. America.

Key to Sections

1. Heads provided with conspicuous involucre made up of more or less connate and variously incised, dentate or rarely herbaceous or membranous bracts; involucre longer than pedicels and sometimes longer than calyx. Stipules mostly dentate or incised, not glandular-toothed. Section Eight. **Involucrarium**
– Heads without involucre, or with short free or slightly connate involucre, and then pedicels much longer than involucre or calyx vesicular or stipules glandular-toothed 2

2. Calyx inflated and vesiculous in fruit; pod included in calyx 3
– Calyx not as above 4
3. Fruiting calyx markedly asymmetric-bilabiate, 10-nerved; teeth of calyx very unequal in fruit. Section Four. **Vesicaria**
– Calyx symmetrical, 20–36-nerved; teeth of calyx equal. Section Three. **Mistyllus**
4. Calyx with 2 very short upper teeth and 3 long lower teeth. Standard persistent, spoon- or boat-shaped, serving as dispersal apparatus for 1-seeded, stipitate pod. Section Five. **Chronosemium**
– Calyx and corolla not as above 5
5. Heads with 2–15 corollate and fertile flowers and 20–50 sterile non-corollate flowers. Section Seven. **Trichocephalum**
– Heads with all flowers corollate, all fertile or very few or only uppermost ones sterile 6
6. Leaflets and stipules serrate, the latter with teeth terminating in glandular tips. Pod 2-seeded, exserted from calyx. Section Two. **Paramesus**
– Stipules not glandular-tipped 7
7. Flowers sessile, ebracteolate almost without exception. Throat of calyx thickened or closed by annular or bilabiate callosity provided with ring of hairs. Pod 1-seeded, membranous with leathery upper part, transversely rupturing when ripe. Section Six. **Trifolium**
– Flowers predominantly pedicellate and bracteolate; rarely bracteoles or pedicels or both lacking but then calyx throat naked and open and pod provided with more or less distinct and thick ventral and/or dorsal suture. Seeds mostly 2–8. Section One. **Lotoidea**

Section One. LOTOIDEA Crantz

Stirp. Austr. 405 (1769)

Subgenus *Lotoidea* Pers., Syn. Pl. 2 : 348 (1807).
Section *Trifoliastrum* Gray, Nat. Arr. Brit. Pl. 2 : 598 (1821).
Section *Trifoliastrum* Ser. in DC. , Prodr. 2 : 198 (1825).
Genus *Amoria* C. Presl, Symb. Bot. 1 : 47 (1830).
Section *Brachydontium* Reichenb., Fl. Germ. Excur. 491 (1832) p. p.
Subgenus *Trifoliastrum* (Ser.) Peterm., Deutschl. Fl. 1 : 140 (1847).
Section *Amoria* (Presl) Lojac., Nuov. Giorn. Bot. Ital. 15 : 125, 220 (1883).
Subgenus *Amoria* (Presl) Hossain, Not. Roy. Bot. Gard. Edinb. 23 : 459 (1961).

Type species : *T. repens* L.

Annuals or perennials. Stems simple or branched, some rhizomatous or scapose. Leaves 3–5(–9)-foliolate. Inflorescences umbellate, capitate or spicate, sometimes scape-like. Bracts entire or bifid or crenulate, rarely inconspicuous or absent (in Subsect. *Neolagopus*). Pedicels long or short, rarely absent. Calyx usually 10-nerved, rarely 5- or 20(–30)-nerved; calyx throat naked, open; calyx teeth generally equal, sometimes unequal. Corolla variously coloured, marcescent. Ovary with 1–12 ovules, sessile, rarely stipitate. Pod indehiscent or opening by ventral suture or also by both sutures.
This Section consists of 99 species arranged in 9 subsections; spread over Europe, N. & S. America, Africa and Asia.

Key to Subsections

1. Leaves (at least part of them) 5–9-foliolate. Pod short-stipitate, hairy towards apex. Subsection III. **Lupinaster**
– Leaves all 3-foliolate. Pod mostly sessile, glabrous or hairy all along 2
2. Petioles adnate for their entire length to stipules so that internodes are sheathed. Subsection IV. **Ochreata**
– Petioles free or almost so 3
3. Fruiting calyx inflated or somewhat accrescent 4
– Fruiting calyx not inflated as above 5
4. Calyx with 10–11 prominent nerves, much inflated in fruit and constricted at throat. Inflorescences dense, many-flowered. Bracts linear, more than twice as long as pedicels. Subsection VIII. **Calycospatha**
– Calyx 15–30-nerved, somewhat accrescent in fruit. Inflorescences 1–12-flowered. Bracts minute. Subsection II. **Loxospermum**
5. Pedicels, all or at least part of them, conspicuous and deflexed in fruit 6
– Pedicels short or absent or sometimes only those of lower flowers deflexed in fruit 8
6. Inflorescences of 2 kinds : aerial and subterranean, the latter 2–8-flowered with long peduncles penetrating soil and fruit ripening there. Subsection VI. **Oxalioidea**

– Inflorescences only aerial 7

7. Pod up to 1 cm long, much exserted from calyx. Inflorescences subsessile or borne on 1–2 cm long peduncles. Subsection I. **Falcatula**

– Pod included in calyx or slightly exserted. Inflorescences usually long-peduncled. Subsection V. **Lotoidea**

8 (5). Pedicels short or only those of lower flowers conspicuous and deflexed in fruit. Bracts conspicuous. Pods 1–2(–4)-seeded. Subsection VII. **Platystylium**

– Pedicels absent. Bracts inconspicuous, cup-like. Pods always 1-seeded. Subsection IX. **Neolagopus**

Synopsis of Species in Section One

Section One. LOTOIDEA Crantz

Subsection I. FALCATULA (Brot.) Aschers. & Graebn.

1. T. ornithopodioides (L.) Sm.

Subsection II. LOXOSPERMUM (Hochst.) Čelak.

2. T. chilaloense Thulin
3. T. decorum Chiov.
4. T. elgonense Gillett
5. T. multinerve A. Rich.
6. T. schimperi A. Rich.

Subsection III. LUPINASTER (Adans.) Belli

7. T. andersonii A. Gray
8. T. gymnocarpon Nutt.
9. T. lemmonii S. Wats.
10. T. lupinaster L.
11. T. macrocephalum (Pursh) Poir.
12. T. polyphyllum C. A. Mey.
13. T. thompsonii Morton

Subsection IV. OCHREATA (Lojac.) Gillett

14. T. cheranganiense Gillett
15. T. cryptopodium A. Rich.
16. T. mauginianum Fiori
17. T. polystachyum Fresen.
18. T. simense Fresen.
19. T. somalense Taub.
20. T. stolzii Harms
21. T. ukingense Harms
22. T. usambarense Taub.
23. T. wentzelianum Harms

Subsection V. LOTOIDEA

Series i. GRANDIFLORA Heller & Zoh.

24. T. alpinum L.
25. T. attenuatum Greene
26. T. eximium Steph. ex Ser.
27. T. nanum Torr
28. T. pilczii Adamović

29. T. uniflorum L.

Series ii. PHYLLODON Heller & Zoh.

30. T. bejariense Moric.
31. T. carolinianum Michx.

Series iii. BRACHYANTHA Heller & Zoh.

32. T. retusum L.

Series iv. LOTOIDEA

33. T. amabile Humb., Bonpl. & Kunth
34. T. angulatum Waldst. & Kit.
35. T. bifidum A. Gray
36. T. bivonae Guss.
37. T. breweri S. Wats.
38. T. burchellianum Ser.
39. T. cernuum Brot.
40. T. euxinum Zoh.
41. T. gillettianum Jac.-Fél.
42. T. gracilentum Torr. & Gray
43. T. hybridum L.
44. T. latifolium (Hook.) Greene
45. T. leibergii Nels. & Macbride
46. T. masaiënse Gillett
47. T. michelianum Savi
48. T. nigrescens Viv.
49. T. pallescens Schreb.
50. T. parnassi Boiss. & Sprun.
51. T. purseglovei Gillett
52. T. radicosum Boiss. & Hohen.
53. T. reflexum L.
54. T. repens L.
55. T. rusbyi Greene
56. T. semipilosum Fresen.
57. T. stoloniferum Muhlenberg
58. T. thalii Vill.
59. T. virginicum Small

Series v. PECTINATA Heller & Zoh.

60. T. ciliolatum Benth.

Series vi. CURVICALYX Heller & Zoh.

61. T. eriocephalum Nutt.

Series vii. PRODUCTA Heller & Zoh.

62. T. beckwithii Brewer
63. T. bolanderii A. Gray
64. T. brandegei S. Wats.
65. T. haydenii Porter
66. T. howellii S. Wats.
67. T. kingii S. Wats.

Series viii. ACAULIA (Baker) Heller & Zoh.

68. T. acaule A. Rich.
69. T. petitianum A. Rich.
70. T. vestitum Heller & Zoh.

Subsection VI. OXALIOIDEA (Gillett) Heller

71. T. polymorphum Poir.

Subsection VII. PLATYSTYLIUM Willk.

Series i. PLATYSTYLIUM

72. T. abyssinicum Heller
73. T. africanum Ser.
74. T. ambiguum M. B.
75. T. baccarinii Chiov.
76. T. bilineatum Fresen.
77. T. isthmocarpum Brot.
78. T. lanceolatum (Gillett) Gillett
79. T. longipes Nutt.
80. T. montanum L.
81. T. owyheense Gilkey
82. T. pichisermollii Gillett
83. T. rueppellianum Fresen.
84. T. spananthum Thulin
85. T. tembense Fresen.

Series ii. MACROCHLAMIS Heller & Zoh.

86. T. dasyphyllum Torr. & Gray
87. T. parryi A. Gray
88. T. riograndense Burkart

Series iii. PHYLLOCEPHALA Heller & Zoh.

89. T. andinum Nutt.

Series iv. MICRANTHEUM (C. Presl) Heller & Zoh.

90. T. glomeratum L.
91. T. pachycalyx Zoh.
92. T. suffocatum L.

Series v. ALTISSIMA (Abrams) Heller

93. T. douglasii House

Subsection VIII. CALYCOSPATHA (Chiov.) Heller & Zoh.

94. T. mattirolianum Chiov.
95. T. pseudostriatum, Baker f.

Subsection IX. NEOLAGOPUS Lojac.

96. T. albopurpureum Torr. & Gray
97. T. dichotomum Hook. & Arn.
98. T. macraei Hook. & Arn.
99. T. plumosum Dougl. ex Hook.

Key to Species

1. Calyx glabrous throughout — 2
– Calyx entirely or partly pubescent — 57
2. Calyx teeth half the length of tube, ovate-triangular, imbricate at base. — **50. T. parnassii**
– Calyx teeth not as above — 3
3. Perennials or rarely annuals with creeping stems rooting at nodes — 4
– Annuals or perennials; stems not creeping and not rooting at nodes — 14
4. Pedicels erect, slightly deflexed or only those of lower flowers deflexed after anthesis — 5

– Pedicels all strongly deflexed after anthesis 11
5. Leaves almost sessile; short petioles almost entirely adnate to sheathing stipules 6
– Upper leaves sometimes subsessile, but stipules never sheathing 7
6. Leaflets obovate-obtriangular, 0.2–1 cm long, rounded or truncate, emarginate and apiculate at apex. **15. T. cryptopodium**
– Leaflets oblong, obovate or rhomboid, 0.7–2 cm long, obtuse or acute at apex. **20. T. stolzii**
7. Annuals. Stems erect or spreading. Upper leaves subsessile 8
– Perennials. Stems densely caespitose or matted. Leaves long-petioled 9
8. Calyx tube 11-nerved; margin of teeth broadly scarious. Corolla about 1 cm long; standard oblong, truncate, recurved at tip. **85. T. tembense**
– Calyx tube 15-nerved; margin of teeth not evident. Corolla 3–4 mm long; standard obovate, rounded at apex. **75. T. baccarinii**
9. Calyx teeth unequal, shorter than tube, the upper ones longer than the lower. **58. T. thalii**
– Calyx teeth equal, as long as tube or longer 10
10. Peduncles much longer than subtending leaves. Leaflets deeply emarginate, long-petioled, up to 4 cm. **69. T. petitianum**
– Peduncles shorter than to as long as subtending leaves. Leaflets rounded or shallowly emarginate, short-petioled, 1–1.5 cm. **68. T. acaule**
11 (4). Corolla at least twice as long as calyx. Peduncles very long. Pedicels often much longer than calyx. **54. T. repens**
– Corolla, peduncles and pedicels shorter than above 12
12. Calyx teeth subulate from base, twice as long as tube. **38. T. burchellianum**
– Calyx teeth triangular up to middle and then subulate or sharp-pointed, as long as or slightly longer than tube 13
13. Calyx teeth equal, slightly ciliate or denticulate, with lower part scarious-margined. **51. T. purseglovei**
– Calyx teeth unequal, neither scarious-margined nor denticulate in lower part, the upper ones longer than the lower. **58. T. thalii**
14 (3). Corolla at least twice as long as calyx 15
– Corolla shorter than above 33
15. Pedicels erect, slightly deflexed or only those of lower flowers deflexed after anthesis 16
– Pedicels all strongly deflexed after anthesis 20
16. Calyx teeth straight 17
– Calyx teeth curved or some of them tortuose 18
17. Bracts oblong, purplish, scarious, outer ones often united, forming involucre. **87. T. parryi**
– Bracts narrowly lanceolate-subulate, conduplicate, greenish, membranous-margined, all free. **74. T. ambiguum**
18. Inflorescences 0.8–1 cm in diam., sessile. **90. T. glomeratum**
– Inflorescences 1.5–3.5 cm in diam., on 3–10 cm long peduncles 19
19. Annuals. Stems many, often fistulose, much branched. Leaflets obovate or obtriangular, cuneate at base. **77. T. isthmocarpum**
– Perennials. Stems few, stout, poorly branched. Leaflets oblong to narrowly elliptical or linear-lanceolate. **93. T. douglasii**
20 (15). Rhachis prolonged beyond flowers, simple or forked, often bearing cluster of sterile, bud-like flowers on upper whorl 21
– Rhachis not as above 26

21. Stems up to 1 m high, fistulose. Leaflets very large, up to 5 cm broad. **66. T. howellii**
– Stems not more than 0.5 m high. Leaflets smaller 22
22. Calyx tube 5-nerved 23
– Calyx tube 10-nerved 24
23. Peduncle slender, geniculate near base of heads. Standard curved, narrowly ovate. **63. T. bolanderi**
– Peduncle thick, erect. Standard straight, obovate to broadly elliptical. **62. T. beckwithii**
24. Leaves mostly basal 25
– Leaves basal and along stems. **67. T. kingii** subsp. **productum**
25. Flowers curved, saccate at base. Calyx teeth shorter than tube. **67. T. kingii** subsp. **rollinsii**
– Flowers straight. Calyx teeth as long as tube. **65. T. haydenii**
26 (20). Calyx teeth at least twice as long as tube 27
– Calyx teeth as long as or somewhat longer or somewhat shorter than tube 28
27. Leaflets linear-lanceolate, to linear or oblanceolate, margin entire or very finely dentate. Bracts short, forming small involucre. **24. T. alpinum**
– Leaflets ovate to obovate, margin denticulate with broad teeth. Bracts scale-like, free. **52. T. radicosum** var. **guestii**
28. Annuals 29
– Perennials 30
29. Inflorescences 6–12-flowered. Calyx greenish, 15–20-nerved. **3. T. decorum**
– Inflorescences 30–50(or more)-flowered. Calyx whitish, with 10 obscure nerves. **40. T. euxinum**
30. Inflorescences 1–8-flowered 31
– Inflorescences 10- to many-flowered 32
31. Leaflets linear to narrowly lanceolate. Calyx teeth lanceolate, subulate. **12. T. polyphyllum**
– Leaflets oblanceolate to obovate. Calyx teeth triangular. **27. T. nanum**
32. Caespitose alpine plants with short, erect or ascending stems and long, scape-like peduncles arising from them. Leaves usually very crowded at base of plants. **49. T. pallescens**
– Stems erect or ascending, rarely caespitose but with well-developed branches. Leaves scattered all along stems. **43. T. hybridum**
33 (14). Leaves sessile; petioles adnate for their entire length to sheathing stipules 34
– Leaves petiolate; stipules not sheathing 35
34. Some leaves 4-foliolate. Calyx 5-nerved (doubtful species). **16. T. mauginianum**
– All leaves 3-foliolate. Calyx 17–20-nerved. **18. T. simense**
35. Pedicels erect, slightly deflexed or only those of lower flowers deflexed after anthesis 36
– Pedicels all strongly deflexed after anthesis 45
36. Calyx tube 15–30-nerved 37
– Calyx tube 5–11-nerved 40
37. Peduncles absent. Flowers axillary, solitary or in pairs with pedicels almost absent. Leaflets up to 7 mm long, cuneate-obcordate or obovate. **4. T. elgonense**
– Inflorescences long-pedunculate. Flowers pedicellate. Leaflets more than 7 mm long 38
38. Leaflets linear. Flowers 1.8–2 cm long. **6. T. schimperi**
– Leaflets oblanceolate. Flowers 0.7–1 cm long 39

39. Peduncles glabrous. Calyx teeth as long as or slightly shorter than tube. **5. T. multinerve**
– Peduncles slightly pilose, especially near base of heads. Calyx teeth twice as long as tube. **2. T. chilaloense**
40 (36). Inflorescences 1–10-flowered 41
– Inflorescences 15- to many-flowered 43
41. Mat-forming perennials. Stipules up to 1.5 cm long, free portion abruptly contracted to fine point. **84. T. spananthum**
– Annuals. Stipules up to 7 mm long, acuminate 42
42. Inflorescences densely crowded, overtopped by broad membranous stipules. Calyx teeth in fruit recurved or falcate. **92. T. suffocatum**
– Inflorescences loosely scattered, not overtopped by leafy stipules. Calyx teeth remain erect in fruit. **1. T. ornithopodioides**
43. Calyx tube tubular, not inflated in fruit; teeth straight in fruit. **83. T. rueppellianum**
– Calyx tube campanulate, inflated in fruit and markedly constricted at throat; teeth strongly recurved in fruit 44
44. Heads sessile, subtended by uppermost leaves. Corolla shorter than to as long as calyx teeth. Bracts linear-lanceolate, up to 5 mm long. **95. T. pseudostriatum**
– Heads pedunculate, not subtended by leaves. Corolla much exserted from calyx. Bracts linear-subulate, up to 3 mm long. **94. T. mattirolianum**
45 (35). Calyx teeth recurved in fruit 46
– Calyx teeth erect in fruit 47
46. Peduncles as long as or shorter than subtending leaves. Calyx somewhat leathery. Corolla shorter than calyx. **32. T. retusum**
– Peduncles longer than subtending leaves. Calyx membranous. Corolla longer than calyx. **48. T. nigrescens**
47. Calyx teeth hyaline, prominently ciliate, dentate to pectinate at margin. **60. T. ciliolatum**
– Calyx teeth neither ciliate, dentate nor pectinate, margin rarely slightly serrate 48
48. Leaflets oblong-cuneate or linear-cuneate, at least some on each plant deeply retuse to deeply bifid. **35. T. bifidum**
– Leaflets not bifid but sometimes slightly retuse 49
49. Leaflets oblong-lanceolate or lanceolate, acute. Calyx teeth membranous-margined. **42. T. gracilentum** var. **palmeri**
– Leaflets obovate to oblong-obovate or broadly elliptical 50
50. Inflorescences (all or at least the upper ones) subsessile. Flowers 4–5 mm long. Standard deeply emarginate. **39. T. cernuum**
– Inflorescences long-pedunculate. Flowers longer 51
51. Calyx teeth (all or part of them) as long as or shorter than tube, rarely slightly longer 52
– Calyx teeth twice as long as tube 53
52. Perennials. Bracts deeply bilobed or dentate. Inflorescences up to 3 cm. Flowers 1–1.2 cm. Pod oblong. **36. T. bivonae**
– Annuals. Bracts oblong, acute or acuminate. Inflorescences 1–2 cm. Flowers 6–9 mm. Pod linear-oblong. **48. T. nigrescens**
53. Inflorescences up to 3 cm in diam. 54
– Inflorescences up to 1.5 cm in diam. 56
54. Inflorescences (2–)5–8(–10)-flowered, loose. Leaves short-petioled, the upper ones subsessile. **28. T. pilczii**

– Inflorescences 10- to many-flowered, dense. Leaves long-petioled, the uppermost ones 1–2 cm long 55

55. Bracts 1 mm, broad-obovate, scarious, truncate to bifid. Standard obovate-oblong, rounded at apex, retuse, apiculate, more or less erose-denticulate. **53. T. reflexum**

– Bracts 2 mm, hyaline, lanceolate. Standard ovate-oblong, acutish, with apical margin entire. **47. T. michelianum**

56 (53). Teeth of calyx equal, subulate. Bracts linear, folded, acute to acuminate. **34. T. angulatum**

– Teeth of calyx unequal, narrowly triangular. Bracts cup-like. **42. T. gracilentum**

57 (1). Heads few-flowered (1–15) 58

– Heads many-flowered (20 to many) 85

58. Pedicels erect, slightly deflexed or only those of lower flowers deflexed after anthesis 59

– Pedicels all deflexed after anthesis 68

59. Inflorescences sessile 60

– Inflorescences distinctly pedunculate 62

60. Calyx about 25-nerved; teeth erect in fruit. Pod 6-seeded. **4. T. elgonense**

– Calyx 10–12-nerved; teeth recurved in fruit. Pod 2-seeded 61

61. Almost stemless minute plants. Heads few-flowered. Flowers erect in fruit. Fruiting calyx prominently nerved, not becoming corky. **92. T. suffocatum**

– Plants with many conspicuous stems. Heads 10–15-flowered. Flowers recurving after anthesis. Fruiting calyx becoming thick, corky on upper face and nerveless in fruit. **91. T. pachycalyx**

62. Leaves almost sessile; short petioles almost entirely adnate to sheathing stipules. **15. T. cryptopodium**

– Leaves long-petioled or only uppermost ones sessile 63

63. Leaves, at least few, 5-foliolate; stipules irregularly dentate. **8. T. gymnocarpon**

– Leaves all 3-foliolate; stipules entire 64

64. Inflorescences usually in pairs, subsessile or sessile, subtended by stipules and upper leaves. Pedicels absent. **98. T. macraei**

– Inflorescences single, pedunculate, not subtended by upper leaves and stipules. Pedicels conspicuous 65

65. Leaflets rounded to acute at apex. Lower bracts forming involucre, sometimes united at base. **86. T. dasyphyllum**

– Leaflets rounded to truncate or emarginate at apex. Lower bracts not forming involucre and not united at base 66

66. Flowers up to 2 cm long. Stems densely covered with sheathing stipules. **72. T. abyssinicum**

– Flowers up to ca. 1 cm long. Stems not densely covered with sheathing stipules 67

67. Annuals. Upper leaves subsessile; petioles sometimes fused with stipules. Standard oblong, truncate, curved downwards at tip. **85. T. tembense**

– Mat-forming perennials. Upper leaves long-petioled. Standard widest above middle and straight. **84. T. spananthum**

68 (58). Peduncles geniculate or bending near base of head 69

– Peduncles erect, neither geniculate nor bending 71

69. Calyx bilabiate, the upper side longer than the lower one; teeth foliaceous with reticulate or branched nerves 70

– Calyx tubular; teeth subequal, subulate. **37. T. breweri**

70. Calyx teeth, except lower one, ovate, reticulately nerved and villous along margin only. **30. T. bejariense**

– Calyx teeth broadly triangular, with slightly branched nerves, villous all along. **31. T. carolinianum**

71. Inflorescences and flowers of two kinds: one terminal, aerial, long-peduncled; the other basal, of subterranean flowers. **71. T. polymorphum**

– Inflorescences and flowers of one kind – aerial only 72

72. Leaves, at least part of them, 5–8-foliolate; petioles very short and united with sheathing stipules. **10. T. lupinaster**

– Leaves all 3-foliolate, petiolate 73

73. Annuals or biennials 74

– Perennials, mostly caespitose 77

74. Margin of calyx teeth ciliate, dentate to pectinate 75

– Margin of teeth not as above 76

75. Leaflets narrow-oblanceolate, rounded and apiculate at apex. Calyx campanulate, green; tube 15–16-nerved. **41. T. gillettianum**

– Leaflets elliptical to cuneate-elliptical, oblong to obovate in lower leaves, rounded, obtuse to retuse. Calyx green and also violet-blue; tube oblique, 10-nerved, upper side longer than lower one. **60. T. ciliolatum**

76. Leaflets narrow-obcordate to oblong-cuneate or linear-cuneate. Corolla 5–7 mm long; standard elliptical to oblong, entire. **35. T. bifidum**

– Leaflets ovate to obovate, broadly oblong. Corolla 0.9–1.4 cm long; standard obovate-oblong, more or less erose-denticulate, flat in fruit. **53. T. reflexum**

77 (73). Calyx teeth less than twice as long as tube 78

– Calyx teeth at least twice as long as tube 82

78. Stoloniferous plants, rooting at nodes. Free portion of stipules falcate. **56. T. semipilosum**

– Plants not as above 79

79. Pedicels thickened and strongly recurved and twisted in fruit. Pod 5–7-seeded. **29. T. uniflorum**

– Pedicels neither thickened nor twisted in fruit. Pod 1–3-seeded 80

80. Leaflets obtriangular, bilobed at apex, mucronate. Bracts 3–6, forming involucre at base of head. **70. T. vestitum**

– Leaflets ovate to obovate or oblong, rounded or emarginate. Bracts minute, not forming involucre 81

81. Leaflets ovate to obovate, cuneate at base. Calyx about 4 mm; teeth lanceolate-subulate. Corolla 8–9 mm long. **52. T. radicosum** var. **radicosum**

– Leaflets oblong to oblong-elliptical. Calyx 7–10 mm; teeth narrow-triangular, acute. Corolla 1.5–1.8 cm long. **64. T. brandegei**

82 (77). Leaflets linear to lanceolate, acute to acuminate. Standard acute. **25. T. attenuatum**

– Leaflets ovate to obovate or oblong, rounded to retuse, rarely acute. Standard rounded, sometimes retuse 83

83. Stipules, at least some of upper ones, with 1 or more lateral teeth or lobes. Leaflets up to 4 × 2.5 cm. **44. T. latifolium**

– Stipules entire or margin slightly serrulate. Leaflets up to 2.2 × 1.4 cm 84

84. Bracts forming scarious, crenate involucre. Corolla 1.5–2 cm long; standard broadly ovate, slightly emarginate. **26. T. eximium**

– Bracts linear, not forming involucre. Corolla 0.6–1.4 cm long; standard obovate-oblong, rounded. **33. T. amabile**

85 (57). Leaves sessile; petioles adnate along their entire length to sheathing stipules 86

– Leaves petiolate; stipules not sheathing 95

86. Leaflets linear or elliptical, tapering at both ends, acute to acuminate, 4–10 times as

long as wide 87
- Leaflets obovate to oblong with obtuse, retuse or rounded apex and cuneate base, sometimes elliptical, 2–3 times as long as wide 91

87. Inflorescences spicate, oblong-cylindrical. **17. T. polystachyum**
- Inflorescences globular or ovoid 88

88. Calyx tube greenish, with less than 15 nerves 89
- Calyx tube whitish, 17–20-nerved; teeth subulate, as long as or somewhat longer than tube. **18. T. simense**

89 Leaflets nearly entire or sometimes with few teeth below middle. Calyx teeth somewhat longer than tube. **21. T. ukingense**
- Leaflets spinulose-dentate or sharply serrulate. Calyx teeth almost twice as long as tube 90

90. Stipules oblong to ovate. Bracts subulate with broad base. Pedicels up to 1 mm long, only lower ones deflexed. **23. T. wentzelianum**
- Stipules lanceolate or linear-lanceolate. Bracts lanceolate. Pedicels 2–4 cm long, all deflexed after anthesis. **10. T. lupinaster**

91 (86). Inflorescences ovoid to elliptical. Calyx conspicuously longer than corolla. **22. T. usambarense**
- Inflorescences more or less globular to semi-globular. Calyx shorter than corolla 92

92. Pedicels 4–6 mm long, deflexed after anthesis 93
- Pedicels up to 2 mm long, none or only those of lower flowers deflexed in fruit 94

93. Roots tuberous and thick. Stems few, erect or ascending. Calyx somewhat shorter than corolla. Pedicels up to 4 mm long. **19. T. somalense**
- Roots stout not tuberous. Stems prostrate, sometimes rooting at nodes. Calyx about half the length of corolla. Pedicels 5–6 mm long. **14. T. cheranganiense**

94. Leaflets obovate-obtriangular, 0.2–1 cm long, rounded or truncate, emarginate and apiculate at apex. **15. T. cryptopodium**
- Leaflets oblong, obovate or rhomboid, 0.7–2 cm long, obtuse or acute at apex. **20. T. stolzii**

95 (85). Leaves digitate, at least few of them 5–9-foliolate 96
- Leaves all 3-foliolate 102

96. Inflorescences up to 7 × (3–)5 cm. Teeth of calyx 2–4 times as long as tube 97
- Inflorescences and teeth of calyx shorter 98

97. Leaflets obovate, 1–2 times as long as wide; stipules oblong, margin irregularly toothed or lobed. **11. T. macrocephalum**
- Leaflets linear to linear-lanceolate, 6–10 times as long as wide; stipules ovate to lanceolate, margin entire. **13. T. thompsonii**

98. Petioles united to and entirely hidden by sheathing stipules. **10. T. lupinaster**
- Petioles free, much exserted from stipules 99

99. Leaflets entire or very obscurely dentate near apex. Pedicels short, erect after anthesis. **7. T. andersonii**
- Leaflets acutely or spinulosely dentate. Pedicels deflexed after anthesis 100

100. Leaflets linear to linear-lanceolate, 6–10 times as long as wide. **13. T. thompsonii**
- Leaflets ovate, obovate to elliptical or oblong, 1–3(–4) times as long as wide 101

101. Stems branching in upper and lower parts. Leaflets obovate, cuneate at base. **9. T. lemmonii**
- Stems short, arising from branching crown. Leaflets ovate to oblong or elliptical. **8. T. gymnocarpon**

102 (95). Pedicels erect, slightly deflexed or only those of lower flowers deflexed after anthesis 103

– Pedicels all deflexed after anthesis 126
103. Leaflets linear, lanceolate, oblanceolate or elliptical, at least 3 times as long as wide 104
– Leaflets obovate, ovate to oblong, no more than twice as long as wide 114
104. Bracts inconspicuous, at most 1–2 times as long as wide 105
– Bracts conspicuous, 4–10 times as long as wide 109
105. Calyx 17–20-nerved 106
– Calyx 5–10-nerved 107
106. Leaflets often folded, falcate. Inflorescences spicate, ovate to cylindrical. Calyx teeth subulate, erect. **99. T. plumosum**
– Leaflets not as above. Inflorescences umbellate, globular. Calyx teeth narrow, triangular to subulate, the lateral and upper ones tortuose and curved downwards and inwards around corolla, the lower one straight. **93. T. douglasii**
107. Inflorescences usually formed of 2 sessile heads, subtended by membranous uppermost stipules and leaves, thus forming pseudo-involucre. **89. T. andinum**
– Inflorescences of heads long-peduncled 108
108. Upper stipules ovate to lanceolate, acuminate, toothed. Calyx pilose; teeth unequal. **79. T. longipes**
– Upper stipules oblong, free portion triangular-lanceolate, entire. Calyx densely patulous-hairy; teeth more or less equal. **73. T. africanum** var. **lydenburgense**
109 (104). Annuals. Stems branching all along 110
– Perennials. Caespitose or stems erect, poorly branching in upper part 111
110. Bracts 5–7 mm long, ovate-lanceolate, 2–3-toothed; lower bracts sometimes united at base forming involucre up to length of inflorescences. **82. T. pichisermolli**
– Bracts up to 3 mm long, linear, entire, lower ones not united and not forming involucre. **78. T. lanceolatum**
111. Bracts membranous-margined, conduplicate 112
– Bracts not membranous, sometimes scarious, not folded 113
112. Leaflets leathery, obtuse or acute, mucronulate. Standard ovate-oblong, recurved, rounded to truncate at apex and slightly apiculate. **80. T. montanum**
– Leaflets thin, rounded at apex. Standard narrowly ovate to elliptical, rounded at apex and slightly retuse. **74. T. ambiguum**
113. Free portion of stipules linear. Leaflets sometimes conduplicate, margin entire. Lower bracts sometimes united, forming involucre. **86. T. dasyphyllum**
– Free portion of stipules triangular-lanceolate. Leaflets not folded, margin obsoletely spinulose-denticulate. Bracts not united and not forming involucre. **73. T. africanum**
114 (103). Annuals. Stems branching all along 115
– Perennials. Caespitose or stems erect, poorly branching in upper part 122
115. Inflorescences spicate. Bracts inconspicuous, cup-like. Pedicels absent 116
– Inflorescences umbellate, mostly globular. Bracts conspicuous, linear, lanceolate. Pedicels conspicuous 118
116. Heads sessile to subsessile, usually in pairs, subtended by upper leaves and their stipules. Calyx tube 5–10-nerved. **98 T. macraei**
– Heads pedunculate, solitary. Calyx tube 20–30-nerved 117
117. Calyx teeth 3–6 times as long as tube. Corolla 4–6 mm, half the length of to as long as calyx. **96. T. albopurpureum**
– Calyx teeth 2–3 times as long as tube. Corolla 1.3–1.7 cm, as long as to more than twice as long as calyx. **97. T. dichotomum**
118. Calyx inflated in fruit, with markedly constricted throat; teeth strongly recurved in fruit. **95. T. pseudostriatum**

– Calyx not as above; teeth not recurved in fruit 119
119. Bracts minute 120
– Bracts 2–7 mm long 121
120. Calyx tube 15-nerved; teeth twice as long as tube. Corolla 3–4 mm long, shorter than calyx. **75. T. baccarinii**
– Calyx tube with 10 obsolete nerves; teeth as long as tube. Corolla 7–9 mm long, longer than calyx. **76. T. bilineatum**
121. Bracts ovate-lanceolate, 2–3-toothed, the lower ones sometimes united at base forming involucre up to length of inflorescences. Standard denticulate at apex. **82. T. pichisermolli**
– Bracts linear or lanceolate, sometimes truncate or bifid, never united at base, not forming involucre. Standard rounded at apex, entire. **83. T. rueppellianum**
122 (114). Outer bracts foliaceous, many, forming involucre up to 1 cm long and up to 6 mm wide, free or irregularly connate. **88. T. riograndense**
– Outer bracts not as above 123
123. Bracts minute 124
– Bracts 2–4 mm long 125
124. Stipules fused at base, broadly obovate, slightly lobed. Leaflets slightly overlapping, ovate, obovate to circular, emarginate. **81. T. owyheense**
– Stipules not fused at base, ovate to lanceolate, entire or toothed. Leaflets mostly lanceolate to elliptical or oblong; cauline leaflets sometimes ovate, apex usually acute. **79. T. longipes**
125. Leaflets leathery, obtuse or acute, mucronulate. Standard ovate-oblong, recurved, rounded to truncate at apex, slightly apiculate. **80. T. montanum**
– Leaflets thin, rounded at apex. Standard narrowly ovate to elliptical, rounded at apex, slightly retuse. **74. T. ambiguum**
126 (102). Perennials with creeping stems rooting at nodes 127
– Annuals or perennials. Stems not creeping and not rooting at nodes 130
127. Inflorescences and flowers of two kinds: a long-peduncled and a basal, short-peduncled on stoloniferous stems with subterranean flowers. **71. T. polymorphum**
– Inflorescences and flowers of one kind – aerial only 128
128. Bracts truncate to bifid. **38. T. burchellianum**
– Bracts lanceolate, acuminate 129
129. Calyx sparingly appressed-pubescent; teeth twice as long as tube. Lateral nerves of leaflets anastomosing near margin. Inflorescences up to 3.5 cm in diam. **57. T. stoloniferum**
– Calyx villous; teeth up to 1.5 times as long as tube. Lateral nerves of leaflets not anastomosing near margin. Inflorescences about 2 cm in diam. **56. T. semipilosum**
130. Leaflets linear, lanceolate, oblanceolate or elliptical, at least 3 times as long as wide 131
– Leaflets obovate to oblong, no more than twice as long as wide 136
131. Annuals 132
– Perennials 133
132. Leaflets narrow-oblanceolate, rounded and apiculate at apex. Calyx tube 15–16-nerved, sides equal; teeth lanceolate, abruptly cuspidate, ciliate at margin. **41. T. gillettianum**
Leaflets narrow-obcordate to oblong-cuneate or linear-cuneate, rounded or mostly truncate, deeply retuse to bifid, mucronate between two lobes. Calyx tube 10-nerved, the upper side slightly larger than the lower one; teeth subulate, margin entire. **35. T. bifidum**

133. Peduncles bent near head – thus heads inverted or horizontal. Calyx tube curved after anthesis; teeth unequal, often curved and twisted. **61. T. eriocephalum**
– Peduncles and calyx not as above 134
134. Standard tapering towards apex, acuminate or rostrate and then also wings and keel rostrate. **55. T. rusbyi**
– Standard elliptical, rounded or truncate, sometimes mucronate or apiculate at apex 135
135. Leaflets very acute or acuminate, margin entire. Inflorescences hemispherical, flowers whorled. **25. T. attenuatum**
– Leaflets rounded or truncate, margin serrulate. Inflorescences spherical, all flowers radiating. **59. T. virginicum**
136 (130). Annuals or rarely biennials 137
– Perennials 142
137. Calyx bilabiate, the upper side longer than the lower one; teeth foliaceous, with reticulate or branched nerves 138
– Calyx and teeth not as above 139
138. Calyx teeth, except lower one, ovate, reticulately nerved and villous along margin only. **30. T. bejariense**
– Calyx teeth broadly triangular, with slightly branched nerves, villous all along. **31. T. carolinianum**
139. Rhachis sometimes prolonged beyond flowers 140
– Rhachis not prolonged 141
140. Upper side of calyx longer than lower one; tube oblique; margin of teeth hyaline and strongly ciliate, dentate to pectinate. **60. T. ciliolatum**
– Calyx sides equal; tube very short; sinus between entire teeth broad. **53. T. reflexum**
141. Bracts 3–7 mm long, oblong, acute to acuminate. Calyx teeth recurved in fruit, as long as or slightly longer than tube. **48. T. nigrescens**
– Bracts minute, lanceolate. Calyx teeth straight in fruit, 2–3 times as long as tube. **46. T. masaiënse**
142 (136). Margin of stipules and leaflets coarsely spinulose-dentate. Densely villous plants. **45. T. leibergii**
– Plants not as above 143
143. Rhachis sometimes prolonged beyond flowers 144
– Rhachis not prolonged 146
144. Peduncles erect, 1 cm long to longer than subtending leaves. Upper side of calyx longer than lower one. **33. T. amabile**
– Peduncles geniculate or bending near base of head – thus flowers inverted. Calyx sides equal 145
145. Stipules acute, entire or 1–3-toothed near apex. Corolla 1.2–1.5 cm long; standard longer than wings and keel. **44. T. latifolium**
– Stipules long-acuminate, entire. Corolla 6–8 mm long; standard as long as wings and keel. **37. T. breweri**
146. Bracts about 1 mm, lanceolate. Sinus between calyx teeth broad. Standard ovate-oblong, slightly denticulate above. **43. T. hybridum**
– Bracts minute, scale-like, oblong, truncate to bifid, slightly denticulate. Sinus between calyx teeth narrow. Standard oblong, somewhat curved at tip. **52. T. radicosum** var. **radicosum**

Subsection I. FALCATULA (Brot.) Aschers. & Graebn.

Syn. Mitteleur. Fl. 6 : 510 (1908)

Genus *Falcatula* Brot., Phyt. Lusit. Select. ed. 2, 1 : 160, t. 68 (1816).
Ornithopoda Malladra, Malphigia 4 : 168 (1890) pro stirpe.
Subsection *Ornithopoda* (Malladra) Taubert in Engl. & Prantl, Naturl. Pflanzenfam. III, 3 : 249 (1893).
Subgenus *Falcatula* (Brot.) Coombe, Fl. Eur. 2 : 160 (1968).

Type species : *T. ornithopodioides* (L.) Sm.

Low annuals. Leaves 3-foliolate, petiolate. Inflorescences 1–4(5)-flowered, all axillary, from subsessile to pendunculate. Peduncles up to 2 cm long. Pedicels erect or rarely deflexed after anthesis. Pod up to 1 cm long, much exserted from calyx, 5–9-seeded, slightly falcate, nerved and dehiscent. One species.

1. T. ornithopodioides (L.) Sm., Fl. Brit. 2 : 782 (1800). *Typus* : England; Hort. Cliff. 376, no. 2 (BM !).

Trifolium Melilotus ornithopodioides L., Sp. Pl. 766 (1753).
Melilotus ornithopodioides (L.) Desr. in Lam., Encycl. Meth. Bot. 4 : 67 (1797).
Trigonella ornithopodioides (L.) Lam. & DC., Fl. Fr. ed. 3, 4 : 550 (1805).
Falcatula falsotrifolium Brot., Phyt. Lusit. Select. ed. 2, 1 : 160, t. 65 (1816), nom. illegit.
Trigonella uniflora Mumby in Bull. Soc. Bot. Fr. 11 : 45 (1864).
Trifolium perpusillum Simk., in Oesterr. Bot. Zeitschr. 40 : 333 (1890).

Icon. : [Plate 1]. Fiori & Paol., Ic. Fl. Ital. 233 (1899).

Low, procumbent, glabrous annual, 5–10(–20) cm. Leaves long-petioled, up to 5 cm, slender; stipules about 8 mm, lanceolate, acuminate; leaflets 0.4–1.4 × 0.3–0.8 cm, obcordate, obovate, cuneate, truncate or retuse, mucronate, finely serrate. Inflorescences 1–4(–5)-flowered, axillary, subsessile to pedunculate. Peduncles up to 2 cm long. Bracts 1 mm, lanceolate, hyaline. Pedicels 1–3 mm, erect or rarely deflexed after anthesis. Calyx 5 mm; tube cylindrical, 10-nerved; teeth triangular, sharp-pointed, longer than tube. Corolla 6–8 mm long, white or pink; standard oblong, rounded at apex, longer than wings and keel. Pod 6–8 (–10) × 3–4 mm, exserted from calyx, linear-oblong, 5–9-seeded, slightly falcate, nerved and hairy, dehiscent. Seed 1–2 mm, ovoid, smooth. Fl. May–June. 2n = 16.

Hab. : Slopes, near coasts and roadsides.

Gen. Distr. : Azores, Baleares, Spain, Portugal, France, Germany, Hungary, Ireland, England, Corsica, Sardinia, Italy, Yugoslavia, Romania, Netherlands, Morocco, Algeria.

Selected specimens : SPAIN : prov. Zamora, Rivadelago, 1948, *Losa* & *Montsarrot* (HUJ). FRANCE : Cherbourg, 1890, *Corbiere* (B). HUNGARY : Kis-Jenö, Com. Arad., 1890, *Simonkai* 2626 (BM, MPU as *T. perpusillum* Simk.). ROMANIA : Oltenia, distr. Craiova, 50 m, 1961, *Pâun* & *Malos* 337 (BM); Banatus, distr. Timis-Torontal, Timisoara, 90 m,

Plate 1. *T. ornithopodioides* (L.) Sm.
(Romania : *Bujorean* 2441, HUJ).
Plant in flower and fruit; flower; fruiting calyx with pod.

1941, *Bujorean* 2441 (HUJ). MOROCCO : Bu-Meziat, 1,700 m, 1927, *Font Quer* 300 (BM). ALGERIA : Djebel-Ksel, prov. Oran, 1856, *Kralik* (BM, MPU).

Subsection II. LOXOSPERMUM (Hochst.) Čelak.

Oesterr. Bot. Zeitschr. 24 : 41 (1874)

Genus *Loxospermum* Hochst. in Flora 29 : 594 (1846).

Type species: *T. schimperi* A. Rich.

Annuals, or rarely perennials, usually glabrous. Leaves 3-foliolate, petiolate; stipules not sheathing. Inflorescences 1–12-flowered, umbellate. Peduncles long or absent. Bracts minute. Pedicels evident or absent, partly deflexed or not deflexed. Calyx 15–30-nerved, somewhat accrescent in fruit. Ovary 3–12-ovuled. Five African species.

2. T. chilaloense Thulin, Bot. Notiser 129 : 169 (1976). *Typus* : Ethiopia : Arussi Prov., Chilalo awraja, near Kersa, ca. 50 km S.W. of Asella, 2,900 m, 12.11.1971, *Thulin* 1642 (UPS, iso. : BR, EA, ETH, FI, HUJ !, K, MO, WAG).

Icon. : [Plate 2]. Thulin, *op. cit.* f. 1 F–J.

Annual, glabrous or sparsely pilose in upper parts, up to 12 cm. Stems few, profusely branching. Leaves on up to 4 cm long petioles; stipules 1–2 cm long, ovate-oblong, with prominent reddish nerves, upper free portion (about one-third) tapering into filiform tip; leaflets 1–2.5 × 0.3–0.7 cm, oblanceolate to very narrowly elliptical, acute, margin finely serrate in upper part. Inflorescences few, umbellate, 5–12-flowered, globular. Peduncles up to 4 cm long, slightly pilose, especially near base of head. Bracts minute, ovate, forming involucre. Pedicels 0.5–1.5 mm long. Calyx 6–8 mm long, glabrous; tube 15–18-nerved; teeth about twice as long as tube, narrowly triangular, subulate at tip, with few hairs at margin. Corolla 7–8 mm long, purplish-red; standard obovate, about as long as wings, longer than keel. Ovary sparsely papillose, 4-ovuled. Pod 5–6 × 3 mm, glabrous, ellipsoidal, 3–4-seeded, with prominently thickened sutures. Seeds 1.6 mm, ovoid, brown, mottled with black spots, finely warty. Fl. October. 2n = 16.

Hab. : Grassland patches in *Hagenia* forest; 2,900 m.

Gen. Distr. : Ethiopia (endemic).

Note : Only the isotype seen.
T. chilaloense differs from *T. multinerve* by its globular heads, slightly pilose peduncles and its calyx teeth, the latter being twice as long as the tube.

3. T. decorum Chiov. in Ann. Bot. Rom. 9 : 56 (1911). *Typus* : Abyssinia: Amhara–Dembia, Gondar, 17.8.1909, *Chiovenda* 1110 (FI ! and syn. : 1117, 1199, 1391 all in FI !).

Plate 2. *T. chilaloense* Thulin
(Ethiopia : *M. Thulin* 1642, HUJ).
Plant in flower and fruit; flower; standard; wing; keel; fruiting calyx.

Icon. : [Plate 3]. Fiori in Nuov. Giorn. Bot. Ital. n. s. 55 : f. 1, 6 (1948).

Annual, glabrous, 20–30 cm. Stems many, erect to ascending, deeply sulcate, leafy, branching all along. Lower leaves long-petioled, upper ones shorter-petioled; stipules about 1 cm long, not sheathing, obovate, free portion ovate-triangular, acuminate; leaflets 0.5–1.8 × 0.3–1 cm, broadly elliptical to obovate, cuneate at base, rounded or truncate at apex, margin denticulate, especially in upper part. Inflorescences 1.5–2.5 cm in diam., umbellate, depressed globular, 6–12-flowered. Peduncles terminal and axillary, 2–6 cm long. Bracts minute, up to 1 mm, ovate to subulate, entire or dentate, forming outer involucre at base of head. Pedicels up to 2 mm long, partly deflexed after anthesis. Calyx 5 mm long, glabrous, membranous; tube becoming campanulate after flowering, with 15–20 distinct nerves, somewhat inflated in fruit; teeth half the length of tube, lanceolate, subulate above. Corolla 1–1.2 cm long, purplish-violet; standard elliptical-spatulate, slightly longer than wings and keel. Pod 6 × 2 mm, ellipsoidal, 3–5-seeded, membranous. Seeds 2 × 1 mm, minutely tuberculate, brown, ellipsoidal. Fl. August.

Hab. : Open, grazed, grassy hillsides, marshy places and streamsides.

Gen. Distr. : Ethiopia (endemic).

Specimens seen : ETHIOPIA : Choke Mts., Gojjam, upper Ghiedeb Valley, 10,000 ft, 1957, *Evans* & *Lythgoe* (K); 6 mi. S. of Dembecka on Addis-Ababa road, 7,100 ft, 1964, *Thomas* 4 (K).

4. T. elgonense Gillett in Kew Bull. 7 : 395 (1952). *Typus* : Kenya : Mount Elgon, Eastern slopes above Japata estate, alpine belt, close to small stream, 3,450 m, 3 March 1948, *O. Hedberg* 266 (UPS, iso. : EA, K !).

Icon. : [Plate 4]. Gillett, *op. cit.* f. 4/9.

Annual, glabrous, 5–20 cm. Stems many, procumbent to ascending, branching. Leaves on 1–3 cm long petioles; stipules about 1 cm long, ovate, free portion triangular-acuminate, membranous, nerved; leaflets 3–7 × 5 mm, obovate to obcordate, cuneate at base, truncate or emarginate at apex, coarsely serrate. Inflorescences axillary, 1–2-flowered. Peduncles absent. Pedicels very short or up to 1.5 mm long. Calyx 5–8 mm long, glabrous or with few hairs; tube inflated, growing in fruit, about 25-nerved; teeth shorter than tube, hairy at margin, broadly triangular, subulate above. Corolla 9 mm long, purplish; standard oblong-obovate, tapering at base, slightly longer than wings and keel. Pod 6–7 mm long, ellipsoidal, 6-seeded, membranous-hyaline, margins incrassate, sutures prominent. Seeds about 1 mm, ovoid, yellowish-brown. Fl. December–March. 2n = 16.

Hab. : Wet open places in upland, evergreen forests, moorlands and alpine zones; 2,700–3,500 m.

Gen. Distr. : Uganda, Kenya, Ethiopia.

Specimens seen : KENYA : Trans Nzoia Distr., Mt. Elgon, Track S. W. of Suam Sawmills, 2,700 m, 1967, *Gillett* 18409 (HUJ). ETHIOPIA : Shoa Prov., Entott Mt., Addis-Ababa, above Italian Embassy, 8,200 ft, 1964, *Meyer* 8620 (NA).

Plate 3. *T. decorum* Chiov.
(Ethiopia : *Chiovenda* 1391, FI).
Plant in flower and fruit; pod; flower; standard; wing; keel.

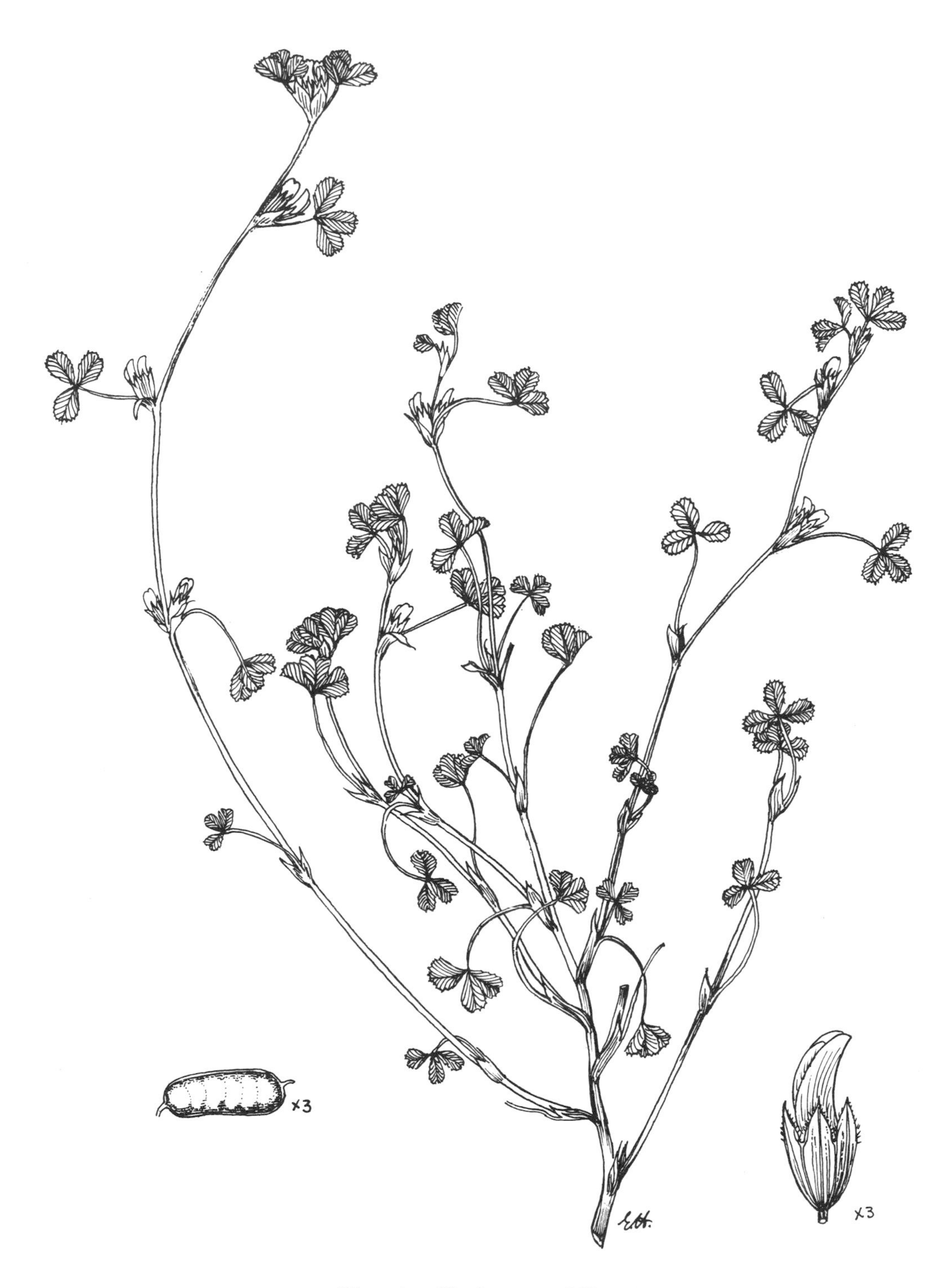

Plate 4. *T. elgonense* Gillett
(Kenya : *Gillett* 18409, HUJ).
Flowering branch; pod (Ethiopia : *Meyer* 8620, NA); flower.

Note: *T. elgonense* is easily distinguished by the absence of peduncles, its 1–2 subsessile flowers and its broadly triangular calyx teeth, hairy at their margin.

5. T. multinerve A. Rich., Tent. Fl. Abyss. 1 : 175 (1847). *Typus*: N. Ethiopia: Locis humidis vallium prope Adoam, 30 Sept. 1837, *Schimper* 300 (P !, syn. K !, isosyn.: B !, G ! W !); Mennessah, *Quartin Dillon* (P !).

Trigonella multinervis Hochst. & Steud. in Sched. Pl. Schimp. Ab. 1 : 300 (1837) nomen.
Trifolium multinerve A. Rich. var. *debilior* A. Rich. *loc.cit.*
Loxospermum multinerve Hochst. in Flora 29 : 595 (1846).

Icon.: [Plate 5]. Gillett, Kew Bull. 7 : f. 5 (1952).

Annual, sometimes perennial, glabrous, 10–20 cm. Stems few, somewhat divaricately branching. Leaves long-petioled; stipules up to 1 cm, ovate-oblong, free filiform portion much shorter than lower part; leaflets 1–2 × 0.4–0.6 cm, oblanceolate-cuneate, rounded or acute, margins obsoletely to strongly dentate-serrate. Inflorescenses 1–6-flowered. Peduncles fairly long. Bracts minute, broadly ovate, mostly dentate, forming outer involucre. Pedicels 1–2 mm long. Calyx 6–8 mm, glabrous; tube tubular, 15–30-nerved, rather inflated in fruit; teeth as long as tube or longer, triangular, acuminate, hirtellous at margin. Corolla 7–11 mm, purple; standard cuneate-obovate. Pod 7–8 mm, glabrous, ellipsoidal, 6–7-seeded. Seeds 1.5 × 0.8 mm, oblong, brown. Fl. May–December. 2n = 16.

Hab.: Short upland grassland and moorland, especially in moist places; 1,800–3,650 m.

Gen. Distr.: Uganda, Kenya, E. Congo, Sudan, Eritrea, Ethiopia.

Selected specimens: KENYA: Eldoret Distr., Elgeyo escarpment, 8,000 ft, 1926, *Harger* (BM); Trans Nzoia Distr., Mt. Elgon, S. W. of Suam Sawmills, 2,700 m, 1967, *Gillett* 18407 (HUJ). ERITREA: Hamasen, Açmara, 2,300 m, 1902, *Pappi* 338 (W); Scimenzana, 2,480 m, 1902, *Pappi* 864 (FI). ETHIOPIA: Scioa, 2,300 m, 1909, *Negri* 87 (FI); Addis-Ababa, 1909, *Philip* 26465b (HUJ).

6. T. schimperi A. Rich., Tent. Fl. Abyss. 1 : 173 (1847). *Typus*: Abyssinia: Locis in cultis vallium prope Adoam, 19 Sept., 1837, *Schimper* 86 (FI ! holo., iso.: BM !, FI !, G !, W !).

Trigonella schimperi Hochst. & Steud. in Sched. Pl. Schimperi (Iter Abyssinicum) Ab. 1 : 86 (1840) nomen.

Icon.: [Plate 6]. Fiori in Nuov. Giorn. Bot. Ital. n. s. 55 : f. 13 (1948).

Annual, glabrous, 15–30 cm. Stems many, erect or ascending, much-branched, grooved. Leaves on 2–5 cm long petioles; stipules 1.5–3 cm long, connate part loosely sheathing, oblong, linear, free portion subulate-cuspidate, about as long as lower part; leaflets 2–5 × 0.2–0.3 cm, linear, tapering at both ends, entire or obsoletely dentate in upper part. Inflorescences 1–3(–5)-flowered, mainly axillary. Peduncles 2–3 cm long, somewhat thickened. Bracts inconspicuous or absent. Pedicels 1–3 mm long. Calyx up to 1.8 cm long, glabrous; tube subcampanulate, with 15 prominent

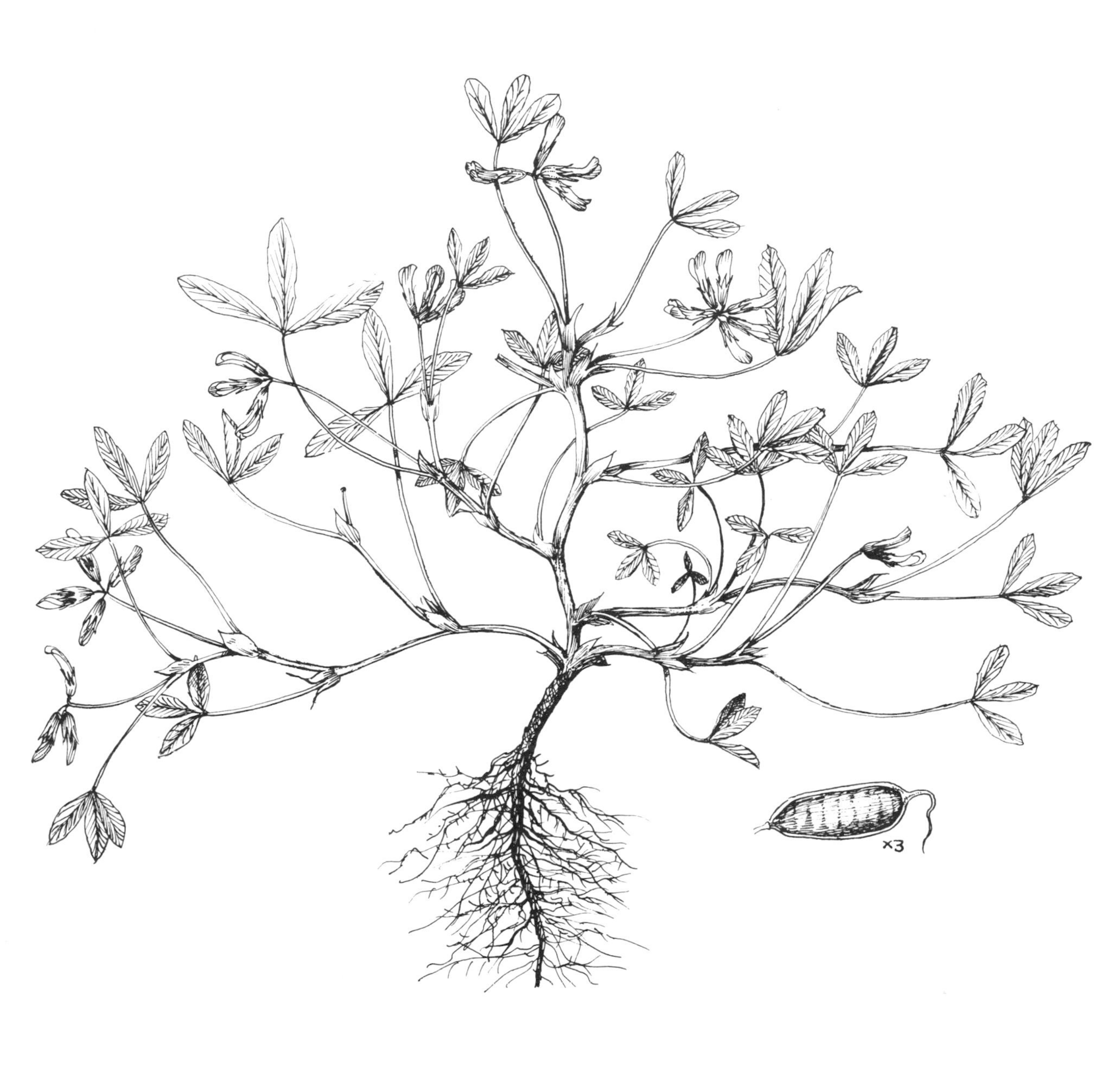

Plate 5. *T. multinerve* A. Rich.
(Ethiopia : *Philip* 26465b, HUJ).
Plant in flower; pod (Eritrea : *Pappi* 864, FI).

Plate 6. *T. schimperi* A. Rich
(Ethiopia : *Schimper* 299, FI).
Plant in flower and fruit.

nerves; teeth at least twice as long as tube, 5-nerved at base, linear-subulate. Corolla 1.8–2 cm long, rose-purple; standard broad-elliptical, tapering at apex. Pod 1.4–1.8 × 0.4–0.5 cm, oblong-linear, 8–10-seeded, with prominent sutures. Seeds 2 × 1 mm, ellipsoidal, minutely tuberculate, brown. Fl. May–September.

Hab. : Valleys.

Gen. Distr. : Ethiopia (endemic).

Specimens seen : ETHIOPIA : Amhara-Dembia, Gondar, 1909, *Chiovenda* 1173 (FI); Scioa: Addis-Ababa, 1939, *Pietro Benedetto* 02 (FI); Jimma, Agr. Tech. School, 1955, *Stewart* A.74 (K); 1854, *Schimper* 299 (FI, G).

Subsection III. LUPINASTER (Adans.) Belli

Mem. Acad. Sci. Torino ser. 2, 44 : 233 (1894) (pro stirpe)

Genus *Lupinaster* Adans., Fam. Pl. 2 : 323 (1763).
Genus *Pentaphyllon* Pers., Syn. Pl. 2 : 352 (1807) "*Pentaphyllum* Pers.", in Spreng., Syst. Veg. 3 : 286 (1826).
Genus *Dactiphyllon* Raf., Am. Monthly Mag. Crit. Rev. 2 : 268 (1818).
"*Dactiphyllum*" Raf., Jour. Phys. Chim. Hist. Nat. 89 : 261 (1819).
Section *Lupinaster* (Adans.) Link, Enum. Hort. Berol. Alt. 2 : 260 (1822); Ser. in DC., Prodr. 2 : 203 (1825) p. p.
Section *Thompsoniana* Morton, Jour. Wash. Acad. Sci. 23 : 271 (1933).
Subsection *Glyzyrrhizum* (Bertol.) Bobrov in Acta Inst. Bot. Acad. Sci. URSS, ser. 1, 6 : 214 (1947) p. p.

Type species : *T. lupinaster* L.

Perennials. Leaves petiolate, rarely sessile, digitate with (3–)5–9 leaflets. Inflorescences often scape-like, subtended by short involucre of connate bracts. Flowers in few whorls, pedicellate. Pod included in or exserted from calyx, shortly stipitate, hairy especially towards apex (sometimes a tuft of hairs). Seeds 2–4(–8). Five N. American and 2 Eurasian species.

7. T. andersonii A. Gray, Proc. Am. Acad. 6 : 522 (1865).

Caespitose, matted perennials, densely pilose, 2–15 cm. Stems many, short, ascending or decumbent, arising from woody thickened base and long taproot, covered with withered stipules. Leaves with 2–5 cm long petioles; stipules 1–2 cm long, scarious, linear to lanceolate, covering base of stem; leaflets 3–7(–8), 5–25 × 2–12 mm, obovate to obovate-elliptical or oblanceolate, often folded, thickened, rounded or acute, obtuse, abruptly mucronate, densely wooly with long hairs on both surfaces, margin entirely or slightly dentate towards apex. Inflorescences 1.5–3 cm in diam., globular, 10- to many-flowered, whorled. Peduncles about as long as or longer than subtending leaves. Bracts minute and scarious, the lower ones arranged in whorls at base of head. Pedicels short, not reflexed. Calyx 8–12 mm long; tube cylindrical, densely pilose; teeth up to twice as long as tube, subulate. Corolla 10–17 mm long; standard pinkish-purple turning brown with age, oblong, rounded or retuse, longer than white wings and keel. Pod small, 1–2-seeded, densely villous towards apex, sessile or very short-stipitate. Seeds ovoid, smooth. Fl. May–July. 2n = 16.

1. Leaflets 9–25 mm long, pilose and spreading-pubescent — 2
– Leaflets 5–11 mm long, appressed-pubescent. — **7b.** subsp. **monoense**
2. Peduncles about as long as subtending leaves. — **7a.** subsp. **andersonii**
– Peduncles longer than subtending leaves. — **7c.** subsp. **beatleyae**

7a. T. andersonii A. Gray Subsp. **andersonii**. *Typus*: Mountains of Nevada, near Carson City, 1862, *C.L. Anderson* (GH, iso.: E !, K !, NY, US).

Icon.: [Plate 7a]. McDerm., N. Am. Sp. Trif. Pl. 72 (1910).

Hab.: Dry slopes and valleys; 1,200–2,500 m
Gen. Distr.: USA: California, Nevada.

Selected specimens: CALIFORNIA: Sierra Co., 1874, *Lemmon* 41 (G). NEVADA: Alum Creek, Washoe Co., 6,000 ft, 1907, *Kennedy* (HUJ, UC); Hills north of Reno, 5,000 ft, 1907, *Heller* 8658 (HUJ).

7b. T. andersonii A. Gray Subsp. **monoense** (Greene) Gillett, Can. Jour. Bot. 50: 1997 (1972). *Typus*: California: White Mountains, Mono Co., 22 July, 1886, *W. H. Shockley* 460 (ND–G, iso.: G !, JEPS, K !, UC).

T. monoense Greene in Erythea 2: 181 (1894).
T. andersonii A. Gray f. *monoense* (Greene) McDerm., N. Am. Sp. Trif. 185 (1910).
T. andersonii A. Gray var. *monoense* (Greene) Isley, Brittonia 32: 55 (1980).

Icon.: [Plate 7b]. McDerm. *op. cit.* Pl.73.

Hab.: Dry rocky mountains and slopes, sagebrush scrub to bristle cone-pine forest; up to 4,000 m. 2n = 16.
Gen. Distr.: USA: California, Nevada.

Selected specimens: CALIFORNIA: McAfee Mdw., White Mts., 11,700 ft, 1930, *V. Duran* 543 (BM, E, H, UC, W); Mono Co., Wyman Creek, 1945, *B. Maguire* & *Holmgren* 26048 (NA). NEVADA: White Mts. near Sunland Mineral Co., 7,250 ft, *Heller* 10498 (E).

Note: The last-mentioned specimens of *Heller* 10498 approach subsp. *beatleyae* Gillett.

7c. T. andersonii A. Gray Subsp. **beatleyae** Gillett, Can. Jour. Bot. 50: 1997 (1972). *Typus*: Nevada: Nye Co., Topopah Spring, S. side of Shoshone Mts., ca. 25 mi. E. of Mercury, 5,800 ft, 23.4.1961, *Cronquist* & *Beatley* 8897 (DAO, seen photo of iso. RSA).

T. andersonii A. Gray var. *beatleyae* (Gillett) Isley, Brittonia 32: 55 (1980).

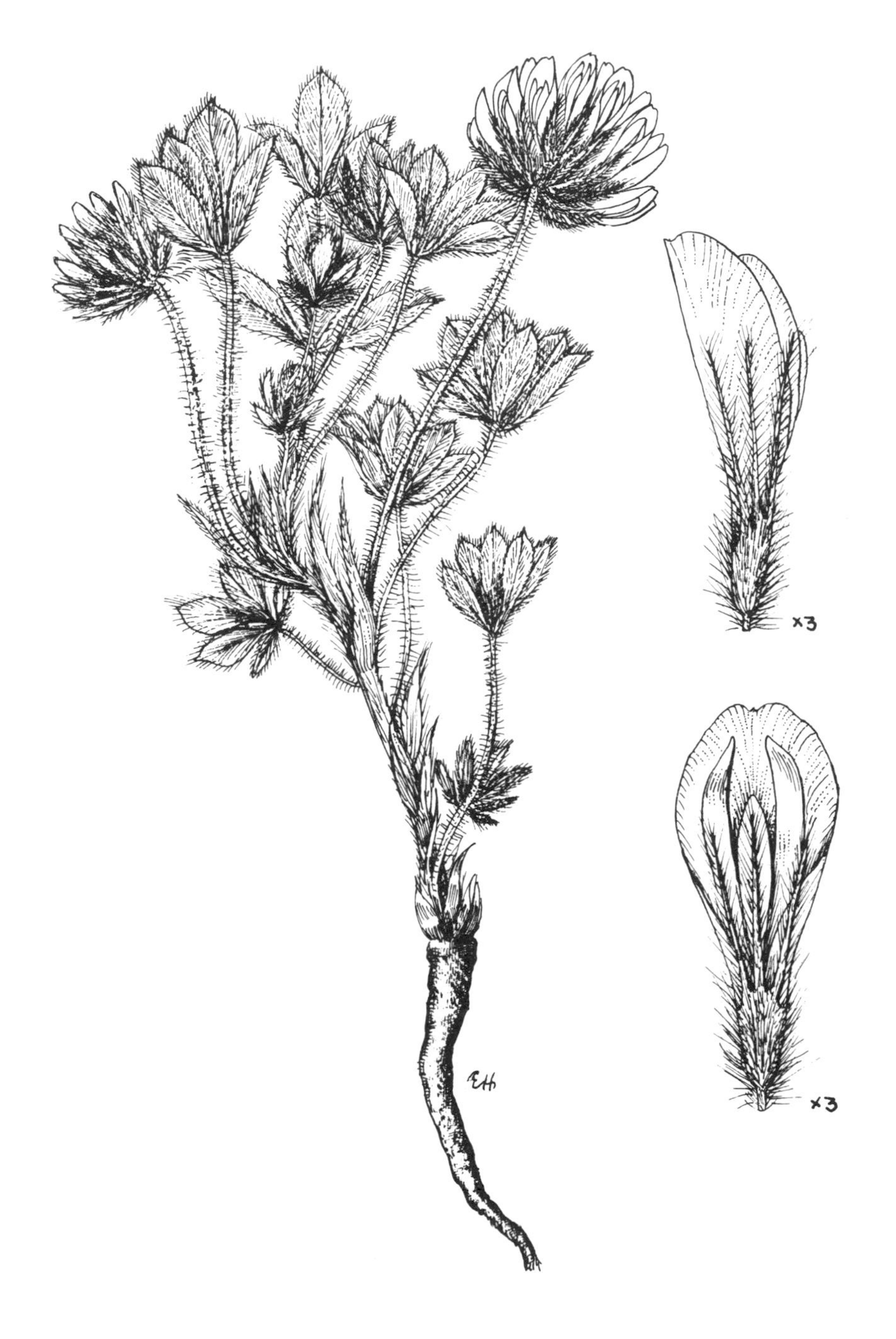

Plate 7a. *T. andersonii* A. Gray Subsp. *andersonii*
(California : *Anderson*, E).
Plant in flower; flowers.

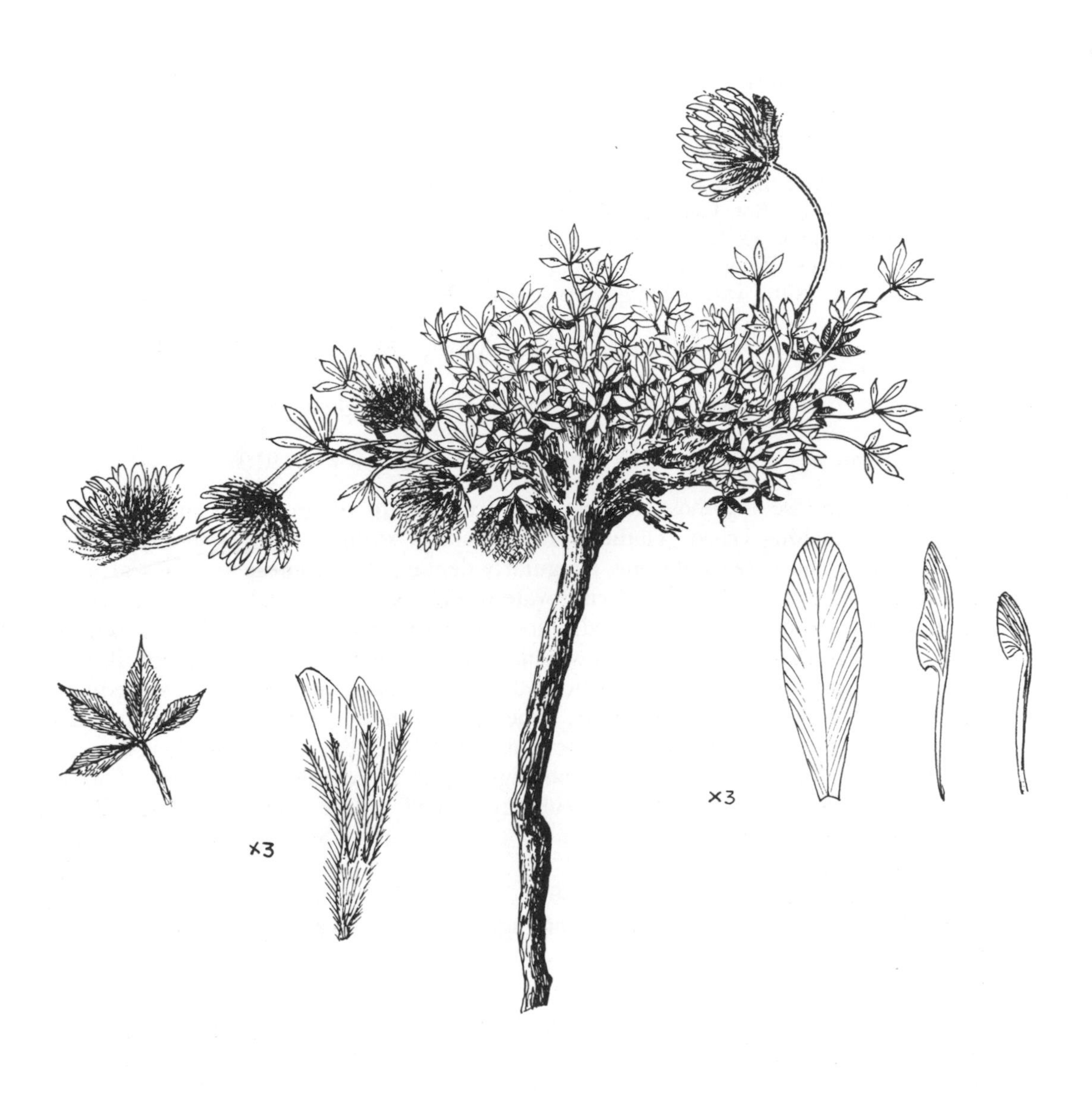

Plant 7b. *T. andersonii* A. Gray Subsp. *monoense* (Greene) Gillett
(California : *V. Duran* 543, E).
Plant in flower; leaf; flower; standard; wing; keel.

Note : This subspecies is, perhaps, an intermediate between subsp. *andersonii* and subsp. *monoense* (Greene) Gillett.

8. T. gymnocarpon Nutt. in Torr. & Gray, Fl. N. Am. 1 : 320 (1838). *Typus* : Wyoming : Dry hills of the Rocky Mts. range, near the source of the Sweetwater of the Platte, *J. Nuttal* (BM !, iso. : GH, K !, NY !, PH).

T. subcaulescens Gray, Ives Rep. Colo. River Bot. 10 (1860).
T. plummerae Wats., Bot. Calif. 2 : 440 (1880).
T. plummeri Lemmon ex Lojac., Nuov. Giorn. Bot. Ital. 15 : 162 (1883) (sphalm.).
T. nemorale Greene, Pittonia 4 : 136 (1900).
T. gymnocarpon Nutt. var. *subcaulescens* (Gray) Nelson, Man. Bot. Rocky Mts. 279 (1909).
T. gymnocarpon Nutt. var. *plummerae* (Wats.) Martin, Bull. Torrey Bot. Club 73 : 368 (1946).
T. gymnocarpon Nutt. subsp. *plummerae* (Wats.) Gillett, Can. Jour. Bot. 50 : 1994 (1972).

Icon. : [Plate 8]. McDerm., N. Am. Sp. Trif. Pls. 74–77 (1910).

Perennial, caespitose or almost stemless, pubescent, up to 10 cm. Stems short, arising from thick branching crown, clothed with persistent stipules. Leaves long-petioled; stipules of very short stems obovate, irregularly dentate, green, those at base scarious, thin; leaflets 3–5, 0.8–2 × 0.2–1.3 cm, ovate to oblong or elliptical, obtuse or acute, spinulose-dentate, prominently nerved, pilose beneath, glabrous above. Inflorescences loose, 1–2 cm in diam., 4–16(–20)-flowered. Peduncles not longer than leaves. Bracts about 1 mm, membranous, ovate. Pedicels 0.5–4 mm long, the lower ones sometimes deflexed. Calyx 4–6 mm long, pilose; tube campanulate, 10-nerved; teeth triangular-subulate, 1–2 times as long as tube. Corolla 0.7–1.4 cm long, whitish-yellow to pinkish-purple; standard oblong or oblong-elliptical, rounded or retuse. Ovary stipitate, 2-ovuled. Pod 1-seeded, reticulately veined, pilose and somewhat villous towards apex, very large, up to 5 mm long at maturity. Seed up to 5 mm long, flattened. Fl. April–June.

Hab. : Plains and slopes; up to 3,200 m.

Gen. Distr. : USA : California, Wyoming, Colorado, Oregon, Nevada, Arizona, Montana, Utah, Idaho.

Selected specimens : CALIFORNIA : Peaks W. of Pyramid Lake, 6,000 ft, *Lemmon & Sara Plummer* (GH, K, isotypes of *T. plummerae* Wats.). WYOMING : Kemmerer, Uinta Co., 1907, *Nelson* 9018 (E). COLORADO : Los Pinos (Bayfield), 1899, *Baker* 446 (E, K, isotypes of *T. nemorale* Greene). OREGON : Trout Creek, Blue Mts., 1885, *Howell* 789 (G, NA). NEVADA : Charlston, M/5, 9,500 ft, 1926, *Edmund C. Jaeger* 6/22 (NA). ARIZONA : Ft. Defiance, 1858, *Newbery* (GH). UTAH : San Juan Co., Navajo Mt. from War God Spring to Summit, 1933, *Benson* 267C (HUJ).

Note : We have not followed the subdivision of *T. gymnocarpon* into two subspecies [subsp. *gymnocarpon* and subsp. *plummerae* (Wats.) Gillett] as proposed by Gillett (1972). The differences between the two taxa appear to be small and there are many intermediates between them.

Plate 8. *T. gymnocarpon* Nutt.
(Oregon : *Howell* 789, NA).
Plant in flower and fruit; flower; fruiting calyx with pod.

9. T. lemmonii S. Wats. in Proc. Am. Acad. 11 : 127 (1876). *Typus* : California : Sierra Valley, Sierra Co., July, *J. G. Lemmon* 978 (GH, iso. : F, G !, JEPS, K !, MO, NA !, PH, POM, UC).

T. gymnocarpon Nutt. var. *lemmonii* (Wats.) McDerm., N. Am. Sp. Trif. 194 (1910).

Icon. : [Plate 9]. McDerm., *loc. cit.* Pl. 78.

Perennial, caespitose, sparingly pubescent, 10–20 cm. Stems arising from crown of thick root, covered at base with withered stipules, branched below and above. Leaves long-petioled; stipules up to 1 cm long, ovate to lanceolate-acuminate, irregularly toothed; leaflets 3–5(–7), 8–17 × 4–8 mm, obovate to broadly elliptical, obtuse, shortly mucronate, cuneate at base, coarsely dentate or serrate. Inflorescences 2–3 cm in diam., many-flowered, whorled, sometimes with few rudimentary flowers. Peduncles much longer than subtending leaves, sometimes bent near heads. Pedicels 2–4 mm long, deflexed after anthesis. Bracts minute. Calyx 3–5 mm long, villous; tube campanulate; teeth subequal, less than twice as long as tube, subulate. Corolla 10–13 mm, recorded as yellow or purplish; standard broadly oblong-elliptical, rounded, retuse, usually apiculate, longer than rostrate wings and keel. Pod 1–2-seeded, large, exserted from calyx, reticulately veined, pubescent in upper part, stipitate. Seeds about 3 mm long, smooth. Fl. June–July.

Hab. : Slopes and valleys; about 1,600–2,300 m.

Gen. Distr. : USA : California (endemic).

Specimens seen : CALIFORNIA : Plumas Co., Red Clover Valley, 1907, *Heller* & *Kennedy* 8758 (E); Webber Lake Valley, 1876, *R. M. Austin* (K); Sierra Co., Dog Valley, 1892, *C. F. Sonne* 10 (HUJ, K); Sierra Valley, 1883, *Lemmon* (K).

Note : It seems that collections available in herbaria are few and restricted to a small number of localities in California. Recently, True (1974) published data on the rediscovery of this species in California.

10. T. lupinaster L., Sp. Pl. 766 (1753). *Typus* : Sibiria, Linn. Herb. 930/13 (LINN; photo HUJ, holo. ?).

T. dimediatum Salisb., Prodr. 333 (1796).
Lupinaster pentaphyllus Moench, Meth. Pl. Suppl. 50 (1802).
Pentaphyllum lupinaster (L.) Pers., Syn. 2 : 352 (1807).
P. amani Ledeb., Ind. Sem. Dorp. Suppl. 5 (1823).
L. purpurascens Fisch. ex DC., Prodr. 2 : 204 (1825).
T. lupinaster L. var. *oblongifolia* Ser. et var. *albiflorum* Ser. in DC., *loc. cit.*
T. lupinaster L. var. *purpurascens* Ledeb., Fl. Alt. 3 : 258 (1831).
T. albens Fisch. ex Loud., Hort. Brit. 300 (1838).
T. lupinastrum St. Lag. in Ann. Soc. Bot. Lyon 7 : 136 (1880).
T. romanicum Brandza, Ann. Acad. Rom. ser. 2, 25 : 153 (1903).
T. pacificum Bobr., Acta Inst. Bot. Acad. Sci. URSS, Fl. Syst. Pl. Vasc., Fasc. 6 : 212 (1947).
T. cisvolgense Sprygin ex Iljin & Trukhaleva in Compt. Rend. Acad. Sci. URSS 82 : 219 (1960).

Plate 9. *T. lemmonii* S. Wats.
(California : *Sonne* 10, HUJ).
Plant in flower; dissected calyx and pod (*Lemmon*, K); flower; standard; wing; keel.

T. litwinowii Iljin in Iljin & Trukhaleva, *loc. cit.*
T. baicalense Belyaëva & Sipl. in Bot. Zhurn. 60 : 819 (1975).

Icon. : [Plate 10]. Reichenb., Ic. Fl. Germ. 22 : t. 81, II 2–6 & 122, II 2–4 (1867–89).

Perennial, glabrous or subglabrous, 15–50 cm. Stems erect or ascending, very poorly branched or only from base. Leaves sessile or very short-petioled and united with lanceolate or linear-lanceolate sheathing stipules; leaflets 3–5(–8), 3–5 × 0.5–1.4 cm, lanceolate or linear-lanceolate, obtuse, margin sharply serrulate with teeth appearing to terminate parallel crowded nerves. Inflorescenses up to 3 cm across, loose, 10–20-flowered. Peduncles as long as subtending leaves or slightly longer. Bracts minute, lanceolate. Pedicels slightly pubescent, 2–4 mm long, deflexed after anthesis. Calyx half the length of corolla, sparsely pubescent; tube cylindrical, 10-nerved; teeth 1–2 times as long as tube, triangular-lanceolate to subulate. Corolla up to 2 cm long, purple or white; standard rhombic, obtuse. Pod about 1 cm long, oblong, stipitate, exserted from calyx, 1–9-seeded. Seeds about 2 mm, reniform, dark brown. Fl. June–July. 2n = 16, 32, 40, 48.

Hab. : Forests to steppe regions.

Gen. Distr. : E. Europe, Central Asia to Siberia, Alaska (introduced ?).

Selected specimens : POLAND : Puszcza Bialowieska, 1929, *Karpinski* 41 (HUJ); Sedlets, pr. Urb. Lukow, 1902, *Puring* 1321a (G). SIBERIA : Birel Altai, 1911, *Meyer* 840 (NA); Nertschinsk, *Steven* s. n. (FI); RUSSIA : Kujbyschev, Elusanj., *Stuchenberg* 3425b (BM, K). MANCHURIA : 2 mi. S. of Hailar, 1934, *MacMillan* & *Stephens* 13 (NA). JAPAN : Honshu, N. of Kose, N. of Karuizawa, 1929, *Beattie* & *Kurihara* 11079 (NA); Minenochaya in Mt. Asama, prov. Asama, 1958, *Ohwi* & *Okamoto* 1794 (NA); Cote de Shiretoko, 1895, *Faurie* 10990 (G). ALASKA : College nr. Fairbanks, 1941, *Anderson* 7325 (US).

Note : Variable in shape of leaflets and colour of corolla. *T. lupinaster* is a most wide-ranging species including several chromosome races which have been treated and reported mainly by Russian botanists (Bobrov, 1947; Igoshima, 1966). Some forms have been recognized as varieties or even as species. We are of the opinion that experimental studies are needed for further understanding of this species.

11. T. macrocephalum (Pursh) Poir., Encycl. Suppl. 5 : 336 (1817). *Typus* : At the headwaters of the Missouri, *M. Lewis* (PH).

Lupinaster macrocephalum Pursh, Fl. Am. Sept. 2 : 479 (1814).
T. megacephalum Nutt., Gen. N. Am. Pl. 2 : 105 (1818).
T. macrocephalum (Pursh) Poir. var. *caeruleomontanum* St. John, Fl. Southeastern Wash. 237 (1937).

Icon. : [Plate 11]. McDerm., N. Am. Sp. Trif. Pls. 79–80 (1910).

Perennial, pubescent, 5–25 cm. Stems usually erect, poorly branched. Lower leaves long-petioled, upper ones shorter-petioled; stipules broadly foliaceous, oblong, the upper ones sometimes free to base, acute or truncate, margin irregularly toothed or lobed, sometimes entire; leaflets 5–9, 1–2.5 × 0.5–1.5 cm, obovate to obovate-cuneate, obtuse, apiculate, denticulate-serrate, often folded, villous beneath,

Plate 10. *T. lupinaster* L. (Japan : *Faurie* 10990, G). Flowering branch; flower; standard; wing; keel.

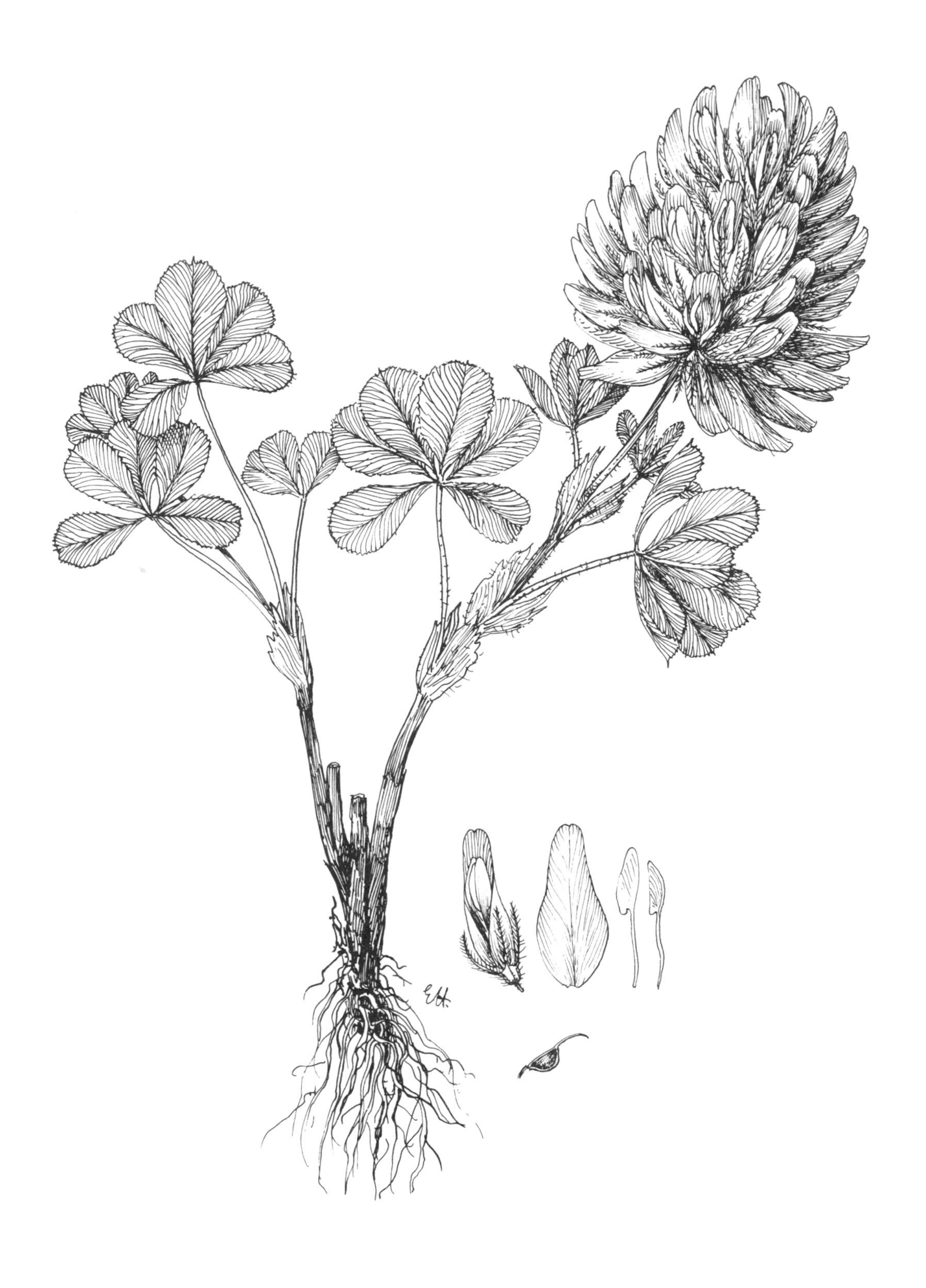

Plate 11. *T. macrocephalum* (Pursh) Poir.
(Washington : *Cronquist* 5838, NA).
Plant in flower; flower; standard; wing; keel; young pod
(Washington : *Hitchcock* & *Muhlick* 8165, NA).

upper surface sometimes glabrous. Inflorescences very large, up to 8 × 6 cm, terminal, many-flowered, whorled. Peduncles longer than subtending leaves. Bracts minute. Pedicels up to 2 mm long, not deflexed. Calyx pubescent; teeth long-villous, 1–1.7 cm long, subulate, 3–4 times longer than membranous, campanulate tube. Corolla 2–2.8 cm long; standard sometimes only one-third longer than calyx, broadly ovate or oblong, ochroleucous or pinkish, longer than purple-tipped wings and keel. Pod about 4 mm, ovoid, membranous, sessile, reticulately veined, villous towards the apex, 1–2-seeded. Seeds about 3 mm, smooth. Fl. April–June. 2n = 32, ca. 84.

Hab. : Dry rocky slopes and valleys; up to 2,500 m. Sagebrush scrub.

Gen. Distr. : USA : Washington, Oregon, Nevada, Idaho, California.

Selected specimens : WASHINGTON : Kittitas Co., Colockum Pass, 1966 *Gillett* & *Crompton* 13086 (HUJ); *ibid.*, 10 mi. N. W. of Ellensburg, 1944, *Hitchcock* & *Muhlick* 8165 (NA); Asotin Co., W. of Anatone Butte, 4,700 ft, 1949, *Cronquist* 5838 (NA); Klickitat Co., near Certerville, 1938, *Thompson* 14274 (NA); Garfield, Umatilla above Turcannon River, 5,000 ft, 1938, *Peters* 353 (NA). OREGON : Wasco Co., between Antelope and Muddy Ranch, 4,000 ft, 1953, *Cronquist* 6934 (H, W); Umatilla Co., 5 mi. N. of Ukiah, 3,600 ft, 1950, *Cronquist* 6362 (NA); Harney Co., 1940, *Sooter* 40 (NA). NEVADA : Washoe Co., Marmol Station, 5,000 ft, 1912, *Heller* (E, R); *ibid.,* Purdy road, N. of Black Springs, 1937, *Henrichs* 31 (NA); *ibid.,* 5 mi. S. W. of Reno, on road to Hunter Creek, 4,800 ft, 1937, *Archer* 5112 (NA). CALIFORNIA : Road to Cedarville, from Alturas, 5,200 ft, 1950, *Balls* 15963 (E); Siskiyou Co., near Yreka, 1876, *Greene* 723 (NA).

12. T. polyphyllum C. A. Mey., Verz. Pfl. Cauc. 139 (1831). *Typus* : Russia : In regione alpine Caucasi occidentalis, alt. 1,200–1,450 hexap., 1829, *C. A. Meyer* (LE, seen photo).

Icon. : [Plate 12]. Bobrov in Fl. URSS 11 : t. 14, f. 1 (1945).

Perennial, glabrous, 4–20 cm. Roots vertical, fusiform. Stems many, covered at base with withered stipules and leaves, flower-bearing ones few, almost leafless. Leaves basal, long-petioled; stipules oblong-linear, membranous, adnate to petiole up to above middle, free portion long-acuminate, with prominent parallel nerves; leaflets 5(–7–9), 1–2(–2.5) × 0.2–0.5 cm, linear to narrowly lanceolate, prominently nerved, sparsely denticulate in upper part, acute to obtuse-mucronulate. Inflorescences umbellate, 3–8-flowered, scape-like, 2-whorled. Peduncles 4–10 cm long, longer than leaves. Bracts about 1 mm, membranous, hyaline, obovate, forming involucre. Pedicels 2–3 mm long, deflexed after anthesis. Calyx about 1 cm long, campanulate, sometimes reddish, 10-nerved; teeth unequal, somewhat longer than tube, lanceolate, subulate. Corolla 1.5–2.5 cm long, pink to purple, violet or yellow to cream-coloured; standard ovate, rounded at apex, or slightly acuminate, usually longer than wings; wings longer than keel. Ovary 2–3-ovuled, lanceolate-linear with curved style. Pod about 1 × 0.5 cm, stipitate, lanceolate, scarious, 1–2-seeded. Seeds 1–2 mm, brown. Fl. June–July. 2n = 16, 18.

Hab. : In steppe-forest, forest clearings on subalpine rocky slopes, scree; 2,300–2,750 m.

Gen. Distr. : Turkey, W. Caucasus, Transcaucasia.

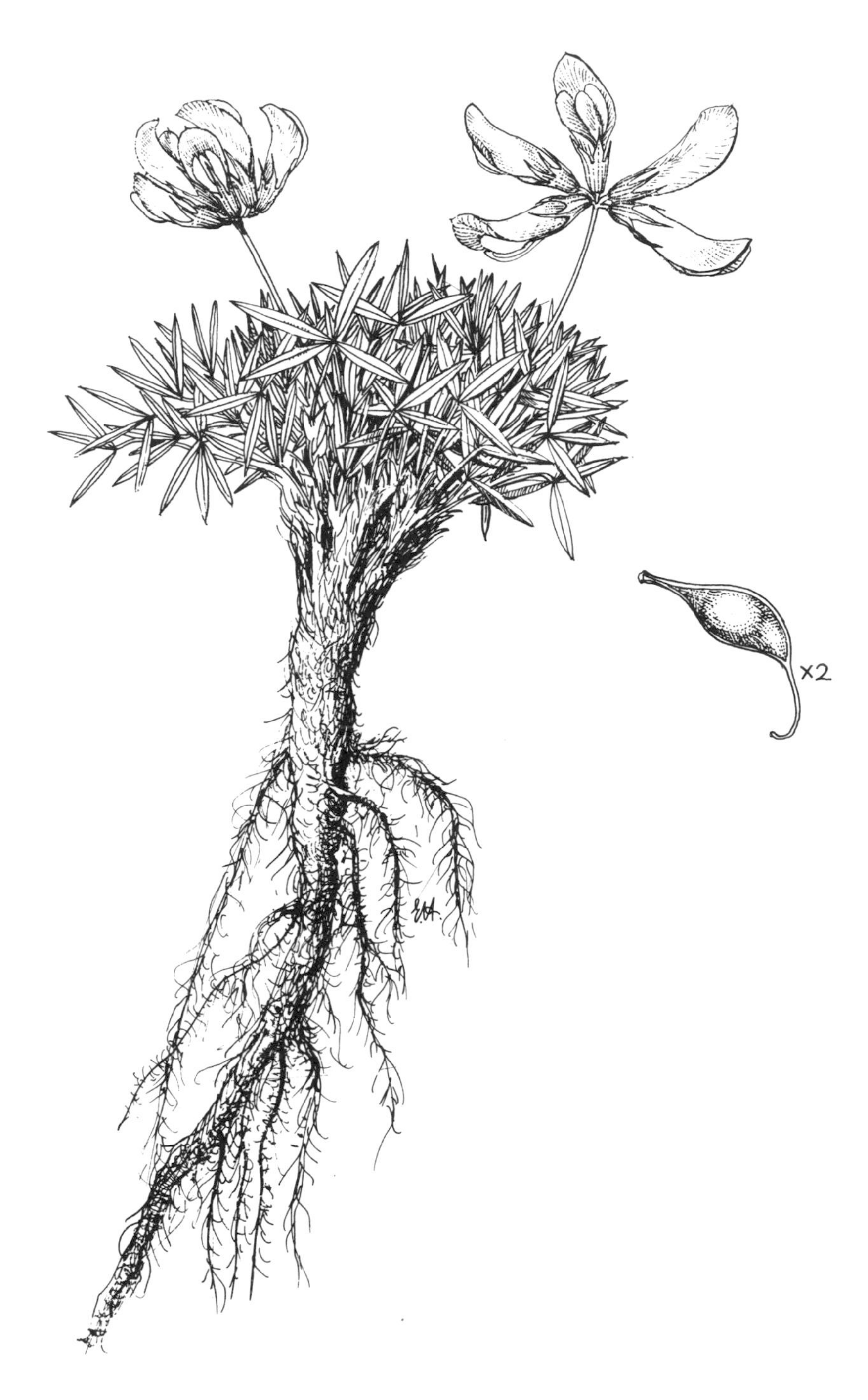

Plate 12. *T. polyphyllum* C. A. Mey.
(Turkey : *Stainton* & *Henderson* 6033, HUJ).
Plant in flower and fruit; pod.

Selected specimens: TURKEY: Prov. Artvin, Magra yalâ, Savval Tepe, Murgul, 1960, *Stainton & Henderson* 6033 (HUJ); Prov. Coruh: Tiryal dag above Murgul, 2,300–2,400 m, 1957, *Davis & Hedge* D. 29955 (E, W); Prov. Gümüsane: N. of Bayburt, Sakarsu, 9,000 ft, 1934, *Balls* 1833 (E). CAUCASUS: Kuban prov., 1909, *N. A. & E. A. Busch* 964 (W); Teberda, 8,000–11,000 ft, 1899, *Dcsoulavy* 660 (K, W). TRANSCAUCASIA: Montkhag, 2,825 m, 1894, *Alboff* 73 (P).

13. T. thompsonii Morton, Jour. Wash. Acad. Sci. 23 : 270 (1933). *Typus*: Washington: near the mouth of Swakane Creek, Chelan Co., June 23, 1932, *J. W. Thompson* 8467 (US, iso.: GH, K.!, MO, NY, UC, US, photo HUJ).

Icon.: [Plate 13]. Hitchcock et al., Vasc. Pl. Pac. N. W. 3 : 371 (1961).

Perennial, tomentose, 5–30(–60) cm, with strong taproots. Stems erect, fistulose, ribbed, poorly branched above. Lower leaves with long petioles (up to 20 cm), upper ones shorter-petioled; stipules up to 3.5 cm long, ovate to lanceolate, acuminate, entire, adnate to petioles for two-thirds of their length; leaflets (4–)5–7, 2–6 × 0.2–0.6 cm, linear to linear-lanceolate, often conduplicate, acuminate at apex, margin spinose-dentate. Inflorescences many-flowered, 3–5 cm across, hemispherical, whorled. Peduncles very long, mainly terminal. Bracts very short, triangular. Pedicels 1–1.5 mm long, those of lower flowers deflexed. Calyx 6–8 mm long; tube campanulate, 10–15-nerved, pubescent; teeth 2–3 times as long as tube, nearly equal, subulate, slightly broadened at base. Corolla 1.8–2.2 cm long; standard broadly ovate-oblong, rounded at apex or retuse, many-nerved, purple or lavender turning reddish-brown with age. Pod about as long as calyx, sessile, villous towards apex, coriaceous, 1–2-seeded. Seeds about 3 mm, smooth. Fl. May–June. $2n = 16$.

Hab.: Grassy hillsides on talus fans.

Gen. Distr.: USA: Washington (endemic).

Specimens seen: WASHINGTON: Chelan Co., Swakane Canyon about 4 mi. W. of Columbia River, 1964, *Hitchcock* 23444 (G, HUJ); *ibid.*, Wenatchee Mts., 1948, *Hitchcock* 17340 (NA); *ibid.*, 27 mi. up Swakane Creek Canyon, 1962, *Gillett & Taylor* 11087 (K).

Subsection IV. OCHREATA (Lojac.) Gillett

Kew Bull. 7 368 (1952)

Section *Ochreata* Lojac. in Nuov. Giorn. Bot. Ital. 15 : 230 (1883).

Type species: *T. polystachyum* Fresen.

Annuals or perennials. Leaves 3- (very rarely 4-) foliolate, sessile; petioles adnate for their entire length to stipules. Inflorescences spicate, globular or ovoid. Calyx 5–20-nerved. Pod 1–2-seeded. Ten African species.

Plate 13. *T. thompsonii* Morton
(Washington : *Hitchcock* 17340, NA).
Plant in flower; flowers.

14. T. cheranganiense Gillett in Kew Bull. 7 : 373 (1952). *Typus* : Kenya : Elgeyo Dist., Cherangani Hills, 2,100–3,000 m, 1951, *S. P. Rawlins* 2 (K ! holo.).

Icon. : [Plate 14].

Perennial, patulous-pubescent, up to 1 m. Stems many, prostrate, leafy, branching, sometimes rooting at nodes. Leaves sessile to subsessile (petioles adnate to stipules all along or for most of their length); stipules green, 1.5–2 cm long, pubescent, sheathing for about half of their length, free portion lanceolate-subulate, long-acuminate; leaflets 1–1.7 × 0.3–0.7 cm, obovate to oblong, cuneate at base, rounded or emarginate at apex, with prominent dense nervation, remotely dentate in lower part, more densely dentate towards apex. Inflorescences 2–2.5 cm in diam., globular, many-flowered, arranged in 2–3 whorls. Peduncles erect, longer than subtending leaves. Bracts about 0.5 mm, lanceolate, membranous, whitish. Pedicels 5–6 mm, strongly deflexed after anthesis. Calyx about 5 mm, pilose; tube tubular to campanulate, with 10 prominent nerves, whitish; teeth longer than tube, lanceolate-subulate. Corolla 8–9 mm long, white or pale pink; standard obovate or broadly elliptical, slightly longer than wings and keel. Pod 3 mm long, 1–2-seeded, scarious, with prominent sutures. Seeds 1.5 mm, ovoid, dark yellow. Fl. August–February. $2n = 16$.

Hab. : Upland grassland, especially where heavily grazed; 2,100–3,100 m.
Gen. Distr. : Uganda, Kenya.

Specimen seen : KENYA : Cult. in Grassland Research Station, 1956, *J. Knight* (HUJ).

15. T. cryptopodium A. Rich., Tent. Fl. Abyss. 1 : 168 (1847). *Typus* : N. Ethiopia, Simen (Simien), Mt. Buahit, Demerki, 8 Aug., 1838, *Schimper* 556 (P !, iso. : G !, K !, W !).

T. kilimandscharium Taub. in Engl. Hochgebirgsfl. Trop. Afr. 254 (1892).
T. cryptopodium A. Rich. var. *kilimandscharium* (Taub.) Gillett in Kew Bull. 7 : 352, f. 1, 16–18 (1952).

Icon. : [Plate 15].

Perennial, glabrous or pubescent to villous in upper part, forming dense mats. Rootstock woody and thick. Stems many, prostrate, rooting at nodes, 10–30 cm long. Leaves almost sessile, the short petioles almost entirely adnate to sheathing stipules; stipules 3–7 mm, membranous, sheathing up to half of their length, free portion ovate-triangular, acute, mucronate; leaflets 0.2–1 × 0.2–0.7 cm, obovate-obtriangular, cuneate at base, rounded or truncate, emarginate and apiculate at apex, coarsely dentate towards apex. Inflorescences 0.6–1.8 cm in diam., ovoid to hemispherical, few- to many-flowered. Peduncles rather long, mostly pubescent. Bracts minute to 2 mm long, short-triangular to subulate, the lower ones usually wide and sometimes divided. Pedicels up to 2 mm long, not deflexed or only the lowermost ones deflexed in fruit. Calyx 4–7 mm, glabrous or pubescent; tube tubular to campanulate, with 10 prominent nerves; teeth 1–2 times as long as tube, subulate, erect or strongly recurved in fruit. Corolla 1–2 times as long as calyx, purple; standard

Plate 14. *T. cheranganiense* Gillett (Kenya : *Knight*, HUJ). Flowering branch; flower.

Plate 15. *T. cryptopodium* A. Rich.
(Kenya : *Gillett* 18412, HUJ).
Plant in flower; flower.

oblanceolate to elliptical, slightly longer than wings and keel. Pod about 2 mm, 1-seeded, oblong, glabrous. Seed about 1 mm, brown. Fl. January–September. 2n = 16.

Hab. : Grassy places, often on rocky ground and sometimes in moist open spots in upland evergreen forest, moorland and alpine zones; 1,800–4,100 m.

Gen. Distr. : Uganda, Kenya, Tanganyika, Ethiopia.

Specimens seen : KENYA : Nanayaki Distr., Mt. Kenya, Sirimon track, 3,150 m, 1969, *Gillett* 18985 (HUJ); Trans Nzoia Distr., Mt. Elgon, S. W. of Suam Sawmills, 2,700 m, 1967, *Gillett* 18412 (HUJ). TANGANYIKA : Kilimanjaro, Bismark hügel, 2,400 m, 1914, *Peter* 49687 (HUJ); Mt. Kilimanjaro, 11,000 ft, 1920, *Swynnerton* 61 (SRGH); *ibid.*, 1,800 m, 1933, *Schlieben* 4313 (G). ETHIOPIA : Scioa: Gafarsa, 2,450 m, 1938, *Borini* 39 (FI, G, W); 1863–8, *Schimper* 1280 (W).

Note : The specimen from Ethiopia, Kawinjiro-Masaidorf Bulbul, *Peter* 42902 (W), differs markedly from other specimens seen from Ethiopia and Tanganyika by the tendency of the lower flowers to deflect, the strongly recurved calyx teeth and the campanulate calyx tube. It represents, most probably, an independent species or it may be a form of *T. cryptopodium* or perhaps a form intermediate between *T. cryptopodium* and *T. usambarense* (see Gillett in Milne-Redhead, Fl. Trop. E. Africa, Leguminosae (Part 4), p. 1025, 1971).

16. T. mauginianum Fiori in Nuov. Giorn. Bot. Ital. n. s. 55 : 343 (1948). *Typus* : Abyssinia : Ist. Agricolo-Coloniale, s. n.

Icon. : Fiori, *op. cit.* f. 21.

Original description: "Planta omnino glabra. Folia ochreata; stipulae petiolo adnatae, pars libera lanceolata, divaricata; foliola terna vel quaterna, oblonga, serrata. Flores racemoso-capitulatis, pedicellatis. Calyx glaberrimus, nervis 5, dentibus lanceolatis, subulatis, tubo subaequalis. Corolla purpurea, 5–6 mm longa".

Note : Gillett in Kew Bull. 7 : 373 (1952) remarks : "Prof. Pichi-Sermolli states in a letter that the type was destroyed in the war. No data are given concerning locality or habitat. The true status of this plant must remain doubtful for the time being; it may be but an aberrant form of *T. polystachyum*".
No further comments can be made about this species until the collection of new specimens confirms its occurrence.

17. T. polystachyum Fresen. in Flora 22 : 50 (1839). *Typus* : N. Ethiopia, *Rueppell* (FR holo.).

T. polystachyum Fresen. var. *psoraloides* Welw. ex Hiern., Cat. Welw. Afr. Pl. 1 : 205 (1896).
T. polystachyum Fresen. var. *contractum* Lanza, Pl. Erythr. t. 4 (1910).
Ochreata polystachya (Fresen.) Bobr., Bot. Zurn. 52 : 1598 (1967).

Icon. : [Plate 16]. Fiori in Nuov. Giorn. Bot. Ital. n. s. 55 : f. 22 (1948).

Plate 16. *T. polystachyum* Fresen.
(Ethiopia : *Schimper* 1071, FI).
Flowering and fruiting branch; flower; pod.

Perennial, glabrous below to sparingly pubescent above, 30–80 cm. Stems many, prostrate, rooting at nodes near base, branching above. Leaves sessile, the short petioles entirely adnate to stipules; stipules up to 2 cm, many-nerved, sheathing, free portion triangular-lanceolate to subulate, less than half the length of lower part; leaflets 1–3 × 0.4–0.8 cm, elliptical to lanceolate or oblanceolate, rounded, acute to obtuse, mucronate, spinulose-denticulate. Inflorescences 1.5–3.5 × 1.8 cm, spicate, many-flowered, lax after flowering, oblong-cylindrical to ovoid. Peduncles rather long. Bracts up to 2 mm, subulate, hairy. Pedicels 1–2 mm, erect, ascending or spreading, hairy. Calyx 4–6 mm, pubescent; tube tubular, 11-nerved; teeth unequal, the upper 2 shorter than the lower ones which are twice as long as tube. Corolla 7–8 mm, purplish; standard obovate-elliptical, rounded to retuse at apex. Pod about 3 mm, obovoid, membranous, glabrous, 1–2-seeded, reticulately nerved. Seeds 1–2 mm, ellipsoidal-reniform, yellow, mottled. Fl. almost all the year round. 2n = 16.

Hab.: Mountain grassland, moist places and forest margins; 1,800–3,000 m.

Gen. Distr.: Ethiopia, Eritrea, Sudan, Kenya, Uganda, Angola, Congo, Zambia.

Selected specimens: ETHIOPIA: prope Adoa, 1838, *Schimper* 1071 (BM, FI, G, MPU, W). ERITREA: Asmara, 1913, *Baldrati* 770 (FI); Scimenzana, Altipiano di Gheleba, 2,480 m, 1902, *Pappi* 142 (W). KENYA: Cherangani Hills, Kaisungor, 7,500 ft, 1959, Mrs. *Symes* et al. 2436 (SRGH); Bolossat, 1936, *Meinertzhagen* 01 (BM). ANGOLA: Benguella, Genda to Caconda, 1,700 m, 1934, *Hundt* 887 (G). CONGO: Prov. Katanga, Elisabethville, 1955, *Schmitz* 5094 (K).

Note: Gillett in Kew Bull. 7 : 371 (1952) considered *T. polystachyum* as comprising 3 varieties (var. *polystachyum*, var. *contractum* and var. *psoraloides*) differing in the length of the corolla and the density of inflorescences. In the specimens seen we found that these diagnostic characters are rather unstable and intergrading.

18. T. simense Fresen. in Flora 22 : 49 (1839). *Typus*: N. Ethiopia, Simen, *Rueppell* s. n. (FR holo.).

Icon.: [Plate 17]. Gillett in Milne-Redhead & Polhil, Fl. Trop. E. Afr. Leg. Papil. 2, f. 143 (1971).

Perennial or annual, glabrous below, pubescent above, 20–60 cm. Stems many, erect or ascending, sparingly branching in upper part. Leaves sessile, with petioles adnate to stipules all along; stipules 1–2(–4) cm, many-nerved, sheathing, free portion much shorter than united sheathing part, subulate-cuspidate; leaflets 1–6 × 0.2–0.4 cm, linear to linear-lanceolate, often cuneate at base, densely nerved, remotely dentate-spinulose towards apex, acute or obtuse, mostly apiculate. Inflorescences 1–1.5 cm in diam., hemispherical, ovoid to globular, many-flowered. Peduncles rather long. Bracts about 1 mm, subulate. Pedicels 1–1.5 mm long, erect. Calyx 3–5 mm, whitish, glabrous to somewhat pilose; tube tubular, 17–20-nerved; teeth as long as or somewhat longer than tube, subulate. Corolla 6–8 mm long, reddish-purple, rarely white; standard ovate to broad-elliptical, rounded to somewhat emarginate. Pod about 2 mm, obovoid, short-stipitate, glabrous, 1–2-seeded, with prominent ventral

Plate 17. *T. simense* Fresen.
(Eritrea : *Pappi* 1621, BM).
Plant in flower; flower; pod (Ethiopia : *Borini* 37, W).

suture. Seeds about 1 mm, ellipsoidal, pale brown. Fl. March–September. 2n = 32.
Hab. : Upland grassland; 1,500–3,000 m.
Gen. Distr. : Uganda, Kenya, Tanganyika, Nyasaland, Congo, Fernando Po, Cameroon, Ruanda Burundi, Sudan, Eritrea, Ethiopia, Zambia, Nigeria.

Selected specimens : KENYA : Trans Nzoia, Cherangani Mts., 3,000 m, *Mass* 6337 (G). TANGANYIKA : Buha Distr.: Kassulo-Tare, 1,315–1,640 m, 1926, *Peter* 38598 (B). NYASALAND : Karonga, Mussissi, 6,500 ft, 1963, *Chapman* 1889 (SRGH). BELGIAN CONGO : Mulingu Distr. (Kivu), 1947, *Hendrickx* 4432 (EA). CAMEROON : Bamenda Div., 2,150 m, 1958, *Hepper* 2150 (K). URUNDI : Murutoke-Niakassu, 1,260–1,900 m, 1926, *Peter* 38098 (B). ERITREA : Ocule Cusai, 2,600 m, 1902, *Pappi* 1621 (BM, W). ETHIOPIA : Addis Ababa, 3,000 ft, 1954, *Mooney* 5781 (SRGH); Scholoda, 1837, *Schimper* 98 (FI, G, W); Scioa, Strada di Debra Breahn, 8 km c. da Aba, 2,450 m, 1938, *Borini* 37 (W).

19. T. somalense Taub. ex Harms in Ann. Ist. Bot. Roma 7 : 90 (1898). *Typus* : Abyssinia (south) : Bidduma- (Galla-Sidamei) Alghae 38–39° E., 5–6° N.; ca. 2,000 m, Sept. 1893, *Riva* 1301 (FI !).

Icon. : [Plate 18]. Gillett in Kew Bull. 7 : f. 2, 8–10 (1952).

Perennial, glabrous below, villous above, 15–30 cm. Roots tuberous; tuber up to 8 mm thick and 8 cm long. Stems few, erect or ascending, branching above. Leaves sessile, with petioles adnate to stipules all along; stipules 1–2 cm, slightly sheathing, linear-lanceolate, free portion triangular and at most one-fourth the length of lower adnate part; leaflets 1–2 × 0.6–0.8 cm, obovate to elliptical or elliptical-oblong, cuneate at base, rounded to mucronate at apex, spinulose-dentate, with many prominent, dense nerves. Inflorescences up to 2 cm in diam., dense, many-flowered, ovoid to globular. Peduncles rather short, sometimes almost absent. Bracts about 1 mm, subulate. Pedicels 1–3(–4) mm, deflexed after anthesis. Calyx 5–6 mm, sparingly villous; tube tubular-campanulate, 11–14-nerved; teeth almost equal, about twice as long as tube, subulate. Corolla 8–10 mm long, white, turning brown; standard obovate, rounded at apex, somewhat tapering at base. Pod about 3 mm, stipitate, linear-oblong, 2-seeded. Fl. November.
Hab. : Open grassland, on sandstone with red sandy soil and on black clay soil.
Gen. Distr. : Ethiopia (endemic).

Specimen seen : ETHIOPIA : Mega, 2,200 m, 1952, *Gillett* 14323 (FI, K).

20. T. stolzii Harms in Feddes Repert. 14 : 257 (1916). *Typus* : Tanganyika : Nyassa Hochland-Station Kyimbila, 1,600 m, 1.5.1912, *A. Stolz* 1231 (B !).

T. wentzelianum Harms var. *stolzii* (Harms) Gillett in Kew Bull. 24 : 217 (1970).

Icon. : [Plate 19]. Gillett in Kew Bull. 7 : f. 2, 11–13 (1952).

Perennial, pilose to glabrescent, up to 50 cm. Root fleshy, almost tuberous. Stems many, creeping, slender, rooting at lower nodes. Leaves sessile, with petioles entirely

Plate 18. *T. somalense* Taub.
(Ethiopia : *Gillett* 14323, FI).
Plant in flower; flower; inflorescence after anthesis.

Plate 19. *T. stolzii* Harms (Tanganyika : *Stolz* 1231, B). Plant in flower; flower.

or very rarely only partly adnate to stipules; stipules 1–1.5 cm long, oblong, sometimes purplish, many-nerved, sheathing, free portion triangular-lanceolate, cuspidate, shorter than to as long as sheathing part; leaflets 0.7–2 × 0.5–1 cm, oblong, obovate or oblong-lanceolate, mostly cuneate at base, sometimes rhomboid, obtuse or acute at apex, margins shallowly denticulate, lateral nerves relatively few, at an angle of 45°. Inflorescences 1.5–2 cm in diam., semi-globular to globular, many-flowered. Peduncles 2–8 cm long, glabrous to glabrescent. Bracts 1–2 mm, subulate, white, the lowest ones truncate, denticulate. Pedicels up to 2 mm, those of lower flowers deflexed. Calyx 4–5 mm long; tube glabrous, subcampanulate, 10–12-nerved, purplish; teeth as long as to longer than tube, subulate, often hairy. Corolla 8–9 mm long, bluish-red; standard obovate to broad-elliptical, rounded to somewhat retuse. Pod 4 mm long, glabrous, 1–2-seeded, oblong. Fl. April–May.

Hab. : Roadsides, pastures, formerly cultivated areas and upland grassland; 1,600–2,800 m.

Gen. Distr. : Tanganyika, Ethiopia.

Specimens seen : TANGANYIKA : Mbeya Peak, 10,000 ft, 1959, *Gaetan* 135 (HUJ); Songea Distr., Matengo Hills, 1,900 m, 1956, *Milne-Redhead* & *Taylor* 10454 (B, K, SRGH); Mbeya Distr., Poroto Mts., 2,340 m, 1957, Mrs. *Richards* 9753 (K).

Note : At first we tended to follow Gillett (*loc. cit.*) in reducing *T. stolzii* to a variety of *T. wentzelianum*. From additional specimens we saw, however, that it does not seem justified to include them in a single taxon, as they differ in several characters: the constitution of the calyx, the shape, nervature and pubescence of the leaflets, the colour and shape of the bracts, and the length of the pedicels.

21. T. ukingense Harms in Bot. Jahrb. 30 : 324 (1902). *Typus* : Tanganyika : Kingagebirge (Ukinga Mts.), Djilula-Berg, 2,400 m, Mai 1899, *Goetze* 923 (B! holo.).

Icon. : Gillett in Kew Bull. 7 : f. 2, 14–16 (1952).

Original diagnosis (translated from Latin) : Perennial, silky-villous or puberulose, up to 20 cm. Stems erect. Leaves sessile with petioles completely adnate to stipules; stipules 1.5–2.5 cm, elongated, sheathing, slightly villous, free portion 0.5–1 cm, lanceolate, acuminate, setose; leaflets 2.5–5 × 0.2–0.4 cm, narrow-lanceolate or linear-lanceolate, tapering towards apex and base, puberulent or glabrescent above, slightly silky-villous beneath, margin often entire, sometimes partly denticulate. Heads subglobular. Peduncles 1.5–2.5 cm. Calyx almost silky-villous; tube up to 2.3 mm, with more than ten nerves; teeth 3–3.2 mm, subulate-linear, somewhat longer than tube. Corolla dark violet. Ovary sparsely pilose towards apex. Fl. March–May.

Hab. : Dry upland grassland on red lateritic soils; 2,100–2,400 m.

Gen. Distr. : Tanganyika (endemic).

Specimen seen : TANGANYIKA : Tandala, 1914, *Stolz* 2591 (B).

22. T. usambarense Taub. in Engl., Pflznw. Ost.-Afr. C. 208 (1895). *Typus*: Tanganyika: Usambara Mts., *Holst* 108 (B! syn.) et Kilimanjaro, Marangu, *Volkens* 578 (B! syn.) et 579 (B! syn., K! isosyn.).

T. polystachyum sensu Baker in Oliv., Fl. Trop. Afr. 2 : 58 (1871) p. p.
T. pseudocryptopodium Fiori in Nuov. Giorn. Bot. Ital. n. s. 55 : 343 (1948).

Icon.: [Plate 20]. Fiori, *op. cit.* f. 20.

Annual or perennial, glabrous below, sparingly pilose above, 20–100 cm. Stems many, erect or ascending, branching mainly in upper part, often rooting at lower nodes. Leaves sessile with petioles adnate to stipules; stipules about 1 cm, herbaceous, with many prominent nerves, oblong, almost open, free portion triangular-lanceolate, cuspidate, somewhat shorter than lower part; leaflets 0.6–1.4(–2.2) × 0.3–0.7 cm, oblong, elliptical to oblanceolate, cuneate at base, rounded and slightly retuse, mucronulate, remotely spinulose-dentate. Inflorescences 0.9–1.8 × 0.7–1 cm, spicate, ovoid to ellipsoidal, terminal and axillary, usually dense, many-flowered. Peduncles 3–6 cm long, pilose above. Bracts about 1 mm, subulate, hairy. Pedicels 1–1.5 mm long, pilose, not deflexed in fruit. Calyx 4–6 mm, pilose; tube tubular, 10–11-nerved; teeth unequal, all considerably longer than tube, subulate, strongly recurved in fruit. Corolla 4–5(–7) mm long, purple or rarely white; standard broad-elliptical. Pod 2–3 mm long, obovoid or broad-ellipsoidal, 2-seeded, membranous, glabrous, with prominent sutures. Seeds about 1 mm, reniform, brown, mottled. Fl.' March–September. 2n = 16.

Hab.: Marshy places and open spots in upland evergreen forest and near streams in scattered tree-grassland; 900–3,000 m.

Gen. Distr.: Uganda, Kenya, Tanganyika, Nyasaland, Fernando Po, Cameroon, E. Congo, Ruanda Burundi, S. Ethiopia, Zambia, Nigeria.

Selected specimens: UGANDA: Mt. Muhavura, 1946, *Williams* B. 5053 (G). KENYA: grown at S. A. L., 5,700 ft, 1961, *Nattrass* 1544 (HUJ). TANGANYIKA: Kilimandscharo, Moschi, 1914, *Peter* 49835 (B); Ngorongoro Crater, 1959, *Neady* 1961 (HUJ). NYASALAND: Kyimbila, 1,350 m, 1913, *Stolz* 2020 (G). CAMEROON: near Ndu, mile 82 from Bamenda, 6,400 ft, 1951, *Keay* & *Russell* 28436 (K). CONGO BELGE: 2392 (FI). RUANDA: Kabaya (Kisenyi), 2,100 m, 1960, *Michel* 6548 (SRGH). S. ETHIOPIA: Keli, E. foot of Amaro Mts., 1,600 m, 1953, *Gillett* 14871 (K).

23. T. wentzelianum Harms in Bot. Jahrb. 30 : 324 (1902). *Typus*: Tanganyika: Livingstone-Gebirge, Masuanu-Berg, 2,400 m, März 1899, *W. Goetze* 825 (B! holo., iso.: B!, BM!, E!, K!).

Icon.: [Plate 21]. Gillett in Kew Bull. 7 : f. 1, 10-12 (1952).

Perennial, densely pubescent, 20–40 cm. Stems erect, sulcate-striate, slightly branching. Leaves sessile with petioles adnate to stipules; stipules 1.5–2 cm, sheathing, oblong to ovate, densely villous, with many prominent nerves, free portion triangular to lanceolate, cuspidate, about as long as sheathing part; leaflets 2–4 × 0.5–1.1 cm, elliptical, lanceolate or oblanceolate, acute to acuminate, cuneate at base, margins spinulose-dentate, with numerous, prominent, anastomising lateral nerves at an an-

Plate 20. *T. usambarense* Taub.
(Kenya : *Nattrass* 1544, HUJ).
Plant in flower; pod (Tanganyika : *Peter* 49835, B);
flower.

Plate 21. *T. wentzelianum* Harms
(Tanganyika : *Goetze* 825, E).
Flowering branch; dissected calyx; flower; standard

gle of 15°–20°. Inflorescences 1.5–2.5 cm in diam., globular to semi-globular, dense, many-flowered. Peduncles rather short, 1.5–3.5 cm, tomentose. Bracts about 1 mm, subulate with broad base. Pedicels up to 1 mm, only those of lower flowers deflexed. Calyx 6–8 mm, villous; tube tubular, with 10–15 prominent nerves; teeth almost twice as long as tube, subulate. Corolla 10–11 mm, purple; standard elliptical, rounded at apex to slightly retuse. Pod 4 mm long, silky-pilose at apex, 1–2-seeded. Seeds 1.5 mm, globular, often mottled. Fl. February–March.

Hab.: Upland grassland; 2,400–2,800 m.

Gen. Distr.: Tanganyika (endemic).

Specimen seen: TANGANYIKA: Njombe Distr., above Matamba, 9,400 ft, 1961, *Procter* 1792 (K).

Subsection V. LOTOIDEA

Subsection *Amoria* (C. Presl) Čelak. in Oesterr. Bot. Zeitschr. 24 : 41 (1874).
Section *Lupinaster* sensu Lojac., Nuov. Giorn. Bot. Ital. 15 : 128 (1883).
Section *Euamoria* Gib. & Belli in Atti Accad. Sci. Torino 22 : 629 (1887).

Type species: *T. repens* L.

Annuals or perennials, sometimes rhizomatous. Leaves all 3-foliolate, petiolate. Inflorescences often umbellate, ovate to globular at anthesis, becoming semi-globular to pyramidal after anthesis; rhachis sometimes prolonged beyond flowers, bearing upper sterile cluster of flowers. Pedicels long, deflexed after anthesis. Calyx 5–10(–20)-nerved. Ovary usually sessile. Pod shorter or longer than calyx, dehiscent or indehiscent, with 2 conspicuous sutures and 2–8(–10) seeds.

Fourty-seven Eurasian, African and N. & S. American species, arranged in 8 series.

Key to Series of Subsection Lotoidea

1. Calyx teeth strongly ciliate, dentate to pectinate at margin. Ser. v. **Pectinata**
– Calyx teeth glabrous or hairy at margin, not dentate or pectinate 2
2. Calyx teeth 2 mm or more, wide, ovate or broadly triangular, foliaceous, with reticulate or branched nerves. Ser. ii. **Phyllodon**
– Calyx teeth lanceolate-linear to subulate, not foliaceous 3
3. Calyx tube curved, with curved and twisted teeth after anthesis. Ser. vi. **Curvicalyx**
– Calyx tube and teeth not as above 4
4. Rhachis of fruiting heads prolonged beyond flowers to form a simple or forked, long cusp; uppermost flowers often rudimentary, bud-like. Ser. vii. **Producta**
– Rhachis not as above; if prolonged – without rudimentary flowers 5
5. Stems densely caespitose or matted. Inflorescences mostly 1–5-flowered, borne on scape-like stems 6

- Stems more profusely branched. Inflorescences many-flowered, usually not borne on scape-like stems 7
6. Flowers 1.5–3 cm long. Ser. i. **Grandiflora**
- Flowers 0.7–1.2 cm long. Ser. viii. **Acaulia**
7. Corolla shorter than calyx. Calyx teeth strongly recurved in fruit, giving the fruiting head a spiny appearance. Ser. iii. **Brachyantha**
- Corolla considerably longer than calyx. Calyx and fruiting head not as above. Ser. iv. **Lotoidea**

Series i. GRANDIFLORA Heller & Zoh. (ser. nov.)*

Section *Glycyrrhizum* Bertol., Fl. Ital. 8 : 101 (1851).
Section *Cryptosciadium* Celak., Oesterr. Bot. Zeitschr. 24 : 42 (1874) p. max. p.

Type species : *T. alpinum* L.

Perennials, densely caespitose. Inflorescences mostly few-flowered, borne on scape-like stems or branches. Flowers large, 1.5–3 cm long. Four Eurasian and 2 N. American species.

24. T. alpinum L., Sp. Pl. 767 (1753). *Typus* : Italy, Hort. Cliff. 499, 23 (BM !).
Lupinaster alpinus (L.) Presl, Symb. Bot. 1 : 47 (1830).

Icon. : [Plate 22]. Hegi, Ill. Fl. Mitteleur. 4 III : t. 163, f. 4 (1923).

Perennial, densely caespitose, glabrous, 5–20 cm, with long taproot. Stem covered at base with fibres of withered stipules and leaf-bases. Leaves radical, with petioles up to 7 cm long; stipules lanceolate, about half the length of petioles and adnate to them for almost their entire length, free portion lanceolate, long-acuminate; leaflets 1–4(–7) × 0.3–0.7 cm, linear-lanceolate, linear or oblanceolate, with many prominent nerves, margin entire or very finely dentate. Inflorescences loose, whorled, scape-like, 3–12(–15)-flowered. Peduncles longer than subtending leaves, up to 10(–15) cm long. Bracts short, forming small involucre subtending each whorl of flowers. Pedicels about as long as calyx tube, deflexed after anthesis. Calyx campanulate, about 1 cm; teeth narrowly triangular at base, linear, subulate above, up to 2.5 times as long as tube. Corolla 1.6–2.5 cm long, pink, purple or rarely white or cream-coloured; standard long-ovate, longer than wings and keel. Pod much exserted from calyx, 1–2-seeded, glabrous, oblong-obovoid, constricted between seeds, somewhat stipitate. Seeds brown, reniform, 3 mm in diam. Fl. June–August. 2n = 16.

Hab. : Stony mountain slopes; meadows and pastures and along alpine streams; mainly 1,700–2,500 m. According to Hegi (1923), it occurs in the Alps on calcium-poor soils otherwise rich in nutritive elements.

Gen. Distr. : N. Spain, C. France, Switzerland, Austria, Italy.

Selected specimens : N. SPAIN : Pyrénées Centrales, Esquierry, 1856, *Zetterstedt* 325 (HUJ). FRANCE : Haute Savoie, 1,700 m, 1894, *Timothes* 4044 (B); Hautes-Alpes,

*See Appendix.

Plate 22. *T. alpinum* L.
(Spain : *Zetterstedt* 325, HUJ).
Plant in flower; reflexed fruiting inflorescence
(France : *Zohary*, HUJ);
fruiting calyx with pod (Switzerland : *Evenari*, HUJ);
wing; keel; flower.

Lautoret, 1869, *Verlot* (R). SWITZERLAND : Ryffelberg, 2,000 m, 1901, *Spencer* s. n. (B); Strelapass-Körbschhorn near Davos, 1958, *Evenari* (HUJ). AUSTRIA : Sulden, 1902, *Sterneck* s. n. (B); Insbruck, Hafelekar, 1950, *Koppel* (HUJ). ITALY : Mont-Cenis, 1863, *L. Netto* 3233 (R).

Note : *T. alpinum* is not widely variable; one variation with white to yellowish flowers (var. *ochroleucum* Gremli) occurs sporadically with normal populations (Hegi, *loc. cit.*).
The flowers are among the largest in the genus and are pollinated by insects with long probosces.
The sweet taproot is used in folk medicine.

25. T. attenuatum Greene, Pittonia 4 : 137 (1900). *Typus* : S. Colorado near Pagosa Peak, 11,500 ft, 6 Aug., 1899, *C. F. Baker* 433 (ND-G lecto., iso. : E !, GH, MO, ND-G, US.

T. bracteolatum Rydb., Bull. Torrey Bot. Club 28 : 500 (1901).
T. lilacinum Rydb., *op. cit.* 37, non Greene (1896).
T. stenolobum Rydb., *op. cit.* 499.
T. petraeum Greene, Pittonia 4 : 307 (1901).
T. dasyphyllum Torr. & Gray var. *stenolobum* (Rydb.) McDerm., N. Am. Sp. Trif. 170 (1910).
T. dasyphyllum Torr. & Gray f. *attenuatum* (Greene) McDerm., *op. cit.* 168.

Icon. : [Plate 23]. McDerm., *loc. cit.* pls. 64–65.

Scapose perennial, pubescent, 5–30 cm. Stems short, covered densely with withered stipules, branching from thickened taproot. Leaves with long petioles (up to 10 cm); stipules up to 2 cm long, scarious, lanceolate; leaflets 1.5–6 × 0.3–1 cm, linear to lanceolate or narrowly elliptical, very acute or acuminate, margin entire. Inflorescences 2.5–4 cm in diam., hemispherical, 10- to many-flowered, whorled. Peduncles 6–20 cm long. Bracts up to 2(–4) mm, scarious, ovate, truncate or sometimes acuminate, lower ones sometimes connate at base. Pedicels 2–4 mm long, deflexed after anthesis. Calyx 8–15 mm long, pubescent; tube campanulate, membranous; teeth up to 3 times as long as tube, subulate, sharp-pointed. Corolla up to 2 cm long, red-purple; standard broadly oblong-elliptical, acute, apiculate, longer than wings and keel. Pod 5–6 mm long, membranous, 1–3-seeded, densely pubescent, sessile or very short-stipitate. Seeds smooth. Fl. June–August. 2n = 16, 48.
Hab. : Open alpine and subalpine slopes; up to 4,000 m.
Gen. Distr. : USA : Colorado, New Mexico, Utah.

Selected specimens : COLORADO : La Plata Mts., Little Kate Basin, 11,500 ft, 1898, *Baker* et al. 631 (W); Custer Co., 1/2 mi. below Hermit Lake, Middle Taylor Peak, Sangre de Cristo Range, 10,600 ft, 1943, *Ewan* 15409 (NA). NEW MEXICO : Santa Fe, Lake Peak, 12,200 ft, 1923, *Renner* 183 (NA); Pecos Baldy, 12,000 ft, 1913, *Stewart* 4 (NA); Bull-of-the-Woods Mts., Taos Co., 11,500 ft, 1968, *J. K. Mitchell* 169 (HUJ). UTAH : La Sal Mts., San Juan Co., 12,000 ft, 1912, *Walker* 276 (E).

Note: Though *T. attenuatum* somewhat resembles *T. dasyphyllum* in its habit, it can be easily distinguished by its smaller bracts, deflexed and larger flowers and usually somewhat

Plate 23. *T. attenuatum* Greene
(New Mexico : *Mitchell* 169, HUJ).
Plant in flower; flower; standard; wing; keel.

lanceolate (not oblanceolate) leaflets. Gillett (1965) recorded the two species as growing together without hybridizing.

We could not confirm that specimens of "*T. dasyphyllum* var *stenolobum*" have "pubescent bracts of involucre" as assumed by McDermott (1910) – not even in the specimen *Baker* 631 which is cited among those she examined.

26. T. eximium Steph. ex Ser. in DC., Prodr. 2 : 203 (1825). *Typus* : W. Siberia, Altai, *Zalezov* (LE).

T. alpinum Pall., Reise 2 : 123 (1773) non L. (1753).
T. eximium Steph. ex Ser. var. *albiflorum* Fisch. ex Ser., *op. cit.* 204.
T. grandiflorum Ledeb. ex Spreng., Syst. Veg. 3 : 218 (1826).
T. albiflorum Fisch. ex Loud., Hort. Brit. 229 (1838).
T. altaicum Vierh., Oesterr. Bot. Zeitschr. 67 : 256 (1918).
Lupinaster eximium (Steph. ex Ser.) Presl, Symb. Bot. 1 : 47 (1830).

Icon. : [Plate 24]. Belli in Mem. Acad. Sci. Torino ser. 2, 44 : 292, t. 2 (1894).

Perennial, pubescent to glabrescent, 5–12 cm. Stems few, ascending or procumbent, branching from straight taproot, covered at base by old stipules and leaves. Leaves short-petioled to sessile; stipules about 1.5 cm long, oblong-ovate, membranous, prominently nerved, adnate to petioles for two-thirds of their length, upper free portion obtriangular, acuminate; leaflets 0.5–2.2 × 0.4–1.4 cm, obovate to oblong-obovate, cuneate at base, rounded to truncate, retuse at apex, prominently nerved, glabrous on upper surface, pubescent beneath, margin dentate. Inflorescences up to 3.5 cm in diam., axillary, 1–6-flowered, loose. Peduncles up to 4 cm long, hairy to subglabrous. Bracts about 1 mm, forming scarious, crenate involucre. Pedicels 1–4 mm long, pubescent, deflexed after anthesis. Calyx 6–8 mm long, pubescent to glabrescent; tube campanulate, with 10 prominent nerves; teeth subequal, 1–2 times as long as tube, broadly triangular, acuminate. Corolla 1.5–2 cm long, reddish, rarely white; standard broadly ovate, slightly emarginate, longer than wings and keel. Pod 1.5 × 0.5 cm, broadly oblong, membranous to scarious, slightly puberulent, falcate, 2–5-seeded, with persistent style up to 0.5 cm long. Seeds about 2 mm, ovoid, dark brown. Fl. June–July. 2n = 16, 36 (?).

Hab. : "Balds, glacial moraines and gravel deposits in the alpine zone; descending over gravelly slopes, taluses and river valleys to lower levels where it often occurs on pebbly or sandy valley accretions" (Bobrov, 1945).

Gen. Distr. : Mongolia, Siberia.

Selected specimens : MONGOLIA : Urga, 1927, *Jettmar* (W); circe lacus Ubsa, 1879, *Potanin* (K, P, W); lac. Kossagol, 1902, *Mikhano* (G). USSR : Ostsiberia circa Ponomarowa ad fl. Ungaram, *Taczanowski* (W); in glareosis subalpinis ad flumen Oka, *Turczaninow* (K, W); Altai, *Politow* (K, W); *ibid.*, *Mayer* (W); *ibid.*, *Ledebour* (K, W).

27. T. nanum Torr., Ann. Lyc. Nat. Hist. N. Y. 1 : 35 (1824). *Typus* : Colorado : James Peak, Rocky Mts., 10,000 ft, *E. James* (NY, seen photo HUJ, iso. (?) K !).

Plate 24. *T. eximium* Steph. ex Ser.
(USSR : *Taczanowski*, W).
Plant in flower and fruit; wing; keel; fruiting calyx with pod.

Icon. : [Plate 25]. Torr., *loc. cit.* t. 3, f. 4.

Perennial, closely matted, caespitose and acaulescent, glabrous. Stems very short, much branched at base from woody, thickened taproot, covered densely with withered stipules and petioles. Leaves abundant, with slender petioles up to several times longer than leaflets; stipules lanceolate, sometimes subulate in upper parts, with scarious margin, entire; leaflets 3–12 × 1.5–5 mm, oblanceolate to obovate, acute to acuminate, margin entire or dentate-serrate. Inflorescences 1–2(–4)-flowered. Peduncles from subsessile to 6 cm long. Bracts inconspicuous, up to 1 mm, triangular, acute, forming involucre. Pedicels about 2 mm long. Calyx 4–7 mm, campanulate; teeth half the length of tube or shorter, triangular, slightly reticulate, sometimes with very narrow, scarious margin. Corolla 1.5–2.5 cm long, purple to violet, sometimes white; standard oblong, longer than wings and keel. Pod about 1 cm long, exserted from calyx, 4–6-seeded, oblong-lanceolate, membranous, reticulately nerved, glabrous. Seeds about 2 mm, orbicular, smooth. Fl. June–July. 2n = 16.

Hab. : Alpine slopes and tundra; 2,600–4,500 m.

Gen. Distr. : USA : Colorado, Montana, Wyoming, New Mexico, E. Utah.

Selected specimens : COLORADO : Clear Creek Co., Grays Peak, 13,300 ft, 1968, *Mitchell* Mt.-156 (HUJ); Independence Pass, Pitkin Co., 12,500 ft, 1938, *Hitchcock* et al. 3992 (NA, UC); Argentine Pass, 12,000 ft, 1878, *Jones* 433 (NA). MONTANA : Carbon Co., Beartooth Mts., on Cooke City – Red Lodge Highway, 1947, *Hitchcock* 16613 (NA). WYOMING : Cooke City Highway, 10,500 ft, 1940, *Swallen* 6319 (NA); Beartooth Pass, 3,200 m, 1966, *Stolz* 1145 (G, W). NEW MEXICO : Taos Co., Mt. Walter, 13,050 ft, 1968, *J. K. Mitchell* 163 (HUJ). UTAH : Grand Co., Mt. Tomasky, 11,600 ft, 1933, *Maguire* 21289 (NA).

28. T. pilczii Adamović, Denkschr. Akad. Wiss. Math.-Nat. Kl. (Wien) 74 : 130 (1904); Vierh., Oesterr. Bot. Zeitschr. 67 : 252 (1918). *Typus* : Macedonia : Babuna-Planina, 50 km N. of Monastir, 1904, *K. Pilcz* (n. v.).

Icon. : [Plate 26]. Adamović, *loc. cit.* t. III, ff. 1–4.

Perennial, caespitose, glabrous or villous in upper parts, 5–8 cm. Stems short, slender, ascending or prostrate. Lower leaves short-petioled, upper ones subsessile; stipules up to 1 cm, triangular, white-membranous, acuminate; leaflets 8–16 × 3–6 mm, obovate to oblong, rounded or cuneate, often retuse at apex, prominently nerved, margin denticulate. Inflorescences (2–)5–8(–10)-flowered, loose, sometimes in two whorls. Peduncles much longer than subtending leaves, villous towards base of head. Bracts minute, hyaline, those of each whorl connate at base. Pedicels up to 2 mm, villous, deflexed. Calyx up to 8 mm long; tube glabrous, campanulate, 10-nerved; teeth up to twice as long as tube, triangular, abruptly acuminate. Corolla up to 1.5 cm, red to purplish; standard broadly oblong, longer than wings and keel. Pod longer than calyx, oblong, 1–2-seeded. Seeds about 2 mm, reniform, light brown. Fl. June–July.

Hab. : Subalpine and alpine slopes; 2,100 m.

Gen. Distr. : Albania, Yugoslavia.

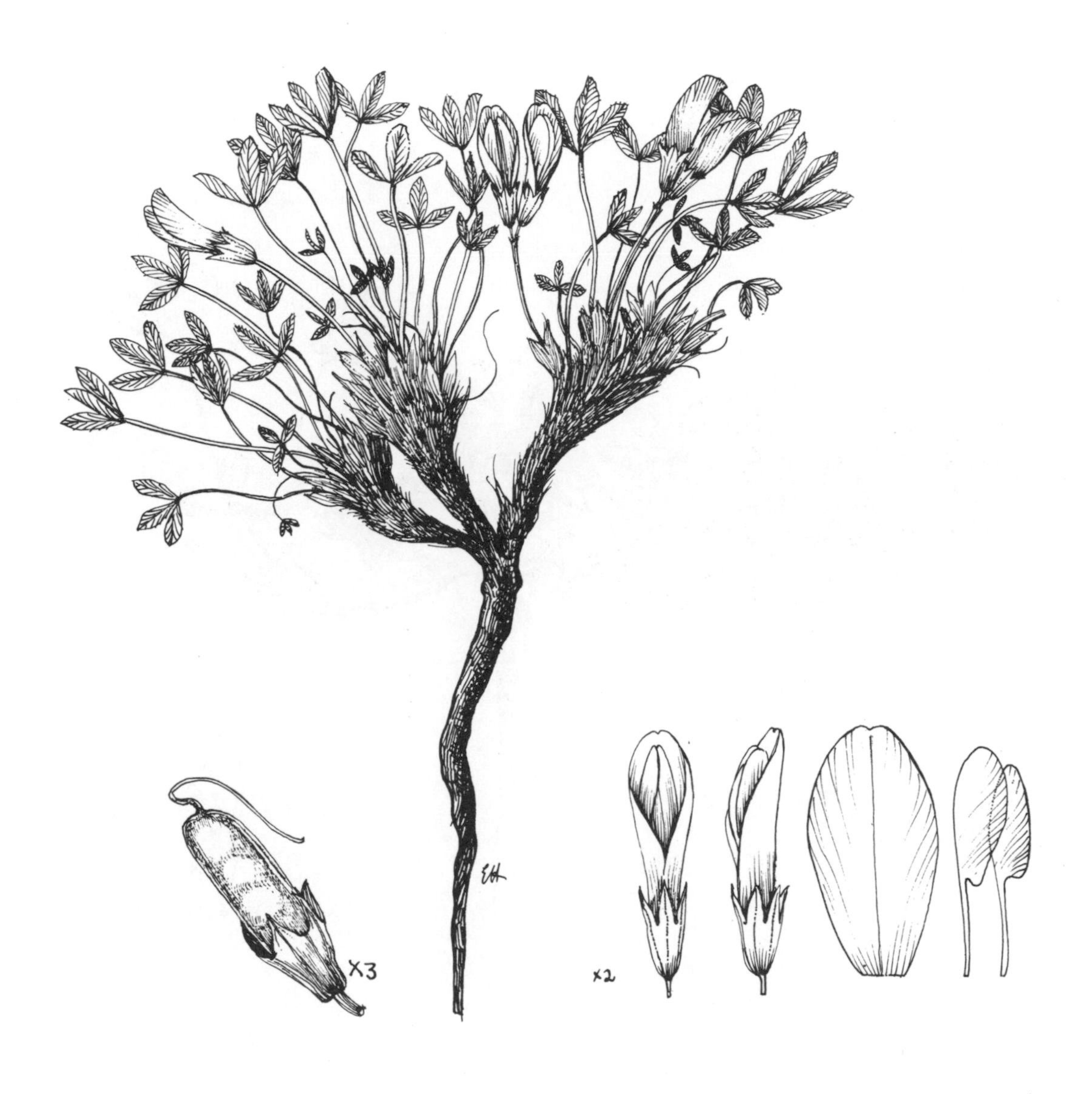

Plate 25. *T. nanum* Torr.
(Colorado : *Mitchell* Mt.- 156, HUJ).
Plant in flower; fruiting calyx with pod
(Utah : *Maguire* 21289, NA);
flowers; standard; wing; keel.

Plate 26. *T. pilczii* Adamović
(Yugoslavia : *Behr*, B).
Plant in flower; flower.

Selected specimens : YUGOSLAVIA : Mazedonien, Galüsia Planina, 2,100 m, 1939, *Qu. E. Behr* (B, W); *ibid.*, Slogowa-Planina, 1,900–2,050 m, 1938, *Lemperg* 741 (E, K, W).

Note : Vierhapper (1918) delimited T. *pilczii* from the other related species.

29. T. uniflorum L., Sp. Pl. 771 (1753). *Typus* : Described from Crete, 930/50 (LINN, photo E !).

T. buxbaumii Sternb. & Hoppe in Denkschr. Bayer. Bot. Ges. Regensb. 1, 2 : 131 (1818).
T. uniflorum L. var. *sternbergianum* Ser. in DC., Prodr. 2 : 203 (1825).
T. savianum Guss., Fl. Sic. Prodr. 2 : 488 (1828).
Lupinaster uniflorus (L.) Presl, Symb. Bot. 1 : 47 (1830).
L. buxbaumii (Sternb. & Hoope) Presl, *loc. cit.*
T. cryptoscias Griseb., Spicil. Fl. Rumel. 1 : 30 (1843).
T. uniflorum L. var. *breviflorum* Boiss., Fl. 2 : 148 (1872).
T. uniflorum L. var. *savianum* (Guss.) Arcang., Comp. Fl. It. 168 (1882).
T. uniflorum L. var. *macrodon* Hausskn. in Mitt. Thür. Bot. Ver. N. F. 5 : 77 (1894).
T. uniflorum L. var. *varians* Vierh. in Verh. Zool.-Bot. Ges. Wien 69 : 206 (1919).

Icon. : [Plate 27]. Reichenb., Ic. Fl. Germ. 22 : t. 114, II9–13 (1867–89).

Perennial, matted-caespitose, sparsely appressed-pubescent, 5–10 cm. Stems many, branching from woody taproot, base covered with withered stipules and petioles. Leaves with very long petioles (up to 10 cm), all radical; stipules crowded, membranous, broadly triangular, terminating in long setaceous tip; leaflets 0.6–1.4 × 0.6–1.2 cm, obovate-cuneate to elliptical, rounded and mostly apiculate, prominently nerved, margin denticulate. Inflorescences 1–3(–5)-flowered. Peduncles very short, hairy, shorter than subtending leaves. Bracts present only when more than 1 flower in inflorescence, hyaline or membranous, cup-shaped or bifid with two lateral cusps. Pedicels shorter than calyx, thickened, and strongly recurved and twisted when fruiting. Calyx 1 cm long, tubular, hairy above, glabrous below; teeth subequal, usually much shorter than tube, triangular at base, lanceolate at tip. Corolla 1.2–3 cm long, white or cream to pink; standard oblong, notched or recurved at apex, much longer than wings and keel. Pod 6–8 mm, linear, appressed-hirsute, beaked, 5–7-seeded. Seeds compressed. Fl. March–May. 2n = 32.

Hab. : Fields, roadsides, maquis and mountain slopes; 0–1,300 m.

Gen. Distr. : S. France, S. Italy, Sicily, Greece, Crete, Libya, Turkey.

Selected specimens : GREECE : Attika, Mt. Parnas, 1933, *Guiol* 23 (HUJ); *ibid.*, Parnis bei Hagia Trias, 950–1,100 m, 1967, *Pfadenhauer* (HUJ); Athenes, 1873, *C. de Candolle* (G); Ins. Scopolos (Sporallis), 1965, *Phitos* 2504 (HUJ). CRETE : 8 km E. of Iraklion, 10 m, 1964, *Zohary* & *Orshan* 29401-39 (HUJ). LIBYA : Cyrenaica, Lamluda-Macchia, 1933, *Pampanini* 4124 (G). TURKEY: Prov. Izmir, Yamanlar Dag, summit, 1,000 m, 1962, *Dudley* D. 34880 (E); Izmir, Kemalpasa, 1966, *Alava, Bocquet* & *von Regel* 4864 (HUJ).

Note : The variability of *T. uniflorum* was summarized by Vierhapper (1919), who divided it into taxa differing mainly in the shape and size of the calyx, the colour and length of the corolla, and the length of the peduncles and pedicels.

Plate 27. *T. uniflorum* L.
(Greece : *Guiol* 23, HUJ).
Plant in flower; inflorescence (Aegean Is. : *Phitos* 2504, HUJ).

Series ii. PHYLLODON Heller & Zoh. (ser. nov)*

Type species : *T. bejariense* Moric.

Annuals, sparingly pubescent to glabrous. Peduncles geniculate near base of heads. Calyx bilabiate, upper side longer than lower one; teeth unequal, foliaceous, with reticulate or branched nerves. Standard slightly dentate or erose at apex. Two N. American species.

30. T. bejariense Moric., Pl. Nuov. Am. 2 (1833). *Typus* : Mexico : prope Bejar, 1828, *D. Berlandier* 1670 (G !).

T. macrocalyx Hook., Ic. Pl. 3 : t. 258 (1840).
T. carolinianum Michx. f. *bejariense* (Moric.) McDerm., N. Am. Sp. Trif. 314 (1910).

Icon. : [Plate 28]. McDerm., *loc. cit.* pl. 131.

Annual, sparingly pubescent, 5–20 cm. Stems erect or ascending, branched from base and above. Leaves long-petioled; stipules 1 cm long, ovate or oblong, long-acuminate, free portion twice as long as lower part, green; leaflets 0.6–1.8 × 0.4–1.0 cm, obovate or obcordate, rounded, sometimes retuse, slightly denticulate. Inflorescences about 2 cm in diam., 5–20-flowered. Peduncles longer than subtending leaves, geniculate at base of heads. Bracts forming small, scarious involucre. Pedicels up to 7 mm, strongly deflexed. Calyx about 6 mm, with prominent reticulate nervation, bilabiate, with upper side longer than lower one, green; teeth unequal, broadly foliaceous, all (except shortest lower one) broadly ovate, rounded or abruptly acute, with prominent reticulate nervation, villous along margins only. Corolla 6–8 mm, yellowish-white, turning reddish-brown with age; standard very broadly ovate, rounded, dentate in upper part, longer than truncate-erose wings and keel; wings awned on one side. Pod longer than calyx, oblong, scarious, reticulately veined, 2–5-seeded. Seeds about 2 mm, pitted, yellow. Fl. March–May. 2n = 16.
Hab. : On open prairies.
Gen. Distr. : USA : E. Texas, W. Louisiana, S. Arkansas. Mexico.

Specimens seen : TEXAS : Austin, 1937, *Tharp* (HUJ); Brazos Co., 2 mi. S.E. of College Station, 1949, *Whisenhunt* Jr. 117 (POM); Limestone Co., 5 mi. E. of Kosse, 1965, *Shinners* 30, 961 (H). ARKANSAS : Pescott, 1900, *Bush* 247 (K). MEXICO : Rio Grande below Donona, *Parry* et al. 247 (P).

Note : The type collection of *T. bejariense* (*Berlandier* no. 1670) seems to be a mixed collection, including specimens of this species and of *T. carolinianum* : two sheets (G and W) named *T. bejariense*, collected by Berlandier (no. 1670), are typical *T. carolinianum*. Many misconceptions relating to these two taxa have their origin in McDermott's treatment.

31. T. carolinianum Michx., Fl. Bor. Am. 2 : 58 (1803). *Typus* : Near Charlestown, South Carolina, *A. Michaux* (P !).

* See Appendix.

Plate 28. *T. bejariense* Moric.
(Texas : grown from seeds, CSIRO 21853, HUJ).
Flowering and fruiting branch; pod; standard; wing; calyx; flower.

T. umbellatum Ser. in DC., Prodr. 2 : 199 (1825).
Amoria caroliniana (Michx.) Presl, Symb. Bot. 1 : 47 (1830).
T. oxypetalum Fisch. & Mey., Ind. Sem. Hort. Petrop. 2 : 51 (1836).
T. saxicolum Small in Bull. Torrey Bot. Club 21 : 302 (1894).
T. arvense Watt., Fl. Carol. 183 (1904).
T. carolinianum Michx. *f. bejariense* (Moric.) McDerm., N. Am. Sp. Trif. 314 (1910) p.p.

Icon. : [Plate 29]. Britten & Brown, Ill. Fl. N. States & Canada ed. 2, 2 : 358 (1913); McDerm., *loc. cit.* pls. 132-133.

Annual, sparingly villous to glabrous, 5–30 cm. Stems procumbent or erect, branched from base and above. Leaves on slender petioles 2–5 cm long; stipules 0.6–1 cm long, lanceolate to ovate, acute or long-acuminate, green, entire or serrate, free portion longer than lower part; leaflets 0.6–2.0 × 0.3–1.2 cm, obovate to obcordate, rounded or retuse and emarginate, margin denticulate. Inflorescences 1–2 cm in diam., globose, 5–40-flowered, loose. Peduncles 5–10 cm long, sharply geniculate near base of heads. Bracts minute, forming small involucre. Pedicels 2–5 mm long, strongly deflexed. Calyx about 4 mm, bilabiate, the upper side longer than the lower one; teeth with 3 somewhat branched nerves, the 2 upper teeth three times as long as tube, broadly triangular, villous on surface and along margins, the 2 lateral ones only twice as long as tube, the lower one shortest. Corolla 4–6 mm, 1–2 times as long as calyx, yellow to brownish-purple; standard broadly ovate, acute or rounded, apiculate, slightly erose at apex, longer than auriculate and acuminate wings. Pod as long as calyx, oblong, reticulately veined, villous near apex, 2–4-seeded. Seeds about 1.5 mm, pitted, brown. Fl. April–May. 2n = 16.

Hab. : Dry fields, open woods and waste land.

Gen. Distr. : USA : E. Kansas, Texas, Florida, Alabama, Arkansas, Louisiana, Carolina, Georgia, New Orleans. Mexico.

Selected specimens : TEXAS : Walker Co., Hundsville, 1918, *Palmer* 13385 (G); Democrat Crossing, 1946, *Parks* (RSA); Austin, 1937, *Tharp* (HUJ); Brazoria Co., Columbia, 1914, *Palmer* 5045 (POM). FLORIDA : Fields near River Junction, 1898, *Curtiss* 6377 (E, K); near Talahassee, 1843, *Rugel* (K). ALABAMA : Montgomery, 1920, *Stone* (NA); Campus, Auburn, 1920, *Pieters* (NA). ARKANSAS : Independence, 1896, *Eggert* (H); *Nuttal* (K). LOUISIANA : Natchitoches, *Palmer* 7351 (K). CAROLINA : Brookgreen Gardens near Georgetown 1946, *Blake* (NA): Hampton Co., E. of Allendale Co. line, 1970, *Leonard & Snow* 3117 (H). GEORGIA : Stone Mt., 1901, *Curtiss* 6776 (E, K, NA, HUJ); Bullock Co., 1964, *Trapell* 46 (POM); Heard Co., 9 mi. W. of Franklin, 1938, *Pyron & McVaugh* 2771 (NA); DeKalb Co., S. Base of Stone Mt., 1,100 ft, 1893, *Small* (NY type of *T. saxicolum* Small). NEW ORLEANS : 1832, *Drummond* 82 (E, K). MEXICO : 1858, *Sumichrash* 724 (G).

Series iii. BRACHYANTHA Heller & Zoh. (ser. nov.)*

Type species : *T. retusum* L.

Annual, glabrous. Stems many, erect or ascending. Inflorescences pedunculate, hemispherical, many-flowered, dense. Calyx somewhat leathery, the upper 2 teeth longer

* See Appendix.

Plate 29. *T. carolinianum* Michx.
(Georgia : *Curtiss* 6776, HUJ).
Plant in flower; pod (Texas : *Tharp*, HUJ); flower.

than others, all teeth strongly recurved in fruit, giving fruiting head a spiny appearance. Corolla shorter than calyx. One Eurasian species.

32. T. retusum L., Demonstr. Pl. 21 (1753). *Typus*: Described from Spain (UPS).

T. parviflorum Ehrh. in Beitr. Bot. 7 : 165 (1792); Boiss., Fl. Or. 2 : 143 (1872).
Amoria parviflora (Ehrh.) Presl, Symb. Bot. 1 : 47 (1830).

Icon.: [Plate 30]. Townsend & Guest, Fl. Iraq 3 : f. 26, 1–9 (1974).

Annual, glabrous, 15–40 cm. Stems many, erect or ascending, obsoletely grooved, profusely branching. Lower leaves long-petioled, upper ones shorter-petioled; stipules 0.7–1.5 cm, membranous, triangular at base, free portion abruptly subulate; leaflets 0.8–1.8 × 0.4–0.9 cm, cuneate at base, dentate-spinulose at margin, rounded to emarginate at apex, prominently nerved, the lower ones obovate to oblong. Inflorescences up to 1 cm, the upper ones subsessile. Bracts longer than pedicels, subulate, folded. Pedicels 1 mm long, curved and deflexed in fruit. Calyx 4–5 mm, green, somewhat leathery, with 10 very prominent nerves; tube obconical, slightly hairy; teeth unequal, longer than tube, subulate-spinulose, the upper two much longer than others, all strongly recurved in fruit, giving fruiting head a spiny appearance. Corolla shorter than calyx, white or pink; standard ovate-cuneate, tapering or rounded and slightly erose-denticulate at apex, not emarginate, longer than wings and keel. Pod slightly exserted from calyx tube, ovate-oblong, membranous, somewhat constricted between 2 seeds, sparingly pilose. Seeds 1 mm, ovoid-reniform, brown, finely granulate. Fl. May–June. 2n = 16.

Hab.: Forest and forest clearings, steppes and pastures.

Gen. Distr.: Spain, Portugal, C. and S. France, Austria, Germany, Czechoslovakia, Hungary, Italy, Yugoslavia, Greece, Bulgaria, Romania, Turkey, Caucasus, Iraq, Crimea, Russia, N. Africa.

Selected specimens: SPAIN: Fuente el Saz, Madrid, 1964, *Bundia* et al. (HUJ). PORTUGAL; Guarda, Torreão, 1951, *Fernandes* et al. 3724 (HUJ). FRANCE: A Moulbrison (Loire), 1858, *Gambey* 2654 (G). ITALY: Pyreneos Orient, *Bonjean* (R). HUNGARIA: Comit. Hajdu. Nagyhortobagy, 1916, *Rapaics* 682 (G, R). ROMANIA: Besarabia, Celatea-Alba, 100 m, 1926, *Rayss* et al. 657 (HUJ). TURKEY: Bingöl, 20 km E. of Bingöl, 1,800 m, 1963, *M. Zohary* 473125 (HUJ); Ankara distr., S. of Ankara, 1963, *Zohary* 2833 (HUJ). CAUCASUS: Daghestan, near Petrovsk, 1910, *Meyer* 486 (G). IRAQ: a single fragment from Penjwin (Townsend, *op. cit.* 187).

Series iv. LOTOIDEA

Series *Repentia* Bobr. in Acta Inst. Bot. Acad. Sci. URSS, ser. 1, 6 : 223 (1947) pro max. parte.

Type species: *T. repens* L.

Annuals or perennials, glabrous or hairy. Inflorescences terminal, scape-like or axillary on branching stems, many-flowered. Calyx teeth equal or unequal. Pod 1- to

Plate 30. *T. retusum* L.
(Turkey : *Zohary* 2833, HUJ).
Plant in flower and fruit; fruiting head; fruiting calyx; standard; wing; keel; young pod; flower.

few-seeded, slightly or much exserted from calyx. Twelve Eurasian, 5 African and 10 N. and S. American species.

33. T. amabile Humb., Bonpl. & Kunth, Nov. Gen. Sp. 6 : 503 (1824).

Icon : [Plate 31].

Caespitose perennials, sparingly villous, 5–30 cm. Stems spreading or decumbent, slender or stout, branching from base and above. Leaves sessile or on 1–8 cm long petioles; stipules up to 1 cm, ovate, acuminate, margin slightly serrulate; leaflets 0.7–1.8 × 0.5–1 cm, ovate to obovate, oblanceolate, rounded or retuse, acute, prominently nerved, margin finely denticulate or serrulate. Inflorescences axillary, 1–3.5 cm in diam., 8- to many-flowered, hemispherical; rhachis sometimes extending beyond flowers. Peduncles 1 cm long to longer than subtending leaves. Bracts 1–2 mm long, linear. Pedicels as long as tube of calyx, strongly deflexed. Calyx 4–5 mm, 10-nerved, sparingly to densely pubescent or pilose, the upper side longer; teeth subequal, triangular to triangular-subulate, 1–2 times as long as membranous tube, sometimes upper one with 2 lateral teeth in upper part. Corolla 0.6–1.4 cm long, purple, rose or white; standard obovate-oblong, rounded, longer than wings and keel. Pod at most as long as standard, oblong, slightly oblique, membranous, sessile, reticulately veined, 2–3-seeded, very sparingly pubescent above. Seeds 1–2 mm, subglobular, smooth, dark brown. Fl. August–October.

Hab. : Moist mountain meadows; 1,600–3,400 m.

Key to Varieties

1. Calyx densely pilose; teeth triangular, about as long as tube. Inflorescences subsessile. — **33c**. var. **pentlandii**
– Calyx sparingly pubescent; teeth triangular-subulate, twice as long as tube. Inflorescences long-peduncled — 2
2. Leaflets broadly obovate, retuse, as long as broad. — **33b**. var. **hemsleyi**
– Leaflets narrowly ovate or oblanceolate, rounded or acute, about 2–3 times as long as broad — 3
3. Leaflets oblanceolate. — **33e**. var. **longifolium**
– Leaflets narrowly ovate — 4
4. Inflorescences up to 1 cm in diam. Leaflets herbaceous, with up to 20 lateral nerves. — **33a**. var. **amabile**
– Inflorescences 2–3.5 cm in diam. Leaflets coriaceous, with lateral nerves denser and more prominent. — **33d**. var. **mexicanum**

33a. T. amabile Humb., Bonpl. & Kunth Var. **amabile**. *Typus* : Mexico : in pratis, alt. 1,380 hex., *Humboldt* & *Bonpland* (P ! photo HUJ).

T. humboldtii Spreng., Syst. 3 : 213 (1826).
Lupinaster amabilis (H., B. & K.) Presl, Symb. Bot. 1 : 47 (1830).
T. reflexum Schlecht. & Cham. in Linnaea 5 : 576 (1830) non L. (1753).

Plate 31. *T. amabile* Humb., Bonpl. & Kunth Var. *hemsleyi* (Lojac.) Heller & Zoh.
(Argentina : *Burkart* 7397, HUJ).
Plant in flower and fruit; pod; flower.

T. pauciflorum Willd. ex Steud., Nom. ed. 2, 2 : 707 (1841).
T. mathewsii A. Gray, Bot. U. St. Expl. Exped. 1 : 398 (1854).
T. schiedeanum S. Wats. in Proc. Am. Acad. 17 : 339 (1883).
T. lozani House in Coult. Bot. Gaz. 41 : 342 (1906).
T. bolivianum Kennedy in Muhlenbergia 7 : 97 (1911).

Icon. : Humb., Bonpl. & Kunth, *loc. cit.* t. 593.

Gen. Distr. : Arizona, Mexico, Peru, Colombia, Bolivia, Guatemala, Chile, Argentina, Ecuador.

Selected specimens : ARIZONA : Bisbee, Tanner Canyon, Huackuco Mts., 1910, *Goodding* 794 (G). MEXICO : Eslava Valley, 7,600 ft, 1901, *Pringle* 9512 (K, type of *T. lozani* House, iso. NA, POM). PERU : Matucana, about 8,000 ft, 1922, *Macbride* & *Featherstone* 354 (G). COLOMBIA : *Hartweg* 940 (W). BOLIVIA : Prov. Lareeaja Ancouma, 3,600–3,800 m, 1860–63, *Mandon* 693 (G). GUATEMALA : near Antigua, 1908, *Kellerman* 7187 (NY). ARGENTINA : Prov. Tucumán, Amaich, 1947, *Schulz* 6656 (SI). ECUADOR : S. Pedro, Loja, 2,200 m, 1946, *Reinaldo Espinosa* 250 (NA).

33b. T. amabile Humb., Bonpl. & Kunth Var. **hemsleyi** (Lojac.) Heller & Zoh. (stat. nov.). *Typus* : Central Mexico : Chiefly in the region of San Luis Potosi, 22° N. Lat., alt. 6,000–8,000 ft, 1878, *C. C. Parry* & *Ed. Palmer* 136 (K ! iso. NY !).

T. hemsleyi Lojac. in Nuov. Giorn. Bot. Ital. 15 : 143, t. IV, f. 1 (1883).
T. macrorrhizum Ulbrich in Fedde Repert. 2 : 2 (1906).

Gen. Distr. : Mexico, Peru, Ecuador, Argentina, Bolivia, Costa Rica, Colombia, Guatemala.

Selected specimens : MEXICO : Temascaltepeç, Cucha, 2,070 m, 1933, *Hinton* 4630 (BM, K). PERU : Pampa Romas, 3,100 m, 1903, *Weberbauer* 3213 (G as *T. macrorrhizum* Ulbr.). ECUADOR :. *Fraser* 1860 (G.) ARGENTINA : Prov. Goidoba, Sierra Grande, Copina, 1,500 m, 1935, *Burkart* 7397 (HUJ, SI). BOLIVIA : Prov. Husillo, km 34, La Rioconada, 3,500 m, 1964, *Hunziker* 8192 (SI). COSTA RICA : Llano Maude, Rancho Redondo, 9,000 ft, 1946, *Barrus* 317 (NA). GUATEMALA : Sierra Madre Mts., Villa Las Cruas, 2,800 m, 1963, *Williams* et al. 22952 (W).

33c. T. amabile Humb., Bonpl. & Kunth Var **pentlandii** Ball in Jour. Linn. Soc. Bot. 22 : 35 (1885). *Typus* : Ex saxosis Andium Peruviae juxta pagum Chicla, 12,000–13,000 ft, April 21–23, 1882, *J. Ball* (K !).

T. chiclense Ball, *loc. cit.*
T. peruvianum Vog. in Nov. Act. Nat. Car. 19 : Suppl. 1 : 12 (1843).
T. titicacense A. W. Hill in Sched. (K).
T. weberbaueri Ulbr., Feddes Repert. 2 : 2 (1906).

Gen. Distr. : Peru, Bolivia.

Selected specimens : PERU : Rio Blanco, about 1,800 ft, 1922, *Macbride* & *Featherstone* 749 (G, W); 80 km S. E. of La Oroya, 4,300 m, 1966, *Bleir* 775 (K). BOLIVIA : Tichicaca, 12,850 ft, *Pentland* (paratype K !); Puna Patanca, 3,300 m, 1904, *Fiebrig* 3432 (W).

33d. T. amabile Humb., Bonpl. & Kunth Var. **mexicanum** (Hemsl.) Heller & Zoh. (stat. nov.). *Typus* : C. Mexico : Chiefly in the region of San Luis Potosi, 22° N. Lat., 6,000–8,000 ft, 1878, *C. C. Parry* & *Ed. Palmer* 137 (K !, iso. : GH, NA !).

T. mexicanum Hemsl., Biol. Centr. Am. Bot. 1 : 233 (1879).
T. potosanum Lojac., Nouv. Giorn. Bot. Ital. 15 : 144 (1883).
T. cognatum House in Coult. Bot. Gaz. 41 : 345 (1906).
T. nelsonii House, *op.cit.* 344.

Gen Distr. : Mexico, Guatemala, Bolivia. 2n = 16.

Selected specimens : MEXICO : State of Hidalgo, Pachuca, 8,500–9,500 ft, 1898, *Pringle* 6933 (K, type of *T. cognatum* House, isotypes : BM, MEXU); Polotitlan, 2,400 m, 1952, *Matuda* & *Colab.* 26556 (MEXU); Flor de Maria, 1890, *Pringle* 3238 (BM, K, MEXU, P, W).

33e. T. amabile Humb., Bonpl. & Kunth Var. **longifolium** Hemsl., Biol. Centr. Am. Bot. 1 : 232 (1879). *Typus* : C. Mexico : Chiefly in the region of San Luis Potosi, 22° N. Lat., 6,000–8,000 ft, 1878, *C. C. Parry* & *Ed. Palmer* 134 (iso. : BM !, NA !).

T. rhombeum S. Schau. in Linnaea 20 : 740 (1847).
T. potosanum Lojac., Nuov. Giorn. Bot. Ital. 15 : 144 (1883) p.p.
T. goniocarpum Lojac., *op. cit.* 145.
T. longifolium (Hemsl.) House in Coult. Bot. Gaz. 41 : 342 (1906).
T. roseum Cervants in Herb.

Gen. Distr. : Mexico.

Selected specimens : MEXICO : Chihuahua, near Colonia Gracia in the Sierra Madres, 7,500 ft, 1899, *Townsend* & *Barber* 177 (BM, MEXU, POM); Monte Rio Frio, 45 km S. E. of Mexico City, 10,000 ft, 1941, *Rollins* 12 (NA); Valle de Mexico, *Schmitz* (W, as *T. roseum* Cervants).

Note: Var. pentlandii differs markedly from the other varieties by its short to subsessile peduncles and shorter, triangular calyx teeth. In this variety one of the upper calyx teeth is sometimes broader and 2 lateral adjacent teeth appear in its upper part. The other 3 varieties are not always so clear cut and are not so easy to distinguish; there are a few intermediate specimens.

34. T. angulatum Waldst. & Kit., Pl. Rar. Hung. 1 : 26 (1800). *Typus* : In Salsis Hungaria, *Waldstein* & *Kitaibel* (PR).

Amoria angulata (Waldst. & Kit.) Presl, Symb. Bot. 1 : 47 (1830).

Icon. : [Plate 32]. Waldst. & Kit., *loc. cit.* t. 27.

Annual, glabrous, 20–30 cm. Stems erect or ascending, grooved, branching in upper part. Leaves long-petioled, the lower ones up to 8 cm long, the upper ones shorter; stipules about 1.5 cm long, broadly ovate, long-acuminate, slender, sharp-pointed; leaflets 0.8–2 × 0.4–1 cm, obovate, cuneate or long obovate-oblong, rounded or slightly retuse, mucronate, margin dentate. Inflorescences 1.2 cm in diam., globular, loose, many-flowered. Peduncles as long as subtending leaves or shorter. Bracts about 1 mm, folded, linear, acute to acuminate. Pedicels as long as or longer than calyx, strongly deflexed. Calyx 3–4 mm long, campanulate; teeth 1–3 times as long as tube, subulate, sharp-pointed. Corolla 6–8 mm long, reddish; standard elliptical-oblong, acuminate, prominently nerved lengthwise, longer than wings and keel; wings with long claw. Pod about as long as standard, oblong to linear, 3–5-seeded, constricted between seeds, glabrous. Seeds 1 mm, rhomboid. Fl. June–August. 2n = 16.

Hab. : Fallow fields, roadsides and salines.

Gen. Distr. : Czechoslovakia, Hungaria, Yugoslavia, Romania, E. Russia. (S. France – introduced.)

Selected specimens : CZECHOSLOVAKIA : Parkan, 113 m, 1935, *Krist* 1075 (HUJ, R). HUNGARIA : prope Puszta Ecseg 1913, *Szartorisz* 679 (K, R): Comit. Báes-Bodrag 1913, *Gy. Prodan* 679 II (R). ROMANIA: Transylvania, distr. Cluj, 304 m, 1938, *Nyárády* & *Todor* 1987 (HUJ); Oltenia, distr. Bals, 80 m, 1960, *Păun* & *Malos* 336 (BM, R).

35. T. bifidum A. Gray in Proc. Calif. Acad. 3 : 102 (1864).

Annual, sparingly pubescent or glabrous, 5–50 cm. Stems many, erect, spreading, usually much branched. Lower leaves long-petioled, upper ones shorter-petioled; stipules 1.2 cm long, ovate to lanceolate, long-acuminate, margin nearly entire; leaflets 1–2.5 × 0.3–0.7 cm, narrow-obcordate to oblong-cuneate or linear-cuneate, rounded at apex to truncate, deeply retuse to bifid, mucronate between two lobes, coarsely serrulate in upper part or entire. Inflorescences 0.8–1.5 cm in diam., 5–30-flowered, whorled, axillary and terminal; rhachis prolonged beyond flowers, simple. Peduncles much longer than subtending leaves, pubescent near head. Bracts inconspicuous, subtending each flower or whorls of flowers. Pedicels 1–3 mm long, all strongly deflexed. Calyx 3.6 mm long; tube campanulate, 10-nerved, sparingly pubescent, the upper side slightly longer than the lower one; teeth subequal, sparingly villous, subulate, 2–4 times as long as tube, the lower ones slightly shorter than the upper ones. Corolla 5–7 mm long, pink to purple; standard elliptical to oblong, rounded at apex, apiculate, longer than wings and keel. Pod about 4 mm long, very short-stipitate, membranous, obovate, 1–2-seeded, glabrous. Seeds 2–3 mm long, smooth. Fl. April–May. 2n = 16.

Plate 32. *T. angulatum* Waldst. & Kit.
(Romania : *Nyárády* & *Todor* 1987, HUJ).
Plant in flower and fruit; pod; standard; flower; keel; wing.

1. Leaflets truncate and more or less deeply bifid. **35a.** var. **bifidum**
– Leaflets, at least some on each plant, deeply retuse with shallow notch at apex. **35b.** var. **decipiens**

35a. T. bifidum A. Gray Var **bifidum**. *Typus*: California: in ravines at Marsh's Ranch near Mt. Diablo, Contra Costa Co., May 29, 1862, *W. H. Brewer* 1184 (GH, iso.: DS, K!, UC!).

Hab.: Grassy fields and open woodlands.
Gen. Distr.: USA: California.

Selected specimens: CALIFORNIA: Sonoma Co., 1919, *Kennedy* (HUJ); Marin Co., Woodacre, 1932, *Lewis S. Rose* 32174 (HUJ).

35b. T. bifidum A. Gray Var. **decipiens** Greene, Fl. Francisc. 24 (1891). *Typus*: California: Berkeley, May, 1884, *E. L. Greene* 38115 (ND-G-lecto.).

T. hallii Howell, Fl. N. W. Amer. 1 : 135 (1903).
T. greenei House, Bot. Gaz. 41 : 334 (1906).

Icon.: [Plate 33]. Hitchcock et al., Vasc. Pl. Pac. N. W. 3 : 359 (1961).

Hab.: As above.
Gen. Distr.: USA: California, Oregon, Washington.

Selected speciments: CALIFORNIA: Mendocino Co., road to Dos Rios from Laytonville, 1966, *Gillett* & *Crompton* 12955A (HUJ); Almeda Co., Berkeley, 1892, *Michener* & *Bioletti* (FI); San Luis Obispo Co., Santa Lucia Range, 1945, *Kappler* 758 (UC). OREGON: Douglas Co., Oakland, 1881, *Howell* (NA, isotype of *T. hallii* Howell); Corvallis, 1922, *Epling* 5527 (UC); 1868–9, *Kellogg* & *Hartford* 133 (BM). WASHINGTON: W. Klickitat Co., Columbia River, 1881, *Suksdorf* (NA).

Note: It appears that in some populations the two varieties occur together and there are some intermediate forms between them. Var. *decipiens* is more common and has a wider range of distribution than the typical variety.
Var. *decipiens* is closely related to *T. gracilentum* but distinguished from it by its pubescent peduncles and oblong-cuneate, retuse leaflets.

36. T. bivonae Guss., Fl. Sic. Prodr. 2 : 512 (1828). *Typus*: Sicily, *Gussone* (NAP).

Amoria calycina Presl, Symb. Bot. 1 : 47 (1830).
T. calycinum (Presl) Steud., Nom. ed. 2, 2 : 705 (1841).

Icon.: [Plate 34]. Fiori & Paol., Ic. Fl. Ital. 238 (1899).

Perennial, glabrous, 10–30 cm. Stems many, erect or ascending, simple, grooved, branched from stout taproot and above. Leaves long-petioled, slender; stipules about

Plate 33. *T. bifidum* A. Gray Var. *decipiens* Greene (California : *Gillett* & *Crompton* 12955A, HUJ). Plant in flower; pod (California : *Michener* & *Bioletti*, FI); standard; wing, keel; flower, calyx.

Plate 34. *T. bivonae* Guss.
(Sicily : *Sommier*, FI).
Plant in flower; flower; standard; wing; keel.

1 cm long, triangular at base, membranous, free portion lanceolate-cuspidate; leaflets 1–2 × 0.5–1 cm, obovate to obovate-elliptical, rounded, obtuse, emarginate, prominently nerved near denticulate margins. Inflorescences up to 3 cm in diam., terminal, hemispherical, many-flowered. Peduncles very long, up to 15 cm. Bracts about 1 mm, white-membranous, deeply bilobed or dentate. Pedicels as long as calyx tube, deflexed after anthesis. Calyx about 7 mm; tube membranous, 10-nerved, tubular; teeth as long as tube, lanceolate, acuminate, the lower ones shorter. Corolla 1.1–1.2 cm, pink; standard elliptical, obtuse, longer than wings and keel. Pod oblong, sessile, 1-seeded. Fl. May–June.

Hab. : Mountain fields.

Gen. Distr. : Sicily (endemic).

Specimens seen : SICILY : Ficuzza, 1895, *Sommier* (FI); Boscodi Ficuzza, 1875, *Lojacono* (W); *ibid.*, 1856, *Huet du Pavillon* 40 (W); Piana del Greci, *Todaro* 594 (K, W).

37. T. breweri S. Wats. in Proc. Am. Acad. 11 : 131 (1876). *Typus* : Sierra Valley, California, 1873, *J. G. Lemmon* (GH).

Icon. : [Plate 35]. McDerm., N. Am. Sp. Trif. pl. 128 (1910).

Perennial, sparingly appressed-pubescent, glaucous, 5–50 cm. Stems erect or decumbent, very slender and diffusely branched. Leaves on slender petioles up to 4 cm long; stipules 0.5–1 cm long, narrowly ovate or lanceolate, long-acuminate, entire; leaflets 0.4–2 × 0.3–1.2 cm, obovate or obcordate, rounded and retuse, apiculate, prominently nerved, margin dentate-serrulate. Inflorescences about 1.5 cm in diam., 5–25-flowered, loose. Peduncles up to 5 cm long, axillary, slender, filiform, geniculate or bending near base of head, thus flowers inverted. Bracts minute, subtending each flower. Pedicels 3–6 mm long, strongly deflexed. Calyx about 5 mm, tubular, pubescent; teeth twice as long as tube, subequal, subulate, with rounded sinuses between them. Corolla 6–9 mm long, pinkish-rose or cream-white; standard ovate-oblong, rounded or retuse, about as long as wings and keel. Pod exserted from calyx, as long as standard, membranous, reticulately veined, short-stipitate, pubescent, 1–2-seeded. Seeds 3 mm, smooth. Fl. May–August. 2n = 16.

Hab. : Wooded slopes and meadows.

Gen. Distr. : USA : California, Oregon.

Selected specimens : CALIFORNIA : Butte Co., Jonesville, 1,550 m, 1929, *Copeland* 376 (H, K, US, W); Siskiyou Mts., 2,000 ft, 1937, *Wolf* 9115 (NA); Tehama Co., above Deer Creek Meadow, 5,000 ft, 1912 *Wilder* (HUJ); Shasta Co., between Trinity Center and Scott Mt., 2,000 ft, 1966, *Gillett* & *Crompton* 12910 (HUJ); Plumas Co., Butterfly Valley, 3 mi. W. of Keddie, 4,000 ft, 1967, *Rose* 67106 (W). OREGON : Joschine Co., 3 mi. W. of O'Brien, 2,500 ft, 1939, *Hitchcock* & *Martin* 5151 (NA); *ibid.*, Siskiyou Mts., 1902, *Cusick* 2920 (W).

Note : *T. breweri* varies in the size of the plant and especially in the dimensions of the leaflets. It is easily distinguished from other species by its inflorescences and the standard which is equal in length to the wings and keel.

Plate 35. *T. breweri* S. Wats.
(California : *Wilder*, HUJ).
Flowering branch; flower; pod
(California : *Gillett* & *Crompton* 12910, HUJ).

38. T. burchellianum Ser. in DC., Prodr. 2 : 200 (1825).

Perennial, glabrous, 5–20 cm. Stems procumbent or prostrate, creeping and rooting at nodes. Leaves long-petioled; stipules up to 2 cm long, triangular, tapering gradually, with free portion about as long as adnate part, the lower ones persistent, covering the stems, with few or many nerves; leaflets 0.8–5 × 0.5–1.5 cm, obovate, obcordate, oblong or elliptical, cuneate, emarginate, truncate or rounded, margin denticulate. Inflorescences up to 3 cm in diam., many-flowered, whorled, globular. Peduncles up to 15 cm, pilose towards base of head. Bracts 2 mm long, lanceolate, truncate to bifid. Pedicels 2.5 mm long, erect or deflexed. Calyx about 6 mm long, glabrous or with few hairs at margin; tube tubular, 10-nerved; teeth twice as long as tube, subulate. Corolla 8–13 mm long, purple; standard broadly obovate, rounded at apex, slightly longer than wings and keel. Pod about 5 mm, ovoid, membranous, 1–2-seeded. Seeds 2 mm long, ovoid, greenish sometimes with dark brown dots. Fl. October–December.

38a. T. burchellianum Ser. Subsp. **burchellianum**. *Typus* : South Africa, near Capetown, Salt River, *Burchell* 516 (holo. G !, iso K !).

Lupinaster ecklonianus Presl, Symb. Bot. 1 : 52 (1832).
L. burchellianus (Ser.) Ecklon & Zeyh., Enum. Pl. Afr. Austr. extra Trop. 222 (1836).
T. subrotundum Baker var. *obcordatum* Welw. ex Hiern., Cat. Welw. Afr. Pl. 1 : 205 (1896).

Icon. : [Plate 36]. Presl, *loc. cit.* t. 32.

Free portion of stipules long and many-nerved; leaflets up to 1 cm broad. Inflorescences rarely more than 2 cm in diam. Standard 8–10 mm long. 2n = 48.
Hab. : Moist forest margins ; 1,800–4,000 m.
Gen. Distr. : Angola, Cape Province, Basutoland, Natal.

Selected specimens : ANGOLA : Huila Humpata, 1,800 m, 1941, *Gossweiler* 12731 (SRGH). S. AFRICA : Cape de Bonne Esperance, *Ecklon* & *Zeyher* 1507 (G); Kapland bei George, 1913, *Peter* 50565 (B); Natal : Distr. Alexandra, Station Dumisa, 850 m, 1911, *Rudatis* 1344 (W).

38b. T. burchellianum Ser. Subsp. **johnstonii** (Oliv.) Gillett in Kew Bull. 25 : 178 (1971). *Typus* : Tanganyika, Kilimanjaro, *Johnston* 114 (holo. K !).

T. johnstonii Oliv. in Trans. Linn. Soc. ser. 2, 2 : 331 (1887).
T. basileianum Chiov. in Nuov. Giorn. Bot. Ital. n. s. 36 : 364 (1929).

Stipules few-nerved; leaflets more than 1 cm long. Inflorescences usually more than 2 cm in diam. Standard 9–13 mm long. 2n = 96.
Gen. Distr. : Uganda, Kenya, Tanganyika, S. Ethiopia.

Selected specimens : KENYA : Bolosat Marsh, 1936, *Meinertzhagen* 01 (BM); Mt. Kenya, clearing in bamboo forest, 10,900 ft, 1936, *Meinertzhagen* (BM). TANGANYIKA : Mt. Meiu, 1966, *Gilbert* N26 (HUJ). S. ETHIOPIA : Sessaganno ridge, 6,300–7,430 ft, 1952, *Gillett* 14775 (FI).

Plate 36. *T. burchellianum* Ser. Subsp. *burchellianum*
(S. Africa : no.335, HUJ).
Plant in flower; fruiting calyx with pod (S. Africa : *Godfrey* 1546, NA);
flower; standard; wing; keel.

Note: Gillett (1952) described an additional variety in subsp. *johnstonii*: var. *oblongum* from specimens from Kenya. No specimen of this taxon has been seen by us. Gillett described this variety as having stout stems (up to 1 m high) and large leaflets (up to 5 cm long). In a later publication (Gillett, 1970), he expressed the view that var. *oblongum* might also be a high polyploid form. On the basis of the description this seems possible, though the possibility of its being a lush form of the subsp. *johnstonii* cannot be excluded.

39. T. cernuum Brot., Phyt. Lusit. ed. 3, 1 : 150 (1816). *Typus*: In vicinis Cintru, *D. Valorado* (LISU).

T. serrulatum Lag., Gen. et Sp. Nov. 23 (1816).
T. parviflorum Perrym., Cat. Pl. Frejus 84 (183) non Ehrh. (1792).
T. minutum Coss., Not. Pl. Crit. 5 (1849).
T. perrymondii Gren. & Godr., Fl. Fr. 1 : 422 (1849).
T. perrymondii Coss. ex Willk. & Lange, Prodr. Fl. Hisp. 3 : 356 (1877).
T. reflexum Pourr. ex Willk. & Lange, *loc. cit.*
T. parviflorum Ehrh. subsp. *cernuum* (Brot.) Vic., Anal. Inst. Bot. Cavanilles 10 : 358 (1952).

Icon.: [Plate 37]. Brot., *loc.cit.* t. 62.

Annual, glabrous, 5–40 cm. Stems many, procumbent or ascending, branching from base and above. Leaves long-petioled; stipules about 1 cm long, triangular-lanceolate, long-acuminate, white-membranous; leaflets 0.4–1.5 × 0.4–1 cm, cuneate, obovate or obcordate, rounded, truncate or emarginate, apiculate, margin coarsely dentate, lower surface prominently nerved. Inflorescences about 1 cm in diam., 8–20-flowered, axillary, more or less loose, umbellate or globular. Peduncles thin, shorter than subtending leaves, the upper ones almost sessile. Bracts about 1 mm, lanceolate. Pedicels 2 mm, as long as calyx tube, strongly deflexed and thickened after anthesis. Calyx tubular, 4 mm long; teeth as long as tube or shorter, subequal, triangular at base, subulate above, the 2 upper ones longer than the lower ones. Corolla 4–5 mm long, pink; standard obovate, spoon-shaped, deeply emarginate, longer than wings and keel. Pod about 4 mm, long-ovoid, exserted from calyx, membranous, opening by prominent sutures, 1–4-seeded. Seeds about 1 mm, ovoid, yellow. Fl. May–July. 2n = 16.

Hab.: Roadsides, grassy (sometimes sandy) places.

Gen. Distr.: Portugal, Spain, France, Corsica, Belgium, Azores, Morocco.

Selected specimens: PORTUGAL: Nespereira, Celarico-Coimbra, 1955, *Fernandes* et al. 5703 (BM); Gavião, 1966, *Fernandes* et al. 9496 (HUJ); Baixo Alentejo, Santiago de Cacém, 55 m, 1968, *Silva* & *Teles* 8142 (HUJ). SPAIN: Galaccia, pr. Orense, 1876, *Hackel* (W); Vallée de Gerte pres Placencia, 1863, *Bourgeau* (W). FRANCE: Prope Ychoux, 1898, *Neyraut* 4043 (W); Gradignan (Gironde), *Maron* (FI). MOROCCO: Tiflet, 350 m, 1936, *Samuelsson* 7299 (B).

40. T. euxinum Zoh. in Not. Roy. Bot. Gard. Edinb. 29 : 321 (1969). *Typus*: Turkey: Kastamonu, Keltepe above Karabük, at Sorgun Yayla, ungrazed pastures, 1,300 m, 4 August, 1962, *Davis* & *Coode*, Cult. Acc. No. X63 1144 (E ! holo.).

Plate 37. *T. cernuum* Brot.
(France : *Maron*, FI).
Plant in flower and fruit; flowers; pod.

Icon. : [Plate 38]. Davis, Fl. Turk. 3 : f. 4, 9 (1970).

Annual, glabrous or glabrescent, 30 cm. Stems few, grooved, sparingly branched above. Lower leaves long-petioled, upper ones almost subsessile; stipules up to 2 cm long, herbaceous, prominently nerved, slightly adnate to petioles, not sheathing, ovate to oblong, gradually terminating in subulate cusp; leaflets of lower leaves 0.6–1.2 cm, obovate, rounded at apex to slightly emarginate, those of upper ones 1–2.2 × 0.8–1.4 cm, rhombic-elliptical, spinulose-dentate, mucronulate. Inflorescences about 2 cm in diam., 30–50 (or more)-flowered, globular. Peduncles longer than subtending leaves. Lower 10–12 bracts minute, forming involucre, ovate-triangular, mucronulate, white; upper bracts whitish, subulate, somewhat folded, about one-third the length of pedicel. Pedicels at least as long as calyx tube, slightly hairy above, deflexed. Calyx 3–5 mm, glabrous; tube obscurely nerved, whitish; teeth longer than tube, greenish, subulate. Corolla twice as long as calyx, pink, becoming flesh-coloured; standard ovate, striate, longer than long-clawed wings and curved, conspicuously beaked keel. Ovary sessile, linear, shorter than almost straight style, 3(–4)-ovuled. Young pod linear, hairy above, with thick sutures and 2–3 seeds.

Hab. : Pastures.

Gen. Distr. : Turkey (endemic).

Note : *T. euxinum* is only known from the type specimen, raised from seeds in the Royal Botanic Garden, Edinburgh.

41. T. gillettianum Jac.-Fél., Adansonia ser. 2, 9 : 553 (1969). *Typus* : Cameroun : Adamaoua, 40 km de Ngaundéré à Belel, plateau laté-ritique de Tournigal, 1,400 m d'altitude, sur dépression humide, 18 Oct. 1967, *H. Jacques-Félix* 8710 (P ! holo., iso. K !).

Icon. : [Plate 39]. Jac.-Fél., *loc. cit.*

Annual, glabrous, 20–30 cm. Stems erect, simple or slightly branched, grooved. Leaves long-petioled, petioles of lower and middle leaves up to 6 cm long, those of upper ones shorter; stipules up to 2 cm long, oblong, adnate to petioles for half of their length, free portion triangular-lanceolate, cuspidate; leaflets up to 1.7 × 0.5 cm, narrow-oblanceolate, rounded and apiculate at apex, serrate-dentate at margin, with many prominent, parallel nerves. Inflorescences 2 cm in diam., globular, 10–20-flowered, arranged in 2 whorls. Peduncles 2–3.5 cm long, sparsely pilose. Bracts almost inconspicuous, linear. Pedicels 1 mm long, deflexed after anthesis. Calyx 6–6.5 mm long, campanulate; tube 15–16-nerved; teeth somewhat longer than tube, lanceolate, 1 mm wide and 3-nerved at base, abruptly cuspidate, ciliate at margin. Corolla about 1 cm, purple; standard elliptical-oblong, as long as wings; keel shorter, about 7 mm. Pod somewhat membranous, with thick sutures, 1–2-seeded. Fl. October.

Hab. : In moist depressions; 1,400 m.

Gen. Distr. : Cameroun (endemic).

Note : Only known from type collection.
T. gillettianum differs from the related *T. rueppellianum* by its narrow oblanceolate leaflets, its 15-nerved calyx and the ciliate margin of the calyx teeth.

Plate 38. *T. euxinum* Zoh.
(Turkey : *Davis* & *Coode* x 631144, E).
Flowering branch; flower; standard; wing; keel.

Plate 39. *T. gillettianum* Jac.-Fél.
(Cameroun : *Jacques-Félix* 8710, P).
Plant in flower; flower; standard; wing; keel; part of branch.

42. T. gracilentum Torr. & Gray, Fl. N. Am. 1 : 316 (1838).

Annual, glabrous, 5–60 cm. Stems erect or ascending, branching all along. Lower leaves long-petioled, upper ones shorter-petioled to subsessile; stipules about 1 cm long, ovate or lanceolate, long-acuminate, margin entire or dentate; leaflets 0.5–3.5 × 0.2–1.5 cm, obovate to obcordate, narrowly elliptical to oblanceolate or lanceolate, cuneate, acute, retuse or apiculate, margin setose-serrulate. Inflorescences 0.5–2 cm in diam., loose, 15- to many-flowered, in whorls; rhachis often prolonged beyond flowers. Peduncles longer than subtending leaves. Bracts minute, cup-like. Pedicels short, erect in flowers, elongating up to 4 mm and deflexed in fruit. Calyx 4–6 mm long, the upper part longer than the lower part; teeth unequal, narrowly triangular at base and gradually setaceously acuminate, 1–2 times as long as tube, margin sometimes scarious and slightly serrulate. Corolla 5–9 mm long, white, pink, red to dark purple; standard broadly ovate or elliptical-oblong, longer than wings and keel. Pod as long as standard or longer, obovoid-oblong, 1–3-seeded, membranous, coriaceous at sutures, reticulately nerved. Seeds about 2 mm, smooth, reniform, brown. Fl. March–May. 2n = 16.

Hab. : Open and grassy places.

42a. T. gracilentum Torr. & Gray Var. **gracilentum**. *Typus* : California, *D. Douglas* (K !, iso. GH).

T. denudatum Nutt., Jour. Acad. Philad. N. S. 1 : 152 (1848).
T. exile Greene, Pittonia 1 : 6 (1887).
T. gracilentum Torr. & Gray var. *inconspicuum* Fern., Zoe 4 : 380 (1894).
T. inconspicuum (Fern.) Heller, Muhlenbergia 1 : 135 (1906).
T. gracilentum Torr. & Gray var. *exile* (Greene) Kennedy, Muhlenbergia 5 : 10 (1909).
T. gracilentum Torr. & Gray var. *reductum* Parish, Bot. Gaz. 65 : 338 (1918).

Icon. : [Plate 40]. McDerm., N. Am. Sp. Trif. pls. 124–125 (1910).

Leaflets obcordate to obovate, retuse.

Gen. Distr. : USA : California, Arizona, Nevada, Oregon, Washington. Mexico.

Selected specimens : CALIFORNIA : San Bernardino Co., 1,000–2,500 ft, 1901, *Parish* 4684 (HUJ); *ibid.*, 1904, *Kennedy* (HUJ); Berkely Hills, 1903, *Mulliken* 40 (HUJ); San Luis Obispo Co., Union, 1925, *Rose* 35045 (B, W – both as *T. variegatum* Nutt.). ARIZONA : San Carlos, Cassadore Spring Canyon, 1935, *Maguire* 10305 (NA as *T. amabile*, H., B et K.); 10 mi. S. of Black River Crossing, 1938, *Gooding* & *Mallery* (NA); Tecata Baja, 1925, *Peirson* 5856 (RSA). OREGON : Oakland Douglas Co., 1881, *Howell* s. n. (K, P). WASHINGTON : Columbia River, W. Klickitat Co., 1881, *Suksdorf* (G). MEXICO : Guadalupe Is., N.E. Anchorage, 1957, *Moran* 5978 (RSA).

42b. T. gracilentum Torr. & Gray Var. **palmeri** (Wats.) McDerm., N. Am. Sp. Trif. 300 (1910). *Typus* : Guadalupe Island, Baja California, 1875, *E. Palmer* 26 (NY, iso. : G !, K !, NA !, P !).

T. palmeri Wats., Proc. Am. Acad. 11 : 132 (1875)

Plate 40. *T. gracilentum* Torr. & Gray Var. *gracilentum* (California : *Mulliken* 10, HUJ).
Plant in flower; pod (California : *Parish* 4684, HUJ); flower.

Leaflets narrowly elliptical to oblanceolate, oblong-lanceolate or lanceolate, acute and apiculate.

Gen. Distr. : USA : California. Mexico.

Selected specimens : CALIFORNIA : Santa Catalina Is., W. of Isthmus Harbor, 50–100 ft, 1932, *Wolf* 3546 (RSA); San Clemente Is., 1903, *Trask* 334 (E); San Nicolas Is., 1897, *Trask* 37 (K). MEXICO : Guadalupe Is., W. Anchorage, 20 m, 1957, *Moran* 5953 (HUJ, POM); *ibid.*, N. Twin Canyon, 100 ft, 1958, *Carlquist* 444 (POM).

43. T. hybridum L., Sp. Pl. 766 (1753).

Perennial, glabrous or glabrescent, 10–70 cm. Stems few or several, sometimes fistulose, erect or ascending, much branched at base, sometimes sparingly branching or almost unbranched or caespitose. Leaves long-petioled; stipules 1–2.5 cm long, obovate to lanceolate, adnate to petioles for about one-third of their length, free portion tapering to subulate or cuspidate tip, more or less herbaceous; leaflets 1–3 × 1–1.8 cm, obovate, rarely rhombic, cuneate at base, emarginate or obcordate, margin entire or serrulate. Inflorescences 1–2.5 cm in diam., axillary, globular, 10- to many-flowered. Peduncles slender, longer than subtending leaves, slightly hairy near head. Bracts about 1 mm, lanceolate. Pedicels up to 4–5 mm long, deflexed after anthesis. Calyx up to 4 mm long, whitish, glabrescent; tube membranous with 5 prominent and 5 inconspicuous intermediate nerves; teeth 1–2 times as long as tube, unequal, lanceolate-subulate, with broad sinuses between them. Corolla 0.7–1 cm long, white to pink or flesh-coloured; standard ovate-oblong, slightly denticulate above. Pod slightly exserted from calyx tube, oblong, stipitate, 2–4-seeded. Seeds ovoid, tuberculate, reddish. Fl. May–September. 2n = 16.

Hab. : Meadows, pastures, roadsides, damp fields, streams.

1. Stems fistulose, erect. Inflorescences 2.5 cm in diam. — **43a.** var. **hybridum**
– Stems not fistulose, procumbent to ascending or caespitose. Inflorescences up to 2 cm in diam. — 2
2. Stems procumbent to ascending, much branched. — **43b.** var. **elegans**
– Stems caespitose, branched at base. — **43c.** var. **anatolicum**

43a. T. hybridum L. Var. **hybridum**. *Typus* : Described from a cultivated specimen (Hb. Linn. 930/15 BM !).

T. fistulosum Gilib., Fl. Lithuan. 2 : 86 (1782).
T. bicolor Moench, Meth. 111 (1794).
T. vaillantii Poir., Encycl. 8 : 4 (1808).
T. intermedium Lapeyr., Suppl. Hist. Pl. Pyr. 115 (1818).
T. michelianum Gaud., Fl. Helv. 4 : 573 (1829) non Savi (1798).
T. caespitosum Eichw., Skizze 166 (1830).
Amoria hybrida (L.) Presl, Symb. Bot. 1 : 47 (1830).

Icon. : [Plate 41]. Reichenb., Ic. Fl. Germ. t. 117 II 12–18 (1867–89).

Plate 41. *T. hybridum* L. Var. *hybridum*
(Turkey : *Zohary* & *Orshan* 2160-2, IIUJ).
Plant in flower and fruit; calyx; flower; young pod; standard; keel; wing.

Gen. Distr. : Probably native only in the Mediterranean countries up to the Caucasus, but widely cultivated.

Selected specimens : TURKEY : Between Muş and Tatvan, 1,300 m, 1964, *Zohary* & *Orshan* 2160-2 (HUJ); Bolu : Abant Göl, 1,400 m, 1962, *Davis* & *Coode* D. 37287 (E); Hakkari : Gevar Ovasi, 2–3 km from Yüksekova, 1,950 m, 1966, *Davis* 45769 (E). ISRAEL : Jerusalem, irrigated garden, 1940, *D. Zohary* (HUJ). IRAN : Azerbaidjan, 2 km E. of Sarab, 1,710 m, 1965, *Danin* et al. 651849 (HUJ).

43b. T. hybridum L. Var. **elegans** (Savi) Boiss., Fl. Or. 2 : 146 (1872). *Typus* : specimen from env. of Pisa (PI ?).

T. elegans Savi, Fl. Pis. 2 : 161, t. 1, f. 2 (1798).
T. hybridum L. var. *pratense* Babenh., Fl. Lusit. 1 : 198 (1839).
Amoria elegans (Savi) Presl, Symb. Bot. 1 : 47 (1830).

Icon. : Reichenb., *loc. cit.* t. 116 II 11–17.

Gen. Distr. : Yugoslavia, France, Italy, Finland, Crete, Turkey, Caucasus, Canada (introduced).

Selected specimens : YUGOSLAVIA : Belgrad, 1888, *Bornmueller* (HUJ). FRANCE : Seine et Loire, 1879, *Verlot* 3232 (R). ITALY : Toscana, Monte Pisano, 1862, *Lais* 3086 (R). FINLAND : Nyland, Mantsala-Ang., 1881, *Stjernvall* (R). TURKEY : betw. Kars and Artvin, Yalnizçam, 2,160 m, 1964, *Zohary* & *Plitmann* 2468-7 (HUJ). CAUCASUS : Helenendorf Georg., 1838, *Hohenacker* 241 (HUJ). CANADA : Nama Creek Road, 1960, *Garton* 8372 (HUJ).

43c. T. hybridum L. Var. **anatolicum** (Boiss.) Boiss., Fl. Or. 2 : 146 (1872). *Typus* : Turkey : Manisa, Tmoli cacumini supra Philadelphiam (Alaşehir), VI 1842, *Boissier* (G ! lecto.).

T. anatolicum Boiss., Diagn. ser. 1, 2 : 31 (1843).
T. parvulum Beck ex Stapf in Denk. Akad. Wiss. Wien Math.-Nat. Kl. 51 : 35 (1886).

Stems caespitose, branched at base. Leaflets up to 1 cm. Inflorescences as in var. *elegans*.

Hab. : Meadows beside streams; 800–2,400 m.

Gen. Distr. : Crete, Bulgaria, Greece, Turkey, Iraq, Iran.

Selected specimens : CRETE : S. of Anoya, 1,200 m, 1964, *Zohary* & *Orshan* 28409-9 (HUJ). TURKEY : Cakeralan, 1,000 m, 1963, *Zohary* (HUJ); Erzurum, Horasan-Karaurgan, 2,000 m, 1957, *Davis* & *Hedge* D. 29518 (HUJ). IRAQ : Hawara Blinda, N.E. of Haji Umran, 1,700–1,900 m, 1959, *Rawi*, *Nuri* & *Kass* 27786 (K). IRAN : Kuh-i-Alwand, Hamadan, 6,000 ft, 1963, *Bowls Bot. Exp.* 1634 (K); 50 km S. of Zirub, Shiblak, 2,350 ft, 1961, *Zohary* & *Orshan* A171/1 (HUJ).

Note : *T. hybridum* is an important fodder crop, widely cultivated as pasture and for cutting. It has been introduced and become naturalized virtually throughout the temperate regions in both hemispheres. It does not suffer from clover sickness and can withstand a considerable soil moisture content. Known as "Alsike clover" or sometimes as "Swedish clover".

44. T. latifolium (Hook.) Greene, Pittonia 3 : 223 (1897). *Typus* : Light pine-woods on the undulating ridge of Coeur d'Aleine Mountains near St. Joseph, *Jane & C.A. Geyer* n. 659 (K ! photo HUJ, iso. : G !, K !).

T. longipes Nutt. *β. latifolium* Hook., London. Jour. Bot. 6 : 209 (1847).
T. aitonii Rydb., Bull. Torrey Bot. Club 34 : 46 (1907).
T. orbiculatum Kenn. & McDerm., Muhlenbergia 3 : 8 (1907).
T. howellii S. Wats. var. *latifolium* (Hook.) McDerm., Ill. Key N. Am. Sp. Trif. 267 (1910).

Icon. : [Plate 42]. McDerm., *loc. cit.* pls. 112–113.

Perennial, appressed-pubescent, 4–40 cm. Roots and rhizomes branching. Stems slender, ascending, spreading, branching from base and slightly above. Lower leaves long-petioled, upper ones shorter-petioled; stipules up to 2 cm, ovate to lanceolate, acute, entire or 1–3-toothed near apex, the upper ones longer and broader than the lower ones; leaflets 1–4 × 0.5–2.5 cm, obovate, elliptical to oblong-ovate, cuneate, rounded to retuse at apex, rarely acute, margin serrulate, sparingly appressed-pubescent beneath, glabrescent above. Inflorescences 2–3 cm in diam., globular, 6–30-flowered; rhachis slightly prolonged beyond flowers. Peduncles longer than subtending leaves, recurved near base of head. Bracts scale-like, minute. Pedicels 2–4 mm long, pilose, strongly deflexed. Calyx 5–8 mm, appressed-pubescent; tube campanulate, 10-nerved; teeth 2–3 times as long as tube, subequal, subulate, sharp-pointed. Corolla 1.2–1.5 cm long, white-yellow or purplish; standard elliptical, tapering above, sometimes retuse, longer than wings and keel. Pod slightly longer than calyx tube, 1–2-seeded, obovoid, short-stipitate, appressed-pilose, reticulately veined. Seeds about 2 mm, flattened, dark brown. Fl. May–July. 2n = 16, 32.

Hab. : Moist meadows, in open woods and on rocky ridges; 800–1,500 m.

Gen. Distr. : USA : Idaho, Montana, Oregon, Washington.

Selected specimens : IDAHO : Lake Santa, 1929, *Epling* (UC); Waha Nez Perces Co., 2,000–3,500 ft, 1896, *Heller* 3273 (E); Kootenai Co., Carlin Bay, 1926, *StJohn* et al. 4229 (UC). MONTANA : near Thompson Falls, Missoula Co., 1925, *Kennedy* 2325 (HUJ). OREGON : near Gibbon station, Umatilla Co., 1910, *Heller* 10180 (LA). WASHINGTON : Asotin Co., 7.4 mi. S. of Anatone, 1962, *Gillett & Taylor* 11007 (HUJ).

Note : *T. latifolium* is closely related to the *T. longipes* complex; some populations show intermediate characters.

45. T. leibergii Nels. & Macbride, Bot. Gaz. 65 : 58 (1918). *Typus* : Oregon : Near Dewsy, serpentine dykes, 1,220 m, June 21, 1896, *J. B. Leiberg* 2344 (GH holo., iso : BM !, F, K !, NY, ORE, P !, POM, UC).

Icon. : [Plate 43].

Perennial, subcaulescent, densely villous, 3–15 cm. Roots large and thick. Stems

Plate 42. *T. latifolium* (Hook.) Greene
(Washington : *Gillett* & *Taylor* 11007, HUJ).
Plant in flower; flower; wing; keel.

Plate 43. *T. leibergii* Nels. & Macbride
(Oregon : *Leiberg* 2344, K).
Plant in flower; flower; standard; wing; keel.

short, branching from base or above, covered at base with persistent stipules. Leaves on 1–2 cm long petioles; stipules about 1 cm, ovate or broadly lanceolate, acuminate, margin irregularly coarsely toothed; leaflets 1–1.5 × 0.7–1.4 cm, obovate-oblong or elliptical, rotundate, rounded, densely villous on both sides, margin coarsely spinulose-dentate from about upper half. Inflorescences 2–3 cm in diam., globular, many-flowered, whorled. Peduncles longer than subtending leaves. Bracts inconspicuous. Pedicels 2–5 mm long, deflexed after anthesis. Calyx 6–7 mm long, densely wooly; teeth about twice as long as tube, subulate, subequal, the upper ones slightly longer. Corolla 1–1.2 cm long, purple; standard broadly elliptical, rounded, longer than wings and keel. Pod longer than calyx tube, ovoid, 1–2-seeded, scarious, reticulately veined, villous towards apex. Seeds about 2 mm, smooth. Fl. June.

Hab. : Serpentine dykes; 1,220 m.

Gen. Distr. : USA : Oregon (endemic).

Note : Known only from type collections.

46. T. masaiënse Gillett in Kew Bull. 7 : 387 (1952).

Annual, pilose to glabrescent, 40–60 cm. Stems erect or ascending, branching from base and above, sometimes lower parts prostrate and rooting at nodes. Leaves on up to 5 cm long petioles; stipules 0.6–2 cm long, ovate-lanceolate, acuminate, membranous; leaflets 18–30 × 10–14 mm, obovate or elliptical, cuneate at base, rounded or emarginate, apiculate, prominently nerved, pilose only on underside of midrib, margin serrulate-dentate. Inflorescences up to 2.5 cm in diam., globular, 20- to many-flowered, arranged in 2 whorls. Peduncles 5–8 cm long, more or less pilose. Bracts minute, hyaline, lanceolate. Pedicels up to 6 mm, pilose, deflexed after anthesis. Calyx 7 mm long, sparsely pilose; tube campanulate, 10-nerved; teeth 2–3 times as long as tube, unequal, triangular at base, gradually narrowing into subulate tip or subulate from base. Corolla about 1 cm long, pink to pinkish-white; standard ovate-elliptical, abruptly narrowing towards deeply retuse apex, longer than wings and keel. Pod longer than calyx tube, 3–4-seeded, constricted between seeds, oblong, glabrous. Seeds about 2 mm, reniform, brown. Fl. April–September. 2n = 16.

46a. T. masaiënse Gillett Subsp. **masaiënse**. *Typus* : Tanganyika : Moister parts of Serengeti plains, Moru, 4,000 ft, 3.6.1932, *H. E. Hornby* H40/32/16 (K ! holo., iso. EA).

Icon. : [Plate 44]. Gillett, *loc. cit.* f. 4/7.

Calyx green or straw-coloured; teeth narrowing gradually, subulate in their upper half.

Hab. : Upland grassland, especially in seasonally wet areas; 1,200–2,100 m.

Gen. Distr. : Tanganyika (endemic).

Selected specimens : TANGANYIKA : Ngorongoro Crater at Haroda, 6,000 ft, 1932, *Burtt* 4191 (K); Masai Distr. at Endulen, 1957, *Bally* 11538 (K); Lerai Descent into Ngorongoro

Plate 44. *T. masaiënse* Gillett Subsp. *masaiënse*
(Tanganyika : *Burtt* 4191, K).
Flowering and fruiting branch; flower; standard; wing; keel; pod;
enlarged part of branch showing stipule.

Crater, 6,800 ft, *Herlocker* H-97 (HUJ); Musoma Distr., Seronera to Simeya Guard Post, 1961, *Greenway* 10, 104 (K).

46b. T. masaiënse Gillett Subsp. **morotoënse** Gillett in Kew Bull. 24 : 218 (1970). *Typus* : Uganda : Karamoja Distr., Mt. Moroto, 2,700 m, grassland, Sept. 1958, *J. Wilson* 536 (K ! holo., iso. EA).

More slender than subsp. *masaiënse*. Flowers smaller. Calyx (at least throat and teeth) purplish; teeth subulate almost from base.

Hab. : Upland grassland with patches of *Juniperus-Podocarpus* forest; 2,100–2,700 m.

Gen. Distr. : Uganda (endemic).

47. T. michelianum Savi, Fl. Pis. 2 : 159 (1798).

Annual, glabrous, 30–60 cm. Stems many, erect or ascending, grooved, fistulose or solid, branching extensively all along. Lower leaves long-petioled, uppermost ones with petioles only 1–2 cm long; stipules 0.5–1.5 cm long, herbaceous, broadly ovate at base, free portion lanceolate-acuminate; leaflets 1–3 × 0.6–2 cm, obovate or oblong, rounded to retuse at apex, those of upper leaves somewhat rhombic, mucronate, margin dentate-spinulose almost all along. Inflorescences up to 3 cm in diam., many-flowered, globular umbellate. Peduncles as long as or longer than subtending leaves. Bracts about 2 mm, hyaline, lanceolate. Pedicels much longer than calyx, strongly deflexed after anthesis. Calyx about 6 mm, glabrous; tube turbinate, 10-nerved; teeth 2–4 times as long as tube, almost equal, subulate; sinuses between teeth rounded. Corolla 0.9–1.3 cm long, pink; standard ovate-oblong, acutish, much longer than linear wings, apical margin entire. Pod exserted from tube, ovoid-oblong, hairy above, 2-seeded, stipitate, ventral suture very prominent, style longer than pod. Seeds 1–2 mm, ovoid-oblong, brown, smooth. Fl. May. 2n = 16.

47a. T. michelianum Savi Var. **michelianum**. *Typus* : Italy : circa Bientinum (Herb. Micheli, FI !).

T. vaillantii Loisel. in Desv., Jour. Bot. 2 : 365 (1809).
Amoria micheliana (Savi) Presl, Symb. Bot. 1 : 47 (1830).
T. macropodum Guss., Fl. Sic. Syn. 2 : 338 (1844).

Icon. : [Plate 45]. Reichenb., Ic. Fl. Germ. 22 : t. 117 I 1–11 (1867–89).

Stems fistulose, erect. Pedicels much longer than calyx. Calyx teeth 4 times as long as tube.

Hab. : Fields, wastes and damp places.

Gen. Distr. : Portugal, Spain, France, Corsica, Sardinia, Italy, Sicily, Yugoslavia, Bulgaria, Romania, Greece, Crete, Turkey.

Selected specimens : PORTUGAL : Prov. Estremadura, 1848, *Welwitsch* 388 (G). FRANCE :

Plate 45. *T. michelianum* Savi Var. *michelianum*
(Romania : *Nyárády* 968, HUJ).
Plant in flower and fruit; flower; pod; standard; wing; keel; calyx.

Saint-François près de Tours, 1854, *Billot* 1649 (G). ITALY : Pisa a Collano (Toscana), 1861, *Billot* 1649 bis (G). YUGOSLAVIA : Bosnia, Dervent, 200 m, 1910, *Prodan* 5249 (G). ROMANIA : Oltenia, distr. Dolj., Tamburesti, 40 m, 1929, *Nyárády* 968 (HUJ). TURKEY : 15 km E. of Mus, *Katznelson* & *Bailey* CPI 45856 (HUJ).

47b. T. michelianum Savi Var. **balansae** (Boiss.) Azn. in Mag. Bot. Lap. 12 : 161 (1913). *Typus* : Turkey : Izmir, Cordilio près de Smyrne, dans les champs en friche, 9 Mai, 1854, *Balansa* 158 (G! iso.: E!, G!, K!, W!,).

T. balansae Boiss., Diagn. ser. 2, 5 : 81 (1856).

Stems solid. Pedicels shorter. Calyx teeth twice as long as tube.
Hab. : As above.
Gen. Distr. : Bulgaria, Turkey.

Selected specimens : BULGARIA : Sofia, 1956, *Velcev* (G). TURKEY : Istanbul : Prinkipo (Büyük Ada), 1909, *Aznavour* (G); Van, 19 km from Ercis to Bendimahi, 1,700 m, 1966, *Davis* 44225 (K); Lydia, Smyrna, 1906, *Bornmüller* 9364 (W).

48. T. nigrescens Viv., Fl. Ital. Fragm. 12, t. 13 (1808).

Annual, glabrous or glabrescent, (5–)10–60 cm. Stems prostrate, erect or ascending, grooved, branching from base and above. Leaves long-petioled, except for uppermost ones; stipules 0.4–2 cm long, triangular-lanceolate, membranous, upper free portion abruptly subulate-cuspidate; leaflets 0.5–2.5(–4) × 0.3–1.5(–2.5) cm, obovate to rhombic or sometimes obtriangular, truncate to rounded or emarginate at apex, sharply denticulate in upper part or all over. Inflorescences 1–2 cm, globular, many-flowered, loose. Peduncles longer than subtending leaves. Bracts as long as or longer than pedicels, oblong, acute or acuminate. Pedicels 3–6 mm long, as long as or longer than calyx tube, deflexed after anthesis. Calyx about 5 mm, glabrous or sparingly hairy; tube white, 5–10-nerved; teeth shorter or longer than tube, greenish, triangular to lanceolate, with or without cuspidate tip, recurved in fruit, subequal. Corolla 6–9 mm, pink or white, turning yellowish or brownish in fruit; standard obovate-oblong, mucronulate or retuse, longer than wings and keel. Pod slightly longer than calyx tube, linear-oblong, 1–5-seeded, margin crenate or constricted between seeds. Seeds ovoid, dark brown. Fl. March–October. 2n = 16, 32.

48a. T. nigrescens Viv. Subsp. **nigrescens**. *Typus* : Italy : Repare in littore Romano, non longe ab Ostia, nunc Flumicino (? Genoa), *Viviani* (GE ?).

T. hybridum Savi, Fl. Pis. 2 : 90 (1798) non L. (1753).
T. vaillantii Ten., Prodr. Fl. Nap. 1 : 44 (1811).
T. pallescens DC., Fl. Fr. Suppl. 555 (1815) non Schreb. (1804).
T. polyanthemum Ten. ex Ser. in DC., Prodr. 2 : 200 (1825).
T. michelianum Koch, Syn. Fl. Germ. ed. 1 : 174 (1835) non Savi (1798).
T. prostratum Griseb., Spicil. Fl. Rumel. 1 : 29 (1843).

T. molybdocalyx Reichenb. f., Ic. Fl. Germ. 22 : 77, t. 2167 (1903).

Icon. : Reichenb., Ic. Fl. Germ. 22 : t. 110 IV 20–35 et 116 I 1–10 (1867–89).

Corolla white to pale pink. Pod crenate at margin. Seeds 3–4.
Hab. : Fields, among shrubs, stony hills, damp places.
Gen. Distr. : Mediterranean countries of Europe, Turkey, Crete, N. Africa.

Selected specimens : PORTUGAL : Coimbra, Cerca de S. Bento, 1948, *Matos* (HUJ). FRANCE : Montpellier, 1930, *Eig* (HUJ). ITALY : Apulien, Mt. Gargano, 1964, *Seitz* (HUJ). GREECE : Attika, Hagia Trias, 950–1,100 m, 1967, *Haesler* 165 (HUJ). TURKEY : Istanbul, San Stefano, 1898, *Aznavour* (G); Sinop, Ayancik Büyükdiz, 300 m, 1970, *Annon.* 240570 (HUJ). CRETE : ruins of Phaestos, 80 m, 1964, *Zohary* & *Orshan* 25400-18 (HUJ). TUNISIA : Ile de Zembra, 1953 (HUJ).

48b. T. nigrescens Viv. Subsp. **petrisavii** (Clem.) Holmboe, Berg. Mus. Skrift. N. R. 1(2) : 106 (1914). *Typus* : Istanbul : in collibus Constantinopolitanus ultra il Gran Campo, *G. Clementi* (BM ?).

T. hygrophilum Boiss., Diagn. ser. 2, 2 : 18 (1856).
T. petrisavii Clem., Sert. Or. 32 (1885).
T. meneghinianum Clem., *op. cit.* 31.
T. thessalonicum Halacsy & Charrel ex Halacsy in Oesterr. Bot. Zeitschr. 41 : 370 (1891).
T. nigrescens Viv. ssp. *petrisavii* (Clem.) Holmboe var. *meneghinianum* (Clem.) Hossain, Not. Roy. Bot. Gard. Edinb. 23 : 468 (1961).

Icon. : [Plate 46]. Clem., *loc. cit.* t. 7, f. 1–2.

Corolla brownish in fruit. Pod often deeply constricted between upper and lower segments. Seed 1 (rarely 2).
Hab. : As in subsp. *nigrescens*; 1,750 m.
Gen. Distr. : Sicily, Yugoslavia, Greece, Aegean Islands, Cyprus, Turkey, Syria, Lebanon, Israel, Iraq, Iran, Caucasus, Armenia.

Selected specimens : SICILY : 35 km S. of Messina, Capo S. Alessio, 1963, *Stud. Biol.* 716 (HUJ). TURKEY : Muğla : Bodrum, 0–10 m, 1966, *Davis* 40868 (E); Siirt, 37 km from Siirt to Baykan, 800–900 m, 1966, *Davis* 43171 (K). SYRIA : Homs, 2,000 ft, 1910, *Haradjian* 3235 (K). LEBANON : Shemlan, 715 m, 1958, *Maitland* 318 (K); Beirut, 1877, *Post* 942 (NA). CYPRUS : Cedars Valley, 3,500 ft, 1967, *Czoyomides* 946 (K). IRAQ : Erbil, Gt. Zab. below Kan Gissik, 250 m, 1948, *Gillett* 10.533 (K). IRAN : Karun, 1919, *Sharples* s. n. (K). ISRAEL : Acco Plain, Haifa to Nesher, 1963, *Amdursky* & *Grizi* 17083 (HUJ); Upper Galilee, betw. Kerem Ben-Zimra and Gush-Halav, 1964, *Zohary* et al. (HUJ); Golan, near Jarba, 1968, *Zohary* (HUJ); Mt. Carmel, near Kishon Bridge, 1965, *Zohary* (HUJ).

49. T. pallescens Schreb. in Sturm., Deutschl. Fl., Abt. 1, Band 4, Heft 15 (1804).

T. glareosum Schleich., Cat. Pl. Helv. ed. 4 : 35 (1821).
Amoria pallescens (Schreb.) Presl, Symb. Bot. 1 : 47 (1830).

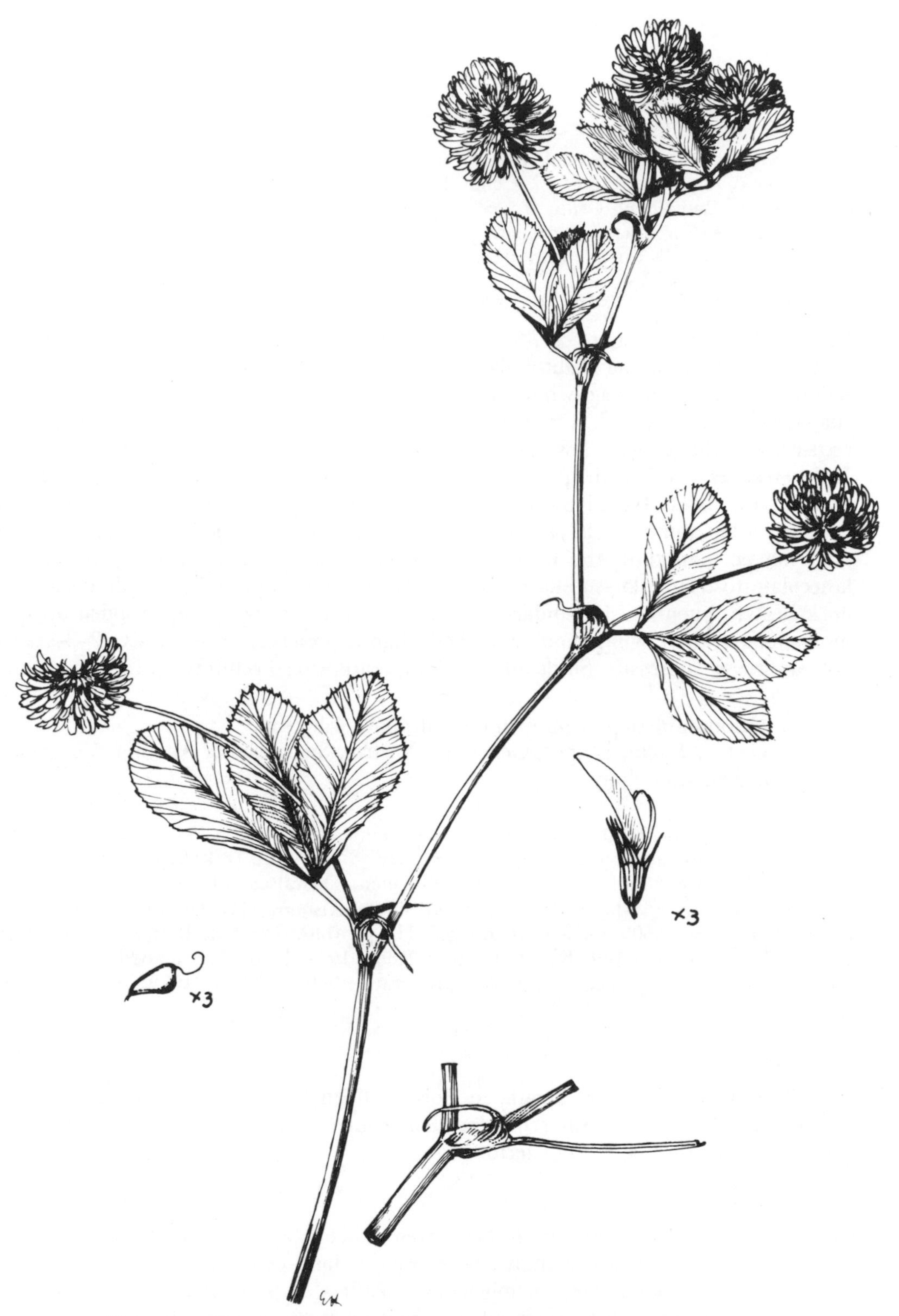

Plate 46. *T. nigrescens* Viv. Subsp. *petrisavii* (Clem.) Holmboe (Golan : *Zohary*, HUJ).
Flowering branch; young pod (Mt. Carmel : *Zohary*, HUJ); flower; enlarged part of branch showing stipule.

T. caespitosum Sturm., Deutschl. Fl, Abt. 3, Band 6, Heft 32 (1851).
T. arvernense Lamotte, Prodr. Fl. Centr. Fr. 1 : 202 (1855).
T. glareosum Boreau, Fl. Centr. Fr. ed. 3 : 158 (1857).
T. glaciale Porc. in Magyar Növényt. Lapok 9 : 126 (1885).
T. pallescens Schreb. var. *glareosum* (Schleich.) Rouy, Fl. Fr. 5 : 80 (1899).

Icon. : [Plate 47]. Sturm., *loc. cit.* t. 17, 4.

Perennial, glabrous, caespitose, 5–20 cm. Stems procumbent or ascending, branching from woody taproot and slightly above. Leaves short- or long-petioled in lower parts; stipules up to 1.5 cm long, white-membranous, broadly lanceolate, long-acuminate to cuspidate; leaflets 0.5–2 × 0.4–1.2 cm, obovate or elliptical, obtuse, rounded or retuse, sometimes with few hairs on nerves of under surface, margin dentate. Inflorescences 1.5–2.5 cm in diam., oblong, many-flowered. Peduncles much longer than leaves, about 10 cm long. Bracts about 2 mm, lanceolate. Pedicels 2–4 mm long, deflexed after anthesis. Calyx about 4 mm, campanulate; teeth unequal, the upper ones longer than tube, the lower ones shorter than or about as long as tube, all lanceolate to triangular-subulate. Corolla 0.6–1.2 cm long, pink or yellowish-white, dark brown when dried; standard elliptical to ovate or spatulate, rounded at apex, much longer than wings and keel. Pod slightly exserted from calyx, 2–3-seeded, oblong-linear, subsessile. Seeds about 1 mm, lentiform to reniform. Fl. July–August. 2n = 16.

Hab.: Mountain slopes, on granite soil above 1,800 m.

Gen. Distr. : Spain, France, Germany, Austria, Switzerland, Albania, Yugoslavia, Bulgaria, Romania, Italy.

Selected specimens : FRANCE : Monétier les Bains, 2,200 m, 1914, *Faure* (MPU); Monte-Dore, Puy de-Dóme, 1,250 m, 1884, *Billiet* 811 (K). GERMANY : St. Bernhard, *Thomas* 1881 (BM, HUJ, K). AUSTRIA : Karnten, Dobralsch bei Villach, 1897, *Dörfler* (K). SWITZERLAND : Matterhorn, 2,200 m, 1960, *Evenari* (HUJ); betw. Zermatt and Schwarzsee, 1,700–2,600 m, 1950, *R. Koppel* (HUJ). ALBANIA : Sala Bergrucken, 1,900 m, 1916 und 1918, *Dörfler* 139 (K). BULGARIA : Rilon, 1921, *Stefanoff* s. n. (BM). ROMANIA : Transsilvania, distr. Hunědoara, Mtibus Rătezat, 1,900–2,100 m, 1929, *Nyárády* 1984 (HUJ).

50. T. parnassi Boiss. & Sprun. in Boiss., Diagn. ser. 1, 2 : 30 (1843) et Fl. Or. 2 : 147 (1872). *Typus* : Am schmelzenden Schnee auf dem Parnaso, 6,800 ft, 15 Juli, 1842, *Spruner* s. n. (G ! lecto.).

Icon. : [Plate 48].

Perennial, caespitose, glabrous, 1–9 cm. Stems rhizomatous, prostrate, branching extensively from base and sparingly above. Leaves long-petioled; stipules 6 mm long, ovate, abruptly acuminate, membranous; leaflets 3–8 × 2–5 mm, obovate or obcordate, cuneate at base, rounded or retuse at apex, prominently nerved, margin serrulate. Inflorescences 0.8–1.4 cm in diam., globular, 8–12(–15)-flowered. Peduncles as long as or longer than subtending leaves. Bracts longer than pedicels, oblong, acuminate. Pedicels about 1 mm, deflexed after anthesis. Calyx 3–4 mm,

Plate 47. *T. pallescens* Schreb.
(Romania : *Nyárády* 1984, HUJ).
Plant in flower and fruit; young pod; calyx; flower; standard; wing; keel.

Plate 48. *T. parnassi* Boiss. & Sprun.
(Greece : *Pichler* 6, 76, **K**).
Plant in flower; pod (Greece : *Heldreich* 93, FI);
flower; keel; wing; standard.

glabrous; tube white, with 10 prominent nerves; teeth half the length of tube, green, ovate-triangular, with membranous margins, acute, imbricate at base. Corolla 5–8 mm long, pink; standard ovate-oblong, rounded or retuse at apex, longer than wings and keel. Pod about 3 mm, ovoid, 2-seeded, included within calyx tube. Seeds 1 mm, reniform, brown. Fl. July–August.

Hab.: Mountain slopes.

Gen. Distr.: Greece (endemic).

Specimens seen: GREECE: Morea: Mts. Taygetos, 1882, *Pichler* 6, 76 (FI, K); Agrapha, Pindi summi Montis Karava, 5,500–6,500 ft, 1885, *Haussknecht* (K); M. Korax, Aetoliae adjectae, 1899, *Chr. Leonis* 304 (B); Tymphresto (nunc Veluchi) Eurythaniae, 5,500–7,140 ft, 1879, *Heldreich* 93 (FI); Parnassi, 1855, *Guicciardi* (MPU); Metsovo, Joannina-Kalabaka road, 1960, *Feinbrun* (HUJ).

51. T. purseglovei Gillett in Kew Bull. 7: 383, f. 4/4 (1952). *Typus*: Uganda, Kigezi District, Kachwekano Farm, Swamp, 1,950 m, Dec. 1949, *J.W. Purseglove* 3139 (K! holo.).

T. rueppellianum Robyns, Fl. Spermat. Parc Nat. Alb. 1: 287 (1948) p. p. et t. 26, non Fresen. (1837).

T. subrotundum Auct. non Hochst. & Steud. ex A. Rich. (1842).

T. johnstonii Auct. non Oliv. (1886).

Icon.: [Plate 49]. Gillett in Fl. Congo Belge et Ruanda-Urundi 4: pl. XVIII (1953).

Perennial, glabrous, 10–30 cm. Stems ascending, prostrate, rooting at nodes, branched, grooved. Leaves long-petioled; stipules 0.8–1.5 cm long, broadly ovate, abruptly narrowing to short acuminate tip; leaflets 1.4–1.7 × 1–1.2 cm, obovate or obcordate to elliptical, cuneate at base, retuse or truncate, emarginate, margin denticulate with broad teeth. Inflorescences 1.5–2 cm in diam., globular, 20–30-flowered, whorled. Peduncles up to 10 cm, sparingly pilose near base of head. Bracts 1.2 mm, subulate-aciculate. Pedicels 2 mm long, more or less deflexed after anthesis. Calyx 4–5 mm long; tube campanulate, 11–13-nerved, white; teeth as long as or longer than tube, narrowly triangular, subulate at tip, scarious-margined and slightly ciliate or denticulate. Corolla 8–9 mm, deep purple; standard broadly oblong, rounded at apex, emarginate, slightly longer than wings and keel. Pod 5.5 mm long, oblong, 2–3-seeded. Seeds about 1.5 mm, ovoid, irregularly flattened, brown. Fl. June–July.

Hab.: Open marshy grassland, on volcanic soil; ca. 2,000 m.

Gen. Distr.: Uganda, Congo, Ruanda.

Specimens seen: UGANDA: Kigezi, Kachwekano Farm, 6,400 ft, 1951, *Purseglove* P. 3613 (FI). CONGO: Sabinio Volcano, below Bamboo zone, 7,500 ft, 1930, *Burtt* 2980 (K).

52. T. radicosum Boiss. & Hohen. in Boiss, Diagn. ser. 1, 9: 27 (1849) et Fl. Or. 2: 146 (1872).

Plate 49. *T. purseglovei* Gillett
(Uganda : *Purseglove* P. 3613, FI).
Plant in flower; flower standard; wing; keel.

Perennial, caespitose, glabrous or sparingly hirsute, 5–25 cm. Stems rhizomatous, with thick vertical rootstock, fusiform, prostrate, branching. Radical leaves numerous, long-petioled, cauline leaves few, short-petioled; stipules 4–7 mm long, whitish, ovate, free portion triangular, acute to muticous; leaflets 0.4–1.2 × 0.3–0.6 cm, ovate to obovate, cuneate at base, rounded or emarginate at apex, obtuse, margin denticulate with broad teeth. Inflorescences about 1.5 cm in diam., globular, few- to many-flowered. Peduncles often longer than subtending leaves. Bracts minute, scale-like, oblong, truncate to bifid, slightly denticulate. Pedicels shorter than calyx tube, deflexed after anthesis, hispid or glabrous. Calyx about 4 mm, appressed-hirsute or glabrous; tube tubular, 10-nerved; teeth somewhat longer or shorter than tube, lanceolate-subulate, equal. Corolla 8–9 mm, purple; standard oblong, somewhat curved at tip, longer than wings; keel straight, tapering into mucro. Pod as long as calyx, obovoid-oblong, 1–2-seeded. Seeds 2 mm in diam., elliptical, smooth, brown. Fl. July–August.

Hab. : Calcareous damp grassy slopes; 3,000–3,600 m.

52a T. radicosum Boiss. & Hohen. Var **radicosum**. *Typus* : Iran : In summo monte Atscha Pasch m. Demawend, 3 Aug., 1843, *Th. Kotschy* 620 (holo. G !, iso. : FI !, K !, P !, W !).

T. tumens Stev. ex M. B. var. *rechingeri* Zoh. in Zoh. & Heller, Israel J. Bot. 19: 322 (1970).

Icon. : [Plate 50]. Townsend in Kew Bull. 21 : f. 1, 13–19 (1968).

Calyx and pedicels hirsute.

Gen. Distr. : Iran (endemic).

Specimens seen : IRAN : Elburz, Gochizor, 50′ N. E. of Karadj, 9,000 ft, 1962, *Furse* 2703 (K, W); Totschal, ca. 3,500–3,600 m, 1902, *Bornmüller* 6600 (K, W); Prov. Mazanderan, distr. Kudjur, monti Ulodj, 3,200–3,400 m, 1948, *K.H.* & *F. Rechinger* 6496 (W).

52b. T. radicosum Boiss. & Hohen. Var. **guestii** (Blakelock) Hossain, Not. Roy. Bot. Gard. Edinb. 23 : 462 (1961). *Typus* : N. Iraq : Algurd Dagh, 3,000 m, on grassy places by a lake, 22.7.1932, *E. Guest* 2854 (K ! holo.).

T. guestii Blakelock, Kew Bull. 3 : 421 (1949).

Icon. : Blakelock, *loc. cit.* f. 2B; Townsend, *loc. cit.* f.1, 1–12.

Calyx and pedicels glabrous.

Gen. Distr. : Iraq (endemic).

Specimens seen : IRAQ : Algurd Dagh, on a damp grassy slope, 3,300 m, 1933, *Guest* 2894 (type locality K).

Note : Townsend (*op. cit.* 439) believes that *T. guestii* Blakelock is a good species and differs from the typical *T. radicosum* Boiss. & Hohen. not only by its glabrous calyx and pedicels, but also by the shape of the calyx teeth.

Plate 50. *T. radicosum* Boiss. & Hohen. Var. *radicosum*
(Iran : *Kotschy* 620, FI).
Plant in flower; flower.

53. T. reflexum L., Sp. Pl. 766 (1753).

Annual or biennial, villous or almost glabrous, 10–60 cm. Stems erect or ascending, many, simple or branched at base and slightly above. Leaves long-petioled; stipules 1–2.5 cm long, foliaceous, broadly ovate, acute or acuminate, margin entire or denticulate; leaflets 1–4.5 × 0.7–2 cm, ovate to obovate, broadly oblong, rounded, acute or emarginate, margin denticulate with broad teeth. Inflorescences 2–4 cm in diam., globular, dense, 10- to many-flowered; rhachis prolonged beyond flowers. Peduncles longer than subtending leaves. Bracts small, about 1 mm, broad-obovate, truncate to bifid, scarious. Pedicels 0.4–1.2 cm long, deflexed after anthesis. Calyx 6–9 mm long, glabrous or pubescent; tube very short, 10-nerved, membranous; teeth 2–3 times as long as tube, narrowly triangular to long-subulate, sharp-pointed with broad sinuses between them. Corolla 0.9–1.4 cm long, pink to dark red, wings and keel light-coloured when young; standard obovate-oblong, rounded at apex, retuse, apiculate, more or less erose-denticulate, flat in fruit, very slightly longer than wings and keel. Pod about 5 mm long, oblong, membranous, 1–6-seeded, stipitate, reticulately nerved, glabrous or sparingly villous, with persistent style about as long as pod. Seeds about 1.5 mm, reniform, yellowish-brown, pitted. Fl. April–June. 2n = 16.

Hab.: Meadows, fields and open woodlands.

53a. T. reflexum L. Var. **reflexum**. *Typus*: The species was described by Linnaeus from a *Clayton* specimen collected in Virginia (perhaps at BM ?).

T. adscendens Hornem., Hort Hafn. 2: 716 (1815).
Amoria reflexa (L.) Presl, Symb. Bot. 1: 47 (1830).
T. platycephalum Bisch., Del. Sem. Hort. Heidelb. (1839).

Icon.: [Plate 51]. Britten & Brown, Ill. Fl. N. States & Canada ed. 2, 2: 357 (1913).

Leaflets finely villous on both surfaces.

Gen. Distr.: USA: Texas, Oklahoma, Kansas, Nebraska, Carolina, Missouri, Alabama, Arkansas, Florida, Louisiana, Georgia, Tennessee, Illinois, Pennsylvania, Virginia, Kentucky. Mexico.

Selected specimens: TEXAS: Grand Saline, 1948, *Bates* 47 (HUJ). N. CAROLINA: Biltmore, 1897, 1310b (UC, W); Chatham Co., 4.4 mi. E. of Siler City, 1970, *Leonard* 3189 (H). MISSOURI: S.E. of Webb City, 1952, *Palmer* 53806 (UC); Phelps Co., Rolla, 1928, *Kellogg* 1817 (UC). ALABAMA: Auburn Lee Co., 1896, *Earle* 2640 (POM). ARKANSAS: Big Island Chute Farm, 1939, *Miller* 150 (NA); Booneville, 1920, *Pieters* (NA). FLORIDA: Leon, Tallahassee 1958, *Godfrey* 56531a (RSA); *ibid.*, 1845, *Rugel* 150 (W). GEORGIA: Mt. Berry, Floyd Co., 1938, *Jones* 318 (NA); De Kalb Co., Stone Mt., 1891, *Eggert* (H). ILLINOIS: Jackson Giant City State Park, 1955, *R. South* (SIU). MEXICO: *Banks* 1553 (W).

53b. T. reflexum L. Var. **glabrum** Lojacono, Nuov. Giorn. Bot. Ital. 15: 150 (1883). *Typus*: Near Augusta, Illinois, June 1848, *S. B. Meed* (?, iso.: NY, PH).

Plate 51. *T. reflexum* L. Var. *reflexum*
(Missouri : *Palmer* 3968, UC).
Plant in flower; young pod (N. Carolina : *Leonard* 3189, H); flower.

Leaflets glabrous, sometimes with few hairs on underside of midrib.
Gen. Distr. : USA : Oklahoma, Kansas, Illinois, Iowa, Kentucky, S.W. Indiana, Ohio, Pennsylvania, Missouri, Arkansas.

Selected specimens : OKLAHOMA : Comancho Co., Panther Creek, Payne Springs, 1941, *McMurry* 909 (NA, RSA). ILLINOIS : Augusta, 1845, *Medd* (K, ILL). KENTUCKY : Trigg Co., 1939, *Clanton* 16 (NA). INDIANA : Posey Co., Mt. Vernon (Erwin Woods), 1935, *Hermann* 6660 (NA). MISSOURI : St. Louis, 1838, *Riehl* 166 (W).

54. T. repens L., Sp. Pl. 767 (1753).

Perennial, glabrous or glabrescent, 10–30 cm. Stems rhizomatous, prostrate, long-creeping and rooting from nodes. Leaves long-petioled, sometimes petioles hairy; stipules broad at base, ovate-lanceolate, with subulate upper part, scarious with reddish or green nerves; leaflets 0.6–2.5(–4) × 1–1.5 cm, broadly obovate to orbicular, obcordate, mostly emarginate, margin sharply serrulate, with parallel, lateral veins (featuring light or dark marks) forked towards margin. Inflorescences 1.5–3.5 cm in diam., umbellate, 20- to many-flowered, rather loose, nearly globular. Peduncles as long as or much longer than subtending leaves, weak. Bracts shorter than pedicels, ovate-oblong, acuminate. Pedicels as long as or longer than calyx tube, somewhat deflexed from early flowering, strongly deflexed after anthesis, glabrous or hairy. Calyx 3–5 mm, glabrous; tube campanulate, 6–10-nerved; teeth about as long as tube, somewhat unequal, triangular-lanceolate. Corolla 4–13 mm long, white, yellow or pink; standard ovate-lanceolate, oblong, rounded at apex; wings somewhat spreading. Pod 4–5 mm, usually 3–4-seeded, linear-oblong, constricted between seeds. Seeds ovoid to reniform, brownish. Fl. March–September. 2n = 16, 28, 32, 48, 64.
Hab. : Damp and swampy soils, also on lawns, grassy places, etc.

1. Petioles and pedicels hairy — 2
– Petioles and pedicels glabrous — 3
2. Corolla white. Plants of Inner Anatolia and Iran. — **54d**. var. **macrorrhizum**
– Corolla pale pink. Plants of Europe. — **54e**. var. **biasolettii**
3. Inflorescences 3–3.5 cm in diam. Leaflets up to 4 cm. — **54b**. var. **giganteum**
– Inflorescences less than 3 cm in diam. Leaflets smaller — 4
4. Corolla yellow or greenish-yellow; standard 3–4 times as long as calyx. — **54g**. var. **ochranthum**
– Corolla cream, white or pale pink; standard shorter — 5
5. Peduncles 1–2 cm, only slightly longer than leaves. Flowers 4–5 mm long. — **54c**. var. **orphanideum**
– Peduncles usually longer than leaves. Flowers more than 5 mm long — 6
6. Calyx with only 6 distinct nerves. — **54h**. var. **orbelicum**
– Calyx with 10 nerves. — 7
7. Leaflets 1 cm or more in length. Inflorescences up to 2.5 cm in diam. — **54a**. var. **repens**
– Leaflets less than 1 cm. Inflorescences less than 2 cm in diam. — **54f**. var. **nevadense**

54a. T. repens L. Var. **repens**. *Typus* : Described from Europe (Hb. Cliff. 375 18, BM !).

Amoria repens (L.) Presl, Symb. Bot. 1 : 47 (1830).
T. luxurians Hort. Par. ex Steud., Nom. ed. 2, 2 : 706 (1841)
T. stipitatum Clos in Gay, Fl. Chile 2 : 71 (1847).
T. nothum Stev. in Bull. Soc. Nat. Mosc. 29, 3 : 137 (1853).
T. limonium Phil. in Linnaea 28 : 679 (1856).
T. umbellatum Losc. ex Willk. & Lange, Prodr. Fl. Hisp. 3 : 355 (1877).
T. repens L. var. *genuinum* Aschers. & Graebn., Syn. Mitteleur. Fl. 6, 2 : 498 (1908).

Icon. : [Plate 52]. Reichenb., Ic. Fl. Germ. 22 : t. 115 I, III, IV, V (1867–89).

Gen. Distr. : Europe, C. and N. Asia. All Mediterranean countries and N. Africa.

Selected specimens : PORTUGAL : Coimbra, Penedo da Meditaçao, 1953, *Matos* (HUJ). FRANCE : Charenton, proximo de Paris, 1876, *Saldanha* 2129 (R). ITALY : Sabaudia, *Bonjean* 68205 (R). FINLAND : Nyland, Mantsala, 1879, *Stjernvall* 68369 (R). HOLLAND : Utrecht, Vechten, 1959, *Mennega* 627 (HUJ). SWEDEN : Tärendö parish, Tärendö Village, 1959, *Alm* 3547 (HUJ). KASAKSTAN : Grib. Aktinsk, 1926, *Knorring* 220 (HUJ). TRANSCAUCASIA : Armenia, distr. Migri, 1934, *Karjagin* (HUJ). RUSSIA: Prope viam, Birštonas, distr. Alytus, 1931, *Sataïte* 9 (HUJ). CRETE : Kissamos, 1932, *Guiol* (HUJ). TURKEY : Bolu, Ala Dag, 700 m, 1960, *Khan* et al. 512 (E). SYRIA : Djebel Druze, env. of Tel Jine, 1,600 m, 1932, *Eig* & *Zohary* (HUJ). LEBANON : Hammana, 1941, *Feinbrun* (HUJ). CYPRUS : Kryos Potumos, 5,100 ft, 1938, *Kennedy* 1289 (K). ISRAEL : Hula Valley, E. of Jahula, *D. Zohary* 944 (HUJ). EGYPT : N. Park, Alexandria, 1944, *Gilbert* 144 (K). IRAQ : Mosul Kiwa, Agra, 1948, *Rawi* 11.404 (K). IRAN : Hamadan, Mt. Alvand, 2,210 m, 1965, *Danin* et al. 65-843 (HUJ). AFGHANISTAN : Hajigak Pass, Koh-l-Baba, 10,000 ft, 1966, *Fursa* 8532 (K). BALUCHISTAN : Kahan, 7,000 ft, 1952, *Crookshank* 463 A (K). ALGERIA : Forêt de M. Sila près El-Ancor, 1923, *Faure* (HUJ).

54b. T. repens L. Var. **giganteum** Lagrèze-Fossat, Fl. Tarn. et Garonne 95 (1847). *Typus* : Described from France.

Icon. : Zoh., Fl. Pal. 2 : t. 231 (1972).

Gen. Distr. : France, Czechoslovakia, Turkey, Lebanon, Israel, Tunisia.

Selected specimens : CZECHOSLOVAKIA : Moravia, Olomouc Larce, 218 m, 1933, *Otruba* 826 (G). TURKEY : Istanbul Thérapia, Parc Amb. d'Allem., 1888, *Aznavour* 675 (G). LEBANON : Hermon, Wadi Sheeba, 1924, *Naftolsky* 17225 (HUJ). ISRAEL : Upper Galilee, Wadi Limun, 1925, *Naftolsky* 17224 (HUJ). TUNISIA : Qued Akmar (Tabarea), 1930, *Eig* (HUJ).

Note : Perhaps only a sporadically occurring form of var. *repens*.

54c. T. repens L. Var. **orphanideum** (Boiss.) Boiss., Fl. Or. 2 : 145 (1872). *Typus* : Greece : In monte Kyllene Peloponnesi supra Trikala, *Orphanides* (G).

T. orphanideum Boiss., Diagn. ser. 2, 2 : 17 (1856).
T. repens L. ssp. *orphanideum* (Boiss.) Coombe, Feddes Repert. 79 : 54 (1968).

Plate 52. *T. repens* L. Var. *repens*
(Sweden : *Alm* 3547, HUJ).
Plant in flower; flower; standard; wing; keel; fruiting calyx with pod
(Portugal : *Matos*, HUJ).

Gen. Distr.: Greece, Crete, Turkey, Lebanon.

Selected specimens: TURKEY: Bursa, Uludag, 1,800–2,180 m, 1956, *Moore* 7294 (E). LEBANON: Shibaniya, 1924, *Eig* (HUJ).

54d. T. repens L. Var. **macrorrhizum** Boiss., Fl. Or. 2 : 145 (1872). *Typus*: Turkey: Ad fontes supra plumbi fodinas in monte Pasch Olug Tauri Cilicici, 2,100 m, *Kotschy* 256 b (K!).

T. macrorrhizum Boiss., Diagn. ser. 2, 5 : 80 (1856) non Waldst. et Kit. (1802).
T. pachypodum O. Schwarz in Feddes Repert. 65 : 140 (1962).

Gen. Distr.: Turkey, Iran.

Selected specimens: TURKEY: Kars, betw. Kars and Ardahan, 2,020 m, *Zohary* & *Plitmann* 2466-7 (HUJ). IRAN: Hamadan, Mt. Alvand, 2,600 m, 1965, *Danin* et al. 65-707 (HUJ).

54e. T. repens L. Var. **biasolettii** (Steud. & Hochst.) Aschers. & Graebn., Syn. Mitteleur. Fl. 6, 2 : 500 (1908).

T. biasolettii Steud. & Hochst., Flora 10 : 72 (1827).
T. prostratum Biasol., Flora 12 : 532 (1829)
Amoria biasolettiana (Steud. & Hochst.) Presl, Symb. Bot. 1 : 47 (1830)
T. biasolettianum Steud. & Hochst. ex Steud., Nomen. ed. 2, 2 : 105 (1842).
T. neglectum Noë & Nym., Consp. 178 (1878).
T. repens L. ssp. *prostratum* (Biasol.) Nym., *loc. cit.*
T. monvernense Shuttlew. ex Rouy & Foucaud, Fl. Fr. 5 : 79 (1899).
T. occidentale Coombe, Watsonia 5 : 70 (1961).
T. repens L. ssp. *occidentale* (Coombe) M. Lainz, Comun. I. N. I. A. Recursos Nat. 2 : 6 (1974).

Icon.: Reichenb., *loc. cit.* t. 115 II.

Gen. Distr.: England, France, Portugal, Spain, Greece, Yugoslavia, Albania, Turkey.

Selected specimens: ENGLAND: Caerthillian Cove, near Lizard Town, W. Cornwall, 20 m, 1959, *Coombe* O/CA (G, isotype of *T. occidentale* Coombe). GREECE: Metsovo, 1960, *Feinbrun* (HUJ). YUGOSLAVIA: Istria, Pomer prope Polam, 2–170 m, 1882, *Pichler* 419 (FI, HUJ). TURKEY: Thracia, 12 km N. of Demirkoy, 400 m, 1963, *Orshan* 55666 (HUJ).

54f. T. repens L. Var. **nevadense** (Boiss.) C. Vicioso in Ann. Inst. Bot. A. J. Cavanilles 10 : 280 (1953). *Typus*: Spain: In humidis glacialibus S. Nevada ad prata Borreguiles dicta, ad Corral de Veleta (G).

T. pallescens Boiss., Voy. Bot. Midi Esp. 2 : 727 (1845) non Schreb. (1804).

T. glareosum Boiss., *loc. cit.* 727 (1845) non Schleich. (1821).
T. nevadense Boiss., Diagn. ser. 2, 2 : 17 (1856).
T. repens L. ssp. *nevadense* (Boiss.) Coombe, Feddes Repert. 79 : 54 (1968).

Gen. Distr. : Spain, Portugal.

Specimens seen : PORTUGAL : Montalegre, Lameiro Grande, 950 m, 1943, *Pedro* & *Myre* 9363 (HUJ).

54g. T. repens L. Var. **ochranthum** Maly ex Aschers. & Graebn., Syn. Mitteleur. Fl. 6, 2 : 500 (1908). *Typus* : Bosnia, auf dem Gipfel der Bjelaśnica, in 2,067 m, *Maly* (GJO).

T. repens L. ssp. *ochranthum* (Maly) Nyar., Bul. Grad. Bot. Cluj 20 : 45 (1940).

Gen. Distr. : Romania.

Specimens seen : ROMANIA : Transsilvania : Braşov, Mt. Bucagi, 2,450 m, 1938, *Buia* et al. 1983 (G, HUJ).

54h. T. repens L. Var. **orbelicum** (Velen.) Fritsch. in Mitteil. Naturw. Ver. Steierm. 47 : 197 (1910).

T. orbelicum Velen. in Abh. Boehm. Ges. Wiss. 2 : 33 (1890).
T. repens L. ssp. *orbelicum* (Velen.) Pawl., Zapiski Fl. Tatr. 4 : 9 (1949).

Gen. Distr. : Poland, Romania, Bulgaria, Yugoslavia.

Specimens seen : ROMANIA : Moldova, Neamtu, Tâmpesti, 220 m, 1937, *Papp* 1982 (HUJ). BULGARIA : Vitosa, 1892, *Jovanović* (W). YUGOSLAVIA : Midjor, 2,140 m, 1896, *Adamović* (W).

Note : *T. repens*, known as "White or Dutch clover", is now extensively cultivated and has been introduced into many countries all over the world, for pasture, fodder and hay. Many cultivars have been selected and naturalized in the tropics of Africa, N., C. & S. America, Canada and elsewhere. The flowers are considered as an important source of honey. They are also used in folk medicine as an anti-rheumatic, anti-scrofulous, etc.
Only some of the varieties of the wild plants are treated above, from which we note the following : in the mountainous plants the stems are short and sometimes subcaespitose, while in the others the stems are long-creeping and root at the nodes. We consider var. *biasolettii* to be identical to *T. occidentale* which was described from dry grassy places near the coasts of S.W. England, N.W. France, N. Portugal and N. Spain. This diploid species was reported to be crossing successfully with the tetraploid *T. repens* (Gibson et al., 1971b). The similarity of their karyotypes (Chen and Gibson, 1971a) and the homology of their chromosomes (Chen and Gibson, 1972b) have also been shown.
T. repens has also been crossed successfully with the annual diploid *T. nigrescens* (by the above-mentioned authors and by others), and the karyotypes of the two species are very similar.

55. T. rusbyi Greene, Pittonia 1 : 5 (1887).

Perennial, villous-pubescent or sparsely pilose, 5–40 cm. Roots sometimes thickened or fusiform. Stems several, branching from crown, ascending, decumbent or erect. Leaves variable in size and shape, basal ones long and slender-petioled, upper ones short-petioled to subsessile ; stipules up to 2 cm long, ovate to lanceolate, acute to acuminate, entire or slightly toothed ; leaflets 0.3–6 × 1.4 cm, elliptical, ovate to oblanceolate or lanceolate, margin entire or serrate to dentate, apex acute. Inflorescences 1–3 cm in diam., obconical, becoming globular or slightly elongated after flowering, 20- to many-flowered, alternating along thick rhachis; rhachis sometimes protruding beyond flowers. Peduncles longer than subtending leaves. Bracts minute, not more than 1 mm or reduced to scarious scales subtending each flower. Pedicels slender or stout, up to 2 mm long, strongly deflexed. Calyx 0.4–1 cm long, pilose to glabrescent, sometimes whitish; teeth 1–4 times as long as tube, linear to subulate. Corolla 1–1.5 cm long, white to cream, purple to pink, becoming dark brown; standard ovate to elliptical, acute to attenuate, sometimes rostrate and then wings and keel also rostrate. Pod slightly exserted from calyx, 1–4-seeded, oblong, sparingly pubescent and reticulately veined. Seeds about 2 mm, smooth, brown. Fl. April–August.

Key to Subspecies

1. Wings blunt, shorter than claw, with free portion of standard shorter than limb. **55g**. subsp. **neurophyllum**
– Wings acute or rostrate, of same length as claw, with free portion of standard as long as limb 2
2. Standard, wings and keel rostrate. **55b**. subsp. **multipedunculatum**
– Standard, wings and keel acute to attenuate 3
3. Rhizomatous perennials 4
– Perennials with thickened or fusiform roots, occasionally with crown of buried stems. **55a**. subsp. **rusbyi**
4. Pedicels slender, 1.5–2 mm long. **55c**. subsp. **oreganum**
– Pedicels stout, about 1 mm long or less 5
5. Corolla usually white to creamy, turning dark brown with age. **55f**. subsp. **reflexum**
– Corolla usually purple to violet-purple 6
6. Inflorescences finally spherical. Flowers up to 1.3 cm long. Calyx greenish. **55d**. subsp. **shastense**
– Inflorescences finally slightly elongated. Flowers up to 1.5 cm. Calyx whitish. **55e**. subsp. **caurinum**

55a. T. rusbyi Greene Subsp. **rusbyi**. *Typus* : Bill Williams Mts., 10 June, 1883, *H. H. Rusby* 557 (US ! lecto., isolecto. : NY !, ORE, US !).

T. longipes Nutt. var. *pygmaeum* Gray in Ives. Rep. Colo. Riv. 4 : 9 (1860).
T. longipes Nutt. var. *brachypus* Wats., Bibl. Ind. N. Am. Bot. 264 (1878).
T. brachypus (Wats.) Blank., Mont. Agr. Coll. Sci. Stud. Bot. 1 : 81 (1905).
T. confusum Rydb., Bull. Torrey Bot. Club 34 : 46 (1907).
T. oreganum Howell f. *rusbyi* (Greene) McDerm., Ill. Key N. Am. Sp. Trif. 260 (1910).

T. oreganum Howell f. *brachypus* (Wats.) McDerm., *op. cit.* 261.
T. longipes Nutt. var. *rusbyi* (Greene) Harrington, Man. Pl. Colo. 332.641 (1954).
T. longipes Nutt. subsp. *pygmaeum* (Gray) Gillett, Can. Jour. Bot. 47 : 102 (1969).

Icon. : [Plate 53]. McDerm., *loc. cit.* pls. 107–108.

Hab. : Open meadows, clay soils rich in humus, in Ponderosa pine forest. 2n = 16.
Gen. Distr. : USA : Arizona, Colorado, New Mexico, Utah, California, Oregon, Washington, Idaho, Montana, Nevada, Wyoming.

Selected specimens : ARIZONA : Coconino Co., Williams, 6,800 ft, 1964, *Gillett* & *Moulds* 12790 (HUJ). NEW MEXICO : Valley, San Antonio Ranger Station, Fres Piedras, 1930, *Talbot* 1292 (NA). UTAH : Pine Valley, 7,000 ft, 1942, *Gould* 1820 (UC); Iron Co., Sevier Forest, 3,000–3,200 m, 1912, *Eggleston* 8431 (NY). IDAHO : Custer Co., Morgan Creek, 1944, *Hitchcock* & *Muhlick* 9399 (NA).

55b. T. rusbyi Greene Subsp. **multipedunculatum** (Kennedy) Heller & Zoh. (comb. nov.). *Typus* : Oregon : Granitic coarse sandy soil near summit of China Cap, Wallowa Mts., 8,600 ft, 30 July, 1907, *W. C. Cusick* 3190 (UC, iso. : DS, E !, F !, G !, GH, ORE, P !, US, W !, WS, WTU).

T. multipedunculatum Kennedy, Muhlenbergia 5 : 59 (1909).
T. longipes Nutt. ssp. *multipedunculatum* (Kennedy) Gillett, *op. cit.* 101.
T. longipes Nutt. var. *multipedunculatum* (Kennedy) Martin ex Isley, Brittonia 32 : 56 (1980).

Icon. : Hitchcock et al., Vasc. Pl. Pac. N. W. 3 : 368 (1961).

Hab. : Meadows, open pine or spruce-fir forest to alpine tundra. 2n = 32, 48.
Gen. Distr. : USA : California, Oregon, Washington.

Selected specimens : E. OREGON, 6,000 ft, 1910, *Cusick* 3352 (NA); Wallowa Co., Wallowa Mts., above Immaha Canyon, 6,300 ft, *Hermann* 18920 (W). WASHINGTON : Kittitas Co., Mission Peak, Wenatchee Mts., 6,000 ft, 1940, *Thompson* 14937 (HUJ, NA).

55c. T. rusbyi Greene Subsp. **oreganum** (Howell) Heller & Zoh. (comb. nov.). *Typus* : Oregon, Eastern base of the Coast Mountain near Waldo, 26 April, 1886, *T. Howell* 1346 (ORE, iso. : JEPS, MO, ND, NY, US, WTU).

T. oreganum Howell, Erythea 1 : 110 (1893).
T. longipes Nutt. ssp. *oreganum* (Howell) Gillett, *op. cit* 103.
T. longipes Nutt. var. *oreganum* (Howell) Isley, Brittonia 32 : 57 (1980).

Icon. : McDerm., *loc. cit.* pl. 106.

Hab. : Forest slopes.
Gen. Distr. : USA : California, Oregon.

Specimen seen : CALIFORNIA : Humboldt Co., Buck Mt. near summit, 5,000 ft, 1913, *Tracy* 4161 (E, UC).

Plate 53. *T. rusbyi* Greene Subsp. *rusbyi*
(Utah : *Gould* 1820, UC).
Plant in flower; flower; young pod (Utah : *Eggleston* 8431, NY).

55d. T. rusbyi Greene Subsp. **shastense** (House) Heller & Zoh. (comb. nov.). *Typus* : California : North side of Mt. Shasta, Siskiyou Co., 1,500–2,700 m, 11–16 June, 1897, *H. E. Brown* 362 (US).

T. shastense House, Bot. Gaz. 44 : 336 (1906).
T. longipes Nutt. var. *shastense* (House) Jeps., Fl. Calif. 2 : 303 (1936).
T. longipes Nutt. ssp. *shastense* (House) Gillett, *op. cit.* 104.

Icon. : McDerm., *loc. cit.* pl. 101.

Hab. : Open gravelly places.
Gen. Distr. : USA : California.

Selected specimens : CALIFORNIA : Siskiyou Co., near the source of Wagon Creek, Mt. Eddy, 8,500 ft, 1915, *Heller* 12222 (HUJ); Crescent City, Grant Pass Road, Smith River Canyon, below mouth of Myrtle Creek, 200 ft, 1937, *Wolf* 8862 (NA); near Pit River Ferry, Shasta Co., 700–900 ft, 1897, *Brown* 362 (NY, photo HUJ – same number as type).

55e. T. rusbyi Greene Subsp. **caurinum** (Piper) Heller & Zoh. (comb. nov.). *Typus* : Washington : Big Creek Prairie, Chehalis Co., Aug. 1897, *F. H. Lamb* 1395 (MO, iso. GH).

T. caurinum Piper, Erythea 6 : 29 (1898).
T. covillei House, Bot. Gaz. 41 : 337 (1906).
T. longipes Nutt. f. *caurinum* (Piper) McDerm., Ill. Key N. Am. Sp. Trif. 250 (1910).
T. oreganum Howell var. *multiovulatum* Hend., Madroño 3 : 231 (1936).
T. longipes Nutt. ssp. *caurinum* (Piper) Gillett, *op. cit.* 104.

Icon. : McDerm., *loc. cit.* pls. 102–103.

Hab. : Alpine slopes. 2n = 48.
Gen. Distr. : USA : Oregon, Washington.

Specimens seen : WASHINGTON : Kittitas Co., Bald Mt., Snoqualmie National Forest, 6,000 ft, 1940, *Thompson* 14790 (HUJ, NA).

55f. T. rusbyi Greene Subsp. **reflexum** (Nels.) Heller & Zoh. (comb. nov.). *Typus* : Wyoming : On the banks of Wind River at the foot of Union Pass, August 9, 1894, *A. Nelson* 918 (RM, iso. : GH (photo !), ND, NY, US).

T. longipes Nutt. var. *reflexum* Nels., 1st Rept. Fl. Wyo. 94 (1896).
T. rydbergii Greene, Pittonia 3 : 222 (1897).
T. oreganum Howell var. *rydbergii* (Greene) McDerm., Ill. Key N. Am. Trif. 261 (1910).
T. longipes Nutt. ssp. *reflexum* (Nels.) Gillett, *op. cit.* 105.

Icon. : McDerm., *loc. cit.* pl. 109.

Hab. : Moist meadows. 2n = 32, 48.
Gen. Distr. : USA : Arizona, Colorado, Idaho, Montana, Nevada, Utah, New Mexico, Oregon, Wyoming.

Selected specimens: COLORADO: Chicken Creek, W. La Plata Mts., 9,500 ft, 1898, *Baker, Earle & Tracy* 342 (G). IDAHO: Fremont Co., 1 mi. W. of Rea, 1939, *Maguire* 17133 (NA). MONTANA: Meagher Co., Highway 89, 5,050 ft, 1963, *Gillett & Mosquin* 11852 (HUJ). UTAH: E. Fork of Bear River, Uinta Mts., Summit Co., 8,100 ft, 1926, *Payson* 4832 (HUJ). WYOMING: Yellowstone Park, 1905, *Setchell* (HUJ).

55g. T. rusbyi Greene Subsp. **neurophyllum** (Greene) Heller & Zoh. (stat. nov.). *Typus*: New Mexico: Middle Fork of Gila River, Mogollon Mts., Socorro Co. (more likely Catron Co.), 8,500 ft, Aug. 17, 1903, *O. B Metcalfe* 532 (ND, iso.: ARIZ, BM !, E !, GH, K !, MO, ND, NY !, UNM, US).

T. neurophyllum Greene, Leafl. Bot. Obs. and Crit. 1 : 154 (1905).
T. longipes Nutt. var. *elmeri* (Greene) McDerm. f. *neurophyllum* (Greene) McDerm., Ill. Key N. Am. Sp. Trif. 257 (1910).
T. longipes Nutt. var. *neurophyllum* (Greene) Martin ex Isley, Brittonia 32 : 57 (1980).

Hab.: Open meadows and in pine forest. 2n = 16.
Gen. Distr.: USA: Arizona, New Mexico.

Note: Seen only isotypes.
Gillett (1969), in a most thorough and widely documented study of the *T. longipes* complex, considered it as comprising *T. longipes* with 11 subspecies and 2 additional, closely related species: *T. latifolium* and *T. neurophyllum*. *T. latifolium* is indeed easily distinguished from all other taxa. The latter comprise plants which have short, erect pedicels (considered here as *T. longipes*) and others which have conspicuous pedicels becoming deflexed after anthesis (separated by us from *T. longipes* as *T. rusbyi*). In the present study, the nature of the pedicel is regarded as a key character of the subsections in Sect. *Lotoidea*. For that reason, the complex is divided by us into two species: *T. rusbyi* included in Subsect. *Lotoidea*, and *T. longipes* in Subsect. *Platystylium*.

56. T. semipilosum Fresen. in Flora 22 : 52 (1839).

Perennial, pilose to glabrescent. Stems caespitose, many, stoloniferous, prostrate, rooting at nodes, branching from strong taproot. Leaves long-petioled; stipules 1–2 cm long, triangular, free portion falcate, triangular-lanceolate, longer than adnate part; leaflets 0.4–1.5(–2) × 0.4–1.1 cm, orbicular, elliptical or oblong-elliptical, cuneate-obovate, rounded, truncate or emarginate, prominently nerved, glabrous above, pilose underneath (sometimes only on midrib), margin toothed. Inflorescences about 2 cm in diam., globose, 10–25-flowered. Peduncles longer than subtending leaves, pilose. Bracts minute, lanceolate, sometimes absent. Pedicels 2–5 mm long, pilose, deflexed after anthesis. Calyx 4–6 mm long; tube campanulate, pilose, with 10 prominent nerves; teeth 1–1.5 times as long as tube, narrowly triangular, abruptly subulate. Corolla 8–10 mm, white or pale pink; standard obovate-oblong, rounded at apex or slightly emarginate, longer than wings and keel. Pod 6 mm, longer than calyx, oblong, 2–6-seeded, stipitate, reticulately nerved, glabrous or pilose at least near apex, sutures prominent. Seeds 1.5 mm, mottled, yellow or light brown. Fl. July–November. 2n = 16.

Hab. : Upland grassland, moist upland evergreen forest; 1,400–3,200 m.

56a. T. semipilosum Fresen. Var. **semipilosum**. *Typus* : N. Ethiopia : Simen (Simien), *Rueppell* s. n. (holo. FR).

T. semipilosum Fresen. var. *microphyllum* Chiov. in Ann. Ist. Bot. Roma 8 : 406 (1908).
T. semipilosum Fresen. var. *kilimanjaricum* Baker f., Legum. Trop. Afr. 81 (1926).
T. africanum sensu Baker f., *op. cit.* 80 p. p. non Ser. (1825).
T. semipilosum Fresen. var. *sennii* Chiov. ex Fiori in Nuov. Giorn. Bot. Ital. n. s. 55 : 342 (1948).
T. brunellii Chiov. ex Fiori, *loc. cit.* f. 12.

Icon. : [Plate 54]. Gillett in Milne-Redhead & Polhill, Fl. Trop. E. Afr., Leg. Papil. 2 : f. 142 (1971).

Leaflets orbicular, elliptical or oblong-elliptical, twice as long as broad, rounded or truncate, silky-pilose underneath. Pod pilose at least at apex.
Gen. Distr. : Yemen, Uganda, Kenya, Tanganyika, Eritrea, Ethiopia.

Selected specimens : YEMEN : Gebel Shibàm, Menacha, 2,300–2,600 m, 1889, *Barbey* 1384 (K). KENYA : Thompson's Falls, 7,800 ft, 1931, *Pierce* 1478 (G); Naivasha Distr., Sisiun, 2,700 m, 1964, *Gillett* 16160 (HUJ, SRGH, W). TANGANYIKA : Masai Distr. E. of Loliondo on Narok Road, 2,200 m, 1964, *Gillett* 16335 (HUJ). ERITREA : Acchéle Guzai, Soyrà 2,800 m, 1902, *Pappi* 339 (G, W). ETHIOPIA : Mega, 2,200 m, 1952, *Gillett* 14204 (FI. HUJ); Malu farm, 50 km S. of Addis Ababa, 8,000 ft, 1951, *Archer* 10074 (NA).

56b. T. semipilosum Fresen. Var. **glabrescens** Gillett in Kew Bull. 7 : 385 (1952). *Typus* : Kenya : S. Kavirondo Distr., near Kisii, Marani, *Greenway* 7871 (K !, iso. EA).

T. repens sensu Baker f., Legum. Trop. Afr. 81 (1926) non L. (1753).
T. johnstonii sensu Edwards in Emp. Jour. Expt. Agr. 3 : 153–9 (1935) non Oliv. (1886).

Leaflets cuneate-obovate, about as long as broad, emarginate, glabrescent underneath with few hairs on midrib. Pod glabrous.
Gen. Distr. : Yemen, Uganda, Kenya, Tanganyika, S. Ethiopia, Malawi (introduced ?).

Selected specimens : YEMEN : Menakha, Djebel Kahel, 2,400 m, 1887, *Deflers* 287 (P). KENYA : Lake Naivasha, 1964, *Agnew* et al. 5567 (HUJ); Grown at S. A. L., 5,700 ft, 1944, *Nattrass* 333 (HUJ). ETHIOPIA : Scioa : Debra Sina, 3,200 m, 1937, *Senni* 1964 (FI); Enschedcap, 1838, *Schimper* 1174 (W).

57. T. stoloniferum Muhlenberg, Catalogue 67 (1813). *Typus* : Ohio : Kent, collector unknown.

Icon. : [Plate 55]. Britten & Brown, Ill. Fl. N. States & Canada ed. 2, 2 : 357 (1913).

Plate 54. *T. semipilosum* Fresen. Var. *semipilosum*
(Ethiopia : *Gillett* 14204, HUJ).
Plant in flower; fruiting calyx with pod (Kenya : *Gillett* 16160, HUJ);
flower.

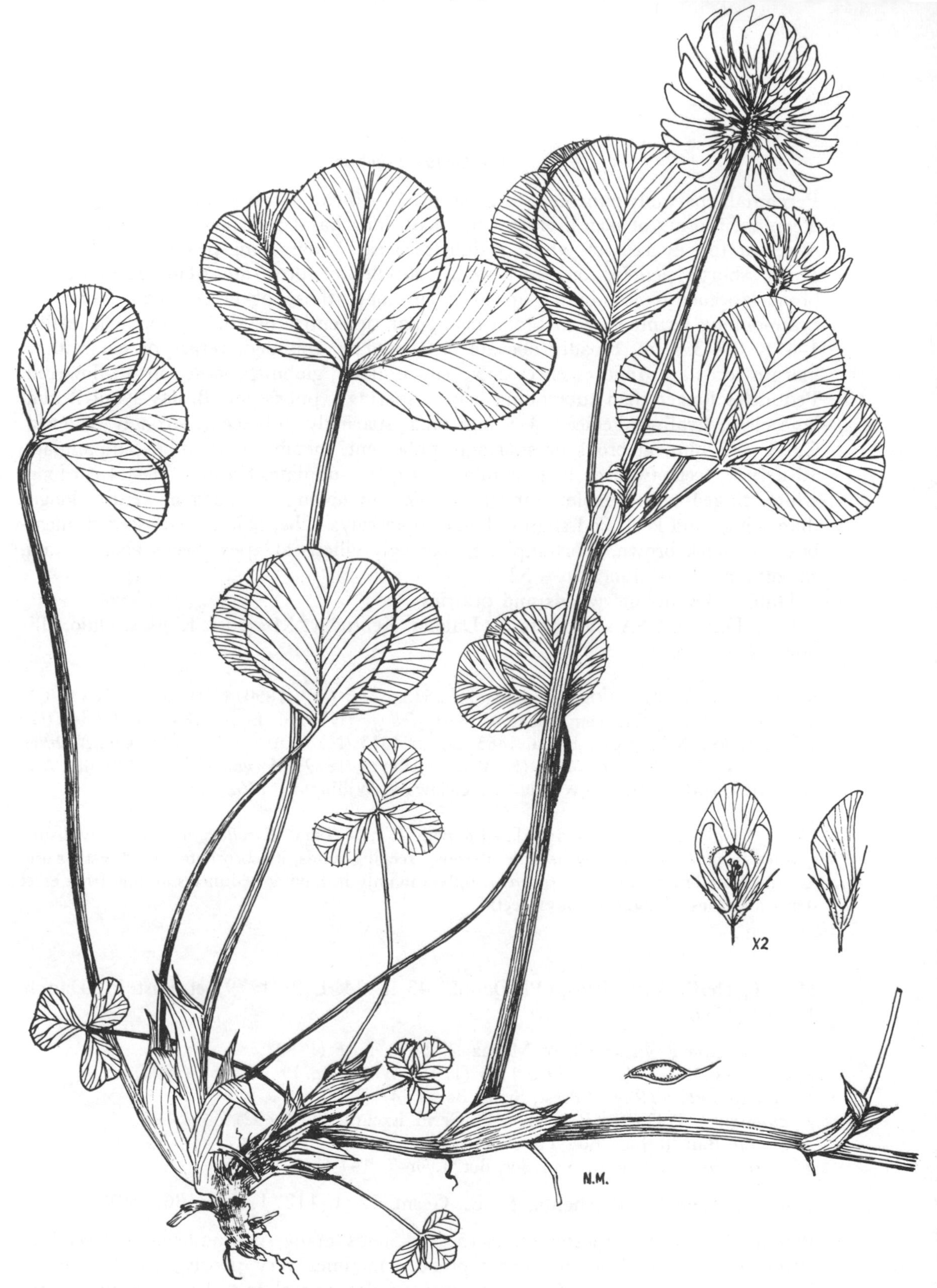

Plate 55. *T. stoloniferum* Muhlenberg
(Kentucky: *Short*, W).
Plant in flower and fruit; flowers; young pod.

Perennial, sparingly pubescent to glabrous, 10–40 cm. Stems stoloniferous, prostrate, creeping, rooting at nodes; erect stems with long basal internode and 2–3 crowded nodes at top in axils of which peduncles arise. Leaves long-petioled on prostrate stems, shorter-petioled on erect stems; stipules of prostrate stems membranous, broad-lanceolate, acuminate, up to 2 cm long, those of erect stems foliaceous, ovate-oblong terminating in broad triangular tip, entire or slightly toothed; leaflets 1–4 × 0.8–3.5 cm, broadly obovate or obcordate, rounded, retuse or emarginate, margin serrulate. Inflorescences 2–3.5 cm in diam., globular, loose, 30–40-flowered. Peduncles longer than subtending leaves, sparingly pubescent. Bracts about 2 mm, lanceolate, hyaline. Pedicels 4–8 mm long, sparingly pubescent, strongly deflexed. Calyx 4–7 mm, glabrous or sparingly pubescent, membranous; tube campanulate, 10-nerved; teeth twice as long as tube, subequal, subulate. Corolla 0.8–1.4 cm long, white, tinged with purple; standard obovate to oblong, rounded or retuse, longer than wings and keel. Pod slightly longer than calyx tube, oblong, 1–3-seeded, membranous, dark brown, short-stipitate, sparingly villous at apex. Seeds about 2 mm, smooth, Fl. May–June. 2n = 32.

Hab.: Dry upland woods and prairies.

Gen. Distr.: USA: Virginia, S. Dakota, Kentucky, Missouri, Kansas, Ohio, Illinois, New Jersey.

Specimens seen: KENTUCKY: Lexington, 1835, *Short* (W); 1850, *C. W. Short* (NA, UC). MISSOURI: Big Spring near Allenton, 1880, *Kellogg* (NY); St. Louis, 1846, *Riehd* 446 (G, W). KANSAS: Miami Co., Paola, 1885, *Oyster* 1513 (NY). OHIO: Hamilton Co., *Matthes* 110 (P, W); North Bend, *Short* (K, W); Cincinnati, 1879, *Morgan* (NY). ILLINOIS: Augusta, 1845, *Mead* (K). NEW JERSEY: Delaware, Willington, 1862, *Canby* (NY).

Note: *T. stoloniferum* is closely related to both *T. repens* and *T. reflexum*. It is easily distinguished from *T. reflexum* by its stoloniferous, creeping stems, its obcordate, broad leaflets and its lanceolate bracts. From *T. repens* it differs mainly in having peduncles arising from erect stems and longer, subulate calyx teeth.

58. T. thalii Vill., Prosp. Pl. Dauph. 43, t. 24, f. 2 (1779) et Hist. Pl. Dauph. 1: 289 (1786).

T. caespitosum Reyn. in Hoepf. Magaz. 2, 2: 78, t. 16 (1788).
T. caespitosum J. F. Gmel., Syst. 1141 (1791); Willd., Sp. Pl. 3: 1359 (1802).
Amoria caespitosa (Reyn.) Presl, Symb. Bot. 1: 47 (1830).
T. angulatum J. Gay ex Reichenb., Fl. Germ. Excurs. 496 (1832).
T. humile Ball in Jour. Bot. 11: 306 (1873).
T. caespitosum St. Lag. in Ann. Soc. Bot. Lyon 7: 144 (1880).

Icon.: [Plate 56]. Reichenb., Ic. Fl. Germ. 22: t. 112, I 1–8 (1867–89).

Perennial, glabrous, caespitose, 5–15 cm. Stems many, procumbent or ascending, branching at base from massive taproot, sometimes very poorly branched above. Leaves with long petioles (up to 10 cm); stipules lanceolate to linear, membranous-whitish, acute to acuminate; leaflets 0.6–1.4 × 0.3–1 cm, elliptical, obovate, cuneate, blunt, rounded or sometimes retuse, margin finely dentate. Inflorescences about

Plate 56. *T. thalii* Vill.
(France : *Verlot*, R).
Plant in flower and fruit; fruiting calyx; flower.

1.5 cm in diam., hemispherical, many-flowered, whorled. Peduncles as long as or longer than petioles. Bracts about 1 mm, ovate, truncate to retuse. Pedicels 1–1.5 mm, only scarcely deflexed after anthesis. Calyx about 4 mm, tubular; teeth unequal, the upper ones longer, but all shorter than tube, triangular, sharp-pointed; fruiting calyx somewhat inflated, about as long as broad, ribbed. Corolla about 8 mm long, rose-coloured; standard elliptical, rounded at apex, longer than wings and keel. Pod somewhat exserted from calyx, oblong, 2–3-seeded, sessile or subsessile. Seeds about 1 mm, brown, smooth, reniform. Fl. July–August. 2n = 16.

Hab.: Grasslands and slopes; up to 3,000 m.

Gen. Distr.: Spain, France, Austria, Germany, Switzerland, Italy, Morocco.

Selected specimens: SPAIN: Cerdagne, Val d'Eype, 2,150 m, 1919, *Sennen* 3675 (K). FRANCE: Savoie, Monte Cenisio, La Rainape, 1863, *Verlot* (R); Lautaret, Hautes Alpes, 1933, *Redgrove* (K). AUSTRIA: Tirol, Valzaregoi, 5,000–6,000 ft, 1870, *Huterk* (K). SWITZERLAND: Eugadiny, 1879, *Scholle* 445 (B). ITALY: Moncenisio, Savalino, 1,900 m, 1937, *Vignolo-Lutati* (B). MOROCCO: Atlantis Majoris, Ait Mesan, 2,00–2,530 m, 1871, *Ball* (FI, syntype of *T. humile* Ball); 70 mi. S. W. of Mident, 9,500 ft, 1968, *Goodchild* 46 (BM).

Note: *T. humile* Ball undoubtedly represents a small, matted form, found in the High Atlas mountains.

59. T. virginicum Small ex Small & Vail, Mem. Torrey Bot. Club 4: 112 (1894). *Typus*: W. Virginia: Kate's Mountain, near White Sulphur Springs, Greenbrier Co., 3,300 ft, May 16, 1892 *J. K. Small* (NY, iso.: G!, GH, NY, PH, UC!).

T. reflexum L. var. *virginicum* (Small) McDerm., N. Am. Sp. Trif. 273 (1910).

Icon.: [Plate 57]. Small & Vail, *loc. cit.* t. 75.

Perennial, pubescent to villous, 2–10 cm. Taproot thick and long, branching. Stems caespitose, spreading, prostrate, very short, branching from crown of taproot, covered at base by persistent stipules. Leaves long-petioled, mostly basal; upper stipules up to 1.5 cm long, ovate, acute, with entire margin, lower ones oblong-lanceolate; leaflets (1–)2–7 × 0.4–1.5 cm, linear-elliptical or oblanceolate, rounded or truncate, acute, glabrous on upper surface, pubescent on lower surface, margin serrulate. Inflorescences 2–3 cm in diam., spherical, 20–40-flowered, all flowers radiating in fruiting heads. Peduncles shorter than subtending leaves. Bracts minute, scarious. Pedicels 3–8 mm long, strongly deflexed. Calyx 4–7 mm, pilose, membranous; teeth 2–3 times as long as tube, triangular-subulate. Corolla 1–1.2 cm long, pink to red or white (?); standard broadly obovate, rounded or truncate, retuse or sometimes mucronate, striate, longer than wings and keel. Pod longer than calyx, long-stipitate, obovoid, membranous, reticulately nerved, 1-seeded, sparingly villous. Seed 2 mm, yellowish-brown, roughened. Fl. May–July. 2n = 16.

Hab.: On barren shale slopes.

Gen. Distr.: USA: Maryland, Pennsylvania, Virginia, W. Virginia.

Selected specimens: MARYLAND: Allegany Co., at base of Polish Mt., 1930, *Freeman*

Plate 57. *T. virginicum* Small
(Virginia : *Heller* 843, E).
Plant in flower; young pod (Maryland : 1930 *Freeman*, NA);
flowers.

(NA); *ibid.*, near Bellegrove, 1946, *Freeman* (NA). VIRGINIA : Frederick Co., 1/2 mi. W. of Gore, 1935, *Freeman* (NA); Detrick, Shenandoah Co., 1940, *Freeman* (NA). W. VIRGINIA : 6 mi. S. W. of Burlington, Mineral Co., 1946, *Freeman* (NA); Kates Mt., Greenbrier Co., 2,500–3,400 ft, 1893, *Heller* 843 (E, POM); Morgan Co., near Largent, 1938, *Freeman* (NA).

Series v. PECTINATA Heller & Zoh. (ser. nov.)*

Type species : *T. ciliolatum* Benth.

Annual. Stems erect. Bracts minute, often united and subtending whorled flowers. Pedicels conspicuous, deflexed after anthesis. Rhachis sometimes prolonged beyond flowers. Calyx teeth unequal, triangular-lanceolate, hyaline, strongly ciliate, dentate to pectinate at margin. One N. American species.

60. T. ciliolatum Benth., Pl. Hartw. 304 (1848). *Typus* : California : In Sacramento Valley, 1848, *Hartweg* 1697 (G !, iso. : G !, GH, K !, NY, W !).

T. ciliatum Nutt., Jour. Acad. Philad. N. S. 1 : 152 (1848) non Clarke (1813–16).
T. ciliatum Nutt. var. *discolor* Lojac., Nuov. Giorn. Bot. Ital. 15 : 146 (1883).

Icon. : [Plate 58]. Nutt., *loc. cit.* t. 23.

Annual, glabrous, slightly hairy or ciliate, 5–50 cm. Stems erect, stout, poorly branched. Lower leaves long-petioled, upper ones short-petioled; stipules 1–1.5 cm long, ovate to lanceolate, acuminate, falcate, margin narrowly hyaline, entire or ciliate; leaflets 0.8–3.5 × 0.5–1.5 cm, elliptical to cuneate-elliptical, oblong to obovate in lower leaves, rounded, obtuse to retuse, glabrous usually on both sides, prominently nerved, margin entire or serrulate. Inflorescences 0.6–1.2 cm in diam., globose, 10- to many-flowered, arranged in whorls; rhachis sometimes prolonged beyond flowers. Peduncles usually longer than subtending leaves, sometimes slightly hairy. Bracts minute, often united and subtending whorled flowers. Pedicels 0.5–6 mm long, short in flowering, later becoming longer and strongly deflexed. Calyx 0.5–1.1 cm long, membranous, green or also violet-blue, the upper side longer than the lower one; tube oblique, 10-nerved, glabrous or with few hairs; teeth unequal, slightly incurved, the upper ones 2–3 times and the lower ones 1–1.5 times as long as tube, triangular-lanceolate, sharp-pointed, margin hyaline and strongly ciliate, dentate to pectinate. Corolla 0.6–1.3 cm, shorter or longer than calyx, white to pink or purple; standard broadly ovate, narrowly rounded at apex, apiculate, longer than wings and keel. Pod as long as corolla or slightly shorter, ovoid, 1–2-seeded, short-stipitate, with reticulate-parallel nervation. Seeds about 3 mm, smooth. Fl. April–June. 2n = 16.

Hab. : Slopes and grassy valleys, or in open forests.
Gen. Distr. : USA : Washington, Oregon, California.

* See Appendix.

Plate 58. *T. ciliolatum* Benth.
(Washington : *Gillett* & *Moulds* 12541, HUJ).
Plant in flower; young pod (California : *Nutting*, HUJ);
flower; calyx.

Selected specimens: WASHINGTON: Klickitat Co., 5 mi. E. of Bingen, 1964, *Gillett & Moulds* 12541 (HUJ); *ibid.*, 1906, *Suksdorf* (UC). OREGON: Wasco Co., Hood River on the Columbia, 1910, *Heller* 10102 (E); Curry Co., Agress above Illinois River, 1953, *Hitchcock* 19905 (POM, RSA); Umpqua Valley, 1881, *Howell* (NA); Douglas Co., N.W. of Roseburg, 1962, *Gillett & Taylor* 11174 (K, small flowers and leaflets). CALIFORNIA: Shasta Co., Redding, *Nutting* (HUJ); Willitts, 1921, *Piper* (HUJ); Santa Clara Co., Mt. Hamilton Range, 8 mi. from Gilroy, 250 ft, 1951, *Raven* 2795 (HUJ); Madera Co., San Joaquin Exp. Range, 1,200 ft, 1937, *Biswell* 150 (HUJ); Contra Costa Co., Crockett, 1895, *Tidestrom* (HUJ); San Bernardino Co., San Bernardino, 1904, *Kennedy* (HUJ, UC).

Note: *T. ciliolatum* is easily distinguished by the hyaline, strongly ciliate, dentate to pectinate margin of the calyx teeth.

Series vi. CURVICALYX Heller & Zoh. (ser. nov.)*

Type species: *T. eriocephalum* Nutt.

Perennial, villous or rarely glabrous. Upper stipules 2–5 cm long, foliaceous, broadly lanceolate. Flowers spirally arranged. Peduncles bent near apex – thus heads inverted or horizontal. Calyx tube curved after anthesis; teeth unequal, often curved and twisted, subulate, the lower ones longer than the upper ones. One N. American species.

61. T. eriocephalum Nutt. in Torr. & Gray, Fl. N. Am. 1: 313 (1838).

Perennial, villous or rarely glabrous, 10–50 cm. Roots thick, sometimes fusiform. Stems erect or decumbent, branching from base and slightly above. Leaves long-petioled; upper stipules 2–5 cm long, foliaceous, broadly lanceolate, acuminate, lower ones smaller and narrower, margins entire or irregularly serrate; leaflets 1–5 × 0.4–1.2 cm, elliptical, oblong, ovate or lanceolate-linear, obtuse, acuminate or acute, often setose and prominently nerved at margin, villous on both sides or glabrous on upper side, margins rarely entire, usually irregularly denticulate or serrulate. Inflorescences 1–3 cm in diam., globose, many-flowered, flowers spirally arranged. Peduncles longer than subtending leaves, terminal, bent near apex – thus heads inverted or horizontal. Bracts small, scarious. Pedicels less than 1 mm, deflexed after anthesis. Calyx 4–11 mm long; tube membranous, villous, curved after anthesis; teeth 2–5 times as long as tube, unequal (to varying degrees in different subspecies), slender, subulate, often curved and twisted, sometimes purplish, long-villous, the lower ones longer than the upper ones. Corolla 8–16 mm, white, yellowish to purple; standard elliptical, ovate or oblong, recurved, rounded or slightly retuse at apex, apiculate, about as long as or slightly longer than wings and keel. Pod slightly longer than calyx tube, 1–4-seeded, oblong, oblique, villous especially towards apex. Seeds about 2 mm, smooth. Fl. May–July. 2n = 16.

* See Appendix.

Key to Subspecies (after Gillett, 1971)

1. Ovary 1–2-ovuled (very rarely 3-ovuled). Calyx teeth straight, usually green, subequal, 3–4 times as long as tube 2
– Ovary 3–4-ovuled. Calyx teeth contorted, usually purple, upper one longer than others, laterals 1–1.5 times as long as tube 3
2. Leaflets of stem leaves elliptical to lanceolate-oblong. Roots cylindrically thickened. **61a.** subsp. **eriocephalum**
– Leaflets of stem leaves narrowly lanceolate to linear, occasionally narrowly elliptical. Roots fusiform, tuberous. **61c.** subsp. **cascadense**
3. Leaflets of basal leaves linear to linear-elliptical, acute. Roots cylindrical, thickened, branched, rhizomatous 4
– Leaflets of basal leaves obovate, rounded to retuse. Roots fusiform, tuberous without rhizomes 5
4. Plants densely villous throughout. Leaflets of median leaves broadly elliptical, serrulate. **61f.** subsp. **villiferum**
– Plants slightly villous throughout. Leaflets of median leaves linear to linear-elliptical, serrate. **61d.** subsp. **cusickii**
5. Leaflets of cauline leaves broadly lanceolate. Flowering heads about 2.5 cm in diam. Peduncles longer than leaves. **61b.** subsp. **arcuatum**
– Leaflets of cauline leaves narrowly lanceolate to linear. Flowering heads about 1 cm in diam. Peduncles not longer than leaves. **61e.** subsp. **martinii**

61a. T. eriocephalum Nutt. Subsp. **eriocephalum**. *Typus* : Oregon : Prairies of the Wahlamet, *T. Nuttall* (BM ! lecto., isolecto. : K !, NY !, PH).

T. scorpioides Blasdale in Erythea 4 : 187 (1896).
T. eriocephalum Nutt. var. *butleri* Jeps., Fl. Calif. 2 : 302 (1936).

Icon. : [Plate 59]. Jeps., Man. Fl. Pl. Calif. 546 (1925).

Hab. : Mountain slopes and meadows.
Gen. Distr. : USA : Washington, Oregon, California.

Selected specimens : OREGON : Josephine Co., Whisky Creek, S. W. of O'Brien, 1964, *Gillett* & *Moulds* 12592 (HUJ); *ibid.*, 4 mi. N. of Grave Creek, 1,200 m, 1942, *Constance* & *Rollins* 2949 (NA). CALIFORNIA : Siskiyou Co., near Yreka, 1876, *Greene* 817 (NA); Shasha Co., 1/4 mi. S. of Burney Falls, 1940, *Hitchcock* 6516 (UC); Mendocino Co., 1866, *Bolander* 6509 (W).

61b. T. eriocephalum Nutt. Subsp. **arcuatum** (Piper) Gillett, Can. Jour. Bot. 49 : 400 (1971). *Typus* : Simcoe Mts., 6 June, 1884, *W. N. Suksdorf* 270 (GH, iso. WS).

T. arcuatum Piper in Bull. Torrey Bot. Club 28 : 39 (1901).
T. eriocephalum Nutt. f. *arcuatum* (Piper) McDerm., N. Am. Sp. Trif. 242 (1910).
T. eriocephalum Nutt. var. *piperi* Martin, Madroño 8 : 154 (1946).

Icon. : Hitchcock et al., Vasc. Pl. Pac. N. W. 3 : 362 (1961).

Plate 59. *T. eriocephalum* Nutt. Subsp. *eriocephalum*
(Oregon : *Gillett & Moulds* 12592, HUJ)
Plant in flower; flowers.

Hab. : Moist mountain meadows and woods.
Gen. Distr. : USA : Oregon, Washington, Idaho.

Specimens seen : E. OREGON : Trout Creek, Wallowa Mts., 6,000 ft, 1900, *Cusick* 2432 (SIU, W); Wallowa Co., Wallowa Mts., Sacajawea Spring, 6,850 ft, 1963, *Hermann* 18934 (W). WASHINGTON : Dry Creek, Yokima, 300 ft, 1932, *Heidenreich* 247 (UC).

61c. T. eriocephalum Nutt. Subsp. **cascadense** Gillett, *loc. cit. Typus* : Oregon : Wasco Co., Warm Spring River at junction with highway 26, meadow at border of stream, under Ponderosa pine. Abundant, 23 June, 1966, *J. M. Gillett* & *C. W. Crompton* 13075 (DAO).

Hab. : Meadows near streams.
Gen. Distr. : USA : Oregon.

Specimens seen : OREGON : Crook Co., Campo Polk, *Cusick* 2811a (UC, W).

61d. T. eriocephalum Nutt. Subsp. **cusickii** (Piper) Gillett, *op. cit.* 402. *Typus* : Oregon : Camp Creek, Maurey's Mts., 2 July, 1901, *W. C. Cusick* 2628 (WS lecto., isolecto. : C, G !, K !, MO, ND-G, NY, ORE, RM, UC !, US, W !).

T. harneyense Howell, Fl. N. W. Am. 1 : 134 (1897).
T. eriocephalum Nutt. var. *harneyense* (Howell) McDerm., N. Am. Sp. Trif. 231 (1910).
T. arcuatum Piper var. *harneyense* (Howell) Piper in McDerm., *loc. cit.* in syn.
T. austinae Rydb. in McDerm., *loc. cit.* in syn. nom. nud.
T. tropicum Nelson, Bot. Gaz. 54 : 409 (1912).
T. eriocephalum Nutt. var. *cusickii* (Piper) Martin in Madroño 8 : 156 (1946).
T. arcuatum Piper var. *cusickii* Piper, Bull. Torrey Bot. Club 29 : 641 (1902).

Icon. : McDerm., *loc. cit.* pl. 99.

Hab. : Moist meadows, open woodland.
Gen. Distr. : USA : E. Oregon, N. Nevada, Idaho, California.

Selected specimens : E. OREGON : Burns, Harney Co., 4,100 ft, 1941, *Hansen* (HUJ); Union Co., Trout Creek, 5,000 ft, 1963, *Hermann* 18865 (W). NEVADA : Nye Co., S. of Dieringer, 1937, *Goodner* & *Henning* 719 (NA). IDAHO : Blaine Co., Corrall, Camas Prairie, *McBride* & *Payson* 2910 (US).

61e. T. eriocephalum Nutt. Subsp. **martinii** Gillett, *op. cit.* 401. *Typus* : Idaho : Valley Co., 2 mi. W. of McCall, 1944, *Hitchcock* & *Muhlick* 8508 (WTU, iso. : DS, F, GH, IDS, MO, NA !, NY, PH, POM, RM, UC, US, UTC, WS).

Hab. : As above.
Gen. Distr. : USA : Idaho, Oregon.

Specimens seen : IDAHO : Adams Co., between New Meadows and McCall, 1939, *Cronquist* 1193 (NA). OREGON : Baker Co., Blue Mts., *Cusick* 2797 (HUJ, UC, W).

61f. T. eriocephalum Nutt. Subsp. **villiferum** (House) Gillett, *op. cit.* 403. *Typus*: Utah: Beaver City, Beaver Co., 1877, *E. Palmer* 91 (US, iso.: GH, MO, NA !, NY).

T. villiferum House, Bot. Gaz. 41 : 335 (1906).
T. eriocephalum Nutt. f. *villiferum* (House) McDerm., N. Am. Sp. Trif. 242 (1910).
T. eriocephalum Nutt. var. *villiferum* (House) Martin, Madroño 8 : 156 (1946).

Icon.: McDerm., *loc. cit.* pl. 98.

Gen. Distr.: USA: Nevada, Utah.

Specimens seen: NEVADA: Elko Co., 25 mi. S. of Wells, H. D. Ranch, 1901, *Jones* (POM).

Note: We have followed here the subdivision of *T. eriocephalum* as proposed by Gillett (1971). It should, however, be pointed out that the delimitation of these infraspecific taxa is sometimes difficult, as characters may overlap to various degrees. This fact was already noted by Gillett himself. It was evident from our herbarium material that in some cases the diagnostic characters used by Gillett may occur in various combinations (e.g., long equal calyx teeth – but purple in colour and contorted).
There is a possibility that further field studies by American botanists will point towards clinal variation of many of the characters studied.

Series vii. PRODUCTA Heller & Zoh. (ser. nov.)*

Type species: *T. kingii* S. Wats.

Perennials, glabrous or sparingly hairy. Stems erect, branching or caespitose. Inflorescences few- to many-flowered, arranged in whorls; rhachis prolonged beyond flowers, simple or forked, often bearing cluster of sterile, bud-like flowers in upper whorl. Six N. American species.

62. T. beckwithii Brewer ex S. Wats. in Proc. Am. Acad. 11 : 28 (1876). *Typus*: On the Sierra Nevada, 22 June, 1854, *James A. Snyder* (NY, iso. GH).

T. altissimum Torr. & Gray, Pacif. Rail. Rep. 2 : 120 (1855) non Dougl. (1831) nec Loisel. (1807).

Icon.: [Plate 60]. McDerm., N. Am. Sp. Trif. pl. 117 (1910).

Perennial, glabrous, 15–50 cm. Stems solitary or few, erect or ascending, arising from thick taproot, sulcate, with 1–2 leaves subtending peduncle. Basal leaves very long-petioled, those below peduncles short-petioled, most of them crowded at base of stem; stipules up to 2 cm, ovate-oblong, free portion triangular-lanceolate; leaflets (1–)2(–6) × 0.5–1.5 cm, broad-elliptical to ovate or oblong, obtuse or acute at apex, margin finely serrulate, the upper ones oblanceolate. Inflorescences 2.5–4 cm across, terminal or axillary, densely whorled, subglobular, many-flowered; rhachis thick, of-

* See Appendix.

Plate 60. *T. beckwithii* Brewer
(California : *Sonne*, HUJ).
Plant in flower; flower; standard; wing; keel.

ten terminating in few sterile flowers. Peduncles very long, 5–15(–20) cm, thick, erect. Bracts 1 mm, subulate to filiform, scarious. Pedicels 2–3 mm, deflexed after anthesis. Calyx 4–6 mm, somewhat gibbous, glabrous, somewhat hyaline or coloured, with 5 obsolete nerves; teeth unequal, the lower ones longer than tube and the 2 upper ones shorter, narrowly triangular-lanceolate to subulate, scarious, with very broad sinus between upper and lower teeth. Corolla 1.3–1.8 cm, pale purple or pink; standard obovate to broadly elliptical, longer than wings and keel. Pod about 5 mm long, short-stipitate, reticulately nerved, 1–4-seeded, membranous, glabrous. Seeds smooth. Fl. May–July. 2n = ca. 48.

Hab.: Valleys and moist meadows in mountains; 1,500–2,000 m.

Gen. Distr.: USA: Nevada, California, Oregon, Idaho, Montana, S. Dakota.

Selected specimens: NEVADA: Deeth, Elko Co., 5,340 ft, 1908, *Heller* 9120 (P). CALIFORNIA: Nevada Co., Sardine Valley, 1887, *Sonne* (HUJ); Plumas Co., 1880, *R. M. Austin* (NA). OREGON: Bear Valley, Southern Blue Mts., 1897, *Cusick* 1696 (E); Harney Co., Silver Creek Valley, 1901, *Cusick* 2607 (W). IDAHO: Corrall, Camas Prairie, Blaine Co., 5,700 ft, 1916, *MacBride* & *Payson* 2915 (E). MONTANA; Anaconda, 4,000 ft, 1906, *Blankinship* 682 (E, H); Armstead, Beaverhead Co., 5,500 ft, *Payson* & *Payson* 1907 (H).

63. T. bolanderi A. Gray in Proc. Am. Acad. Sci. 7: 335 (1868). *Typus*: California: Westfall's Meadow, above the Yosemite Valley, 1866, *H. N. Bolander* 4967 (GH, iso.: MO, US).

Icon.: [Plate 61]. McDerm., N. Am. Sp. Trif. pl. 118 (1910).

Glabrous perennial, 15–30 cm. Stems many, branching from stout taproot, spreading, slender. Basal leaves long-petioled, those on stems shorter-petioled; stipules ovate, free portion acute, entire, the lower ones persistent, covering stem; leaflets 0.6–1.8 × 0.4–1.2 cm, obovate to obcordate or elliptical, rounded or retuse, apiculate, margin serrulate. Inflorescences up to 2 cm in diam., 15- to many-flowered, some sterile flowers at end of rhachis. Peduncles up to 20 cm, slender, geniculate near base of heads. Bracts minute. Pedicels 1 mm long, slightly deflexed after anthesis. Calyx gibbous, 3–5 mm long, scarious, dark purple; tube 5-nerved; teeth about as long as or shorter than tube, subequal, triangular-subulate. Corolla 10–12 mm long, pale purple to rose, curved; standard narrowly ovate, rounded at apex, apiculate, longer than wings and keel. Young pod glabrous, stipitate, 2-seeded, reticulately veined. Fl. June–July.

Hab.: Wet meadows; about 2,500 m.

Gen. Distr.: USA: California (endemic).

Specimens seen: CALIFORNIA: Mariposa Co., Yosemite Valley, Peregoy Meadow, 7,000 ft, 1936, *Sharsmith* 2194A (H, UC); Sierra Nevada Mts., North fork of Kings River, 7,000 ft, 1900, *Hall* & *Chandler* 424 (E); Westfall meadows, head of Old Creek, Wawona head, 7,100 ft, 1913, *Kennedy* & *Behrend* 3020 (HUJ, UC).

Plate 61. *T. bolanderi* A. Gray
(California : *Sharsmith* 2194A, UC).
Plant in flower; flower; standard; wing; keel.

64. T. brandegei S. Wats. in Proc. Am. Acad. 11 : 130 (1876). *Typus* : On the northwestern border of New Mexico, 1875, *T. S. Brandegee* (GH, seen photo HUJ, iso. : MO, NY, US).

T. kingii S. Wats. var. *brandegei* (S. Wats.) McDerm., N. Am. Sp. Trif. 284 (1910).

Icon. : [Plate 62]. McDerm., *loc. cit* pl. 120.

Perennial, caespitose and acaulescent, glabrous (peduncle, rhachis and calyces sparsely hairy), 5–15 cm. Stems short, arising from slender, branched rhizome, covered at base with few persistent stipules. Leaves clustered; petioles up to 12 cm long; stipules broadly lanceolate, acute or rounded, scarious; leaflets 1.5–3 × 0.5–1.5 cm, oblong to oblong-elliptical, rounded, glabrous, margin with very prominent nerves sometimes terminating in shallow teeth. Inflorescences loose, about 2.5 cm in diam., 4–15-flowered, arranged in 2–3 whorls; rhachis prominently prolonged beyond flowers. Peduncles up to 20 cm longer than subtending leaves. Bracts minute, subtending flowers. Pedicels 1–2 mm, strongly deflexed after anthesis. Calyx 7–10 mm long, hairy, green to purple when dry; tube campanulate, membranous: teeth longer than tube, narrow-triangular, acute. Corolla 1.5–1.8 cm long, purple, darkening with age, violet when dry; standard broadly ovate, obtuse to retuse, longer than wings and keel. Pod about 7 mm, oblong, glabrous, 1–3-seeded. Fl. July–August. 2n = 16.

Hab. : Alpine and subalpine slopes in open mountain forests; about 3,500–4,000 m.

Gen. Distr. : USA : Colorado, New Mexico, Utah (?).

Selected specimens : COLORADO : Near Pagos Peak, 1899, *Baker* 444 (E); Mineral Co., N. of Wolf Creek Pass, 10,800–12,000 ft, 1963. *Gillett* & *Mosquin* 12215 (W). NEW MEXICO : Wheeler Peak, Taos Co., 13,161 ft, 1968, *J.K. Mitchell* Mt.-162 (HUJ).

65. T. haydenii Porter, Annual Rep. U. S. Geol. Surv. Mont. 1871 : 480 (1872). *Typus* : Montana : In the mountains, south of Virginia City, June 25–30, 1871, *G. N. Allen* & *R. Adams* (PH, lecto., iso. : GH, ND-G, NY).

T. idahoense Henderson, Madroño 3 : 1 (1936).
T. haydenii Porter var. *barnebyi* Isley, Brittonia, 32 : 56 (1980).

Icon. : [Plate 63]. Hitchcock et al., Vasc. Pl. Pac. N. W. 3 : 365 (1961).

Perennial, matted, caespitose or subcaulescent, glabrous, 5–15 cm. Stems from branched rhizomes very short, covered at base with withered stipules and leaf bases. Leaves mostly basal, sometimes single leaf on flowering branch, with petioles up to 5 cm long, often shorter; stipules scarious, those of lower part obtuse, entire, those of upper part acute or toothed; leaflets 0.5–2 × 0.3–1 cm, ovate or obovate to subrotund, apiculate, prominently nerved, margin denticulate. Inflorescences about 2–2.5(–3) cm in diam., 5–20-flowered, arranged in whorls; rhachis up to 1 cm long, usually prolonged beyond flowers. Peduncles longer than subtending leaves, sometimes bent near top of head and then flowers horizontal. Bracts minute, truncate, subtending flowers. Pedicels 1–2 mm, deflexed after anthesis. Calyx about 6 mm long, green or purple; tube membranous; teeth as long as tube, narrowly

Plate 62. *T. brandegei* S. Wats.
(New Mexico : *Mitchell* Mt.-162, HUJ).
Plant in flower and fruit; young pod; flower; standard; wing; keel.

Plate 63. *T. haydenii* Porter
(Montana : *Gillett* & *Taylor* 11622, E).
Plant in flower; flower; wing; keel; young pod
(Montana : *Blankinship* 125, H).

triangular-subulate, sharp-pointed. Corolla 1–1.7 cm, cream to pink; standard broadly oblong-elliptical, rounded, sometimes obtuse or emarginate, apiculate, longer than wings and keel. Ovary stipitate, 3–4-ovuled. Pod obovoid, membranous, glabrous, reticulately veined, stipitate. Mature seeds not seen. Fl. July–August. 2n = 16.

Hab. : Alpine and subalpine slopes and ridges; 2,200–3,800 m.
Gen. Distr. : USA : Montana, Wyoming, Idaho.

Selected specimens : MONTANA : Park Co., Boulder River Canyon, 1947, *Hitchcock* 16416 (NA); Lone Mountain, 11,000 ft, 1906, *Blankinship* 125 (E); Carbon Co., 23 mi. from Red Lodge, 10,000 ft, 1955, *Cronquist* 8058 (S, H, a small-leafed form). IDAHO : Lost River Mts., 10,200 ft, 1895, *Henderson* 3962 (BM, isolectotype of *T. idahoense* Henderson).

66. T. howellii S. Wats., Proc. Am. Acad. Sci. 23 : 262 (1888). *Typus* : In the Siskiyou Mountains, Southern Oregon, July 11, 1887, *T. Howell* 671 (GH, iso. ORE).

Icon. : [Plate 64]. McDerm., N. Am. Sp. Trif. pls. 110–111 (1910).

Perennial, glabrous, 30–80(–100) cm. Rootstocks stout. Stems erect, simple, fistulose. Leaves short-petioled; stipules 1–4 cm long, ovate, acute, foliaceous, margin serrulate or slightly lobed; leaflets 4–9 × 2–5 cm, elliptical to ovate or obovate, rounded or often apiculate, thin, margin denticulate. Inflorescences 2–2.5 cm wide and up to 4 cm long, oblong, loose, 20- to many-flowered, whorled; rhachis prolonged beyond flowers, with cluster of small sterile flowers. Peduncles much longer than subtending leaves, slender and erect. Bracts minute, hyaline. Pedicels 1–2 mm long, strongly deflexed after anthesis. Calyx 4–7 mm, glabrous; tube membranous, 10-nerved; teeth as long as or slightly longer than tube, subulate, subequal, separated by broad sinuses. Corolla 1–1.5 cm long, yellowish-white or somewhat pinkish-tinged; standard obovate to broadly elliptical-oblong, rounded or sometimes apiculate, longer than wings and keel. Pod small, somewhat longer than calyx tube, obovate, short-stipitate, membranous, 1–3-seeded, reticulately veined; persistent style short, about 1 mm. Seeds about 2 mm, smooth. Fl. June–August.

Hab. : Wet or shady places in mountains, springs and woods; 1,100–2,000 m.
Gen. Distr. : USA : Oregon and California.

Selected specimens : OREGON : Cow Creek, 1887, *Howell* (NA, P). CALIFORNIA : Siskiyou Co., Shackleford Creek, 3,200–4,900 ft, 1926, *Kennedy* & *Reimer* (H, HUJ, R).

Note : *T. howellii* is easily distinguished by its large, thin leaflets (perhaps the largest within the genus) and stipules.

67. T. kingii S. Wats., Bot. King's Rep. 5 : 59 (1871).

Perennial, glabrous or sparingly pilose, 5–50 cm. Stems several, erect, branched from crown of stout taproot. Lower leaves long-petioled, upper ones short-petioled;

Plate 64. *T. howellii* S. Wats.
(California : *Kennedy* & *Reimer*, R).
Flowering branch; flower; standard; wing; keel.

stipules about 1 cm, ovate or lanceolate, acute, margin entire or serrate; leaflets 1–7 × 0.5–2 cm, lanceolate or elliptical to ovate, acute, long-mucronate, lower leaflets shorter and broader, ovate to obovate, rounded, margins of all leaflets coarsely and unevenly dentate or serrate, rarely entire. Inflorescences 1.5–3 cm in diam., obconical, 15- to many-flowered, whorled; rhachis often prolonged beyond flowers, simple or forked, with cluster of small sterile flowers; sometimes upper flowers erect or curved at top and then heads inverted. Peduncles much longer than subtending leaves. Bracts minute, the upper ones setaceous, 1–3 mm long. Pedicels 1 mm long, deflexed early together with flowers which are curved at base. Calyx 3–6 mm long, glabrous or sparingly pilose; tube unequal, membranous, 5–10-nerved, sometimes purplish; teeth slightly shorter than to twice as long as tube, triangular-subulate, sharp-pointed, subequal. Corolla 1.1–1.8 cm long, whitish or yellowish-white, purple-tipped; standard broadly oblong-obovate, broadly rounded at apex, notched, sometimes apiculate, longer than wings and keel. Pod obovoid, long-stipitate, glabrous, reticulately veined, 1–3-seeded, membranous. Seeds 2–3 mm, smooth. Fl. May–August.

Key to Subspecies

1. Calyx sparsely pilose 2
– Calyx glabrous 3
2. Inflorescences about as long as wide. Calyx teeth about as long as tube. **67a.** subsp. **kingii**
– Inflorescences longer than wide. Calyx teeth shorter than tube. **67c.** subsp. **macilentum**
3. Rhachis prolonged for up to 1.5 cm beyond flowers, forked. **67b.** subsp. **productum**
– Rhachis only slightly prolonged beyond flowers, entire, never forked 4
4. Nearly all leaves basal; leaflets obovate to rhombic, cuneate, margins dentate. Calyx purplish; teeth shorter than tube. **67e.** subsp. **rollinsii**
– Stems leafy. Basal leaflets ovate-elliptical, cauline ones lanceolate to linear, margins remotely serrate. Calyx green; teeth about twice as long as tube. **67d.** subsp. **dedeckerae**

67a. T. kingii S. Wats. Subsp. **kingii.** *Typus* : Utah : In a damp canyon above Parley's Park in the Wahsatch, 1868, *S. Watson* 239 (GH, iso. US).

Icon. : McDerm., N. Am. Sp. Trif. pl. 119 (1910).

Hab. : Open forested slopes; 1,100–3,200 m.
Gen. Distr. : USA : Utah, Nevada, Arizona.

Selected specimens : UTAH : San Juan Co., La Sal Mts., Geyser Pass, 10,500 ft, 1967, *Wiens* & *Arnow* 4140 (HUJ); Grand Co., La Sal Mts., Lake Oowah, trail to Mt. Tomasaki, 1962, *Gillett* & *Taylor* 11424 (HUJ).

67b. T. kingii S. Wats. Subsp. **productum** (Greene) Heller (stat. nov.). *Typus* : California : King's Valley, Aug. 1882, Mrs. *R. M. Austin* (ND-G, photo HUJ, iso. : CAN, NA !, US).

T. productum Greene, Erythea 2 : 181 (1894).
T. kingii S. Wats. var. *productum* (Greene) Jeps., Fl. Calif. 2 : 304 (1936).

Icon. : Hitchcock et al., Vasc. Pl. Pac. N. W. 3 : 371 (1961).

Hab. : Stream banks to fairly dry forest soil or open ridges; 1,100–2,800 m.
Gen. Distr. : USA : Oregon, California, Nevada.

Selected specimens : OREGON : Crater Lake, National Park, 5,900 ft, 1916, *Heller* 12627 (G). CALIFORNIA : Siskiyou Co., 1908, *Butler* 377 (HUJ); Plumas Co., Little Grizzly Creek below Genasse, 1907, *Heller* & *Kennedy* 8761 (G); Butte Co., Jonesville, 1,700 m, 1931, *Copeland* 653 (H, UC, W); Stanislau Co., 1 mi. S. W. of Sonora Pass, 1966, *Gillett* & *Crompton* 13028 (HUJ); Sierra Co., Salmon Lakes, 6,700 ft, 1929, *Kennedy* 10328 (HUJ); Shasta Co., Summit Lake, 6,700 ft, 1939, *Heller* 15419 (HUJ); Nevada Co., Sagchen Creek, 1957, *Alava* 1066 (W); Sierra Co., W. of Gold Lake, 7,000 ft, 1934, *Rose* 34381 (W).

67c. T. kingii S. Wats. Subsp. **macilentum** (Greene) Gillett, Can. Jour. Bot. 50 : 2007 (1972). *Typus* : Southern Utah, Northern Arizona, 1877, *E. Palmer* 90 (US, iso. : MO, NA !, NY, photo HUJ).

T. macilentum Greene, Pittonia 3 : 223 (1897).
T. kingii S. Wats. var. *macilentum* (Greene) Isley, Brittonia 32 : 56 (1980).

Icon. : [Plate 65].

Gen. Distr. : Utah (endemic).

Specimen seen : S. UTAH : 1874, *C. C. Parry* 34 (NA).

67d. T. kingii S. Wats. Subsp. **dedeckerae** (Gillett) Heller (stat. nov.). *Typus* : California : Inyo Co., along Horseshoe Meadows Road on a ridge S. of Caroll Creek, 7,500 ft, 1968, *DeDecker* 1899 (RSA, photo HUJ).

T. dedeckerae Gillett, Madroño 21 : 451 (1972).

Icon. : Gillett, *loc. cit.* f. 1.

Hab. : Steep exposed slope of decomposed granite.
Gen. Distr. : California (endemic).

Note : Seen only phototype.

67e. T. kingii S. Wats. Subsp. **rollinsii** (Gillett) Heller (stat. nov.). *Typus* : Nevada : Nye Co., Toiyabe Mts., 10,500 ft, 1938, *Rollins* & *Chambers* 2526 (DS, photo HUJ).

T. rollinsii Gillett, Madroño 21 : 453 (1972).

Icon. : Gillett, *loc. cit.* f. 2.

Plate 65. *T. kingii* S. Wats. Subsp. *macilentum* (Greene) Gillett (isotype : *Palmer* 90, NA). Plant in flower; flower; wing; keel.

Hab. : Steep W. talus slope, rocky clay soil.
Gen. Distr. : Nevada (endemic).

Note : Seen only phototype.
Our concept of *T. kingii* is broader than that of Gillett (1972 a, b), who included only subsp. *macilentum* with the typical subspecies. However, he pointed out the close resemblance of *T. kingii* to *T. productum* and of the latter to his two new species.
It seems to us that these five taxa form a complex which we prefer to consider under a single species.

Series viii. ACAULIA (Baker, f.) Heller & Zoh. (stat. nov.)

Subsection *Acaulia* Baker f., Leg. Trop. Afr. 75 (1926)

Type species : *T. acaule* A. Rich.

Perennials, glabrous to glabrescent. Stems densely caespitose or matted, ascending or creepingly rooting at nodes. Peduncles 1 cm to longer than subtending leaves. Inflorescences 1–3(–6)-flowered. Bracts 3–6, separate, lanceolate to subulate, subtending flowers, forming involucre at base of head. Two African and 1 S. American species.

68. T. acaule A. Rich., Tent. Fl. Abyss. 1 : 169 (1847). *Typus* : N. Ethiopia : Semien, Mt. Buahit, Entchetcab, 2 July, 1838, *Schimper* 587 (holo. P !, iso. : BM !, K !).

T. acaule Steud. in Sched. Pl. Schimp. Ab. 2 : no. 587, nomen.

Icon. : [Plate 66]. Fröman, Ill. Guide Past. Leg. Ethiopia, Rur. Dev. Stud. 3 : pl.52 (1975).

Perennial, glabrous, low-matted. Stems short, prostrate, densely caespitose, rooting at nodes, base covered with withered stipules and leaves. Leaves with 1–1.5 cm long petioles; stipules up to 8 mm long, lanceolate, long-acuminate; leaflets 3–6 × 2–2.5 mm, obovate to oblong, sometimes somewhat cuneate, shallowly emarginate, apiculate, prominently nerved, margin crenate or dentate. Inflorescences axillary, 1–4(–5)-flowered. Peduncles about 1 cm long. Bracts up to 2 mm long, scarious, lanceolate or truncate and often connate at base. Pedicels 2–3 mm long, erect or sometimes slightly deflexed after anthesis. Calyx 3–4 mm long, glabrous; tube campanulate, 11-nerved; teeth about as long as tube, triangular, slightly scarious-margined. Corolla 7–9 mm long, pale blue to mauve; standard oblong, broadly rounded at apex, longer than wings and keel. Ovary 2–4-ovuled, glabrous. Pod not seen. Fl. May–December.

Hab. : High mountain and alpine region, in short grass and rock crevices; 2,600–4,100 m.

Gen. Distr. : N. Ethiopia, Uganda, Kenya.

Plate 66. *T. acaule* A. Rich.
(Ethiopia : *Schimper* 587, BM).
Plant in flower; flower.

Specimens seen : ETHIOPIA : Scioa, 2,600 m, 1909, *Negri* 295 (FI); Yaramba, Yajasa (C. 6°45' N, 38°42' E), 1958, *Smeds* 1160 (K); Addis-Ababa, Entoto Hills, 1948, *Scott* 18 (K). KENYA : Nzoia Distr., Mt. Elgon, Path W. of Kipserehill, 3,300–3,750 m, 1968, *Gillett* 18478 (HUJ).

Note : *T. acaule* is related to *T. petitianum* (see observation there). Its prostrate growth, with almost sessile inflorescences (see Hedberg, 1973), is of particular interest.

69. T. petitianum A. Rich., Tent. Fl. Abyss. 1 : 168 (1847). *Typus* : Ethiopia : Crescit in provincia Ouodgerate (Ant. Petit.), prope Demerki, 14 Aug., 1838, *Schimper* 1389 (FI !, syn. : BM !, FI !, G !, W !).

Trigonella caespitosa Steud. in Sched. Pl. Schimp. Ab. 2 : no.1389, nomen.

Icon. : [Plate 67]. Fröman, Ill. Guide Past. Leg. Ethiopia, Rur. Dev. Stud. 3 : pl. 63 (1975).

Perennial, glabrous, with thick woody rootstock. Stems loosely caespitose, prostrate, creeping and rooting at nodes. Leaves on slender, ca. 4 cm long petioles; stipules up to 7 mm long, lanceolate to narrowly lanceolate, long-acuminate; leaflets 2–8(–12) × 2–5(–7) mm, obovate-cuneate to obcordate, deeply emarginate, apiculate, prominently nerved, margin denticulate only in upper part. Inflorescenes axillary, (1–)3–6-flowered. Peduncles slender, much longer than subtending leaves, sometimes with few curled hairs. Bracts about 1 mm, scarious, often connate at base, lanceolate. Pedicels 2–4 mm long, erect or deflexed after anthesis. Calyx 3–5 mm long, glabrous; tube campanulate, 10-nerved; teeth shorter than tube, triangular-lanceolate, acuminate. Corolla 1–1.2 cm, pale purple; standard obovate-oblong, rounded at apex, longer than wings and keel. Pod oblong, sessile, 1–3-seeded, exserted from calyx. Seeds about 1 mm, smooth, brown. Fl. August–December.

Hab. : High mountain slopes and escarpments; 2,800–3,100 m.

Gen. Distr. : Ethiopia, Eritrea.

Specimens seen : ETHIOPIA : Amhara-Semien, 1909, *Chiovenda* 1048 (FI); Simien between Debbivar, 7,200 ft, and top of Walchefet Pass, 9,900 ft, 1952, *Scott* 345 (K); *ibid.*, 9,000–10,000 ft, 1862, *Schimper* 320 (BM, K); *ibid.*, 1853, *Schimper* 924 (FI, G); *ibid.*, 1853, *Schimper* 117 (FI, G, K). ERITREA : Acchele Guzai, monti Soyra, 2,800 m, 1902, *Pappi* 1084-336 (BM).

Note : *T. petitianum* and *T. acaule* are closely related species and some forms are even difficult to place in either of them. This was pointed out by J . B. Gillett (1952, p. 384) : "It must be left to future research... to decide whether there are two species hybridizing or one very variable species".

Three specimens seen [Ethiopia : Shoa Province, Entotto Mt., 2,800 m, 1961, *F. G. Meyer* 7568 (K); Berg. Gunna, 10,000 ft, Sept. 1863, *Schimper* 320 (K); Choke Mts., Gojjom, 11,500 ft (?), 1957, *Leaky* & *Cythgol* 635 (FI)] show some characters intermediate between *T. acaule* and *T. petitianum*. Without studying local populations, it is difficult to determine whether these are lush forms of *T. acaule* (at least the *Meyer* specimen was collected on "moist bank by edge of stream") or hybrids, or should be considered as a separate species (*T. aethiopicum* Zoh. & Heller in Herb.).

Plate 67. *T. petitianum* A. Rich.
(Ethiopia : *Schimper* 1389, FI).
Plant in flower; flower; leaf.

As may be seen, the collection *Schimper* 320 includes plants collected in October 1862, which are typical for *T. petitianum*, while those from September 1863 belong to these intermediate forms.

70. T. vestitum Heller & Zoh. (sp. nov.).**Typus* : Chile : Coquimbo, Cordellera Ovalle, Gordito, 2,900 m s. m., 30.1.1954, *G. Jiles* P. 2549 (SI ! holo.).

T. volekmanni Phil. in Herb.

Icon. : [Plate 68].

Perennial, glabrous or glabrescent, 7–10 cm. Stems many, procumbent, ascending, caespitose, densely covered by persistent stipules for up to 2–3 cm of their length, branching from woody crown. Leaves on long, slender petioles; stipules oblong, membranous, upper free portion triangular, acute to acuminate; leaflets 3–7 × 3–4 mm, obtriangular, bilobed, mucronate, deeply and sharply dentate at apex, entire in lower part, prominently nerved, glabrous. Inflorescences umbellate, on slender hairy peduncles longer than leaves, 2–5-flowered. Bracts up to 2 mm, oblong-elliptical, hyaline, prominently nerved, acute to acuminate, 3–6 bracts forming involucre at base of head. Pedicels 3–6 mm long, hairy, deflexed after anthesis. Calyx 4–5 mm long, hairy; tube 10-nerved, campanulate; teeth slightly unequal, triangular-lanceolate, acuminate, membranous-margined at base, 1-nerved, the longest ones longer than tube. Corolla 7–9 mm long, flesh-coloured (when dry); standard broad, elliptical, rounded at apex to slightly retuse, longer than wings and keel; wings auriculate, claw 7 mm long. Ovary 2–3 mm, sessile, linear, 3-ovuled; style 3 mm long. Pod and seeds unknown. Fl. January.

Gen. Distr. : Chile (endemic).

Specimens seen : CHILE : Cordillera de Santiago, 1876, *Philippi* 2028 (B, photo G as *T. volekmanni* Phil.); Cordillera de Las Aranôs, prov. Santiago (FI as *T. volekmanni* Phil.).

Subsection VI. OXALIOIDEA (Gillett) Heller (comb. et stat. nov.)

Section *Oxalioides* Gillett in Report of the *Trifolium* research work conference (U. S. Dept. Agric.), 39 (1970).

Type species : *T. polymorphum* Poir.

Perennial, sparingly pubescent to densely villous. Stems prostrate, stoloniferous, creeping, rooting at nodes. Leaflets broadly obcordate or obovate-cuneate, retuse to deeply emarginate. Inflorescences and flowers of two kinds : one terminal, long-peduncled, the other basal, axillary, short-peduncled and on stoloniferous stems. Basal inflorescences of 1–8 cleistogamous and subterranean flowers. One American species.

* See Appendix.

Plate 68. *T. vestitum* Heller & Zoh.
(Chile : *Jiles* P. 2549, SI).
Plant in flower; flower; leaf.

71. T. polymorphum Poir. ex Lam. & Poir., Encycl. Meth. Bot. 8 : 20 (1808). *Typus* : Dict. Ic. Magellan, 92 (Herb. de Lamarck, P !, photo HUJ).

T obcordatum Desv., Jour. Bot. 3 : 76 (1814).
T. grandiflorum Hook. & Arn., Bot. Beech. Voy. 16 (1830).
T. mirabile Hook. inedit.
Amoria polymorpha (Poir.) Presl, Symb. Bot. 1 : 47 (1830).
A. obcordata (Desv.) Presl, *loc. cit.*
T. amphianthum Torr. & Gray, Fl. N. Am. 1 : 316 (1838).
T. megalanthum Steud., Nom. ed. 2, 2 : 706 (1841).
T. rivale Steud., *loc. cit.*
T. crossneri Clos in C. Gay, Fl. Chil. 2 : 68 (1847).
T. indecorum Clos in C. Gay, *op. cit.* 67.
T. rivale Clos in C. Gay, *op. cit.* 70.
T. simplex Clos in C. Gay, *op. cit.* 69.
T. reflexum Rowm. ex Scheele in Linnaea 21 : 460 (1848).
T. roemerianum Scheele, *loc. cit.*
T. brevipes Phil., Linnaea 28 : 680 (1856).
T. concinum Phil., Fl. Atac. 14 (1860).
T. nivale Arvet-Touvet, Ess. Pl. Dauph. 23 (1871) (Quid ?).
T. amphicarpum Phil., Anal. Univ. Santiago 84 : 11 (1894).
T. argentinense Speg. in Comm. Mus. Buenos Aires 1 : 49 (1898).

Icon. : [Plate 69]. Burkart in Physis 9 : 270 (1928).

Perennial, sparingly pubescent to densely villous. Stems prostrate, stoloniferous, creeping, rooting at nodes. Leaves long and slender-petioled; stipules ovate to lanceolate, scarious, acute to acuminate; leaflets 0.5–1.8 × 0.5–1.5 cm, broadly obcordate or obovate-cuneate, retuse to deeply emarginate, prominently nerved, margin entire or irregularly denticulate in upper part. Inflorescences and flowers of two kinds : one–terminal, long-peduncled, and the other–basal, axillary, short-peduncled and on stoloniferous stems. Terminal inflorescences : 1–2.5 cm in diam., globular, 5–20-flowered, arranged in whorls. Peduncles much longer than subtending leaves, erect. Bracts 2–4 mm, lanceolate, separate ones forming involucre. Pedicels 2–7 mm long, initially erect, later strongly deflexed. Calyx 3–5 mm, sparingly to densely pubescent, green; teeth 1–1.5 times as long as tube, narrowly triangular to lanceolate-subulate, sharp-pointed, subequal, sinus between 2 upper ones much shallower than others. Corolla 0.8–1.5 cm long, pink to rose; standard very broadly elliptical-obovate or subrotund, broadly rounded at apex, retuse, slightly longer than wings and keel. Pod slightly longer than calyx, elliptical, stipitate, 2–4(–6)-seeded, villous towards apex, reticulately nerved. Seeds about 1.5 mm, obovoid, smooth. Basal inflorescences : flowers axillary, solitary or 2–8 from nodes of stoloniferous stems. Pedicels 0.4–2 cm long, slender. Calyx 1.5–2.5 mm long, irregular; teeth shallowly triangular, much shorter than tube, sometimes inconspicuous. Corolla 4–5 mm long; standard hooded, enclosing shorter wings and keel. Pod about 3 times as long as calyx, subglobose, coriaceous, 2–3-seeded, puberulent. Seeds about 2 mm, obovoid to reniform, smooth or somewhat scabrous. Fl. March–June. 2n = 16, 32.

Hab. : Low open prairies.

Plate 69. *T. polymorphum* Poir.
(Texas : *Tracy* 9086, E).
Plant in flower and fruit; flower; standard; wing; keel.

Gen. Distr. : USA : E. Texas, W. Louisiana; Chile, Argentina, Peru, Uruguay, Paraguay, Brazil.

Selected specimens : TEXAS : *Drumond* in 1833 (G, W, type of *T. amphianthum* Torr. et Gray); College station Brazos, 1946, *Parks* (RSA); Victoria, 1905, *Tracy* 9086 (BM, E, W). CHILE : Porillos, *Stolp* 4076 (SGO, photo HUJ, type of *T. brevipes* Phil.); Prov. Coquimbo, Cordilleras de Ovalle, 2,700 m, *Gay* 592 (P, type of *T. indecorum* Clos); Concon prope Valparaiso, 1884, *Philippi* 2167 (SGO, photo HUJ, type of *T. amphicarpum* Phil.); Valdivia, ensenada bei Corral, 1902, *Buchtien* (BM, W). ARGENTINA : Isla, Martin Garcia, 1934, *Pastore* 6978 (SI); Prov. Buenos Aires, La Plata, 1950, *Cabrera* 10651 (K). PERU : Prov. Moquegua, Mostorilla Ilo, 1949, *Vargas* 8566 (HUJ). URUGUAY : San José, Arazati, *Rosengurtt* 78/a (SI); Concepcion, 1875, *Lorentz* (W). PARAGUAY : Villa-Aica, 1874, *Balansa* 1518 (G); Caaguazuensis, Yhú, 1905, *Hassler* 9423 (W). BRAZIL : Arroyo near Pelotas, 1936, *Archer* 4278 (R); Rio Grande do Sul, Porto Alegro, 1931, *Rambo* 1367 (SI).

Note : *T. polymorphum* is a most variable species and we were unsuccessful in our attempt to divide it into infraspecific taxa. Characters proved to be inconstant. Plants from S. America seem to be, in general, more pubescent than those from N. America. While plants from Texas named *T. amphianthum* are usually quite glabrous, except for the peduncle near the base of the head, it was found that in young specimens the peduncles are more densely pubescent. Different combinations of characters are encountered : large flowers may occur together with large leaflets *(T. concinum)* or with very small leaflets *(T. grandiflorum)*.

Subsection VII. PLATYSTYLIUM Willk.

in Willk. & Lange, Prodr. Fl. Hisp. 3 : 353 (1877)

Pseudolupinaster Lojac. in Nuov. Giorn. Bot. Ital. 15 : 247 (1883).
Subsection *Amoria* Gib. & Belli in Atti Accad. Sci. Torino 22 : 628 (1887).

Type species : *T. montanum* L.

Annuals or perennials, caulescent or acaulescent. Leaves 3-foliolate. Inflorescences pedunculate or sessile, few- to many-flowered. Bracts conspicuous. Pedicels short or virtually absent, not deflexed or only those of lower flowers deflexed after anthesis. Pod almost sessile, 1–2(–4)-seeded, glabrous or slightly hairy. Twenty-two Eurasian, African and N. & S. American species; arranged in 5 series.

Key to Series of Subsection Platystylium

1. Inflorescences involucrate (involucre consisting of large bracts or stipules of uppermost leaves) 2
– Inflorescences not involucrate 3
2. Inflorescences subtended by uppermost stipules and leaves, thus forming a pseudo-involucre. Series iii. **Phyllocephala**

– Inflorescences subtended by a row of large bracts, about one-third to one-half the length of heads. Series ii. **Macrochlamis**
3. Calyx tube 17–20-nerved; lateral and upper teeth tortuose and curved downwards and inwards around corolla, lower tooth straight. Series v. **Altissima**
– Calyx tube 5–10-nerved; teeth neither tortuose nor curved inwards 4
4. Inflorescences sessile. Pedicels absent. Calyx teeth almost equal, spreading or recurved outwards in fruit. Series iv. **Micrantheum**
– Inflorescences pedunculate. Flowers pedicellate. Calyx teeth remain erect in fruit. Series i. **Platystylium**

Series i. PLATYSTYLIUM

Type species : *T. montanum* L.

Annuals or perennials, glabrous or pubescent. Inflorescences pedunculate, few- to many-flowered. Bracts conspicuous or minute, subtending each flower. Pedicels short, erect or only those of lower flowers deflexed after anthesis. Three Eurasian, nine African, and two N. American species.

72. T. abyssinicum Heller (nom. nov.). *Typus* : Abyssinia (North), *Rueppell* (FR).

T. calocephalum Fresen. in Flora 22 : 50 (1839) non Nutt. ex Torr. & Gray (1838).
Loxospermum calocephalum (Fresen.) Hochst. in Flora 29 : 595 (1846).

Icon. : [Plate 70]. Fröman, Ill. Guide Past. Leg. Ethiopia, Rur. Dev. Stud. 3 : pl. 56 (1975).

Annual or sometimes perennial, glabrous, 10–30 cm. Stems densely caespitose, diffuse, erect or procumbent, rooting at nodes, branching, grooved. Leaves long-petioled, slender; stipules up to 2.5 cm, broadly sheathing, scarious, free portion only one-third the length of adnate part, broadly triangular to lanceolate; leaflets 1–2.5 × 0.4–0.8 cm, oblong to obovate-cuneate, rounded or retuse to emarginate at apex, margin denticulate. Inflorescences 3–4 cm in diam., 3–8-flowered. Peduncles much longer than subtending leaves. Bracts 2–4 mm long, hyaline, lanceolate. Pedicels 2–4 mm long, erect. Calyx 0.7–1 cm, glabrous; tube campanulate, with 10 obsolete nerves; teeth as long as or longer than tube, triangular, membranous-margined, slightly ciliate at base, subulate above. Corolla 1.5–2 cm long, purple; standard broadly elliptical, retuse to emarginate at apex, slightly longer than wings and keel. Pod about as long as calyx, elliptical, rounded at apex, up to 7-seeded. Fl. March–May.
Hab. : Mountain streams and damp grass; about 2,700 m.
Gen. Distr. : Ethiopia (endemic).

Specimens seen : ETHIOPIA : Scioa, 2,600 m, 1909, *Negri* 533 (FI); Taccasse River, 2,500 m, 1947, *Hall* 138 (BM); Addis Ababa, 8,000 ft, 1951, *Curle* 17 (BM); *Schimper* 1279 (W); Arussi Prov., Chilalo awraja, 10 km N. of Koffale, 2,700 m, 1971, *Mats Thulin* 1469 (HUJ); 1853, *Schimper* 328 (G).

Plate 70. *T. abyssinicum* Heller
(Ethiopia : *Negri* 533, FI).
Plant in flower; flower.

73. T. africanum Ser. in DC., Prodr. 2 : 200 (1825).

Perennial, glabrescent to pilose, 10–25 cm. Stems many, erect or ascending, sometimes prostrate and rooting at nodes. Leaves with petioles about as long as leaflets; stipules oblong, many-nerved, tapering at apex, free portion shorter than sheathing lower part, triangular-lanceolate; leaflets usually 1–2(–4) × 0.4-0.8 cm, oblanceolate, oblong-lanceolate to lanceolate, obtuse to acute or slightly emarginate, obsoletely spinulose-denticulate. Inflorescences 1–1.5(–2) cm across, semi-globular, many-flowered. Peduncles often rather thick, usually 5–15 cm long. Outer bracts as long as tube, inner ones shorter, all subulate or ovate, truncate. Pedicels very short, up to 1.5 mm, horizontal to erect. Calyx 5–8 mm long; tube tubular to campanulate, densely patulous-hairy, 10-nerved; teeth as long as or somewhat longer than tube, more or less equal, subulate. Corolla 8–10 mm, purple to brick red; standard obovate, longer than wings and keel, apex slightly emarginate. Pod long, protruding from calyx tube, glabrous, broad-elliptical, 2-seeded. Seeds about 2 mm, ovoid, finely tuberculate, yellowish-green. Fl. October–November. 2n = 32.

73a. T. africanum Ser. Var. **africanum.** *Typus*: Cape Province (Catalogus Geographicus Plantarum Africano Australes Extratropicae), *Burchell* no. 2817 (holo. G !).

T. hirsutum Thunb. sec. E. Mey., Comm. Pl. Afr. Austr. 91 (1836).
T. hirsutum Thunb. var. *glabellum* E. Mey., *Loc. cit.*
Lupinaster africanus (Ser.) Ecklon & Zeyh., Enum. Pl. Afr. Austr. Extra Trop. 223 (1836).
T. africanum Ser. var. *glabellum* (E. Mey.) Harvey, Flor. Cap. 2 : 159 (1862).
T. burchellianum Ser. var. *africanum* (Ser.) O. Ktze., Rev. Gen. 3, 2 : 73 (1898).

Icon. : [Plate 71]. Fiori in Nuov. Giorn. Bot. Ital. n. s. 55 : f. 7 (1948).

Leaflets 2–4 times as long as wide. Bracts up to 3 mm long, subulate. Calyx densely hirsute.
Hab. : Waste places.
Gen. Distr. : Cape Province, Orange Free State, Transvaal, Natal, Basutoland.

Specimens seen : ORANGE FREE STATE : Paul Roux. Distr., waste places in town, 5,200 ft, 1946, *Acocks* 13188 (HUJ). TRANSVAAL : Pietersburg, 1 km N. of Haenertsburg, 5,000 ft, 1954, *Codd* 8423 (SRGH).

73b. T. africanum Ser. Var. **lydenburgense** Gillett in Kew Bull. 7 : 380 (1952). *Typus* : Habitat republic Transvaal. Distr. Lydenburg. Bai der Stadt, Lydenburg. An feuchten Orten, Oct. 1895, *F. Wilms* 288 (holo. K !, iso. G !).

T. lydenburgense Harms, ms.

Icon. : t'Mannetje, E. Afr. Agr. & For. J. 31 : f. 3 (1966).

Leaflets 4–10 times as long as wide. Bracts inconspicuous or up to 1 mm broad, ovate, truncate. Calyx sparingly villous.

Plate 71. *T. africanum* Ser. Var. *africanum* (Orange Free State : *Acocks* 13188, HUJ). Plant in flower and fruit; flower; young pod.

Hab. : Moist grassy places and marshes; 600–2,000 m.
Gen. Distr. : Transvaal, Cape Province.

Specimens seen : TRANSVAAL : Johannesburg, 1925, *Moss* 11168 (BM); Pretoria Distr., 5,000 ft, 1946, *Story* 1474 (HUJ); Wakkerstroom, Oshoek, 6,400 ft, 1960, *Devenish* 284 (SRGH); Lake Chrisste, 1929, *Moss* 16327 (BM).

Note : Some forms of *T. africanum* show a superficial resemblance to *T. burchellianum* but can easily be distinguished by their oblanceolate to lanceolate leaflets and short and erect pedicels.

74. T. ambiguum M. B., Fl. Taur.-Cauc. 2 : 208 (1808). *Typus* : Caucasus : Habitat frequens Tauriae et Caucasi pratis, *M. Bieberstein* (LE photo. E!).

T. vaillantii M. B. ex Fisch. Cat. Hort. Gorenk. 111 (1808).
T ruprechtii Tomasch. & Fedorov, Jour. Bot. URSS 34 : 164 (1949).
T. ambiguum M. B. var. *majus* Hossain, Not. Roy. Bot. Gard. Edinb. 23 : 464 (1961).

Icon. : [Plate 72]. Bobrov, Acta Inst. Bot. Acad. Sci. URSS ser.1,6 : pl. 3 (1947).

Perennial, glabrous or sparingly hairy, 8–40(–60) cm. Roots thick and rhizomatous. Stems many, leafy, procumbent to ascending, grooved to angled, branched from base and slightly above. Leaves long-petioled, the uppermost ones sometimes subsessile; stipules broadly ovate, membranous, long-adnate to petioles, upper free portion lanceolate-subulate; leaflets 1–3.5(–7) × 0.6–2(–3) cm, obovate, elliptical to broadly elliptical, rounded at apex, prominently nerved, the upper ones coarsely serrate-dentate. Inflorescences 1–2 cm, many-flowered, capitate, solitary or in pairs, ovoid in flowering, later elongating, spicate, up to 3.5 cm. Peduncles 4–20 cm long, longer than subtending leaves. Bracts longer than pedicels, as long as calyx tube, narrowly lanceolate-subulate, conduplicate, membranous-margined. Pedicels, 1–2 mm long, those of the lower flowers deflexed. Calyx 4-5 mm long; tube campanulate, whitish, hirsute, in upper part obscurely nerved; teeth considerably shorter than to about as long as tube, subequal, narrowly triangular-subulate, membranous-margined at base. Corolla 1–1.3 cm, white, flesh-coloured after anthesis; standard narrowly ovate to elliptical, rounded at apex, slightly retuse, much longer than wings and keel. Pod 3 mm long, ellipsoidal, glabrous, sessile, 1–2-seeded. Seeds 1.2 mm across, lenticular. Fl. June–August. 2n = 16, 32, 48.

Hab. : Steep scree slopes, forest borders, eroded banks, stream sides, fields, etc.;1,700–2,750 m.

Gen. Distr. : Romania, Russia (Crimea, Caucasus, Transcaucasus, Armenia), Turkey, Iran, Iraq.

Selected specimens : ROMANIA : Bessarabia, distr. Tighina, Zloti, 1924, *Savulescu* & *Rayss* 556 (HUJ, K). RUSSIA : Odessa, *Lang* & *Szovile* 148 (G); Caucasus, Karasubazar, Barultscha, *Callier* 579 (K); Transcaucasus, Azerbajdzhan, distr. Zokataly, 1935, *Beideman* (HUJ); Armenia, prope pag. Gedjolan, 6,400 ft, 1929, *Schelkovnikov* & *Kara-Murza* (HUJ). TURKEY : Lazistan, Chokje, 9,000 ft, 1934, *Balls* 1915 (E); N.E. Anatolia, betw.

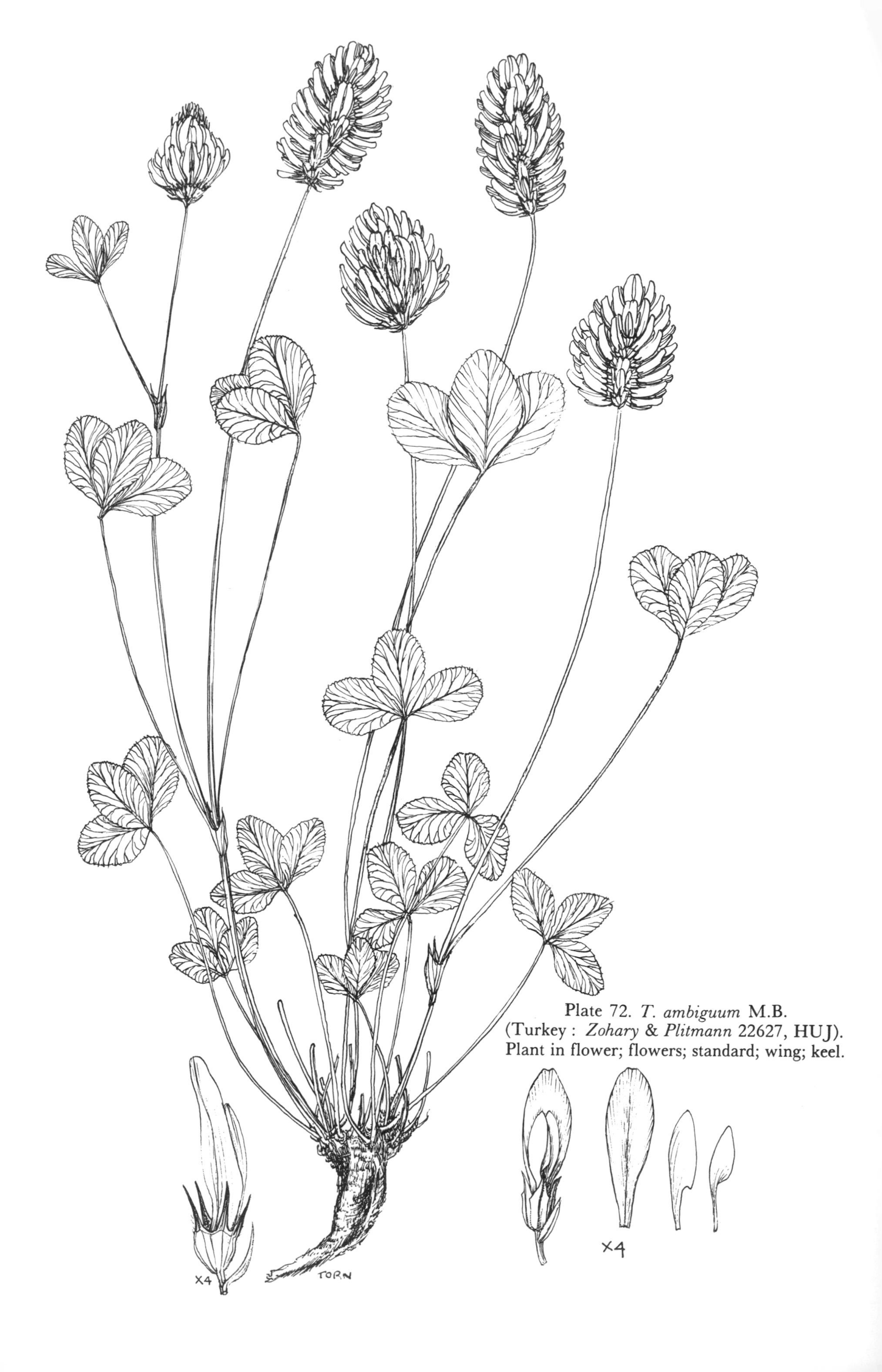

Plate 72. *T. ambiguum* M.B.
(Turkey : *Zohary* & *Plitmann* 22627, HUJ).
Plant in flower; flowers; standard; wing; keel.

Agri and Dogobayazit, 1,840 m, 1964, *Zohary* & *Plitmann* 22627 (HUJ). IRAN: Kuh-i-Savalan, 20 W. of Ardabil, 7,000 ft, 1962, *Furse* 2431 (K). IRAQ: Helgord (Algurd Dagh), 2,000 m, 1954, *Rawi* 13750 (K).

Note: *T. ambiguum* is very variable in size and indumentum. In our view it includes *T. ruprechtii*, which seems to be a small alpine form of *T. ambiguum*, and var. *majus* Hossain, on the basis of glabrous taller plants.
The common name used is "Kura or Caucasian clover". This species is considered as a valuable pasture plant even for cutting in perennial pastures.

75. T. baccarini Chiov., Ann. Bot. Roma 9 : 57 (1911). *Typus*: E. Ethiopia, Dembia, *Chiovenda* 1774 et 2155 (FI !).

Trifolium subrotundum A. Rich., Tent. Fl. Abyss. 1 : 172 (1847) p. p. non Hochst. & Steud. (1842).
Trigonella marginata Bak. in Flora Trop. Afr. 2 : 50 (1871) nom. nud.
Trifolium rueppellianum Chev. in Rev. Bot. Appl. 14 : 439 (1934) non Fres. (1839).
Trifolium rueppellianum Chev. f. *minor* Chev., *op. cit.* 440, Fig. 4/1.
Trifolium marginatum (Bak.) Cufod., Bull. Jard. Bot. Brux. 25 : suppl. 251 (1955).

Icon.: [Plate 73]. Fiori in Nuov. Giorn. Bot. Ital. n. s. 55 : f. 4 (1948).

Annual, glabrous or sparingly hairy in upper parts, 20–50 cm. Stems erect, ascending or prostrate and sometimes rooting at nodes in overgrazed areas, grooved, branched. Lower leaves long-petioled, upper ones short-petioled to subsessile; stipules about 1 cm long, ovate, triangular at apex, free portion of upper ones long-cuspidate; leaflets 0.7–1.5 × 0.4–1.4 cm, obovate or elliptical, cuneate, rounded to emarginate, apiculate, margin dentate. Inflorescences about 1 cm in diam., 5–20-flowered, globular. Peduncles up to 4 cm long, longer than subtending leaves. Bracts inconspicuous. Pedicels 1–2 mm long, more or less erect after flowering. Calyx about 5 mm, glabrous or sparingly hairy; tube 1.5 mm, tubular, 15-nerved; teeth twice as long as tube, triangular, tapering gradually to subulate upper part. Corolla 3–4 mm long, purple or whitish; standard obovate, rounded at apex, longer than wings and keel. Pod 4 mm long, slightly hairy, with prominent sutures, ovoid, 2-seeded, beaked. Seeds 1 mm, ovoid, with brown-black spots. Fl. September. [Recorded as cleistogamous (Pritchard & t' Mannetje, 1967)]. 2n = 16.
Hab.: Pastures in areas with fairly heavy rainfall; 1,200–2,100(–2,500) m.
Gen. Distr.: Uganda, Kenya, Tanganyika, Cameroon, E. Congo, N. Nigeria, Rwanda-Burundi, Ethiopia.

Specimens seen: TANGANYIKA: Aruscha-Moschi, 1926, *Peter* 42466 (B). KENYA: 1961, *Bogdan* 5296 (HUJ, K). BELGIAN CONGO: Kabare, 1947, *Hendrick* 4429 (HUJ). ETHIOPIA: Gimma loc. Malco, 1,850 m, 1939, *Ufficio Agrario* (FI, G).

76. T. bilineatum Fresen. in Flora 22 : 52 (1839). *Typus*: N. Abyssinia, *Rueppell* (FR).

T. costatum Hochst. ex Engl., Hochgebirgsflora Trop. Afr. 252 (1892).

Plate 73. *T. baccarini* Chiov.
(Kenya : *Bogdan* 5296, HUJ).
Fruiting branch; fruiting calyx with pod; flowering branch; flower.

Icon. : [Plate 74]. Fiori, Nuov. Giorn. Bot. Ital. n. s. 55 : f. 8 (1948).

Annual, patulous-pubescent, 25–50 cm. Stems many, herbaceous, erect or ascending, leafy all along, branching, sulcate. Lower leaves long-petioled, upper ones short-petioled to subsessile; stipules about 1 cm long, ovate, free portion triangular-lanceolate, cuspidate, herbaceous; leaflets 1–1.5(–2.5) × 0.8–1.5 cm, ovate, obovate-cuneate to broad-elliptical, obtuse to acute, usually rounded, sometimes short-mucronate, prominently nerved, dentate. Inflorescences 0.8–1.5 cm in diam., globular, dense, many-flowered. Peduncles slender, much longer than subtending leaves. Bracts minute, ovate, truncate to bifid or with two teeth. Pedicels up to 0.5 mm long, those of lower flowers slightly deflexed after anthesis. Calyx 4–6 mm, pubescent; tube with 10 obsolete nerves, tubular to campanulate; teeth as long as tube, subequal, subulate. Corolla 7–9 mm long, light purple to pink; standard broad-elliptical, retuse, longer than wings and keel. Pod about 3 mm, obovoid, membranous, 1-seeded, glabrous. Seed up to 2 mm, brown, ellipsoidal. Fl. July–October.

Hab. : Stony pathways.

Gen. Distr. : Ethiopia, Eritrea.

Specimens seen : ETHIOPIA : Gafat, 1863–8, *Schimper* 1264 (W); Scioa, Strad di Ficca, 2,650 m, 1938, *Borini* 21 (G); Bahor Dor, 6,100 ft, 1964, *Thomas* 2 (K); Malu Farm, 50 km S. of Addis Ababa, 8,000 ft, 1951, *Archer* 10073 (NA); Amhara-Dembia, Gondar, 1909, *Chiovenda* 1198 (FI). ERITREA : 1943, *Isaia Baldrati* 4248 (FI).

77. T. isthmocarpum Brot., Phyt. Lusit. ed. 3, 1 : 148 (1816). *Typus* : Described from Portugal (LISU).

Amoria isthmocarpa (Brot.) Presl, Symb. Bot. 1 : 50 (1832).
T. mutelii Gren. in Acad. Sci. Besanç. Seances Publ. 127 (1838).
T. strangulatum Huet du Pav., Pl. Sic. exsicc. (1855).
T. jaminianum Boiss., Diagn. ser. 2, 2 : 19 (1856).
T. rouxii Gren. in Mém. Soc. Emul. Doubs. ser. 3, 2 : 413 (1858).
T. mauritanicum Salzm. ex Ball in Jour. Linn. Soc. 16 : 420 (1878).
T. rubicundum Schousb. ex Ball, *loc. cit.*
T. isthmocarpum Brot. ssp. *jaminianum* (Boiss.) Murb., Lunds Univ. Arsskr. 33 : 67 (1897).

Icon. : [Plate 75]. Brot., *loc. cit.* t. 61.

Annual, glabrous, 30–60 cm. Stems many, erect or ascending, rarely procumbent, often fistulose, grooved, branching from base and above. Lower leaves with long petioles (up to 10 cm), uppermost ones subsessile; stipules 0.6–2 cm long, ovate, abruptly narrowing to a long filiform cusp, membranous, whitish, sheathing; leaflets 1–3.5 × 0.5–2 cm, broadly elliptical to obovate, rounded at apex, cuneate at base, margin entire in lower part and coarsely dentate-serrate in upper part. Inflorescences 1.5–2.5 cm wide, broadly ovoid in flower, becoming globular or cylindrical in fruit, compact, many-flowered. Peduncles all on upper part of stems, 3–10 cm long. Bracts almost as long as calyx tube, membranous, subulate, folded. Pedicels shorter than calyx tube. Calyx 4–6 mm long, white, glabrous; tube campanulate, 10-nerved; teeth shorter than to almost as long as tube, almost equal, triangular-lanceolate, long-acuminate, straight or curved. Corolla 0.8–1.2 cm long, pink,

Plate 74. *T. bilineatum* Fresen.
(Eritrea : *Baldrati* 4248, FI).
Plant in flower; flowers.

Plate 75. *T. isthmocarpum* Brot.
(Portugal : *Matos* & *Matos*, HUJ).
Plant in flower and fruit; pod; standard; wing; keel; flower with bract.

becoming brownish in fruit; standard oblong, rounded to slightly retuse at apex, longer than wings and keel. Pod about 4 mm long, membranous, oblong, 2-seeded, constricted between seeds, with slightly twisted style longer than seed-bearing part. Seeds about 1.2 mm, ovoid-reniform, finely tuberculate, brown. Fl. April–June. 2n = 16.

Hab.: Hills, moist meadows.

Gen. Distr.: Corsica, Portugal, Spain, Italy, Sicily, Turkey, Algeria, Morocco, Tunisia.

Selected specimens: PORTUGAL: 5 km para além. de Castelo Branco, 1959, *Matos & Marques* 6677 (HUJ); Fugueira da Foz, 1950, *J. Matos & A. Matos* (HUJ). SPAIN: Algeciras, 1887, *Reverchon* 141 (HUJ); Lagos, Algarue, 1853, *Bourgeau* 1831 (G). TURKEY: Gallipoli near Helles, 1923, *Ingoldby* 101 (K). ALGERIA: Bougie, 1896, *Reverchon* 24 (HUJ). MOROCCO: Forest of Marmora, nr. Rabat, 1939, *Irvine* 3284 (BM); inter Azrou et Ito, 1,200–1,300 m, *Maire* (HUJ). TUNISIA: Infructicetis Djebel Zaghouan, 1854, *Kralik* (FI).

78. T. lanceolatum (Gillett) Gillett in Kew Bull. 24: 219 (1970). *Typus*: Kenya: Nakuru Distr., Molo, 2,800 m, July 1931, *Edwards* 1808 (K! holo., iso. EA).

T. rueppellianum Fresen. var. *lanceolatum* Gillett in Kew Bull. 7: 390 (1952).

Icon.: [Plate 76]. Fröman, Ill. Guide Past. Leg. Ethiopia, Rur. Dev. Stud. 3: pl. 60 (1975).

Annual, glabrous except in upper parts, 15–40 cm. Stems erect or ascending, grooved, branching from base and above. Lower leaves long-petioled, upper ones subsessile; stipules up to 2 cm long, adnate to petioles for about two-thirds of their length, free portion broadly lanceolate, acuminate, margin laciniate or sometimes 3–4-toothed; leaflets 1–3 × 0.4–0.7 cm, lanceolate to elliptical, acute, prominently nerved, margin coarsely dentate, teeth terminating in subulate tip. Inflorescences 0.8–1.4 cm in diam., globular, 15–30-flowered. Peduncles much longer than subtending leaves, sparingly pilose above. Bracts up to 3 mm long, linear. Pedicels up to 2 mm long, pilose, those of lower flowers slightly deflexed. Calyx 3–5 mm long, pilose, tubular; tube 10-nerved; teeth about as long as to longer than tube, subulate. Corolla 5 mm long, shorter than calyx, purple; standard oblong, rounded at apex, slightly longer than wings and keel. Pod 3–4 mm long, ovoid, 2-seeded, with short persistent style, membranous, sutures prominent. Seeds about 1 mm, reniform, brown. Fl. May–October. 2n = 16.

Hab.: Upland grassland and as a weed in cultivation; 2,400–2,800 m.

Gen. Distr.: Kenya, Tanganyika, Ethiopia (possibly introduced).

Specimens seen: KENYA: Nakuru Distr., Molo, 8,300–9,000 ft, 1951, *Bogdan* AB 3170 & 3175 (K); Grown at KGRS from seeds collected at Molo, 8,000 ft, 1964, *Bogdan* AB 5740 (HUJ). ETHIOPIA: Harar Distr., near Dire Dawa, 1958, *Imp. Coll. Agric.* C-90 (K).

Note: *T. lanceolatum* differs from *T. rueppellianum* in its laciniate stipules, lanceolate, acute leaflets and pilose calyx.

Plate 76. *T. lanceolatum* (Gillett) Gillett
(Kenya : *Bogdan* AB5740, HUJ).
Flowering and fruiting branch; flower; fruiting calyx with pod.

79. T. longipes Nutt. in Torr. & Gray, Fl. N. Am. 1 : 314 (1838).

Perennial, pubescent to pilose, 5–50 cm. Stems several, erect or ascending, branching from long woody root and above. Leaves extremely variable in size and shape, long-petioled in lower parts, shorter-petioled in upper parts; stipules 0.5–2.5 cm long, ovate to lanceolate, acuminate, entire or toothed; leaflets 1–6 × 0.2–1.6 cm, mostly lanceolate to elliptical or oblong, the cauline ones shorter and wider, sometimes ovate, entire or serrate to dentate, apex usually acute. Inflorescences 1–3.5 cm in diam., 10- to many-flowered, rounded or elongated. Peduncles longer than subtending leaves, erect. Bracts minute, 0.5 mm, reduced to scarious scales, subtending flowers. Pedicels about 2 mm long, pilose, remaining erect or only those of lower flowers deflexed after anthesis. Calyx 0.4–1 cm long, pilose to glabrous; tube membranous, 5–10-nerved; teeth shorter than to six times as long as tube, unequal, linear or subulate or sometimes broadened at base, sharp-pointed. Corolla 1–1.8 cm long, white, purple to pink or flesh-coloured; standard ovate to elliptical, rounded to acute or obtuse, rarely emarginate, longer than wings and keel. Pod exserted from calyx, oblong, 1–4-seeded, membranous, sparingly pubescent, reticulately veined. Seeds about 3 mm, flattened, smooth, dark brown. Fl. May–July.

Key to Subspecies

1. Calyx, leaves and peduncles densely appressed-pilose; upper calyx teeth much reduced, less than 2 mm long. **79b.** subsp. **atrorubens**
– Calyx and leaves sparingly pubescent to glabrous; upper calyx teeth more than 2 mm long 2
2. Standard and wings slender-attenuate. **79a.** subsp. **longipes**
– Standard and wings acute, obtuse or rarely emarginate 3
3. Calyx teeth often somewhat rigid, about 6 times as long as tube, usually prominently nerved. **79c.** subsp. **elmeri**
– Calyx teeth lax, usually only slightly longer than tube 4
4. Plants with fusiform or thickened roots. Stems ascending from crown. **79d.** subsp. **pedunculatum**
– Plants rhizomatous. **79e.** subsp. **hansenii**

79a. T. longipes Nutt. Subsp. **longipes.** *Typus* : Valleys of the central chain of the Rocky Mountains range and on the moist plains of the Oregon as low as the Wahlamet, *Nuttall* (K ! lecto.–selected by Kennedy, 1910, iso. : BM !, GH, NY, PH).

Icon. : [Plate 77]. McDerm., N. Am. Sp. Trif. pl. 100 (1910).

Hab. : Moist places below 3,000 m, meadows, open prairies, river banks and mostly montane coniferous forest. n = 8.

Gen. Distr. : USA : Washington, N. Oregon.

Specimens seen : WASHINGTON : Kittitas Co., Bald Mt., 5,900 ft, 1940, *Thompson* 14841 (HUJ, NA); Wenatchee Mts., 5,500 ft, 1963, *Hermann* 18996 (W); Chelan Co., Camas Land, Wenatche Mts., 3,000 ft, 1935, *Thompson* 11762 (NA).

Plate 77. *T. longipes* Nutt. Subsp. *longipes*
(Washington : *Thompson* 14841, HUJ).
Plant in flower; flower; standard; wing; keel; young pod
(Washington : *Thompson* 11762, NA).

79b. T. longipes Nutt. Subsp. **atrorubens** (Greene) Gillett, Can. Jour. Bot. 47 : 108 (1969). *Typus* : Bluff Lake, San Bernardino Co., California, 21–27 June, 1895, *J. B. Parish* 3745 (US, lecto., iso. : GH, MO, ND, NY, UC).

T. rusbyi Greene var. *atrorubens* Greene, Erythea 4 : 66 (1896).
T. atrorubens (Greene) House, Bot. Gaz. 41 : 336 (1906).
T. longipes Nutt. var. *elmeri* (Greene) McDerm. f. *atrorubens* (Greene) McDerm., Ill. Key N. Am. Sp. Trif. 256 (1910).
T. longipes Nutt. var. *atrorubens* (Greene) Jeps., Fl. Calif. 2 : 303 (1936).

Hab. : As above. 2n = 48.
Gen. Distr. : USA : California.

Specimens seen : S. CALIFORNIA : San Bernardino Co., Bear Valley, 5,000 ft, 1887, *Parish* 1809 (HUJ); Tulare Co., Monache Meadow, S. Fork of Kern River, 8,050 ft, *Munz* 15079 (NA).

79c. T. longipes Nutt. Subsp. **elmeri** (Greene) Gillett, *op. cit.* 111. *Typus* : N.W. California : S. Fork, Trinity River, near Grouse Creek, 1888, *V. K. Chestnut* & *E. R. Drew* (ND, iso. : JEPS, UC).

T. elmeri Greene, Pittonia 3 : 223 (1897).
T. longipes Nutt. var. *elmeri* (Greene) McDerm., *op. cit.* 253.

Icon. : McDerm., *loc. cit.* pl. 104

Hab. : As above. n = 8.
Gen. Distr. : USA : California, Oregon.

Specimens seen : CALIFORNIA : Siskiyou Co., Sanger Peak Trail, 13 mi. S. of Takilma, 3,500 ft, 1939, *Hitchcock* & *Martin* 5275 (NA); N. Fork, Smith River, Adams Station, 1926, *Doris K. Keldale* 2330 (HUJ). OREGON : Whiskey Creek, near Waldo, Josephine Co., 1925, *Kennedy* (HUJ); Josephine Co., Bank of Illinois River, 1902, *Cusick* 2909 (W).

79d. T. longipes Nutt. Subsp. **pedunculatum** (Rydb.) Gillett, *op. cit.* 109. *Typus* : Long Valley, Idaho, 1895, *Henderson* 3096 (US, iso. : F !, NY, ORE, UC).

T. pedunculatum Rydb., Bull. Torrey Bot. Club 30 : 254 (1903).
T. longipes Nutt. var. *pedunculatum* (Rydb.) Isley, Brittonia 32 : 57 (1980).

Hab. : As above. n = 8, 24.
Gen. Distr. : USA : Idaho.

Specimens seen : IDAHO : Valley Co., Round Prairie near Alpha, 1937, *Christ* & *Ward* 8141 (HUJ); Camas Co., 15 mi. E. of Featherville, 4,400 ft, 1944, *Hitchcock* & *Muhlick* 8775 (NA).

79e. T. longipes Nutt. Subsp. **hansenii** (Greene) Gillett, *op. cit.* 109. *Typus* : High Sierra in Alpine Co., Calif., 1892, *Geo. Hansen* (ND lecto.).

T. hansenii Greene in Erythea 3 : 17 (1895).
T. longipes Nutt. f. *hansenii* (Greene) McDerm., *op. cit.* 250.
T. longipes Nutt. var. *nevadense* Jeps., Fl. Calif. 2 : 303 (1936).
T. longipes Nutt. var. *hansenii* (Greene) Jeps., *op. cit.* 302.

Hab. : As above. 2n = 24.
Gen. Distr. : USA : California, Nevada, Oregon.

Specimens seen : CALIFORNIA : Alpine Co., Monitor Pass, 7,000 ft, 1964, *Gillett* & *Moulds* 12710 (HUJ); Placer Co., Squaw Valley, 1916, *Smith* 511 (HUJ); Northern Lake Co., 1902, *Mackie* (HUJ). NEVADA : Galena Creek, Washoe Co., 8,000 ft, *Kennedy* 1216 (HUJ). OREGON : Deschutes Co., E. of Little Cultus Lake, 4,800 ft, 1939, *Hitchcock* & *Martin* 4928a (NA).

Note : For the treatment of this species by Gillett (1969), see note under *T. rusbyi.*

80. T. montanum L., Sp. Pl. 770 (1753).

Perennial, pubescent or glabrous, with woody stock, 15–80 cm. Stems many or few, erect or ascending, grooved, poorly branching above. Lower leaves with long petioles (up to 20 cm), upper ones short-petioled to almost sessile, all sparsely scattered along stem; stipules about 1 cm, membranous, ovate, long-adnate to stem, the lowest ones sheath-like, upper free portion acuminate to subulate; leaflets 2–7(–10) × 0.8–2.5 cm, elliptical, ovate to oblong-lanceolate, rather leathery, obtuse or acute, mucronulate, hairy or glabrous, prominently nerved especially near serrulate margin. Inflorescences 1.5–3 cm in diam., capitate, 2–3 together, dense, many-flowered, globular to ovoid, later elongating into spike (up to 3 × 2 cm). Peduncles 4–15 cm long. Bracts somewhat shorter than to about as long calyx tube, membranous or scarious-margined, lanceolate, acuminate. Pedicels as long as or shorter than calyx tube, slightly deflexed in fruit of only lower flowers. Calyx 6–8 mm long, campanulate, sparingly hairy to glabrescent; tube 10-nerved, whitish; teeth as long as or slightly longer than tube, unequal to subequal, 3-nerved at triangular base, subulate at tip. Corolla up to 1.5 cm long, white or yellowish, rarely pink; standard ovate-oblong, recurved, rounded to truncate at apex, slightly apiculate, much longer than wings and keel. Pod 4–6 mm, ovoid-oblong, slightly oblique, with short, persistent style, 1–2-seeded. Seeds 2 mm, oblong-ovoid, brown. Fl. June–August. 2n = 16, 32.

Key to Subspecies

1. Stems and leaves glabrous (upper part sparsely pubescent). Flowers 1–1.5 cm long. Bracts 3–4 mm long, membranous, with scarious margin.
80b. subsp. **humboldtianum**

– Stems and leaves pubescent. Flowers shorter. Bracts 1 mm long, membranous, without scarious margin 2
2. Calyx teeth subequal, longer than the tube, triangular. Corolla 1 cm. **80c.** subsp. **rupestre**
– Calyx teeth unequal, about as long as tube or shorter, subulate. Corolla 7–9 mm. **80a.** subsp. **montanum**

80a. T. montanum L. Subsp. **montanum.** *Typus* : Habitat in pratis siccis Europaeae septentrionalis (930/56, LINN ?).

T. subulatum Gilib., Fl. Lithuan. 2 : 90 (1782).
T. odoratum Schrank, Baier. Fl. 2 : 286 (1789).
T. tenoreanum Buck, Ind. ad DC., Prodr. 1 : P. VII (1824).
T. enderssi J. Gay ex Gren. & Godr., Fl. Fr. 1 : 417 (1849).
T. bifurcatum Pourr. ex Willk. & Lange, Prodr. Fl. Hisp. 3 : 353 (1877).
T. celtiberneum Pau, Not. Bot. Fl. Espan. 1 : n. 7 (1887) et 4 : 28 (1891).

Icon. : Sturm., Fl. Deutschl. ed. 2, 9 : 153 (1901).

Hab. : Dry grassy places.

Gen. Distr. : Europe, absent in the West, the Mediterranean region and the North.

Selected specimens : FRANCE : H-te Savoie, Fillière, 900 m, 1931, *Eig* (HUJ). BELGIUM : Env. de Goé et de Dolhain, prov. de Liège, 1866, Cent. II no. 128 (HUJ). SPAIN : Alemanha, Arnstadt, 1911, *Krahmer* (R). POLAND : Silesia Orient., Strzemieszyce Wielkie, 1952, *H. Błaszczyk* 436 (HUJ). CZECHOSLOVAKIA : Moravia Merid., Uh. Ostroh, Borsice, 400 m, 1933, *Podpera* 830 (HUJ). ROMANIA : Transilvania distr., Turda, 600 m, 1938, *Cupcea* & *Todor* 1993 (HUJ).

80b. T. montanum L. Subsp. **humboldtianum** (A. Br. & Aschers.) Hossain, Not. Roy. Bot. Gard. Edinb. 23 : 463 (1961). *Typus* : Described from a plant cultivated in Berlin, raised from seed collected in Tiflis (Georgia) by *Noodt* (B).

T. humboldtianum A. Br. & Aschers. in Index Sem. Hort. Berol. 24 (1868).
T. montanum L. var. *grandiflorum* A. Br., *ibid.* 17 (1868) nomen.
T. montanum L. var. *humboldtianum* (A. Br. & Aschers.) Gib. & Belli in Atti R. Accad. Sc. Torino 22 : 451 (1887).
T. bordzilowskyi Grossh. in Jour. Soc. Bot. Russe 14 : 311 (1929).
T. elizabethae Grossh., *ibid.* 16 : 309 (1930).

Icon. : [Plate 78]. Bobrov, Acta Inst. Bot. Acad. Sci. URSS. ser. 1, 6 : pl. 2, f. 1 & 3 (1947).

Hab. : Mountain slopes, 1,800–2,400 m.
Gen. Distr. : Turkey, Caucasus, Armenia.

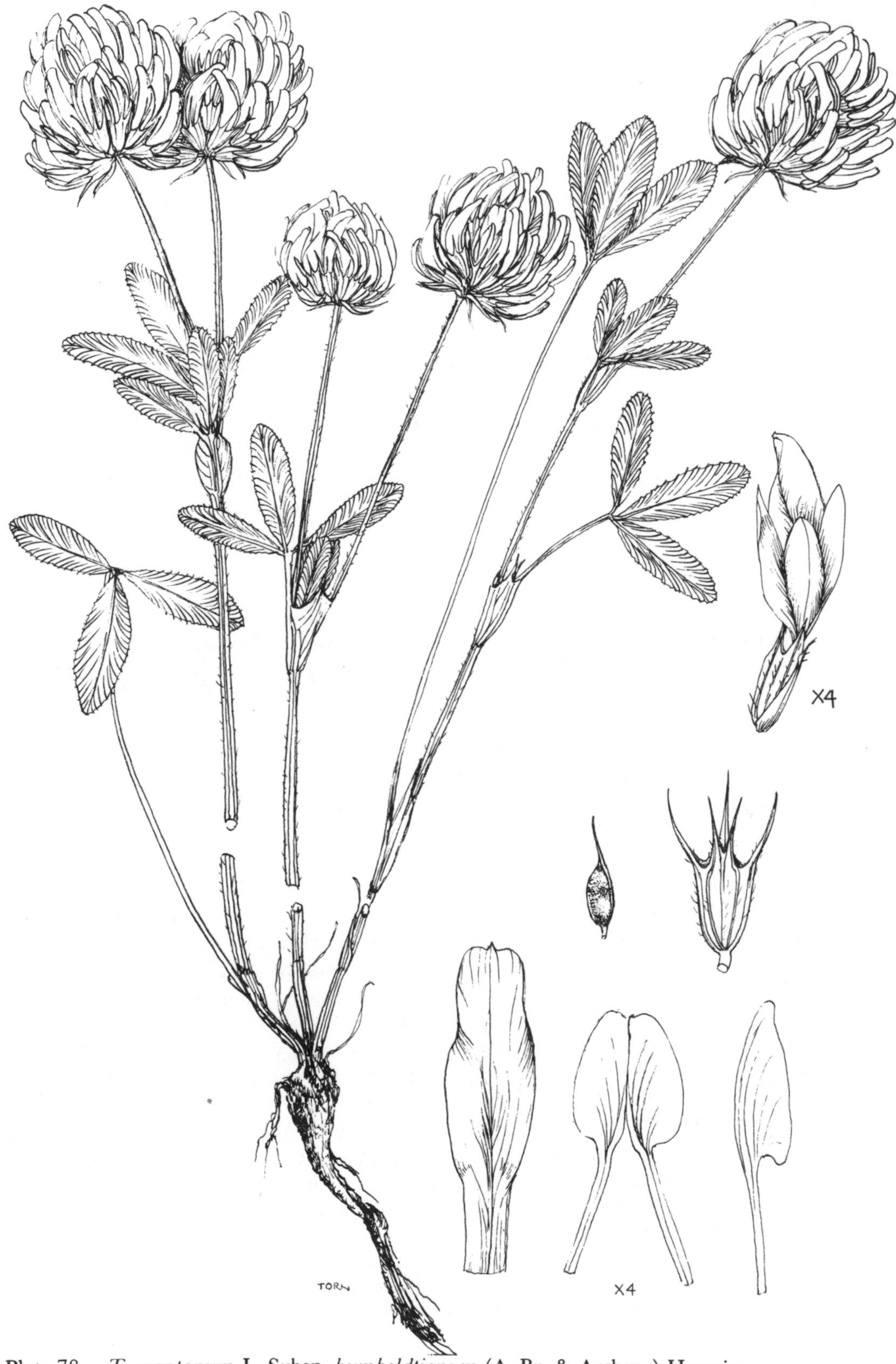

Plate 78. *T. montanum* L. Subsp. *humboldtianum* (A. Br. & Aschers.) Hossain (Turkey: *Zohary* 870228, HUJ).
Plant in flower and fruit; flower with bract; pod; calyx; standard; keel; wing.

Selected specimens: TURKEY: Van, 5 km N. E. of Baskale, 2,150 m, 1966, *Davis* 45941 (K); 3 km N. of Kars, 1,760 m, 1964, *Zohary* & *Plitmann* 2461–21 (HUJ); Erzurum, 33 km N. of Hinis, 1,750 m, 1963, *Zohary* 870228 (HUJ). ARMENIA: Martuni Mts., 1955, *Gabrielian* (HUJ).

80c. T. montanum L. Subsp. **rupestre** (Ten.) Pignatti, Giorn. Bot. Ital. 107 : 217 (1973).

T. rupestre Ten., Prodr. Fl. Nap. 1 : 43 (1811).
T. balbisianum Ser. in DC., Prodr. 2 : 201 (1825).
T. montanum L. var. *rupestre* (Ten.) Fiori in Fiori & Paoletti, Fl. Anal. Ital. 2 : 63 (1900).

Icon.: Reichenb., Ic. Fl. Germ. 22 : t. 109, I 1–7 (1867–89).

Hab.: Mountain slopes.
Gen. Distr.: S. France, N. W. Italy.

Selected specimens: FRANCE: St.-Michel sur Annot, 1885, *Reverchon* 48 (W); Basses-Alpes, Fugeret, 1888, *Reverchon* & *Derbez* 48 (E, FI, K, W). ITALY: Madonna delle Finistre, 1,880 m, 1890, *Bernoulli* (FI); Gende, 1854, *Huet de Pavillon* (G).

81. T. owyheense Gilkey in Madroño 13 : 169 (1956). *Typus*: Oregon: ridge above Sucker-Creek, Malheur Co., May 1954, *B. F. Murphy* (OSC, photo HUJ !, iso.: DAO, DS, GH, NY, ORE, WS).

Icon.: Gilkey, *loc. cit.* f. 1.

Perennial, glaucous, glabrous, up to 20 cm tall. Stems few, spreading, branching at base and slightly above from fleshy, deeply penetrating roots. Leaves long-petioled; stipules 1–2 cm long, fused at base, broadly obovate, slightly lobed; leaflets 1–2 × 0.7–2.3 cm, ovate, obovate to circular, emarginate, slightly overlapping, sparingly dentate at margin. Inflorescences 2.5–4 cm in diam., 20–30-flowered. Peduncles 3–5 cm long, longer than subtending leaves. Bracts very small, cupulate. Pedicels very short to subsessile, only those of lowermost flowers slightly deflexed. Calyx 0.9–1.2 cm long, conical, membranous, pilose; teeth subequal, about as long as tube, the lower one longest, all subulate. Corolla 2–2.3 cm long, pale to dark violet; standard pale violet, tubular for most of its length, longer than dark violet-tipped wings and keel. Ovary finely pilose at apex below long style, 1–2-ovuled. Fl. May.
Hab.: Dry shale hillsides, on blue-grey diatomaceous earth; 400 ft.
Gen. Distr.: Oregon (endemic).

Note: Only seen the phototype and drawing accompanying diagnosis; description based on the original publication and Gillett, Can. J. Bot. 50 : 1988 (1972).

82. T. pichisermollii Gillett in Kew Bull. 7 : 388 (1952). *Typus*: Central Abyssinia: Scioa: Ambossa, ca. 2,300 m, 15 Apr., 1909, *G. Negri* 85 bis (FI !).

Icon. : [Plate 79].

Annual, pubescent, up to 25 cm. Stems many, ascending, branching all along. Petioles of lower leaves up to 3 cm, those of upper ones 0.5 cm; stipules hyaline-membranous, with prominent sometimes purplish nerves, lower adnate part up to 5 mm long, ovate, upper free portion up to 3 mm long, lanceolate-subulate; leaflets 1–2 × 0.4–0.7 cm, oblanceolate, markedly serrate-dentate, rounded or acute at apex, narrowly cuneate at base. Inflorescences about 1.2 × 0.7 cm, almost semi-globular, about 20-flowered, arranged in 2–3 whorls. Peduncles up to 4 cm long. Bracts green, 5(–7) × (1.5–)2–3 mm, 2–3-toothed, 1–3-nerved, ovate-lanceolate, the lower ones sometimes united at base forming involucre up to length of inflorescence. Pedicels up to 1 mm long, not deflexed after anthesis or only deflexed in lower flowers. Calyx about 5 mm long, densely pilose, with 10 prominent nerves; teeth about as long as tube, triangular at base, subulate at tip with thickened margin, dark purple after drying. Corolla 6–8 mm long, purple; standard 2 mm broad, tapering, denticulate at apex, longer than wings and keel. Pod 2 mm long, (1–)2-seeded, papery, oblong, with slightly thickened margin. Seeds about 1.5 cm in diam., flattened, brown. Fl. April–September. 2n = 16.

Hab. : Mountains; up to 2,700 m.

Gen. Distr. : Central Ethiopia (endemic).

Specimens seen : ETHIOPIA : Scioa : Ambò-Uorkà, 2,350 m, 1938, *Borini* 23 (FI); Mulu, N. of Col. Sandford's Farm, 8,000 ft, 1967, *Bally* B 12969 (G, K); 64 km N.E. of Addis-Ababa on Dessie road, 2,700 m, 1953, *Gillett* 14814 (K); plants grown from seeds : the road Addis-Bahar, 2,700 m, 1967, *Zohary* & *Ladijinsky* (HUJ).

Note : The size of the bracts and the extent to which they are united at the base are varying characters in *T. pichisermolli* : the specimen *Gillett* 14814 is a non-bracteate form (other ebracteate specimens are cited by Gillett, *op. cit.*); the lower bracts of the plants of *Zohary* & *Ladijinsky* form an involucre about as long as or even longer than the inflorescence.

83. T. rueppellianum Fresen. in Flora 22 : 51 (1839).

Annual, glabrous, 20–60 cm. Stems erect or less often prostrate, branched, grooved. Lower leaves on 5 cm long petioles, upper ones very short-petioled to sessile; stipules about 1.5 cm, free portion one-third of length, lanceolate, acuminate; leaflets 1–2.4 × 0.8–1.4 cm, ovate or oblong to cuneate-obovate, rarely broadly lanceolate, rounded or truncate, rarely acute, sometimes emarginate, margin dentate. Inflorescences about 1.5 cm in diam., globose, 15–30-flowered. Peduncles much longer than subtending leaves, pilose especially above. Bracts up to 3 mm long, linear or lanceolate, sometimes truncate or bifid. Pedicels up to 2 mm long, glabrous or pilose, sometimes those of lower flowers slightly deflexed after anthesis. Calyx 3–6 mm, glabrous or very sparingly hairy at margin; tube with 11 prominent nerves; teeth as long as to more than twice as long as tube, subulate. Corolla 4–8 mm long, purple or rarely white; standard oblong, rounded at apex, longer than wings and keel. Pod 3–4 mm long, oblong, glabrous, 2–3-seeded. Seeds about 1 mm, ovoid, brown. Fl. May–October. 2n = 16.

Plate 79. *T. pichisermollii* Gillett
(Ethiopia : *Bally* B12969, G).
Plant in flower; flower.

83a. T. rueppellianum Fresen. Var. **rueppellianum.** *Typus* : N. Ethiopia, Simen (Simien), *Rueppell* (FR holo.).

T. subrotundum A. Rich., Tent. Fl. Abyss. 1 : 172 (1847), pro majore parte.
T. rueppellianum Fresen. var. *preussii* (Bak. f.) Gillett in Kew Bull. 7 : 389 (1952).
T. preussii Bak. f., Leg. Trop. Afr. 82 (1926).

Icon. : [Plate 80]. Fröman, Ill. Guide Past. Leg. Ethiopia, Rur. Dev. Stud. 3 : pl. 67 (1975).

Leaflets up to 2.4 cm long, cuneate-obovate or cuneate-ovate, rounded at apex or rarely truncate or retuse. Pedicels up to 2 mm long. Calyx tube 1.5–2.2 mm long. Standard 5–8 mm long.

Hab. : Upland grassland, moorland, tracks through upland forest and as a weed in cultivated areas, usually in rather wet places; 1,600–3,600 m.

Gen. Distr. : Uganda, Kenya, Tanganyika, Fernando Po, E. Congo, Sudan, Eritrea, Ethiopia, Cameroun.

Selected specimens : KENYA : Trans Nzoia Distr., Mt. Elgon, Track S.W. of Suam Sawmills, 2,700 m, 1967, *Gillett* 18411 (HUJ). TANGANYIKA : Masai Distr., Ngorongoro crater, 1960, *Bally* B 12130 (HUJ); Mbulu, 1,685 m, 1926, *Peter* 4394 (B). ERITREA : Scimezana, altipiano di Gheleba, 1902, *Pappi* 340 (W). ETHIOPIA : Shoa Prov., Entotto Mt., Addis Ababa. above Italian Embassy, 8,200 ft, 1964, *Meyer* 8599 (FI, NA); 20 km N. of Gondar, 2,900 m, 1969, *D. Zohary* (HUJ). CAMEROUN: Kamerun Mt., *Preuss* 972 (W, iso. of *T. preussii* Bak. f.)

83b. T. rueppellianum Fresen. Var. **minimiflorum** Gillett in Kew Bull. 7 : 390 (1952). *Typus* : Tanganyika, near Kilimanjaro, Machame (Mashami), *Haarer* 353 (K ! holo.).

Leaflets up to 1 cm long, rounded at apex. Pedicels 1 mm long, pilose. Calyx tube about 1.2 mm long. Standard about 3–4 mm long. Cleistogamous and self-fertile.

Hab. : Grassland; 1,000–1,800 m.

Gen. Distr. : Tanganyika (endemic).

Note : Seen only the type specimen.

84. T. spananthum Thulin, Bot. Notiser 129 : 167 (1976). *Typus* : Ethiopia : Arussi Prov., Chilalo awraja, near Bekoji, ca. 60 km S. of Asella, 2,700 m, 5.10.1971, *M. Thulin* 1355 (UPS holo., iso. : BR, EA, ETH, FI, HUJ !, K, MO, WAG).

Icon. : [Plate 81]. Thulin, *loc. cit.* f. 1A-E (1976).

Perennial, glabrous, 3–20 cm. Stems many, prostrate or ascending, branching from taproot, often mat-forming. Leaves (even upper ones) on up to 4 cm long petioles; stipules up to 1.5 cm long, adnate to petioles for two-thirds of their length, often

Plate 80. *T. rueppellianum* Fresen. Var. *rueppellianum* (Kenya : *Gillett* 18411, HUJ).
Plant in flower; pod (Ethiopia : *D. Zohary*, HUJ); flowers.

Plate 81. *T. spananthum* Thulin
(Ethiopia : *M. Thulin* 1355, HUJ).
Flowering and fruiting branch; flower; pod.

reddish-nerved, free portion abruptly contracted to fine point; leaflets 1–2 × 0.4–0.8 cm, elliptical or obovate, cuneate at base, acute, rounded, truncate or slightly emarginate, dentate-serrate at margin. Inflorescences about 1–1.5 cm in diam., 1–6-flowered. Peduncles up to 3 cm long. Bracts about 1 mm long, oblong, usually bifid. Pedicels up to 1.5 mm long, erect after anthesis. Calyx 4–6 mm long, somewhat inflated in fruit; tube campanulate, 9–10-nerved; teeth about as long as tube, narrowly triangular, with subulate tips, sparsely hairy, margin scarious. Corolla 0.6–1.1 cm long, purple; standard oblong, broadest above middle, longer than wings and keel. Ovary 4–5-ovuled, sparsely hairy or papillose. Pod 4–6 × 3 mm, oblong, glabrous, 3–5-seeded, sutures prominent, thickened. Seeds about 2 mm, ovoid, brown, usually mottled with purple, smooth. Fl. October. 2n = 16.

Hab.: Grassland, often in heavily grazed pastures or on disturbed or base ground along tracks; 2,300–3,900 m.

Gen. Distr.: Ethiopia (endemic).

Note: Only seen the isotype specimen.
According to Thulin (1976), *T. spananthum* is very similar to *T. tembense*. It differs from the latter, firstly, by its perennial and mat-forming habit (though certain specimens of the annual *T. tembense* sometimes root from the nodes and the plant *Gillett* no. 18466/A from Kenya, Mt. Elgon, seems to be mat-forming). Its upper leaves are always long-petioled, while in *T. tembense* the upper leaves are subsessile and sometimes the petioles are fused with the stipules. The tube of its calyx is 9–10-nerved instead of 11-nerved as is usual in *T. tembense*. Its standard is broadest above the middle and straight, while that in *T. tembense* is oblong, truncate and curved downward at the tip.

85. T. tembense Fresen. in Flora 22 : 51 (1839). *Typus*: N. Ethiopia, Tembien Prov., *Rueppell* (FR, holo.).

T. umbellulatum A. Rich., Tent. Fl. Abyss. 1 : 172 (1847).
T. goetzenii Taub. in von Goetzen, Durch Afrika von Ost nach West, 376–8 (1895).
T. calocephalum Fresen. var. *parviflorum* Chiov. in Ann. Ist. Bot. Rom. 8 : 405 (1908).
T. subrotundum Harms in Engl., Pflanzenwelt Afrikas 3(1) : 568 (1915).

Icon.: [Plate 82]. Fröman, Ill. Guide Past. Leg. Ethiopia, Rur. Dev. Stud. 3 : pl. 73 (1975).

Annual, glabrous or subglabrous, 20–50 cm. Stems many, erect or spreading, sometimes rooting at lower nodes, striate, branching. Leaves on up to 5 cm long petioles, upper ones subsessile; stipules about 1.5 cm long, adnate to petioles for two-thirds of their length, free portion ovate, cuspidate; leaflets 1–2 × 0.6–1 cm, elliptical or obovate, cuneate at base, acute, rounded at apex or truncate to slightly emarginate, margin dentate. Inflorescences about 2.5 cm in diam., globose, 3–10(–15)-flowered. Peduncles up to 7 cm long, longer than subtending leaves. Bracts about 2 mm long, oblong to broadly ovate, truncate to bifid. Pedicels up to 3 mm long, glabrous, erect after anthesis. Calyx 5–7 mm long, sparingly pilose at margin; tube campanulate, with 11 prominent nerves; teeth up to twice as long as tube, gradually tapering from broad triangular base, margins broadly scarious. Corolla about 1 cm, purple; standard oblong, truncate, recurved at tip, longer than wings and keel. Pod 5–8 × 3 mm,

Plate 82. *T. tembense* Fresen.
(Ethiopia : *Chiovenda* 1496, FI).
Plant in flower and fruit; fruiting calyx with pod; fruiting calyx with open pod.

slightly longer than calyx teeth, glabrous, opening along prominent sutures, oblong, 2–4-seeded. Seeds about 2 mm, ovoid, brown, flattened. Fl. April–September. 2n = 16.

Hab.: Wet places in upland grassland, forest, moorland and alpine zones – 2,000–3,800 m; rarely near water – as low as 1,400 m.

Gen. Distr.: Uganda, Kenya, Tanganyika, E. Congo, Ethiopia, Eritrea.

Selected specimens: KENYA: Trans Nzoia Distr., Mt. Elgon, Path W. of Kipsare Hill, 3,300 m, 1967, *Gillett* 18466/A (HUJ). EHTIOPIA: Chilalo awraja, near Bekoji, ca. 60 km from Asella, 2,700 m, 1971, *Thulin* 1355 (HUJ); Shoa Prov., Entotto Mt., Addis-Ababa, above Italian Embassy, 8,200 ft, 1964, *Meyer* 8600 (NA); Amhara-Dembia, Gondar, 1909, *Chiovenda* 1496 (FI). ERITREA: Hamasien, Asmara, 2,300 m, 1938, *Zenmaro* 44/3 (G).

Series ii. MACROCHLAMIS Heller & Zoh. (ser. nov.)*

Type species: *T. parryi* A. Gray.

Perennials, glabrous or pubescent to villous. Stems always caespitose or creeping, rooting at nodes, branched from large taproot, covered by persistent stipules. Inflorescences subtended by row of conspicuous and large bracts forming involucre; bracts sometimes slightly united at base, upper ones smaller. Three N. and S. American species.

86. T. dasyphyllum Torr. & Gray, Fl. N. Am. 1: 315 (1838).

Perennial, caespitose-matted, pubescent, 5–20 cm. Taproot thick, woody, deeply penetrating. Stems many, short, leafy, covered with persistent stipules. Leaves short- or long-petioled; stipules adnate to petioles for more than half of their length, scarious, often purple, glabrous or pubescent in linear, green, free portion; leaflets 0.6–3 × 0.2–0.6 cm, sometimes conduplicate, linear to narrowly elliptical or oblanceolate, sometimes obovate, rounded to acute, margins entire. Inflorescences 1.5–3.5 cm in diam., scape-like, subglobose, 5–20-flowered. Peduncles longer than leaves. Lower bracts 2–6 mm long, forming involucre, scarious, lanceolate to subulate, sometimes united at base, upper ones reduced to entire or toothed scales. Pedicels up to 2 mm, erect. Calyx 6–9 mm long, pubescent; tube campanulate, 10-nerved; teeth up to twice as long as tube, unequal, subulate. Corolla 1.2–1.6 cm long, one-coloured (violet or rarely white) or two-coloured (standard cream to pale violet), with or without purple tips; wings and keel purple-tipped; standard broadly elliptical, ovate or oblong, rounded at apex, apiculate. Pod about 6 mm long, oblong, appressed-pubescent, 1–4-seeded. Seeds dark brown. Fl. June–August.

* See Appendix.

Key to Subspecies (after Gillett, 1965)

1. Leaflets oblong-elliptical, occasionally somewhat oblanceolate, acute. Flowers two-coloured; standard and wings cream to pale violet; at least keel tips and frequently entire keel strong purplish-violet. **86a.** subsp. **dasyphyllum**
– Leaflets tending to oblanceolate, sometimes markedly obovate, obtuse. Flowers usually uniformly reddish-violet or violet-purple or if standard is pale, then keel is violet-purple 2
2. Upper surface of leaflets densely appressed, silvery-pubescent. Diameter of inflorescence 1.8–2.8 cm. **86b.** subsp. **anemophilum**
– Upper surface of leaves pubescent to glabrate. Diameter of inflorescence 1.2–2.2 cm. **86c.** subsp. **uintense**

86a. T. dasyphyllum Torr. & Gray Subsp. **dasyphyllum.** *Typus*: Rocky Mts., *Dr. James* (NY).

T. lividum Rydb., Bull. Torrey Bot. Club 30 : 254 (1903).

Icon.: [Plate 83]. F. E. & E. S. Clements, Rocky Mt. Fl. t. 28 (1914).

Hab.: Alpine meadows and rocky slopes; 3,000–4,000 m. 2n = 16, 24.
Gen. Distr.: USA: Colorado, Wyoming.

Selected specimens: COLORADO: Pikes Peak Region, 12,000 ft, 1891, *Penard* 189 (W); *ibid.*, Bald Mt., 12,000 ft, 1920, *Johnston* 2536 (HUJ). WYOMING: Albany Co., Medicine Bow Peak, 12,013 ft, 1968, *Mitchell* Mt-174 (HUJ); *ibid.*, Libby Creek, 10,000 ft, 1935, *Louis Williams* 2395 (H, HUJ); Fremont Co., Dickinson Peak, 9,500 ft, 1965, *Scott* 500 (HUJ).

86b. T. dasyphyllum Torr. & Gray Subsp. **anemophilum** (Greene) Gillett in Brittonia 17 : 125 (1965). *Typus*: Wyoming: Foothills southeast of Laramie, 25 May, 1894, *A. Nelson* 68 (ND-G, photo HUJ !, iso.: NY,US).

T. anemophilum Greene, Pittonia 4 : 137 (1900).
T. scariosum A. Nels., Bull. Torrey Bot. Club 29 : 401 (1902)
T. dasyphyllum Torr. & Gray var. *anemophilum* (Greene) Isley, Brittonia 32 : 55 (1980).

Icon.: McDerm., N. Am. Sp. Trif. pl. 63 (1910).

Hab.: As above, 2,000–2,800 m. 2n = 16.
Gen. Distr.: Wyoming.

Specimens seen: WYOMING: Albany Co., Laramie Hills, 1903, *Nelson* 8944 (E); Sherman Hill, 8,500 ft, 1935, *Louis Williams* 2232 (H, HUJ).

86c. T. dasyphyllum Torr. & Gray Subsp. **uintense** (Rydb.) Gillett, *loc. cit.* *Typus*: Uintaha, N. Utah, 11,000–12,000 ft, Aug. 1869, *S. Watson* 241 (NY, iso.: UC, US).

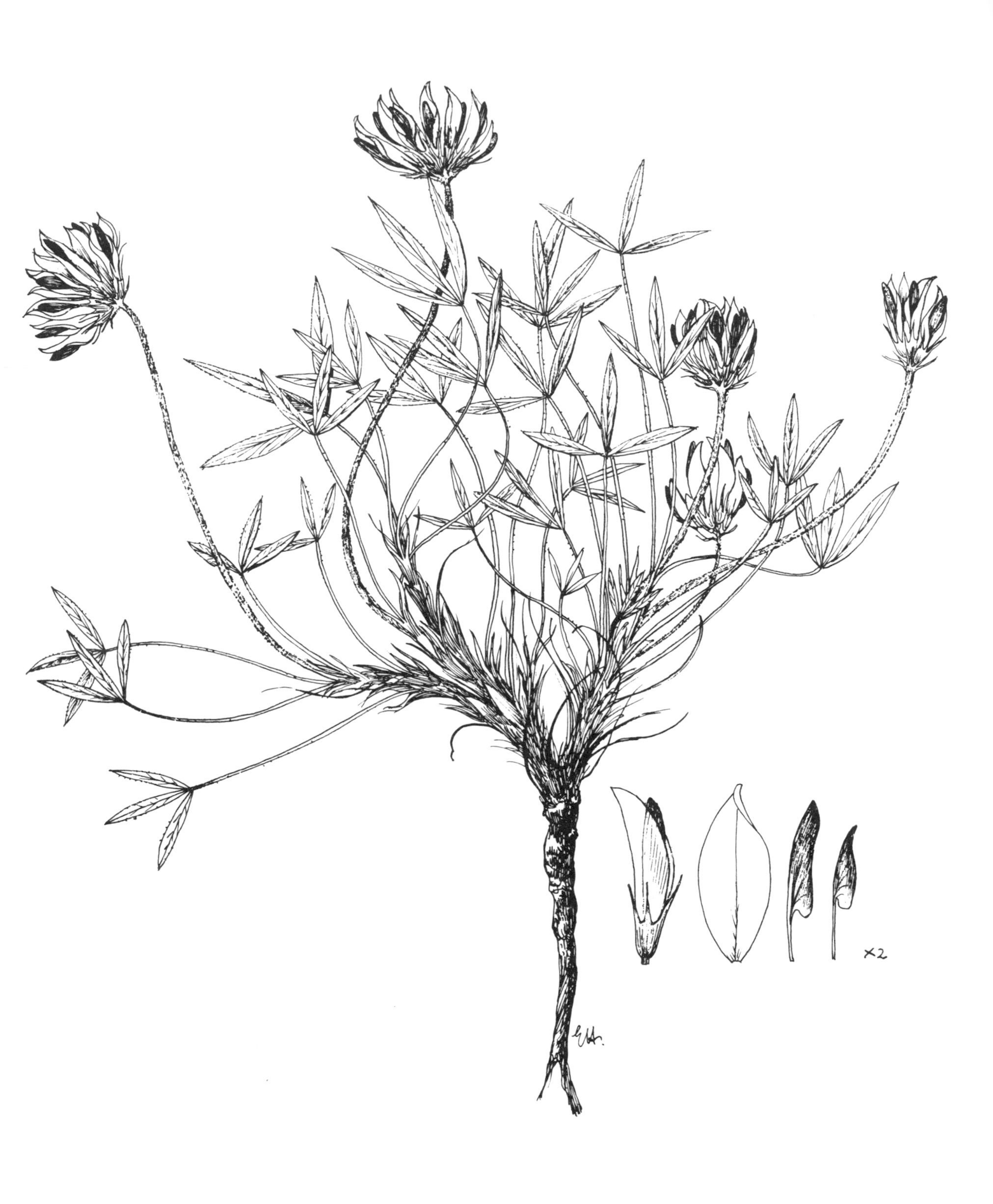

Plate 83. *T. dasyphyllum* Torr. & Gray Subsp. *dasyphyllum*
(Wyoming : *Mitchell* Mt.- 174, HUJ).
Plant in flower; flower; standard; wing; keel.

T. uintense Rydb., Bull. Torrey Bot. Club 34 : 47 (1907).
T. dasyphyllum Torr. & Gray f. *uintense* (Rydb.) McDerm., Ill. Key N. Am. Sp. Trif. 168 (1910).
T. dasyphyllum Torr. & Gray var. *uintense* (Greene) Martin ex Isley, Brittonia 32 : 56 (1980).

Hab. : As above. 2n = 16.
Gen. Distr. : Montana, Utah, Wyoming, Colorado.

Selected specimens : MONTANA : Carbon Co., Beartooth Pass, 10,300 ft, 1968, *Mitchell* Mt.-179 (HUJ); Carbon Co., Cook City, Red Lodge Highway, 1947, *Hitchcock* 16605 (NA, UC). UTAH : San Juan Co., Horsehead Peak, Abajo Mts., 11,000 ft, 1942, *Ellison* 842-70 (NA); *ibid.*, La Sal Mts., summit of Mt. Pool, 13,089 ft, 1933, *Maguire* et al. 21295 (NA, UC). COLORADO : Park Co., Mt. Bross above Alma, 13,000 ft, 1940, *Ewan* 12569 (NA).

Note : Gillett (1965) pointed out the overlapping of the diagnostic characters of the subspecies of *T. dasyphyllum* in some populations. In certain specimens studied by us, the combination of characters is intermediate between the subspecies (e.g., the inflorescence of the specimen of subsp. *uintense* – *Hitchcock* 16605 – resembles that of subsp. *dasyphyllum*).

87. T. parryi A. Gray, Am. Jour. Sci. Arts II, 33 : 409 (1862).

Perennial, glabrous, often with scattered brownish scales or with hairs especially on peduncles, 10–20 cm. Stems with long rhizomes; aerial stems few, scape-like, leafy only at base. Leaves almost all basal, long-petioled; stipules crowded at base of stem, persistent, obliquely ovate, membranous, obtuse or acute, often coloured, some of those from previous years brown and withered; leaflets 1–4 × 0.5–1.5 cm, broadly elliptical to obovate or oblong, entire to dentate at margin, rounded to acute at apex. Inflorescences 1.5–3(–3.7) cm across, (4–)10–20(–30)-flowered, subglobular to globular. Peduncles long, up to 20 cm, much longer than leaves. Bracts half the length of to longer than calyx, purplish, scarious, oblong, acute or obtuse, sometimes bifid or irregular, 0.6–1 cm wide, often united in pairs, outer, lower ones forming involucre, upper ones smaller, free. Pedicels 1–2 mm long, erect. Calyx 6–8 mm long, scarious; tube reddish, (5–)6–9-nerved, campanulate, glabrous to hairy; teeth unequal, the upper 2 shorter than the lower ones, from half to twice the length of tube, subulate to triangular-lanceolate. Corolla 1–2(–2.2) cm long, dark purple; standard up to 9 mm wide, broadly elliptical or oblong, with apex rounded or rarely retuse or acute, longer than wings and keel. Pod about 7 mm, fusiform, oblong, short-stipitate, glabrous, 1–4-seeded. Seeds 2 mm long, smooth. Fl. June–August.
Hab. : Alpine tundra and meadows above timberline, on grassy slopes and in open areas of subalpine or montane forest; 9,000–13,500 ft.

Key to Subspecies

1. Inflorescences longer than wide, 3–3.7 cm long. Leaves thick and fleshy; stipules 1.5–2 cm long. Peduncles thickened. **87c.** subsp. **salictorum**
– Inflorescences globular, less than 3 cm in diam. Leaves fleshy but thin; stipules 0.5–1 cm long. Peduncles usually slender, sometimes thickened 2

2. Involucral bracts large and acute. Inflorescences 2–2.9 cm long. Flowers 1.4–2.2 cm long. **87a.** subsp. **parryi**
– Involucral bracts shorter, obtuse. Inflorescences 1.4–2.4 cm long. Flowers 1.2–1.7 cm long. **87b.** subsp. **montanense**

87a. T. parryi A. Gray Subsp. **parryi.** *Typus* : Rocky Mountains, Colorado Territory, 1862, *Parry* 178 (GH, iso. : G !, K !, MO).

Icon. : [Plate 84]. Hitchcock et al., Vasc. Pl. Pac. N. W. 3 : 368 (1961).

Hab. : As above. 2n = 16.
Gen. Distr. : USA : Colorado, New Mexico, Wyoming.

Selected specimens : COLORADO : Larimer Co.–Jackson Co. Line, Cameron Pass, summit of P 52, 11,852 ft, 1968, *Mitchell* Mt-176 (HUJ); Argentine Pass, 12,000 ft, 1878, *Jones* 419 (BM, NA). NEW MEXICO : Mora Co., Pecos Baldy, 1908, *Standley* (UC). WYOMING : Albany Co., Medicine Bow Peak, 11,400 ft, 1968, *Mitchell* Mt -174 (HUJ); *ibid.*, 10,000 ft, 1960, *Porter* & *Marjorie W. Porter* 8194 (H).

87b. T. parryi A. Gray Subsp. **montanense** (Rydb.) Gillett in Brittonia 17 : 132 (1965). *Typus* : Montana : Old Hollowtop, near Pony, 9,000 ft, 7 July, 1897, *Rydberg* & *Bessey* 4461 (NY, iso. : E !, GH, K !, MONT).

T. montanense Rydb., Mem. N. Y. Bot. Gard. 1 : 263 (1900).
T. inaequale Rydb., Bull. Torrey Bot. Club 34 : 47 (1907).
T. parryi A. Gray var. *montanense* (Rydb.) Isley, Brittonia 32 : 57 (1980).

Hab. : As above.
Gen. Distr. : USA : Idaho, Montana, Utah, Wyoming.

Selected specimens : MONTANA : Carbon Co., Beartooth Mts., 24 mi. S. of Red Lodge, 10,000 ft, *Cronquist* 8046 (H). UTAH : Summit Co., Bald Mountain Pass, 10,678 ft, 1968, *Mitchell* Mt-182 (HUJ). WYOMING : Park Co., Beartooth Lake, 8,960 ft, 1968, *Mitchell* Mt-180 (HUJ); Albany Co., Brookline Lake, 10,500 ft, 1935, *Williams* 2393 (H).

87c. T. parryi A. Gray Subsp. **salictorum** (Greene ex Rydb.) Gillett, *loc. cit.* *Typus* : Carson, Colorado, *Baker* 307 (ND, iso. : DS, K !, MO, UC).

T. salictorum Greene ex Rydb., Fl. Colo. 202 (1906) nom. nud.; Rydb., Fl. Rocky Mts. and Adj. Plains 475 (1917).

Icon. : McDerm., N. Am. Sp. Trif. pl. 70 (1910).

Hab. : As above. 2n = 32.
Gen. Distr. : Colorado.

Specimens seen : COLORADO : Twin Lakes, 1873, *Wolf* 3878 (R); Marshall Pass, 10,000 ft, 1901, *Baker* (UC).

Plate 84. *T. parryi* A. Gray Subsp. *parryi*
(Wyoming : *Mitchell* Mt.- 174, HUJ).
Plant in flower; young pod (Colorado : *Baker*, UC);
flower with bract; dissected calyx.

Note: We have here adopted the subdivision of *T. parryi* as proposed by Gillett (1965). Subsp. *salictorum* appears to be a well-defined subspecies, perhaps a spontaneous tetraploid form of subsp. *parryi* with which it is sympatric. The delimitation of the other two subspecies, which are geographically separated, is based on quantitative characters, not always making possible a sharp distinction.

88. T. riograndense Burkart in Darviniana 3 : 421 (1939).

Perennial, pubescent to villous, 10–30 cm. Stems rhizomatous, creeping, rooting and branching from nodes. Leaves very long-petioled, arising from creeping stems; stipules 4–8 mm long, oblong-lanceolate, cuspidate, adnate to petioles for up to half their length; leaflets 0.5–2.2 × 0.4–1.8 cm, obovate to obcordate, cuneate at base, deeply notched to emarginate at apex, densely dentate-serrate. Inflorescences 1–1.5 cm across, semi-globular to globular, 20–40(–70)-flowered. Peduncles axillary, erect, much longer than petioles. Bracts hyaline, outer ones foliaceous, many, forming involucre up to 1 cm long and up to 6 mm wide, free or irregularly connate, ovate to elliptical, acute to acuminate, inner ones gradually decreasing in size, the innermost being about 3 mm long, lanceolate. Pedicels about 2 mm long, not deflexed in fruit. Calyx about 6 mm long, membranous, appressed-villous; tube campanulate; teeth longer than tube, subequal, triangular, lanceolate to triangular. Corolla 6–8 mm long, pink to purple or white; standard obovate-oblong, broadly rounded at apex, longer than wings and keel. Pod about 3 mm long, included in calyx, obovoid-oblong, 1–2-seeded, glabrous, brown. Seeds 1–1.3 mm, yellowish. Fl. November–February. 2n = 16.

Key to Subspecies

1. Inflorescences semi-globular, 20–40-flowered. Outer bracts almost free. Calyx about 6 mm long; teeth lanceolate to triangular, acute. **88a.** subsp. **riograndense**
– Inflorescences globular, 40–70-flowered. Outer bracts irregularly connate. Calyx 4–4.5 cm long; teeth triangular, acuminate. **88b.** subsp. **pseudocaliculatum**

88a. T. riograndense Burkart Subsp. **riograndense**. *Typus*: Brasil: Rio Grande do Sul, Tupaceretàn, XII, 1934, *A.A. de Araujo* 151 b (SI, iso. HUJ !).

Icon.: [Plate 85]. Del Puerto, *loc. cit.* 3, f. M, N.

Hab.: Meadows.
Gen. Distr.: Brazil, Argentina, Uruguay.

Specimens seen: BRAZIL: Santa Catarina, Campos dos Curitilanos, 1876, *Müller* 191 et 1993 (R); *ibid.*, Mun. Lajes, between Palmeiras and Lajes, 800–900 m, 1956, *Smith* & *Klein* 8056 (MEXU); Rio Grande do Sul, 1945, *Schultz* 4 (SI). ARGENTINA: B. de Irigoyen, Misiones, 1958, *Gamerro* & *Toursarkissian* 101 (SI).

88b. T. riograndense Burkart Subsp. **pseudocaliculatum** Del Puerto in Com. Bot. Mus. Hist. Nat. Montevideo 55(4) : 1 (1972). *Typus*: Quebrada de los

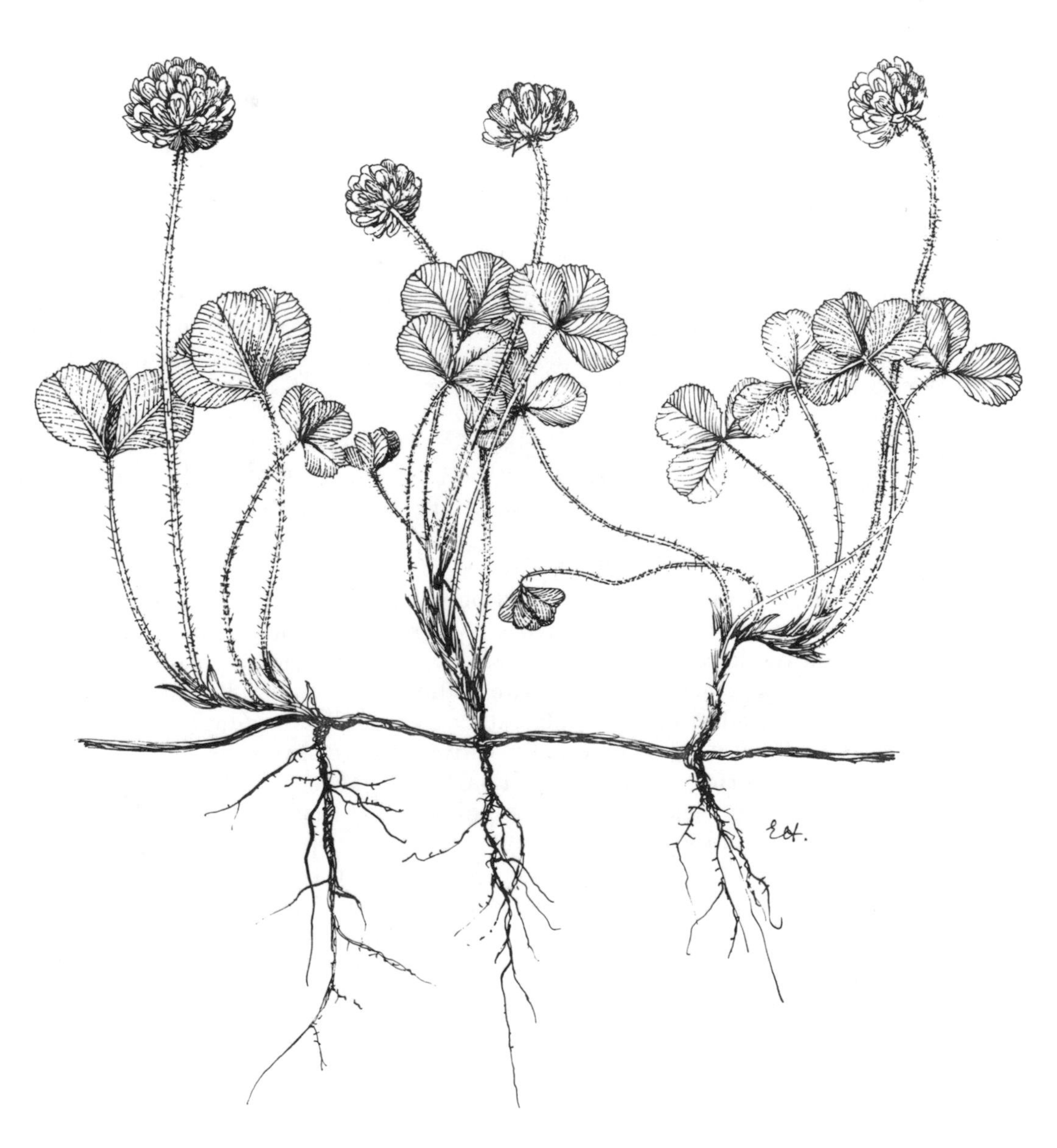

Plate 85. *T. riograndense* Burkart Subsp. *riograndense*
(Brazil : *Müller* 191, R).
Plant in flower; flower with bract; flower.

Cuervos, Dpto. Treinta y Tres, Uruguay, 8.12.1966, *Rosengurtt, Izaguirre* & *Del Puerto* 10213 (MVFA).

Icon. : Del Puerto, *loc. cit.* 3, f. A–L

Hab. : Low inundated meadows.
Gen. Distr. : Uruguay.

Note : No specimen seen.

Series iii. PHYLLOCEPHALA Heller & Zoh. (ser. nov.)*

Type species : *T. andinum* Nutt.

Perennial, appressed-pubescent. Stems densely caespitose, covered with persistent stipules. Inflorescences axillary, usually formed of 2 heads, sessile, subtended by membranous uppermost stipules and leaves, thus forming pseudo-involucre. One N. American species.

89. T. andinum Nutt. in Torr. & Gray, Fl. N. Am. 1 : 314 (1838). *Typus* : Rocky Mountains (?), *Th. Nuttall* (BM !, iso. : GH. K !, NY, PH).

Icon. : [Plate 86]. McDerm., N. Am. Sp. Trif. pl. 71 (1910).

Perennial, densely caespitose, appressed-pubescent, 5–15 cm. Stems many, erect or ascending, lower part densely covered with persistent membranous stipules. Leaves long-petioled, the lower ones almost sessile with petioles adnate to stipules; stipules membranous, many-nerved, fused with petioles for over half their length, oblong-lanceolate; leaflets 0.6–1.8 × 0.2–0.5 cm, sometimes conduplicate, elliptical to oblanceolate, acute, apiculate, usually entire. Inflorescences usually formed of 2 heads, 1–2 cm in diam., semi-globular to almost globular, 15–25-flowered, sessile in axils of uppermost leaves; upper stipules broad-ovate, membranous, forming involucre-like structure. Bracts minute, scarious, cupulate to varying degrees. Pedicels up to 1 mm. Calyx 6–9 mm long, villous, membranous; tube tubular-campanulate, usually with 10 prominent nerves; teeth 1–1.5 times as long as tube, subulate. Corolla 1–1.5 cm; standard violet, oblong, rounded or truncate at apex, as long as or slightly longer than ochroleucous keel and wings. Pod small, scarious, oblong, 1–2-seeded, densely villous towards apex. Seeds 2–3 mm, smooth or minutely roughened. Fl. May–July. 2n = 16.
Hab. : Dry rocky and sandy hillsides.
Gen. Distr. : USA : Wyoming, Utah, Arizona.

Selected specimens : WYOMING : Uinta Co., Junction of Hams Fork and Black Fork Rivers, S.E. of Lyman, 6,500 ft, 1937, *Rollins* 1663 (G); Fremont Co., Beaver Hill, 6,100 ft, 1955,

* See Appendix.

Plate 86. *T. andinum* Nutt.
(Wyoming : *Rollins* 1663, G).
Plant in flower; flower.

Porter 6665 (W); W. of Fort Bridger, Highway, 1963, *Gillett* & *Mosquin* 12006 (HUJ). UTAH : Daggett Co., Flaming Gorge, S.E. of Manila, *Rollins* 2277 (US).

Note : The involucre-like structure formed by the uppermost stipules is also present in two other species of Sect. *Trifolium* : *T. cherleri* L. and *T. hirtum* All.

Series iv. MICRANTHEUM (C. Presl) Heller & Zoh. (stat. nov.)

Genus *Micrantheum* C. Presl, Symb. Bot. 1 : 47 (1830).
Section *Micrantheum* (C. Presl) Gib. & Belli, Mem. Accad. Sci. Torino ser. 2, 41 : 197 (1891).
Subsection *Micrantheum* (C. Presl) Čelak. in Oesterr. Bot. Zeitschr. 24 : 41 (1874).

Type species : *T. glomeratum* L.

Annuals, glabrous or very sparingly hairy. Inflorescences axillary, sessile or almost so. Calyx 10(–12)-nerved; teeth almost equal, recurved in fruit. Pod sessile to subsessile, 2-seeded, included in calyx. Three Eurasian species.

90. T. glomeratum L., Sp. Pl. 770 (1753). *Typus* : Described from England, Hort. Cliff. 373, 3 (BM !).

Micrantheum glomeratum (L.) Presl, Symb. Bot. 1 : 48 (1830).
T. duodecimnerve Losc. ex Willk. & Lange, Prodr. Fl. Hisp. 3 : 357 (1877).
T. axillare Phil., Anal. Univ. Santiago 84 : 10 (1894).

Icon. : [Plate 87]. Bonnier, Fl. Compl. Fr. Suisse & Belg. 3 : t. 135 (1914).

Annual, glabrous, 10–30 cm. Stems slender, few, branching, procumbent or decumbent to ascending. Leaves varying from long-petioled to sessile; stipules about 1 cm long, ovate, white, membranous, longitudinally nerved, adnate to petioles for half their length, free portion ovate with subulate-setaceous tip; leaflets 0.6–1.5 × 0.4–1 cm, obovate, cuneate at base, retuse, spinulose-serrulate. Inflorescences 0.8 (–1) cm in diam., axillary, globular, dense, many-flowered, subtended by 1 or 2 opposite leaves. Peduncles absent. Bracts membranous, lanceolate, acuminate, up to one-quarter the length of calyx tube. Pedicels very short to absent. Calyx 3–4 mm long, tubular-obconical, with 10 (–12) prominent nerves; teeth about half the length of tube, triangular-ovate, with 3 anastomosing nerves, somewhat auriculate at base, spreading or recurved in fruit, terminating in spinulose awn. Corolla 6–8 mm long, pink; standard oblong-ovate, folded; wings long-clawed. Pod 2–3 mm long, membranous, obovoid with oblique mucro, 1–2-seeded. Seeds about 1 mm, reniform, brown, tuberculate. Fl. May–June. 2n = 14, 16.

Hab. : Fallow fields, dry river banks and among rocks.

Gen. Distr. : E. England, Germany, Portugal, Spain, S. France, Corsica, Sicily, Italy, Albania. Yugoslavia, Bulgaria, Greece, Aegean Islands, Crete, Turkey, Cyprus, Israel, Iran, Caucasus, Azores, Canary Islands, Algeria, S. Africa, Chile (intr.?).

Plate 87. *T. glomeratum* L.
(Corsica : *Reverchon*, HUJ).
Plant in flower and fruit; pod; flowers.

Selected specimens: SPAIN: Nalon prope Penâflor, 1835, *Durieu* 354 (G); Fuente El Saz Madrid, 1964, *Biundia* 213 (HUJ). FRANCE: Ilha du Titan, *Montagne* 3287 (R); Montpellier, Doscaries, 1930, *Eig* (HUJ). CORSICA: Bonifacio, 1880, *Reverchon* (HUJ). GREECE: Evia, 56 km from Loutra to Halkis, 1975, *Danin* & *Heyn* 75632 (HUJ). TURKEY: Istanbul, Alemitehir, Adampol, 1896, *Aznavour* 673 (G); Sinop, Ince Burun, 1966, *Tobey* 1618 (K). CYPRUS: Platres, 3,750 ft, 1938, *Kennedy* 1281 (K). ISRAEL: Golan, clearing in forest near Mas'ada, 1968, *Zohary* (HUJ). IRAN: Lahijan, Guilan, 1937, *Lindsay* 841 (G). CANARY ISLANDS: Teneriffe, 1845, *Bourgeau* 336 (G). S. AFRICA: Kapstadt, Sea Point, 1913, *Peter* 51291 (B). ALGERIA: Maison-Caroie, 1880, *Meyer* 78 ter (G). CHILE: Santiago, La Reina, Santiago, 1951, *Frödin* 28 (BM).

91. T. pachycalyx Zoh. in Not. Roy. Bot. Gard. Edinb. 29 : 321 (1970). *Typus*: Turkey: Istanbul, Champs de lin, entre Chichli (Şişli) & Kiathané Keuy, 23 Mai, 1892, *Aznavour* (G ! holo., iso.: G !, HUJ !).

Icon.: [Plate 88]. Davis, Fl. Turk. 3 : f. 4/27 (1970).

Annual, glabrous, 5–15 cm. Stems many, procumbent to ascending, mostly simple, somewhat flexuose, leafy. Leaves with very long, filiform petioles scattered all along stems; stipules 0.7 mm long, broad-ovate, membranous, lower part adnate to petioles, free portion lanceolate, longitudinally nerved; leaflets 0.8–1 × 0.3–0.8 cm, obovate-cuneate, truncate to retuse, serrate-dentate in upper part, glabrous. Inflorescences 0.7–1 cm across, globular, 10–15-flowered, sessile in axils of stipules and largely overtopped by subtending leaves. Pedicels absent. Bracts minute, about 1 mm, membranous, ovate-oblong, mucronate. Calyx 5–6 mm, cylindrical, irregular, becoming bilabiate in fruit; tube sparingly appressed-puberulent, usually 12-nerved; teeth almost equal, about as long as tube, triangular-lanceolate, subulate at tip, narrowly membranous at margin. Corolla 3–4 mm long, white, shorter than calyx, somewhat longer then calyx tube; standard elliptical, long-clawed, longer than wings and keel. Fruiting calyx deflexed, lower face strongly gibbous, tridentate, upper face straight, indurated-spongy, nerveless, with 2 triangular-lanceolate teeth wider than those of lower face; throat naked. Pod 3 mm long, membranous, subsessile, ellipsoidal, 2-seeded, constricted between seeds; beak up to one-half the length of pod. Seeds globular, reniform, brownish. Fl. May.

Hab.: Meadows and grassy places near sea level.

Gen. Distr.: Turkey (endemic).

Selected specimens: TURKEY: Istanbul distr., entre Chichli et Kiathané, 1891, *Aznavour* (G); pelouses entre Chichli et Flamour, 1904, *Aznavour* (G); Proti Kinaliada, 1938, *Post* (G).

Note: *T. pachycalyx* superficially resembles *T. cernuum* Brot., but differs distinctly from the latter in several characters: heads sessile not pedunculate, minute bracts one-fifth the length of the calyx, gibbous calyx, bilabiate fruiting calyx with upper face strongly indurate-spongy, nerveless.

92. T. suffocatum L., Mantissa Alt. 276 (1771). *Typus*: Sicily (930/68, LINN ?).

Plate 88. *T. pachycalyx* Zoh.
(Turkey : *Aznavour*, G).
Plant in flower and fruit; flowers; standard; wing; keel; dissected calyx; fruiting calyx; fruiting calyx cut; pod.

Micrantheum suffocatum (L.) Presl, Symb. Bot. 1 : 48 (1830)

Icon. : [Plate 89]. Reichenb., Ic. Fl. Germ. 22 : t.110 I–III 1–19 (1867–89).

Annual, procumbent, sometimes almost stemless plant, glabrous or very sparingly hairy, 3–10 cm. Leaves very long-petioled, erect or ascending; stipules ovate, acuminate, scarious or membranous and thin; leaflets 3–8 × 2–6 mm, obovate-cuneate, obcordate, truncate and notched at apex, finely serrate in upper part, prominently nerved on lower surface. Inflorescences 5–8 mm across, few-flowered, densely crowded, sessile, axillary, overtopped by broad membranous stipules. Flowers sessile or subsessile. Bracts minute, lanceolate. Calyx 3–5 mm long, sparingly pilose or glabrous; tube with 10 prominent nerves; teeth lanceolate, as long as ovoid calyx tube, 3-nerved, recurved or falcate in fruit. Corolla 3–4 mm long, white : standard obovate, longer than wings and keel. Pod 3–5 mm long, membranous, ovoid, obliquely mucronate, 2-seeded, somewhat constricted between seeds. Seeds yellowish, lenticular to reniform, obscurely tuberculate. Fl. March–April. 2n = 16.

Hab. : Grazed places and roadsides.

Gen. Distr. : England, France, Spain, Portugal, Sardinia, Corsica, Sicily, Italy, Greece, Albania, Yugoslavia, Romania, Crete, Bulgaria, Turkey, Caucasus, Cyprus, Syria, Lebanon, Israel, Cyclades, N. Africa, Canary Islands, Madeira, Azores, Baleares.

Selected specimens : PORTUGAL : Coruche, 1963, *Paiva* et al. 8769 (HUJ). FRANCE : Manche, Grainville d'Avranches, 1955, *Brechier* 13 (G); Marseille, Mazarg, 1930, *Eig* (HUJ). ITALY : pr. Romam, 1922, *Lusina* (HUJ). GREECE : Attika, Acropolis, 1877, *Heldreich* & *Holzmann* 79 (G). TURKEY : Istanbul distr., Ichen Guelkuey, 1908, *Aznavour* (G). CYPRUS : Platanisso, Karpas, 150 ft, 1950, *Chapman* 258 (K). ISRAEL : Upper Galilee : Kerem-Ben-Zimra, 1963, *Zohary* 9 (HUJ); Sharon Plain : Tel-Aviv, 1927, *Eig* (HUJ). ALGERIA : Sahel de Collo, 1944, *Faurel* (MPU).

Note : Some plants of *T. suffocatum* have inflorescences only at their base. Whether exclusive basiocarpy is dependent on conditions prevailing in the habitat, should be elucidated experimentally.
A parallel form is found in *T. tomentosum* var. *chtonocephalum*, and there are also some basiocarpic species in *Trigonella* and *Astragalus.*

Series v. ALTISSIMA (Abrams) Heller (stat. nov.)

Altissima Abrams, Ill. Fl. Pac. States 2 : 522 (1944) pro min. p. (as a group ?).

Type species : *T. douglasii* House.

Perennial, glabrous or sparingly pubescent in upper parts. Root thick, vertical or fleshy-tuberous. Stipules up to 7 cm long, free portion leaf-like. Calyx membranous, slightly oblique, 17–20-nerved; teeth unequal, the lateral and upper ones tortuose and curved downwards and inwards around corolla, the lower one straight. One N. American species.

Plate 89. *T. suffocatum* L.
(France : *Eig*, HUJ).
Plant in flower and fruit; part of branch with enlarged stipule and leaf; pod; flower; corolla; standard; wing; keel.

93. T. douglasii House in Bot. Gaz. 41 : 335 (1906). *Typus* : Washington : between the Spokane River and Kettle Falls of the Columbia, *D. Douglas* (K !, photo HUJ !).

T. altissimum Douglas ex Hook., Fl. Bor. Am.1 : 130 (1831) non Loisel. (1807).

Icon. : [Plate 90]. Douglas ex Hook., *loc.cit.* t. 48.

Perennial, glabrous or sparingly pubescent in upper parts, 20–80 cm. Roots thick, vertical or fleshy-tuberous. Stems several, stout, erect, striate to grooved, many-leaved, poorly branched. Lower leaves with petioles up to 10 cm, upper ones much shorter-petioled to subsessile; stipules up to 7 cm long, the upper free portion (one-third) leaf-like, ovate to lanceolate, acuminate, prominently nerved, irregularly serrulate; leaflets 3–9 × 0.6–2 cm, oblong to narrowly elliptical or linear-lanceolate, acute or rarely rounded, apiculate, glabrous or sparingly pubescent beneath, prominently nerved, margin denticulate or serrulate. Inflorescences 2–3.5 cm in diam., many-flowered, globular. Peduncles longer than subtending leaves, up to 10 cm long, erect, hairy below inflorescences. Bracts inconspicuous, truncate, hyaline, cup-shaped. Pedicels about 1 mm, straight or deflexed only in lowest flowers after anthesis. Calyx 6–9 mm long, glabrous or very sparingly villous; tube campanulate, membranous, slightly oblique, 17–20-nerved; teeth 2–2.5 times as long as tube, unequal, the lower one straight, the lateral and upper ones shorter, tortuose and curved downwards and inwards around corolla, narrowly triangular to subulate, spiny-tipped, membranous-margined, entire or slightly serrulate, sometimes purplish at apex. Corolla 1.4–1.8 cm, reddish-purple, straight; standard narrowly oblong, acute or obtuse, longer than wings and keel. Pod small, ovoid, 1–2-seeded, sessile, membranous, smooth, glabrous or with few hairs. Seeds about 3 mm, smooth. Fl. June–July. 2n = 16.

Hab. : Meadows and along streams; 650–1,500 m.

Gen. Distr. : USA : Washington, Oregon, Idaho.

Selected specimens : WASHINGTON : Spokane Co., Spokane, 1884, *Suksdorf* 647 (P); Whitman Co., Palonse Creek, 1896, *Elmer* 56 (K). OREGON : E. Oregon, wet meadows, 1897, *Cusick* 1912 (E); Clear Water, *Spalding* (K). IDAHO : Lewis Co., N.W. Grangeville, 3,300 ft, 1940, *Constance* et al. 2739 (K, NA); Kootenai Co., Coeur d'Alene Mts., Santianne Creek, 1895, *Leiberg* 1033 (K, UC, US); Little Potlach River, 1892, *Sandberg* et al. 334 (K, US); Nez Perce Co., Lake Waha, 2,000–3,500 ft, 1896, *A.A.* & *E.G. Heller* 3382 (E, UC).

Note : *T. douglasii* is a very distinct species easily recognizable by its many-flowered and large heads and its strongly incurved calyx teeth.

Subsection VIII. CALYCOSPATHA (Chiov.) Heller & Zoh. (stat. nov.)

Section *Calycospatha* Chiov. in Ann. Bot. Rom. 9 : 57 (1911).

Type species : *T. pseudostriatum* Baker, f.

Annuals, glabrous to glabrescent. Inflorescences semi-globular to globular, many-

Plate 90. *T. douglasii* House
(Idaho : *Heller* 3382, E).
Plant in flower; flower; standard; wing; keel.

flowered. Bracts linear, conspicuous, more than twice as long as erect or horizontal pedicels. Calyx tube with 10–11 prominent nerves, much inflated and constricted at throat, splitting at maturity; teeth strongly recurved in fruit. Two African species.

94. T. mattirolianum Chiov. in Ann. Bot. Rom. 9 : 57 (1911). *Typus* : Abyssinia : Dembia (Gondar), *Chiovenda* 1422 et 1706 (FI !).

T. subrotundum Steud. & Hochst. var. *stipuli incissis* Rich., Tent. Fl. Abyss. 1 : 172 (1847).

Icon. : [Plate 91]. Fiori, in Nuov. Giorn. Bot. Ital. n. s. 55 : f. 5 (1948).

Annual, glabrous, (15–)20–40 cm. Stems many, erect to ascending, branching all along, deeply grooved. Leaves long-petioled in lower part, shorter-petioled in upper part; stipules 0.6–1 cm long, broad-ovate, adnate to petioles for about half their length, upper free portion ovate, cuspidate, herbaceous, prominently nerved, margin slightly laciniate; leaflets 0.8–1.5 × 0.4–1.2 cm, obovate, cuneate, apex rounded or slightly retuse, apiculate, margins spinulose-denticulate, prominently nerved. Inflorescences 1–1.5(–2) cm across, semi-globular to globular, many-flowered. Peduncles much longer than subtending leaves. Bracts 2–3 mm long, linear-subulate, subulate to aciculate. Pedicels up to 1 mm, erect. Calyx 5–6 mm, campanulate-urceolate, membranous; tube with 10–11 prominent nerves, strongly constricted at throat, inflated in fruit, splitting at maturity; teeth shorter to longer than tube, subulate, recurved, almost spinulose, one of them sometimes splitting into two equal or unequal lobes. Corolla up to 1 cm long, purplish-violet; standard oblong, elliptical, tapering at apex, longer than wings and keel. Pod about 3 mm long, obovate, concealed in calyx tube, whitish, leathery, 1–2-seeded, indehiscent; style long; dorsal suture thickened. Seeds about 1 mm, reniform, yellowish-brown. Fl. October. 2n = 16.

Hab. : Among field crops and in open grassland.

Gen. Distr. : Ethiopia (endemic).

Specimens seen : ETHIOPIA : Gojjam Prov., 36 km N.W. of Debra Marcos, 6,000 ft, 1964, *Meyer* 8625 (K, NA); Gimma loc. Malco, 1850, 1939, *Ufficio Agrario* 10 (G, W); Mattu near Gore, 4,500 m, 1958, *Mooney* 7547 (K).

95. T. pseudostriatum Baker, f., Leg. Trop. Afr. 83 (1926). *Typus* : Uganda : Kigezi Distr. Rukiga (Ruchiga), 7,000 ft, 1903–4, *Bagshawe* 413 (BM ! holo.).

Icon : [Plate 92].

Annual, glabrous to glabrescent, 10–60 cm. Stems many, erect to ascending, branching all along, grooved. Leaves short-petioled, upper ones subsessile; stipules 1–2 cm long, lanceolate, not sheathing, long-acuminate, free portion longer than lower part, somewhat diverging from stem, herbaceous; leaflets 1–2.5 × 0.6–1.1 cm, obovate, oblong to elliptical, cuneate at base, rounded to truncate at apex, apiculate-mucronate, margins acutely dentate. Inflorescences 1–1.5 cm in diam., semi-globular to globular, terminal and lateral, dense, many-flowered, sessile and subtended by 1–2

Plate 91. *T. mattirolianum* Chiov.
(Ethiopia : *Meyer* 8625, NA).
Plant in flower and fruit; flower; pod.

Plate 92. *T. pseudostriatum* Baker, f.
(Tanganyika : *Milne-Redhead* & *Taylor* 10155, B).
Plant in flower; flower.

leaves. Bracts up to 5 mm long, linear-lanceolate, subulate to acicular. Pedicels 1 mm or less, mostly erect. Calyx 5–7 mm long, glabrous to glabrescent; tube campanulate, whitish, with 10–11 prominent nerves, inflated in fruit with strongly constricted throat; teeth as long as or longer than calyx, triangular at base, later subulate and often hairy all along, strongly recurved in fruit. Corolla 6–8 mm long, purple; standard oblong-elliptical tapering at apex, as long as wings or longer. Pod about 3 mm, obovate, thickened at sutures, 1–2-seeded. Seeds 1.2–1.3 mm, yellowish-brown, reniform. Fl. January–May. 2n = 16.

Hab.: Pastures and short grassland; 1,400–2,200 m.

Gen. Distr.: Uganda, Tanganyika, E. Congo, N. Malawi, Ruanda-Urundi.

Selected specimens: UGANDA: Ankole, Busharyi, 5,250 ft, 1966, *Harrington* 23 (HUJ). TANGANYIKA: Mbeya Distr., Airfield, 1,650 m, 1956, *Milne-Redhead* & *Taylor* 10155 (B, K). CONGO (BELGE): Kizozi, 1939, *Lejume* 329 (G); 10 mi. N.E. of Rutshuru, betw. Lake Edward and Lake Kivu, 4,300 ft, 1924, *Chapin* 283 (NY). URUNDI: Env. Karuzi, Nyarusange, 1,500 m, 1958, *Van Der Ben* 2087 (K).

Subsection IX. NEOLAGOPUS Lojac.

Nuov. Giorn. Bot. Ital. 15 : 139, 275 (1883).

Type species: *T. macraei* Hook. & Arn.

Annuals or perennials, villous. Inflorescences many-flowered, mostly spicate, pedunculate or sessile. Bracts inconspicuous, cup-like, arranged in whorls along rhachis. Pedicels absent. Pod membranous, 1-seeded, enclosed in calyx. Four N. & S. American species.

96. T. albopurpureum Torr. & Gray, Fl. N. Am. 1 : 313 (1838) *Typus*: California, 1833, *D. Douglas* (GH, iso.: K !, NY. !, photo HUJ !).

T. neolagopus Lojac., Nuov. Giorn. Bot. Ital. 15 : 194 (1883).
T. columbinum Greene, Pittonia 1 : 4 (1887).
T. olivaceum Greene, *loc.cit.*
T. macraei Hook. & Arn. var. *albopurpureum* (Torr. & Gray) Greene, Fl. Francisc. 26 (1891).
T. columbinum Greene var. *argillorum* Jeps., Fl. W. Mid. Calif. 307 (1901).
T. columbinum Greene var. *olivaceum* (Greene) Jeps., *loc.cit.*
T. albopurpureum Torr. & Gray var. *neolagopus* (Lojac.) McDerm., N. Am. Sp. Trif. 209 (1910).
T. albopurpureum Torr. & Gray var. *neolagopus* McDerm. f. *argillorum* (Jeps.) Jeps. in McDerm., *op.cit.* 213.
T. olivaceum Greene f. *columbinum* (Greene) McDerm., *op.cit* 213.
T. helleri Kennedy, Muhlenbergia 9 : 25 (1913).
T. insularum Kennedy, *op.cit.* 29.
T. pseudoalbopurpureum Kennedy, *op.cit.* 20.

T. olivaceum Greene var. *columbinum* (Greene) Jeps., Man. Fl. Pl. Calif. 546 (1925).
T. olivaceum Greene var. *griseum* Jeps., *loc.cit.* (1925).
T. albopurpureum Torr. & Gray var. *olivaceum* (Greene) Isley, Brittonia 32 : 55 (1980).

Icon. : [Plate 93]. Muhlenbergia 9 : t. 1–5 (1913).

Annual, villous, 15–45 cm. Stems erect, dichotomously branching in upper part. Lower leaves long-petioled, upper ones shorter-petioled; stipules 0.8–1.5 cm, ovate, free portion shorter than lower part, triangular, cuspidate; leaflets 1.5–2.5 × 0.8–1.5 cm, obovate to obovate-oblong, cuneate at base, rounded to truncate, apiculate, margins denticulate. Inflorescences 1.5–2 × 1.3–1.5 cm, broad-ovate to globular, dense, many-flowered. Peduncles rather long. Bracts minute, membranous, cup-like, arranged in whorls along rhachis. Pedicels absent. Calyx 0.9–1.5 cm long, villous; tube 2 mm long, tubular, 20–30-nerved; teeth (3)4–6 times as long as tube, subulate-setaceous, plumose. Corolla 4–6 mm long, purple; standard oblong to ovate, rounded. Pod 3 mm long, broad-elliptical, glabrous or sparingly hairy, membranous, parallely nerved, 1-seeded. Seed 2 mm, ellipsoidal, dark brown. Fl. April–May. 2n = 16.

Hab. : Foothills and valleys; 50–1,000 ft.

Gen. Distr. : USA : California, Washington, Arizona, Oregon.

Selected specimens : CALIFORNIA : Napa Co., Near Napa, 1902, *Heller* & *Brown* 5359 (G); Jehama Co., Red Bluff, 1928, *Kennedy* (R); Ventura Co., between Calif. Prep. School and Ojai, 1949, *Pollard* (B); Vacaville, 1886, *Greene* 37925 (HUJ, photo of lectotype of *T. olivaceum* Greene). ARIZONA : Camp Creek, 1928, *Peebles* et al. 5140 (UC); 10 mi. S. of Black River crossing Highway 60, 1938, *Gooding* & *Mallery* (NA). OREGON : Josephine Co., Wonder, 1964, *Gillett* & *Moulds* 12583 (HUJ); Douglas Co., 1881, *Howell* (NA).

97. T. dichotomum Hook. & Arn., Bot. Beech. Voy. 330 (1841). *Typus* : California, 1833, *D. Douglas* (K ! iso. G !).

T. macraei Hook. & Arn. var. *dichotomum* (Hook. & Arn.) Brewer ex Wats., Proc. Am. Acad. 11 : 129 (1876).
T. amoenum Greene, Fl. Francisc. 27 (1891).
T. dichotomum Hook. & Arn. var. *turbinatum* Jeps., Fl. W. Mid. Calif. 306 (1901).
T. petrophilum Greene ex Baker, West. Am. Pl. 2 : 18 (1903) nom. nud.; Greene ex Heller in Muhlenbergia 2 : 298 (1907).
T. dichotomum Hook. & Arn. f. *petrophilum* (Greene ex Heller) McDerm., N. Am. Sp. Trif. 221 (1910).
T. californicum Jeps. in McDerm., *op.cit.* 215.
T. californicum Jeps. f. *turbinatum* (Jeps.) Jeps. in McDerm., *op. cit.* 217.
T. albopurpureum Torr. & Gray var. *dichotomum* (Hook & Arn.) Isley, Brittonia 32 : 55 (1980).

Icon. : [Plate 94]. Jeps., Man. Fl. Pl. Calif. 545 (1925).

Annual, villous, 15–60 cm. Stems many, erect, branching especially in upper part, sulcate. Lower leaves long-petioled, upper ones subsessile; stipules 1–2 cm long, ovate, not sheathing, free portion as long as lower part, acuminate, herbaceous, en-

Plate 93. *T. albopurpureum* Torr. & Gray
(California : *Pollard*, B).
Plant in flower; flower; standard; wing; keel.

Plate 94. *T. dichotomum* Hook. & Arn.
(California : *Smith* 1468, HUJ).
Plant in flower; flower; pod (California : *Baker* 2797, K).

tire or toothed; leaflets 1.5–4 × 0.5–2 cm, obovate to oblanceolate, cuneate at base, rounded at apex, denticulate to serrate. Inflorescences 2–3 × 1.5–2 cm, ovoid to globular, terminal and lateral, dense, many-flowered, whorled. Peduncles up to 10 cm long. Bracts minute, cup-like in several whorls along rhachis. Pedicels absent. Calyx 0.8–1.3 cm, villous; tube campanulate, 20–30-nerved; teeth about 3 times as long as tube, subulate. Corolla 1.3–1.7 cm long, dark purple, lighter-tipped : standard obovate-oblong, rounded, often retuse. Pod 4 mm long, obovoid, membranous, villous towards apex, 1-seeded. Seed 2–3 mm long, smooth. Fl. May and June. 2n = 32.

Hab : Low meadows; 10–4,500 ft.

Gen. Distr. : USA : California, Washington, Vancouver Island.

Selected specimens : CALIFORNIA : Solano Co., near Vanden Station, 1902, *Heller* & *Brown* 5593 (G, P, W); Santa Clara, Mt. Hamilton, 4,000 ft, 1908, *Smith* 1468 (HUJ); *ibid.*, Gilroy, 1903, *Elmer* 4909 (E, G, W); Alameda Co., Midway, 100 ft, 1938, *Rose* 38203 (B, W); Byron, Countra Costa Co., 1903, *Baker* 2797 (B, K, W). WASHINGTON : Columbia River, W. Klickitat Co., 1881, *Suksdorf* (NA).

98. T. macraei Hook. & Arn. in Hook., Bot. Misc. 3 : 179 (1833). *Typus* : Chile : Baths of Collina, 1825, *Macrae* (iso. K !, photo HUJ !).

T. catalinae Wats., Proc. Am. Acad. 25 : 128 (1890).
T. bicephalum Elmer, Bot. Gaz. 41 : 312 (1906).
T. macraei Hook. & Arn. f. *catalinae* (Wats.) McDerm., N. Am. Sp. Trif. 201 (1910).
T. mercedense Kennedy in Muhlenbergia 9 : 17 (1913).
T. traskae Kennedy, *op. cit.* 19.

Icon. : [Plate 95]. McDerm., *loc. cit.* pls. 81–82.

Annual, hairy to various degrees, usually villous, 10–30 cm. Stems few to many, mostly dichotomously branching from base or all along, erect, ascending to prostrate. Leaves short-petioled, uppermost ones sessile; stipules herbaceous, ovate to oblong, acuminate to cuspidate; leaflets 1–2 × 0.3–1.5 cm, narrowly to broadly obovate or elliptical, rounded, often slightly retuse, denticulate to serrate, rarely almost entire. Inflorescences 0.5–2 cm in diam., usually in pairs, subsessile to sessile, 10- to many-flowered, ovoid to semi-globular, subtended by stipules and upper leaves. Peduncles absent or very short. Bracts minute, up to 2 mm, connate at base forming inconspicuous involucre, upper one cup-like or tooth-like, forming several whorls along rhachis. Pedicels absent. Calyx 4 mm long, villous; tube 5–10-nerved; teeth subequal, as long as to shorter than tube, subulate. Corolla 5–7 mm, purplish to pink or white; standard obovate-oblong, rounded and slightly denticulate at apex. Pod 3 mm, oblong, glabrous to sparingly villous towards apex, parallel-nerved, 1-seeded. Seed 2 mm, smooth, brown. Fl. May–June.

Hab. : Grassy fields and sandy ocean bluffs; 50–2,000 ft.

Gen. Distr. : USA : California, Oregon. Chile.

Selected specimens : CALIFORNIA : San Mateo Co., San Pedro, 1903, *Elmer* 4812 (US, W, isotypes of *T. bicephalum* Elmer); San Francisco, Presidio, N. of Golf Link, 250 ft, 1965,

Plate 95. *T. macraei* Hook. & Arn.
(California : *Rose* 65064, H).
Plant in flower and fruit; flower; pod.

Rose 65064 (H). OREGON : Corvallis, 1922, *Epling* 5528 (UC). CHILE : prov. Colchgua, 1862, *Briges* & *Cumming* 191 (NY); Santiago, *Philippi* 651 (K).

99. T. plumosum Dougl. ex Hook., Fl. Bor. Am. 1 : 130 (1831).

Perennial, densely villous, 15–60 cm. Stems few, rhizomatous, with deeply penetrating, thickened taproot, sparingly branched above. Lower leaves long-petioled, upper ones short-petioled to subsessile; stipules 1.5–3.5 cm long, closely sheathing, oblong to lanceolate, densely nerved, free portion about as long as lower part, lanceolate, acute to acuminate, margin entire or serrulate; leaflets 3–10 × 0.2–1.6 cm, often folded, falcate, linear-lanceolate, elliptical or lanceolate-elliptical, acute to acuminate, margins dentate-serrulate or entire, prominently nerved. Inflorescences 2.5–7 × 2–3 cm, spicate, dense, many-flowered, arranged in whorls, ovate to cylindrical. Peduncles rather short, 1.5 cm long. Bracts minute, membranous, whitish, broad-ovate. Pedicels absent. Calyx 1–1.4 cm long, silky-villous; tube tubular, 20-nerved; teeth subequal, 1–2 times as long as tube, subulate. Corolla 1.2–2.2 cm long, cream to crimson; standard oblong, often folded, acute and curved at tip, longer than wings and crimson keel. Pod 4 mm, thin-membranous, obovoid, 1(-2)-seeded. Seed 1–2 mm, brown, smooth. Fl. June–July. 2n=32.

99a. T. plumosum Dougl. ex Hook. Subsp. **plumosum.** *Typus* : N. W. American, Blue Mts., in alluvial soil, *D. Douglas* (K !, iso. GH).

Icon. : Hook., *op. cit.* t. 49 (1831)

Plants 20–35 cm tall. Leaflets of basal leaves 2–5(–7) mm wide, acuminate.
Hab. : Dry sagebrush slopes, meadows and open pine woods; 500–1,500 m.
Gen. Distr. : USA : Washington, Oregon, Utah.

Selected specimens : WASHINGTON : Walla Walla Co., above Mill Creek, 1,700 ft, 1937, *Moore* 162 (NA). OREGON : Grande Ronde Valley, mouth of Lodd canyon, 1899, *Cusick* 2348 (E, W); Union Co., Umatilla, Starkey Range, 4,800 ft, 1941, *Keil* 1125 (NA). UTAH : Wasatch Co., E. of Kamas, 7,300 ft, 1968, *Farrer* 40 (HUJ).

99b. T. plumosum Dougl. ex Hook. Subsp. **amplifolium** (Martin) Gillett, Can. J. Bot. 50 : 1981 (1972). *Typus* : Idaho : Washington Co., Salmon Meadows, 4,000 ft, June 22, 1899, *M.E. Jones* 6254 (POM, iso. : DS, NY, ORE, OSC, POM, US).

T. plumosum Dougl. ex Hook. var. *amplifolium* Martin, Bull. Torrey Bot. Club 73 : 369 (1946).

Icon. : [Plate 96].

Plants up to 60 cm tall. Leaflets of basal leaves (8–)9–16 mm wide, acute.
Hab. : On dry hilltops of stony loams and sagebrush slopes.
Gen. Distr. : Idaho (endemic).

Plate 96. *T. plumosum* Dougl. ex Hook. Subsp. *amplifolium* (Martin) Gillett
(Idaho : *Gillett* & *Mosquin* 11921, HUJ).
Plant in flower; flower.

Specimens seen: IDAHO: Lake Waha, Nez Perce Co., 2,000–3,500 ft, 1896, *Heller* 3398 (E); N. of White Bird, Arid Transition Life Zone, 1938, *Sharsmith* 3619 (NA); Idaho Co., 4.6 mi. N.W. of Grangeville, 2,950 ft, 1963, *Gillett* & *Mosquin* 11921 (HUJ).

Doubtful Names from Section LOTOIDEA

T. bobrovii Chalilov in Not. *Syst.* Herb. Inst. Bot. Acad. Sci. URSS 12 : 119 (1950) – Azerbaijan – related to *T. ambiguum* M.B.
T. coccineum Steud., Nom. ed. 2. 2 : 705 (1841) – Chile
T. coeruleum Viv., Pl. Aeg.-Arab. Dec. 15 (1831) – Egypt
T. densiflorum Phil. in Linnaea 28 : 619 (1856) – Chile
T. erythraeum Schausb. ex Ball in Jour. Linn. Soc. 16 : 420 (1878) = *T. isthmocarpum* Brot.
T. flavum Larranaga, Escritos D. A. Larranaga 2 : 232 (1923) – Uruguay
T. issajevii Chalilov in Dokl. Akad. Nauk Azerbaijan SSR 22(11) : 66 (1966 – Azerbaijan)
T. lasiocephalum Link, Enum. Hort. Berol. 2 : 262 (1822) = *T. africanum* Ser.
T. littorale Larranaga, Escritos D. A. Larranaga 2 : 232 (1923) – Uruguay
T. rotundifolium Sibth. & Sm., Fl. Grace. 8 : 35, t. 747 (1832) – Greece
T. rouyanum Bonnier, Fl. Compl. France, Suisse et Belgique 3 : 40 (1914) = *T. montanum* L. var. *flaviflorum* Rouy
T. rubrum Larranaga, Escritos D. A. Larranaga 2 : 232 (1923) – Uruguay
T. volubile Lour., Fl. Cochinch. 542 (1793) – E. Africa

Section Two. PARAMESUS (C. Presl) Endl.

Gen. Pl. 1268 (1840); Griseb., Spicil. Fl. Rumel. 1 : 28 (1843); Godr. in Gren. & Godr., Fl. Fr. 1 : 416 (1849)

Genus *Paramesus* C. Presl, Symb. Bot. 1 : 45 (1830).
Section *Melilotea* Bertol., Fl. Ital. 7 : 99 (1850).
Subgenus *Paramesus* (C. Presl) Hossain, Not. Roy. Bot. Gard. Edinb. 23 : 444 (1961).

Type species : *T. strictum* L.

Annuals. Stipules denticulate. Heads terminal and axillary. Bracts minute. Flowers sessile. Calyx tube 10-nerved, with open throat; teeth unequal. Petals connate below. Pod 2-seeded, somewhat exserted from calyx tube.

Key to Species

1. Corolla 9–12 mm long. — **100. T. glanduliferum**
   Corolla 6–8 mm long. — **101. T. strictum**

100. T. glanduliferum Boiss., Diagn. ser. 1, 2 : 30 (1843) et Fl. 2 : 141 (1872).

Annual, glabrous or slightly hairy, 10–25 cm. Stems erect or ascending, rarely prostrate, simple or branching, terete. Leaves short-petioled; stipules adnate to petioles up to more than half of their length, scarious with green margin, semi-ovate, obtuse, many-nerved, sharply dentate with teeth often ending in glands; leaflets 1–2(–3) cm, oblong, elliptical to linear, cuneate at base, obtuse, mucronate, the lower ones obovate, all prominently nerved, sharply dentate with or without sessile or stipitate glands on or between teeth. Peduncles axillary, much longer than leaves. Heads ovoid to globular with fairly conspicuous involucre or without involucre. Bracteoles reduced to very minute, tridentate scales. Flowers sessile. Calyx tube white, obconical, glabrous or rarely with some scattered hairs, with 10 prominent nerves; teeth green, unequal, mostly longer than tube, lanceolate-subulate, keeled-nerved, spreading or deflexed in fruit. Corolla 2–3 times as long as calyx, pink to flesh-coloured (rarely white); standard erect, linear, twice as long as keel; keel somewhat shorter than wings. Pod stipitate, oblique, longer than calyx tube with long-exserting style, 2-seeded. Seeds 1–1.2 mm, brown, ovoid to oblong. Dispersal through rupture of upper, hardened part of pod from lower, membranous one. Fl. March–April (May). 2n = 16.

Hab. : Fields, sandy places, among scrub and forests.

Gen. Distr. : Albania, Chios, Greece, Turkey, Syria, Lebanon, Israel.

1. Heads subtended by row of connate, involucral, glandular bracts. Leaflets, stipules and calyx teeth abundantly glandular. Calyx white between prominent nerves.
100a. var. **glanduliferum**
– Involucral bracts 0 or almost so. Leaflets and stipules sparingly or not at all glandular. **100b.** var. **nervulosum**

100a. T. glanduliferum Boiss. Var. **glanduliferum.** Zoh., in Davis, Fl. Turkey 3:414 (1970); Zoh., Fl. Pal. 2:175 (1972). *Typus*: Yachawichlar-Keui, à Zlieues au d'Ouchak (Phrygie) 1857, *Balansa* 310 (G lecto.).

T. glanduliferum Boiss. var. *tmoleum* (Boiss.) Boiss., Fl. 2:141 (1872).
T. tmoleum Boiss. in Balansa, Pl. exs. 175 (1854).

Selected specimens: TURKEY: Prov. Izmir: Sinus Smyrnaeus, Yamanlardagh, 900m, 1906, *Bornmüller* 7352 (E). Yaila de Bozdagh (Tmolus occidental) dans les prairies, 1854, *Balansa* 175 (type of *T. tmoleum* G, E). ISRAEL: Hula Plain: E. of Mishmar Hayarden, 1963, *Zohary* 6/17 (HUJ).

100b. T. glanduliferum Boiss. Var. **nervulosum** (Boiss. & Heldr.) Zoh., in Davis, Fl. Turkey 3:414 (1970); Zohary, Fl. Pal. 2:175 (1972). *Typus*: In viridis succis collium Alaya Pampiliae, 1845, *Heldreich* (K); in arenosis ad Gaza Palaestinae, *Boissier* (G lecto).

T. nervulosum Boiss. & Heldr. in Boiss., Diagn. ser. 1, 9:25 (1849).
T. galilaeum Boiss., *op. cit.* 26.
T. nervulosum Boiss. & Heldr. var. *galilaeum* (Boiss.) Boiss., Fl. 2:142 (1872).

Icon.: [Plate 97].

Selected specimens: TURKEY: Prov. Bursa: Renkoei: in monte Ulu Dagh, 1883, *Sintenis* 366 (E); Prov. Antalya: Adalia, 1860, *Bourgeau* (E); Manavgat-Kara point, 3m, 1956, *Davis* & *Polunin* 25.837 (E); Alanyabay, 2m, 1956, *Davis* & *Polunin* 25.916 (E). SYRIA: 74 km S. W. of Damascus, 950 m, 1933, *Eig* & *Zohary* 7 (HUJ). ISRAEL: Coastal Galilee, N. of Nahariya, 1955, *Zohary* 17023 (HUJ); Sharon Plain, env. of Wadi Faliq, 1956, *Zohary* 17022 (HUJ).

Note: After an extensive study of this species and comparison with the types of T. *glanduliferum* Boiss. and *T. nervulosum* Boiss. & Heldr. in Herbarium Boissier, we concluded that the above two taxa cannot be retained at a specific level and that the whole complex is made up of a series of intergrading forms. While the typical specimens of *T. glanduliferum* from Lydia have rather conspicuous involucres, the ones from Palestine have smaller but still discernible involucres which are densely glandular. A step further leads to a form where the involucre is altogether lacking but the leaflets, stipules and calyx teeth are still sparsely but consistently glandular. In var. *nervulosum* the leaves and stipules are, according to Boissier, not glandular, but one also encounters here and there specimens regarded by Boissier as *T. nervulosum* possessing sparse glands. The length of the calyx teeth and the presence of hairs between them are very weak and unreliable characteristics.

Plate 97. *T. glanduliferum* Boiss. Var. *nervulosum* (Boiss. & Heldr.) Zoh.
(Palestine : *Naftolsky* 17028, HUJ).
Plant in flower; leaflet showing glands along margin; flower; flowering branch showing glandular stipules.

101. T. strictum L., Cent. Pl. 1 : 24 (1755). *Typus*: Described from Italy.

T. laevigatum Poiret, Voy. Barb. 2 : 219 (1789) et Desf., Fl. Atl. 2 : 195 (1799).
Paramesus strictus (L.) Presl, Symb. Bot. 1 : 46 (1830).

Icon.: [Plate 98].

Annual, glabrous, (3–)10–40 cm. Stems few, erect or ascending, grooved, dichotomously branching in upper part. Lower leaves with rather long petioles, upper ones short-petioled to subsessile; stipules 1–1.5 cm, hyaline, ovate or rhombic with glandular-denticulate margins, the lower ones as long as internode; leaflets 0.8–2 × 0.2–0.7 cm, linear to linear-elliptical, prominently nerved, obtuse, with glandular-denticulate margin, the lower ones lanceolate to narrowly obovate. Inflorescences 0.7–1 cm in diam., ovoid, 10–20-flowered. Peduncles axillary or terminal, longer than subtending leaves, slightly hairy at top. Bracts 2–4 mm long, hyaline, ovate, connate at base forming involucre. Pedicels about 1 mm. Calyx 3–5 mm long; tube campanulate, with 10 prominent nerves swollen in fruit; teeth unequal, the lower 3 longer than tube and upper ones, all triangular at base and 3-nerved, abruptly cuspidate, spreading in fruit. Corolla 5–7 mm, pink; standard oblong, rounded or truncate, folded, long-clawed. Pod 3–4 mm long, almost globular to orbicular, exserted from calyx tube, dorsally gibbose, 2-seeded, curved and beaked, membranous to coriaceous in upper part, dehiscent only at base. Seeds about 1 mm, ovoid to lentiform. Fl. May–June. 2n = 16.

Hab.: Fields and meadows.

Gen. Distr.: England, Spain, Portugal, France, Sardinia, Corsica, Italy, Germany, Czechoslovakia, Bulgaria, Hungary, Romania, Greece, Turkey, Morocco, Algeria.

Selected specimens: PORTUGAL: Estrada Macedo-Bragança, a 35 km de Bragança, 1958, *Fernandes* et al. 6295 (HUJ). FRANCE: Nantes, Loire, 1844, *Llyod* 2216 (R). HUNGARY: Comit. Hajdu, Nagyhortobàgy, 1916, *Rapaica* 6801 (R). BULGARIA: M. Rodope, Rastaka prope Madan, 250 m, 1953, *Stojanov* et al. 162 (HUJ). ROMANIA: Banatus, dist. Caras-Severin, 140 m, 1923, *Nayrady* 562 (HUJ). TURKEY: Istanbul, Sinekli, Büyük-han, *Davidov* (fide Davis, Fl. Turkey 3 : 415, 1970). ALGERIA: environs d'Alger, 1861, *Bourlier* 421 (P). MOROCCO: prope oppidum Boulhaut, 1924, *Maire* (HUJ).

Plate 98. *T. strictum* L.
Plant in flower and fruit;
fruiting calyx with corolla; pod.

Section Three. MISTYLLUS (C. Presl) Godr.

in Gren. & Godr., Fl. Fr. 1 : 415 (1849)

Genus *Mistyllus* C. Presl, Symb. Bot. 1 : 49 (1830).
Section *Vesicastrum* Koch, Syn. Fl. Germ. Helv. 1 : 149 (1835); Boiss., Fl. 2 : 112 (1872).
Section *Vesicaria* Richt.. Cod. Bot. Linn. 745 (1835–39) p. p.
Section *Trigantheum* Gib. & Belli, Mem. Accad. Sci. Torino ser. 2, 42 : 3 (1891).

Lectotypus : *T. spumosum* L.

Annual, glabrous herbs with spuriously terminated, caespitose heads. Flowers sessile or subsessile. Bracts large, many-nerved. Calyx with 20 or more nerves, regularly becoming more or less inflated after flowering; throat not closed. Petals clawed, persistent, becoming scarious after flowering; standard free. Pod sessile, 2–4-seeded, long-beaked, included within calyx. Seeds verruculose.

Note : A natural and more or less uniform group, with no clear connection to other sections except for a slight resemblance to certain species of Subsect. *Physosemium* of Sect. *Involucrarium*. Its developed bracts and pods indicate some relationship to Sect. *Lotoidea*. The most puzzling feature of this Section is its distribution, namely of the nine species three are Afro-alpine while the other six are Mediterranean and this is perhaps indicative of its African origin where it is sympatric with Subsect. *Physosemium*.

Synopsis of Species in Section Three

Section Three. MISTYLLUS (C. Presl) Godr.

102. T. spumosum L.
103. T. vesiculosum Savi
104. T. aintabense Boiss. & Hausskn.
105. T. setiferum Boiss.
106. T. argutum Sol.
107. T. mutabile Portenschl.
108. T. lugardii Bullock
109. T. steudneri Schweinf.
110. T. quartinianum A. Rich.

Key to Species

1. Tube of fruiting calyx longitudinally nerved with lateral anastomoses between nerves (the anastomoses sometimes not very prominent) 2
– Tube of fruiting calyx with longitudinal nerves only 3
2. Leaflets obovate or obtriangular. Bracts shorter than calyx. Corolla short-exserting, straight. **102. T. spumosum**

– Leaflets elliptical. Bracts as long as calyx. Corolla long-exserting, deflexed. **103. T. vesiculosum**

3 (1). Calyx tube with two longitudinal rows of hairs 4

– Calyx tube glabrous 5

4. Calyx teeth as long as or longer than tube. Corolla 1–1.2 cm. Bracts lanceolate, acuminate. **105. T. setiferum**

– Calyx teeth one-half to three-quarters the length of tube. Corolla 6–7 mm. Bracts oblong-cuneate to ovate, mucronate. **106. T. argutum**

5 (3). Calyx teeth longer than tube 6

– Calyx teeth shorter than or as long as tube 7

6. Leaflets narrowly-lanceolate, 5 times as long as broad. Peduncles pilose near top. Inflorescences about 30-flowered. Corolla shorter than calyx. **108. T. lugardii**

– Leaflets about 3 times as long as broad. Peduncles glabrous. Inflorescences less than 20-flowered. Corolla longer than calyx. **110. T. quartinianum**

7 (5). Leaflets lanceolate-linear. Peduncles very short. **109. T. steudneri**

– Leaflets obovate-elliptical or obovate, cuneate at base, not exceeding 1 cm in length 8

8. Calyx tube about 40-nerved, slender. Pod 1-seeded. Standard, ovate-obtuse, rounded. **104. T. aintabense**

– Calyx tube about 24-nerved, slender. Pod 2–3-seeded. Standard lanceolate, acute to acuminate. **107. T. mutabile**

102. T. spumosum L., Sp. Pl. 771 (1753). *Typus*: Described from Europe (Hort. Cliff. 373, 7).

T. apulum Horst ex All. in Misc. Taur. 5 : 76 (1774).
T. folliculatum Lam., Fl. Fr. 2 : 599 (1778).
Mistyllus spumosus (L.) Presl, Symb. Bot. 1 : 49 (1830).

Icon.: [Plate 99].

Annual, glabrous, 10–40 cm. Stems few or many, striate, prostrate to erect, diffusely branching. Lower leaves long-petioled, upper ones short-petioled; stipules membranous, white, ovate, abruptly long-cuspidate, prominently nerved; leaflets 1–2.5 cm, obovate to rhombic, cuneate at base, often truncate to retuse at apex, denticulate. Peduncles as long as or longer than subtending leaves. Flowering heads 1.5–2 cm, many-flowered, broadly ovoid to globular; fruiting heads 2–4 cm, globular, ovoid to cylindrical. Bracts somewhat shorter than calyx, oblong (outermost ones ovate), long-mucronate to subulate. Pedicels short, thick. Calyx about 1 cm, membranous; tube ellipsoidal, inflating considerably and becoming ovoid or globular in fruit, with 20 longitudinal and numerous transverse veins between nerves; teeth about half the length of tube, subulate-linear, recurved in fruit. Corolla 1–1.5 cm long, purple or reddish; standard with persistent, ovate limb, longer than divergent wings. Pod 1–1.5 cm, including ensiform beak, sessile, 2–4-seeded, somewhat constricted between seeds. Seeds 1.6–1.8 mm, ovoid, light brown, minutely warty. Fl. March–May. 2n = 16.

Hab.: Damp fields, among scrub, roadsides, grassy places, etc.

Gen. Distr.: England, Spain, S. France, Corsica, Sardinia, Sicily, Greece, Aegean Islands, Cyprus, Turkey, Syria, Lebanon, Israel, Iraq, Morocco.

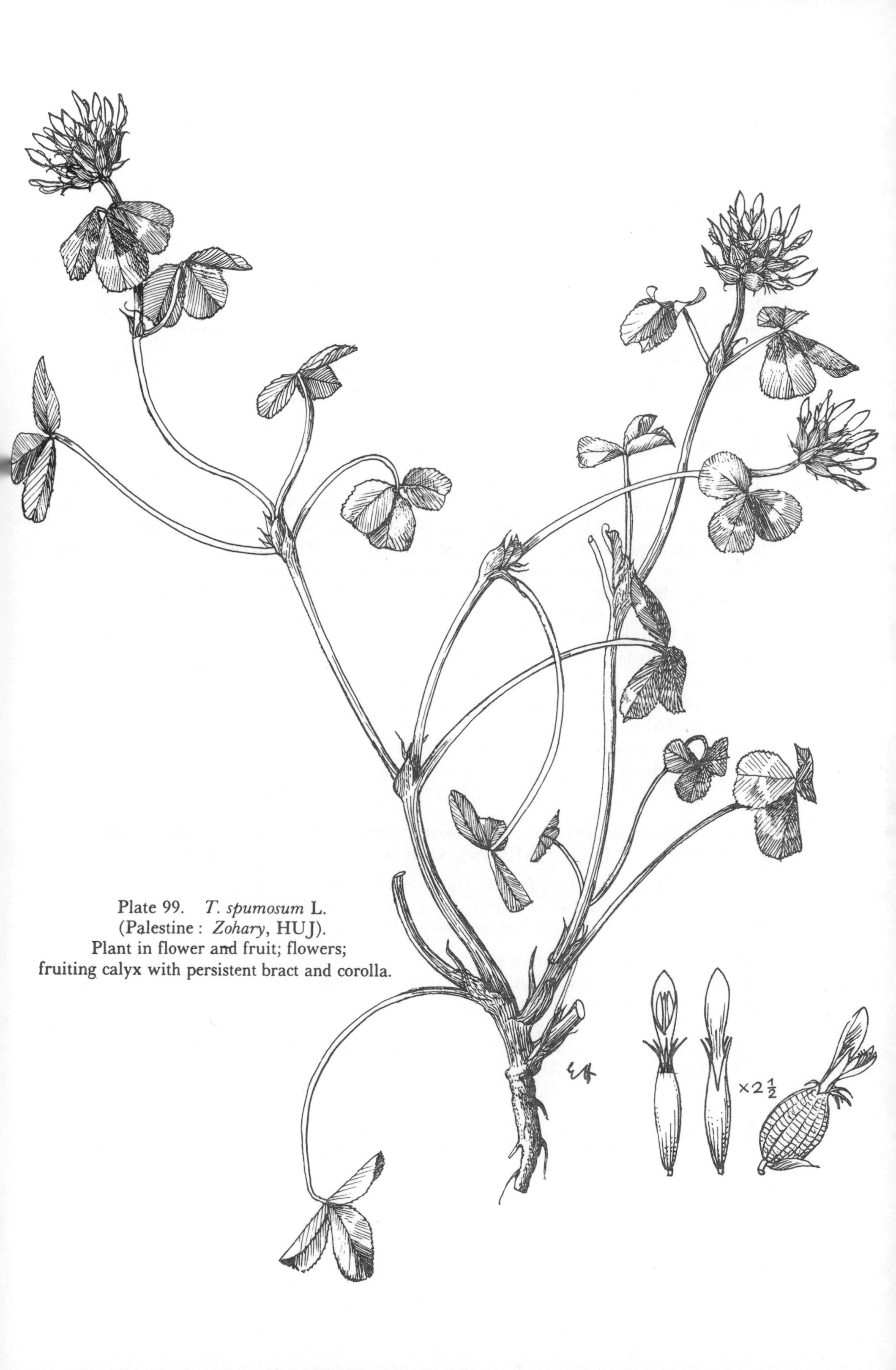

Plate 99. *T. spumosum* L.
(Palestine : *Zohary*, HUJ).
Plant in flower and fruit; flowers;
fruiting calyx with persistent bract and corolla.

Selected specimens: ENGLAND: Tip. Kettlewell, Lances, 1967, *Pearce* (K). SPAIN: Ciudad Universitaria, Madrid, 1948, *Monasterio* (HUJ). FRANCE: d'Agde (Hérault), 1884, *Nayra* 1961 (K). CORSICA: Algaiola, près de Lalvi, 1867, *Mabille* 222 (K). SARDINIA: Santa Teresa Caltura, 1881, *Reverchon* 192 (K). SICILY: Sopra Monreale, *Todaro* 794 (K). GREECE: Dodecanese, Patmos, 1969, *Townsend* 69/68 (K). CYCLADES: Naxos nr. Chora, 1940, *Davis* 1574 (K). CYPRUS: Ayios, Epakchtos, 50 ft, 1967, *Crodforton* 324 (K). TURKEY: Istanbul distr.: Cimet. Jura, aux Petits Champs, Pera, 1889, *Aznavour* 633 (G); Prov. Çanakkale: Erenköy, 1883, *Sintenis* 170 (E). LEBANON: Beyrut, 1933, *Menertzhagen* s. n. (K). SYRIA: Banias, 1863–4, *Lowne* (E). ISRAEL: Philistean Plain: Env. of Wadi Rubin, 1954, *Sheinkar* 17322 (HUJ); E. Upper Galilee: Yiftah, 1954, *Sheinkar* 17306 (HUJ). IRAQ: Serderian near Aqra, 1948, *Memaryan* 10789 (K). MOROCCO: Massif des Beni-Snassen, 1929, *Faure* (HUJ).

103. T. vesiculosum Savi, Fl. Pis. 2 : 165 (1798); Boiss., Fl. 2 : 139 (1872).

Annual, glabrous, 15–60 cm. Stems many, grooved, poorly branching, erect or ascending. Leaves rather large, lower ones long-petioled, upper ones short-petioled to subsessile; stipules lanceolate, white-membranous, lower part adnate to petiole, free portion long, subulate-setaceous; leaflets up to 3 × 1.2 cm, those of lower leaves obovate, the rest elliptical, acute, serrulate-dentate, mucronulate. Peduncles thick, often very short. Heads axillary and terminal, ovate to globular, elongating in fruit up to 6 × 2 cm, many-flowered, often subtended by upper leaves. Bracts lanceolate, acuminate, as long as calyx. Pedicels 0. Calyx 6–7 mm long; tube tubular, glabrous, 24-nerved with less prominent, rarely obsolete, transverse anastomoses; teeth almost equal, about as long as tube, lanceolate-subulate, erect, later recurving. Corolla 1.2–1.5 cm long, pink to purple; standard lanceolate with ovate, acuminate limb and broad, somewhat tapering claw; wings lanceolate, acute; keel mucronate; keel and wings much shorter than standard. Fruiting calyx urceolate. Pod membranous, 2–3-seeded. Seeds 1–1.2 mm, ovoid, brown, granulate. Fl. April–July. 2n = 16.

103a. T. vesiculosum Savi Var. **vesiculosum.** *Typus*: Italy: "Clairières de bois Pisa Coll. P. San Julliet" (M lecto.); Coombe, Fl. Eur. 2 : 164 (1968).

T. recurvum Waldst. & Kit., Pl. Rar. Hung. 2 : 179, t.165 (1805).
T. turgidum M. B., Fl. Taur.-Cauc. 2 : 216 (1808) et Suppl. 511.
Mistyllus turgidus (M.B.) Presl, Symb. Bot 1 : 49 (1830).

Calyx with prominent transverse veins between nerves, strongly inflated in fruit.
Gen. Distr.: Albania, Bulgaria, Hungary, Yugoslavia, Romania, Greece, Italy, Sicily, Corsica, Crimea, Russia.

Selected specimens: ITALY: Pisa (Toscana), Palaretto, 1857, *Savi* 242 (HUJ). CORSICA: 1919, *Consturier* (HUJ).

103b. T. vesiculosum Savi Var. **rumelicum** Griseb., Spicil. Fl. Rum. 1 : 35 (1843). *Typus*: In Rumelia orientalis Friv. (GOET). Zoh. in Davis, Fl. Turkey 3 : 407, f. 5 (1970).

Plate 100. *T. vesiculosum* Savi Var. *rumelicum* Griseb.
(Turkey: *Orshan* 55116, HUJ).
Flowering and fruiting branch; bract; flower with persistent bract; fruiting calyx with persistent bract and corolla.

T. multistriatum Koch, Syn. ed. 1, 2 : 190 (1844); Vis., Fl. Dalm. 3 : 300 (1850).

Icon. : [Plate 100]. Reichenb., Fl. Germ. 22 : 2154, f.5 (1903).

Differs from the type by less prominent transverse nerves of calyx tube. Fruiting calyx less inflated.

Gen. Distr. : Greece, Italy, Corsica, Albania, Serbia, W. Turkey.

Selected specimens : GREECE : Thessalnia superior : Penei infra Kun Kunovo, 2,000 ft, 1885, *Heldreich* 840 (E). ITALY : Calabria, 1899, *Fiori* (FI). TURKEY : Prov. Kirklareli : Derenköy near the pontice guard house, 630 m, 1961, *Karamanoğlu* (E); Thracia : 16 km N. of Tekirdag (road to Hayrabolu), 120 m, 1963, *M. Zohary* 55110 (HUJ); Istanbul distr. : Yorgandjibagtché, 1893, *Aznavour* 634 (G). CORSICA : Paturages, Porto Vecchio, 1849, *Kralik* 541 (V). YUGOSLAVIA : Vranya, 800 m, *Adamowić* 1900 (E).

104. T. aintabense Boiss. & Hausskn. in Boiss., Fl. 2 : 148 (1872); Hossain, Not. Roy. Bot. Gard. Edinb. 23 : 457 (1961); Zoh. in Davis, Fl. Turkey 3 : 407, f. 5 (1970). *Typus* : Prov. Gaziantep : "In graminosis calcareis ad Ispadrul prope Aintab", 10 June, 1865, *Haussknecht* (G holo.).

Icon. : [Plate 101].

Annual, glabrescent to glabrous, 25–50 cm. Stems slender, many, poorly branched, erect to ascending, with striate grooves. Leaves scattered, few and small; petioles rather short, those of lower leaves longer; stipules 0.3(–1) cm, membranous, longitudinally nerved, ovate, sheathing for almost half their length, free portion lanceolate; leaflets 0.6–1.4 × 0.3–0.8 cm, all obovate, mostly cuneate at base, dentate-spinulose all around. Peduncles short, up to 3–4 cm, sometimes 0.5–1 cm, prominently ribbed. Heads about 2 × 1.5 cm, ovate, 60–100-flowered. Bracts 3.5 mm, ovate-oblong, mucronate, scarious, very densely striate. Pedicels 0. Calyx 3–4 mm, tubular; tube about 40-nerved, white, with slightly hyaline base; teeth subulate with triangular base, greenish, one-third to one-half the length of tube. Corolla up to 8 mm long, pale pink, drying yellowish-brown; standard with ovate-obtuse limb and oblong, broad, slightly tapering claw; wings elliptical, acute, much shorter than standard, almost equal in length to acute keel. Ovary subsessile, ovate-oblong, scabrous, hairy above, 1–2-ovuled; styles 3 times as long as ovary, curved above. Fruiting calyx with strongly inflated tube and erect or slightly recurved teeth. Pod about 2 mm, ovate, 1-seeded, with long style. Seed 1.5 mm, ovoid, granulate, yellowish-brown. Fl. April–May.

Hab. : Field edges and among shrubs.

Gen. Distr. : S. and E. Turkey (endemic).

Selected specimens : TURKEY : S. Anatolia, 7 km E. of Adana, 1962, *M.* & *D. Zohary* 3812 (HUJ); Prov. Adana : Karatas, 1961, *Karamanoğlu* 20/1961 (E); Prov. Hatay : S. of Paias, near Alexandretta, 1931, *Eig* & *Zohary* (HUJ); Alexandretta, 1931, *Zohary* (HUJ); Prov. Maraş : 10 km S. E. of Maraş, 1962, *M.* & *D. Zohary* 27214 (HUJ); Bingöl distr., 22 km E. of Bingöl, 1,800 m, 1963, *Zohary* 473126 (HUJ). SYRIA : Eastern ranges of Cassius Mts., S. of the village of Urdu, 1931, *Zohary* (HUJ).

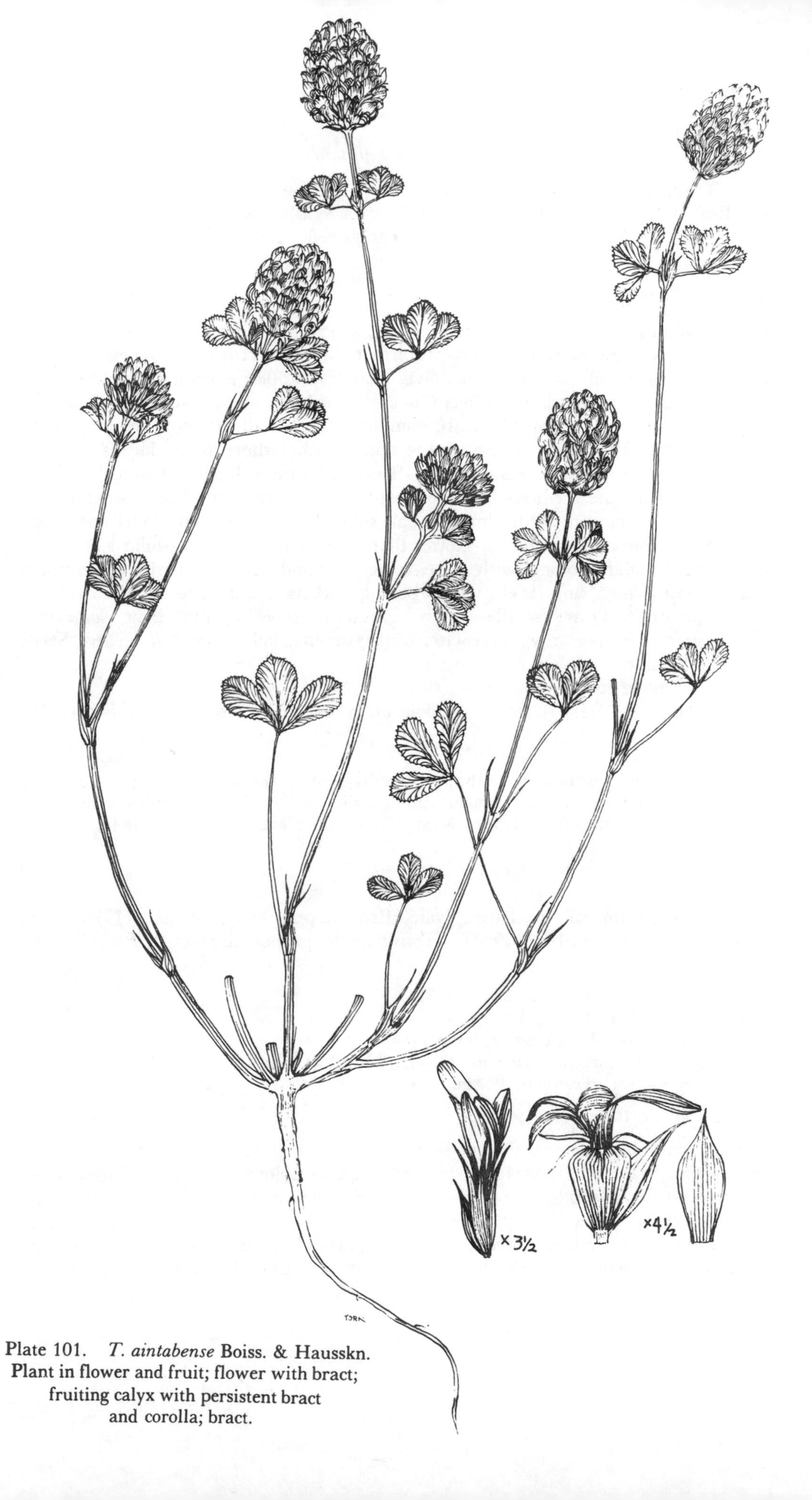

Plate 101. *T. aintabense* Boiss. & Hausskn.
Plant in flower and fruit; flower with bract;
fruiting calyx with persistent bract
and corolla; bract.

105. T. setiferum Boiss., Diagn. ser.1, 2 : 32 (1843) et Fl. 2 : 139 (1872); Hossain, Not. Roy. Bot. Gard. Edinb. 23 : 457 (1961 excl. syn.); Zoh. in Davis, Fl. Turkey 4 : 408 (1970). *Typus* : (Prov. Izmir) Montagne de Jenidje (Yenice), May 1842, *Boissier* (G ! holo.).

Icon. : [Plate 102].

Annual, glabrous, up to 50 cm. Stems many, procumbent, branching, striate, somewhat flexuose. Leaves rather short-petioled; petioles decreasing in length towards apex; stipules 1.5–2 cm, short-adnate to petioles, upper portion lanceolate, long subulate-acuminate, caudate; leaflets 0.8–2 × 0.4–1 cm, obovate-cuneate (a few of the uppermost ones almost elliptical), spinulose-dentate all around with long spiny mucro at apex. Peduncles axillary and terminal, rather short, thick. Heads 2–2.5 × 1.5–2 cm, ovate-globular, dense, many-flowered. Bracts longer than calyx tube, lanceolate-acuminate. Pedicels 0. Calyx slightly shorter than corolla; tube white, tubular, about 36-nerved; teeth almost equal, subulate-setaceous with lanceolate base, ciliolate in lower part, somewhat shorter than to as long as tube. Corolla 1.2–1.5 cm long, pink; standard oblong, with ovate, acute limb and oblong, tapering claw, much longer than wings and keel; wings oblong, acutish; keel shorter than wings, acute-mucronate. Ovary sessile, linear, gradually tapering into long, somewhat curved style. Fruiting calyx urceolate; teeth curved. Pod ovate, 1–2-seeded. Seeds brown. Fl. June.

Hab. : Among scrub.

Gen. Distr. : S. Italy (Calabria), Bulgaria, E. and S. Greece, Albania, Serbia, W. Turkey.

Selected specimens : BULGARIA : Silvan, Bekerdze, 1906, *Schneider* 455 (K). TURKEY : Lydia : in monte Güme Dagh (Mesogis), supra oppidum Tire (Tyrrha), 300–600 m, 1906, *Bornmüller* 9368 (E); Prov. Izmir : Mesogis inter Dervent et Alaşehir, 1842, *Boissier* (syntype : E).

106. T. argutum Sol. in Russ., Nat. Hist. Aleppo, ed. 2, 2 : 260 (1794); Eig, Journ. Bot. Lond. 75 : 188 (1937). *Typus* : Syria prope Aleppo, *P. Russell* (BM holo.).

T. ornatum Clarke, Travels 2, 2 : 336 (1814).
T. xerocephalum Fenzl, Pug. 1 : 5 (1842).
T. moriferum Boiss., Diagn. ser. 1, 9 : 28 (1849).
T. xerocephalum Fenzl var. *cruentum* Bornm., Ver. Zool.-Bot. Ges. Wien 48 : 581 (1898).
T. argutum Sol. var. *cruentum* (Bornm.) Eig, *loc. cit.*

Icon. : [Plate 103].

Annual, glabrous or sparingly pubescent, 10–30 cm. Stems few or many, erect or procumbent, branched. Leaves long-petioled; stipules oblong, with lanceolate-subulate free portion; leaflets 0.6–1.2 × 0.3–0.8 cm, obovate-cuneate to oblong, obtuse and mucronulate at apex, sharply dentate. Heads short-peduncled or sessile and involucrate by upper leaves, ovoid and about (0.8–) 1 cm in flower, somewhat elongating in fruit, many-flowered. Bracts as long as or longer than calyx tube, scarious,

Plate 102. *T. setiferum* Boiss.
Flowering and fruiting branches; fruiting calyx with persistent corolla; bract; dissected calyx; flower with bract.

Plate 103. *T. argutum* Sol.
(Palestine : *Eig*, *Zohary* & *Feinbrun*, HUJ).
Flowering and fruiting branch; fruiting calyx with persistent bract and corolla; flower.

oblong-cuneate to ovate, mucronate, membranous-margined. Calyx about 3 mm, white, becoming somewhat inflated and top-shaped in fruit, villous along 2 rows, about 36-nerved; teeth one-half to three-quarters the length of tube, more or less equal, spreading, subulate with broader base. Corolla 6–7 mm, persistent, scarious, white, pink or reddish, multi-striate; limb of standard erect, ovate to oblong, longer than erect keel and divergent or deflexed wings, turning pink or dirty white in fruit. Pod sessile, 1-seeded, with beak 3–4 times as long as fruit. Seed about 1 mm, ovoid, brown. Dispersal: entire heads separating from peduncles at maturity. Fl. January–May. 2n = 16.

Hab.: Roadsides, fields and among scrub.

Gen. Distr.: Greece, Aegean Islands, Rhodes, Turkey, Cyprus, Syria, Lebanon, Israel, Sinai, Egypt.

Selected specimens: GREECE: Dodecanese, Patmos, 1969, *Townsend* 61/64 (K). AEGEAN ISLANDS: Cos, 1802, *Clarke* s. n. (as *T. ornatum* Clarke, type BM). TURKEY: Mersin distr., Bouloukli, près de Mersine (Cilicia), 1855, *Balansa* 453 (E); Içel: Mut to Silifke, 5 mi. S. of Mut, 400 m, 1965, *Coode* & *Jones* 1024 (E). CYPRUS: Kissonera, 10 ft, 1955, *Marton* 2123 (K). SYRIA: 74 km S. W. of Damascus, 956 m, 1933, *Eig* & *Zohary* (HUJ); Sueda Jebel Druz, 1931, *Zohary* (HUJ). LEBANON: Beirut, 1931, *Zohary* (HUJ). ISRAEL: Sharon Plain: Jaqum, 1950, *Segal* 17443 (HUJ); Judean Mts.: Env. of Deir-esh-Sheikh, *N. Feinbrun* 17435 (HUJ); Upper Jordan Valley: Yarmouq bridge, 1930, *Naftolsky* 17447 (HUJ). SINAI: S. E. of El-Arish, 1925, *Eig* (HUJ).

Note: Var. *cruentum* (Bornm.) Eig, *loc. cit.*, is not a form but a state in which the flowers turn pink after anthesis.

107. T. mutabile Portenschl., Enum. Pl. Dalmat. 16, t. 12, f. 1 (1824). *Typus*: In Insula Lissa, in campo grande 4.

T. leiocalycinum Boiss. & Sprun. in Boiss., Diagn. ser. 1, 2: 31 (1843).
T. paleaceum Portenschl. (nomen in herb. 1824).

Icon.: [Plate 104].

Annual, glabrous, 15–50 cm. Stems many, erect or ascending, striate, branched from base and above. Lower leaves rather long-petioled, upper ones short-petioled to subsessile; stipules membranous, lanceolate, long subulate-acuminate; leaflets 1–3 × 0.4–1.5 cm, elliptical, acute, the lower ones obovate, all prominently nerved, serrulate-dentate, each tooth terminating in cusp, the cusp at apex up to 2 mm long. Inflorescences ovoid, many-flowered, elongated in fruit up to 9 × 2.5 cm. Peduncles terminal and axillary, often longer than subtending leaves, sometimes shorter. Bracts as long as calyx tube, lanceolate, long-acuminate, with prominent longitudinal nerves. Pedicels 0. Calyx 0.8–1.1 cm long; tube cylindrical, with 24 faint, slender, longitudinal nerves and no transverse nerves, not inflated in fruit; teeth about as long as tube or longer, lanceolate-subulate, 3-nerved at base, slightly hairy at margin, recurved in fruit. Corolla 1.4–2 cm long, purple, with prominent longitudinal nerves; standard lanceolate, acute to acuminate. Pod about 3 mm, oblong-elliptical, membra-

Plate 104. *T. mutabile* Portenschl.
Flowering branch; flower.

nous, 2–3-seeded, constricted between seeds. Seeds 1–1.3 mm, subglobular, brown. Fl. June–July. 2n = 16.
Hab. : Dry grassy places.
Gen. Distr. : Italy, Sicily, Albania, Yugoslavia, Greece, Turkey.

Selected specimens : ITALY : Apulia, Bari, Spinazzola, 500 m, 1913, *Fiori* 2106 (FI). YUGOSLAVIA : Dalmatia in Lesina, *Botteri* s. n. (K). GREECE : prope Anterrhion, 1893, *Halacsy* (HUJ). TURKEY : 4 km E. of Sarai towards Istanbul, *Crowford* (CPI 45853).

108. T. lugardii Bullock in Kew Bull., 1932 : 494 (1932). *Typus* : Kenya, Mt. Elgon, *Lugard* 97 (K holo.).

Icon. : [Plate 105].

Annual, subglabrous, up to 60 cm. Stems few, erect, dichotomously branched, grooved. Lower leaves long-petioled; stipules up to 3 cm, free portion about as long as sheathing part, triangular-lanceolate, sharp-pointed; leaflets 1.5–4.5 × 0.5–0.9 cm, lanceolate to narrowly oblong, obtuse or acute at apex, with many prominent nerves, margin denticulate, each tooth terminating in small cusp, teeth at apex longer, up to 2 mm. Inflorescences about 2 cm in diam., globular, many-flowered. Peduncles 5–6 cm long, pilose near base of head. Bracts scarious, the outer ones up to 12 mm long, forming involucre, broadly lanceolate, subulate at tip, 5–10-nerved, sometimes toothed at margin; inner bracts about 6 mm long, linear-lanceolate. Pedicels about 2 mm long, glabrous, erect after anthesis. Calyx 1–1.2 cm long, glabrous or with few hairs; tube about 30-nerved, whitish, strongly inflated and irregularly splitting in fruit; teeth twice as long as tube, triangular-subulate, with 5-nerved base. Corolla 9–10 mm long, purplish. Pod about 4 mm, glabrous, oblique, 4-seeded. Seeds about 2 mm, oblong, yellowish-brown. Fl. August–November. 2n = 16.
Hab. : Upland grassland and forest margins, especially in damp places; 1,800–2,550 m.
Gen. Distr. : Kenya, Uganda.

Specimen seen : KENYA : Trans Nzoia Distr., 7,500 ft, 1956, *Irwin* 307 (HUJ); *ibid.* , Mt. Elgon, forest on eastern slopes, 8,500 ft, *I. Knight* 40 1956 (K).

109. T. steudneri Schweinf. in Verh. Zool.-Bot. Ges. Wien 18 : 652 (1868). *Typus* : N. Ethiopia : bei Abbena am Ataba, ca. 6,000 ft, 8 Jänner, 1862, *Steudner* 159 (B holo.).

T. bellianum Chiov. in Ann. Ist. Roma 8 : 407 (1908).
T. schimperianum Hochst. ex E. G. Baker, Legum. Trop. Afr. 82 (1926).

Icon. : [Plate 106].

Annual, glabrous, 20–50 cm. Stems many, erect to spreading, branched from below and above, grooved. Leaves rather long-petioled, the upper ones shorter-petioled; stipules up to 2.5 cm long, free portion one-quarter of length, lanceolate-acute; leaflets 1.5–5 × 0.3–0.8 cm, linear-lanceolate, terminating in long cusp at apex, margin

Plate 105. *T. lugardii* Bullock
(Kenya : *P. H. Irvin* 307, HUJ).
Flowering and fruiting branch; flower.

Plate 106. *T. steudneri* Schweinf.
(Kenya : *R. Strange* 168, HUJ).
Plant in flower; flower; keel; wing; standard.

denticulate. Inflorescences 1.3 cm in diam., globular or nearly so. Outer bracts up to 7 mm long, oblong, abruptly and sharply pointed; inner bracts 3 mm long, oblong, green-tipped. Pedicels 1 mm long or shorter, erect after anthesis. Calyx 5–7 mm long; tube campanulate, white, membranous, about 40-nerved, inflated in fruit; teeth about as long as tube or shorter, triangular with few hairs at margin. Corolla 7–8 mm, purple; standard oblong, obtuse. Pod about 4 mm, obovoid, oblique, 2-seeded. Seeds 1.2 mm, ovoid, light brown. Fl. August–October. 2n = 16.

Hab.: Upland grassland and forest margins, especially in damp places; 1,800–2,400 m.

Gen. Distr.: Uganda, Kenya, Ethiopia, Eritrea.

Selected specimens: KENYA: Kitale Distr., 1951, *Strange* 168 (HUJ). ERITREA: Sarae: Gaza Gobo, 1,900 m, 1902, *Pappi* 342 (FI). ETHIOPIA: Amhara Bacino del Tana Piana di Guramba, 2,080 m, 1939, *Agr. di Gondar* II (FI).

110. T. quartinianum A. Rich., Tent. Fl. Abyss. 1 : 169 (1847). *Typus*: N. Ethiopia, near Tchalatchekanne, *Quartin Dillon* s. n. (P holo.).

Icon.: [Plate 107].

Annual, subglabrous, 20–40 cm. Stems few, erect or spreading, stout, branching from base and above, grooved. Leaves rather long-petioled in lower part and shorter-petioled to subsessile in upper part; stipules 1–2 cm long, upper third free, lanceolate, lower adnate part ovate, membranous; leaflets 1.5–5.2 × 0.6–1.6 cm, elliptical to lanceolate, rounded to acute, margin with fine teeth terminating in cusp. Inflorescences about 2 cm in diam., globular, many-flowered. Peduncles longer than subtending leaves. Outer bracts up to 7 mm long, ovate, lobed or toothed, scarious-margined, with many prominent nerves, imbricate and forming involucre. Pedicels about 1 mm, glabrous, erect after anthesis. Calyx 0.9–1.1 cm long, glabrous or nearly so; tube about 30-nerved, campanulate, strongly inflated in fruit, membranous; teeth about as long as tube, triangular at base, with scarious margin, overlapping when mature, subulate above. Corolla 1.1–1.2 cm long, purple. Pod about 4 mm, oblong, pointed, (1–2–)4–5-seeded, with wide dorsal suture. Seeds 1.2–1.8 mm, elliptical-oblong, golden-brown. Fl. August–October.

Hab.: Short grassland; 1,500–2,600 m.

Gen. Distr.: Eritrea, N. Ethiopia, Uganda.

Selected specimens: ERITREA: Oculé Cusai: Loggo Sarda, Deggahenm, 2,600 m, 1902, *Pappi* 1369 (FI). ETHIOPIA: 1853, *Schimper* 642 (FI).

Plate 107. *T. quartinianum* A. Rich.
Plant in flower and fruit:
fruiting calyx with persistent corolla;
fruiting calyx with persistent bract and corolla.

Section Four. VESICARIA Crantz.

Stirp. Austr. ed. 2 : 404 (1769) pro max part.; Savi in Bertol., Fl. Ital. 8 : 185 (1850) p. p.; Zoh. & Heller, Israel J. Bot. 19 : 316 (1970)

Section *Vesicastrum* Ser. in DC., Prodr. 2 : 202 (1825) p. p.
Genus *Galearia* Presl, Symb. Bot. 1 : 49 (1830).
Section *Fragifera* Koch, Syn. Fl. Germ. Helv. 171 (1835).
Section *Galearia* (Presl) Godr. in Gren. & Godr., Fl. France 1 : 413 (1848); Boiss., Fl. 2 : 111 (1872); Gibelli & Belli, Mem. Accad. Sci. Torino ser. 2, 41 : 3 (1890).
Subgenus *Galearia* (Presl) Hossain, Not. Roy. Bot. Gard. Edinb. 23 : 446 (1961).

Lectotypus : *T. fragiferum* L.

Annuals or perennials. Heads pedunculate or sessile, fruiting ones globular or ovoid or stellate. Flowers many, sessile or pedicellate, bracteolate, sometimes lower bracts forming involucre. Calyx bilabiate; upper lip bidentate, tube growing in fruit into inflated, vesicular, reticulately nerved body, with teeth becoming setaceous; lower lip tridentate, unchanged in fruit; throat of fruiting calyx narrowed or closed. Corolla resupinate or not, marcescent or deciduous; standard free or more or less connate with other petals. Ovary elliptical or ovate, with straight or geniculate style. Pod 1–2-seeded, enclosed in scarious or membranous, inflated, hairy or glabrous calyx tube that generally functions as dispersal unit.
Seven Eurasian Species.

Synopsis of Species in Section Four

Section Four. VESICARIA Crantz
- 111. T. fragiferum L.
- 112. T. physodes Stev. ex M. B.
- 113. T. tumens Stev. ex M. B.
- 114. T. resupinatum L.
- 115. T. clusii Godr. & Gren.
- 116. T. tomentosum L.
- 117. T. bullatum Boiss. & Hausskn.

Key to Species

1. Perennials with creeping or caespitose stems. Petals not resupinate — 2
– Annuals. Petals resupinate — 4
2. Heads distinctly involucrate by relatively larger bracts of lower flowers. — **111. T. fragiferum**
– Heads not at all or rarely very inconspicuously involucrate — 3
3. Flowers 1–1.5 cm. Flowering heads 2–2.5 cm. — **112. T. physodes**
– Flowers 0.6–0.8(–1) cm. Flowering heads 1.5(–2) cm. — **113. T. tumens**

4 (1). Fruiting heads star-shaped, i.e., forming group of stellately spreading calyces. Fruiting calyx ellipsoidal or pyriform, with open throat and erect or divergent teeth. Flowers 0.6–1.2 cm, usually fragrant. Leaflets of uppermost leaves oblong-elliptical to rhombic. **114. T. resupinatum**

– Fruiting heads not star-shaped. Fruiting calyx not as above 5

5. Fruiting heads with marcescent corollas, all borne on fairly long peduncles deflexed or bent near top. Flowering heads about 6–9 mm. Leaflets obovate, never truncate or notched at apex. Upper calyx teeth long and well developed in fruit. **115. T. clusii**

– Fruiting heads without corollas. Peduncles, if present, erect or regularly bowed. Flowering heads smaller than above. Leaflets, at least those of upper leaves, truncate or notched at apex, usually cuneate at base 6

6. Flowering heads 3–5 mm, few-flowered. Lower calyx teeth 1–2 mm, triangular. Fruiting heads never cottony, usually glabrous or glabrescent, 4–7(–8) mm, all borne on long, regularly bowed peduncles. Fruiting calyx often almost globular, with upper teeth very short or reduced to callosities. **117. T. bullatum**

– Flowering heads 5–7 mm, many-flowered. Lower calyx teeth 2–3 mm, usually lanceolate or linear. Fruiting heads generally hairy, cottony or with calyces showing prominent reticulate nervation, 0.7–1.2 cm, sessile or part of them borne on erect or bowed peduncles. Fruiting calyx mostly ovoid, with conspicuous, recurved or divaricate upper teeth. **116. T. tomentosum**

111. T. fragiferum L., Sp. Pl. 772 (1753).

Perennial, branching from neck, with prostrate or creeping stocks, 5–50 cm long, rooting from nodes. Leaves 2–20 cm, with erect, hairy or glabrous petioles; stipules up to 2 cm long, linear to lanceolate-linear, membranous, with acuminate or subulate free portion; leaflets 0.5–3 × 0.3–1.5 cm, ovate to obovate-elliptical, rarely ovate or suborbicular, cuneate at base, obtuse or retuse, spinulose-toothed, glaucous or glaucescent, glabrous or hairy along nerves and margin. Peduncles axillary, terete, furrowed, villous or glabrescent. Flowering heads hemispherical to ovoid, 0.8–2 cm, 10–30-flowered, subtended by a 2–6 mm long involucre made up of united, entire or toothed bracts of lower flowers and often concealing calyx tubes of latter. Flowers 6(–8) mm long, short-pedicelled, each subtended by entire or bifid, more or less ciliate bract. Calyx tubular-obconical, bilabiate, often reddish, mostly pilose to woolly on upper side, rarely all over; teeth unequal, lanceolate, the upper ones becoming subulate-aristate, erect or recurved, shorter than or almost as long as or longer than tube, the 3 lower ones somewhat broader and shorter than upper ones, often ciliate. Corolla whitish, pink or flesh-coloured, considerably longer than calyx, marcescent in fruit, not resupinate (rarely slightly resupinate); standard with marginate or crenulate limb, free or slightly connate with other petals or with staminal tube; wings somewhat shorter than standard and somewhat longer than keel, oblong to oblong-lanceolate, auricled; keel obtuse, subapiculate, without auricles. Ovary substipitate, linear, with erect or slightly curved style. Fruiting head 1–2.5 cm, ovoid or globular; fruiting calyx deflexed, inflated, reticulately nerved, pilose to hispid above or all over, with marcescent corolla largely or slightly protruding from calyx. Pod ovoid to globular, 1–2-seeded. Seeds reniform, brown, 1–1.4 mm long. Fl.: April–November. 2n = 16.

Hab.: Grassy and muddy places, sometimes on saline ground.

Gen. Distr. : All European countries (incl. British Isles), W. Asia (incl. W. Siberia), all Mediterranean countries, Anatolia, Iraq, Caucasus, Iran, Afghanistan, Pakistan, Middle Asia, Turkestan. Also recorded from Ethiopia. Cultivated in some countries, especially in Australia.

1. Fruiting heads 1.4–2.5 cm across. Plants up to 50 cm. **111b.** var. **majus**
– Fruiting heads up to 1.5 cm across. Plants usually smaller 2
2. Heads few-flowered, loose. Dwarf, alpine, caespitose plants. **111d.** var. **modestum**
– Heads usually many-flowered, more or less compact 3
3. Stems short, woody, congested, with long scarious stipules covering internodes. Leaves crowded; leaflets small to minute. Peduncles 2–5 cm 4
– Stems longer, not congested. Leaflets about 1–3 cm. Peduncles 10 cm or more. **111a.** var. **fragiferum**
4. Teeth of calyx very unequal, the lower ones at most half the length of upper ones; fruiting calyx about 5 mm. **111c.** var. **pulchellum**
– Teeth of calyx almost equal, all about 2 mm; fruiting calyx about 8 mm. **111e.** var. **orthodon**

111a. T. fragiferum L. Var. **fragiferum**. *Typus*: Herb. Linn. 930/54 (holo); Boiss., Fl. 2 : 135 (1872); Gib. & Belli, Mem. Accad. Sci. Torino, ser. 2, 41 : 22 (1890); Reichenb., *loc. cit.*, Fl. Germ. Helv. 22 : /1 (1903) incl. var. *ericetorum* Reichenb. f. in Reichenb., *loc. cit.*

T. bonanni Presl, Delic. Prag. 51 (1822).
Galearia fragifera (L.) Presl, Symb. Bot. 1 : 50 (1831).
T. congestum Link, Linnaea 9 : 584 (1835) non Guss. (1831).
T. neglectum C. A. Mey., Suppl. ad Ind. Sem. Hort. Petrop. 21 (1843).
T. fragiferum L. var. *bonanni* (Presl) Gib. & Belli, *op. cit.* 23.

Icon. : [Plate 108]. Gib. & Belli, *loc. cit.* t. 1, f. 3; Reichenb., *loc. cit.* 106 et 113; Hegi, Ill. Fl. Mitteleur. 4, 3 : t. 164 (1923).

Selected specimens : SWEDEN : Gotland, *Wisby* 1981 (W). FINLAND : Alandia, *Seppäla* 7637 (P). NETHERLANDS : prov. Friesland n. Oosterend, 1951, *Kramer-Lindman* (W). BRITISH ISLES : Outer Hebrides, S. Mist, *Meinertzhagen* (W). GERMANY : Rheinprovinz, bei Köln, Rheinufer, 1934, *Hupke* (W). ESTONIA MARITIMA : parish of Kihelkonna, S. of village Austla, 1933, *Saarsoo* (HUJ). LATVIA : Dunnenmunde, haud procul a Riga, 1897 (P). AUSTRIA : Burgenband, 1926, *Ferny* (W). HUNGARY : near Neusiedler See, 1908, *Veller* (W). RUSSIA MERIDIONALIS : Amu Darya, *Kavalkov* & *Krause* (W). PORTUGAL : *Welwitsch* 203 (G). SPAIN : Aragon, 1872, *Bordere* (G). FRANCE : Près de Nancy, 1848, *Mathieu* 344 (G); Bois de Boulogne, 1822, *Choisy* (G). ITALY : Sicily, Palermo, 1899, *Ross* 226 (G). YUGOSLAVIA : Dalmatien, Golf von Cattaro, 1926, *Korb* (W). BULGARIA : Mt. Rhodope, 1954, *Stojanoff* et al. 362 (W). TURKEY : Mersine, 1931, *Eig* & *Zohary* (HUJ). CYPRUS : Platres, 1931, *Naftolsky* (HUJ). SYRIA : betw. Selemie and Hama, env. of 'Ein Kasrein, 1932, *Eig* & *Zohary* (HUJ). LEBANON : Hamana, 1941, *Feinbrun* (HUJ). PALESTINE : n. Haifa, 1923, *Eig* 813 (HUJ) (also collected from Sharon Plain, Philistean Plain, Upper and Lower Galilee, Esdraelon Plain, Dan Valley, Hula Plain and elsewhere). EGYPT : in incultis ad vias Nougzap et Sidi Gaber pr. Ramle, 1880, *Letourneux* 242 (W). IRAQ : Deltawah, 1932, *Guest* 30785 (K). IRAN : S.W. ca. 70 km S. of Dehbid n.

Plate 108. *T. fragiferum* L. Var. *fragiferum*
(Palestine : *Grizi* 533, HUJ).
Plant in flower and fruit; fruiting calyx with persistent corolla.

Passargad, 1960, *M.* & *D. Zohary* (HUJ). AFGHANISTAN : Khamabad, *Volk* 662 (W). TRANSCASPIA : Kisil Arvat in Valle Yeldere, *Freyn* 1923 (G). Also seen from other countries of the Mediterranean region as well as from Poland, Czechoslovakia, Romania, Kashmir and the Himalayas.

111b. T. fragiferum L. Var. **majus** Rouy, Fl. Fr. 5 : 91 (1899). *Typus* : Russia, Prov. Voronesh, vic. Bobrov, 1904, *B. Maximov* 164 (HUJ lecto.)

T. ampulescens Gilib., Fl. Lith. 89 (1782).
T. fragiferum sensu Bobrov, Acta Inst. Bot. Acad. Sci. URSS ser. 1,6 : 250 (1947).

Selected specimens : GERMANY : Breslau, Güterbahnhof, 1936, *Meyer* (HUJ). SWITZERLAND : Sion Vallis, 1945, *Kichler* (G). RUSSIA AUSTRALIS : Azov, 1887, *Laupman* (W).

111c. T. fragiferum L. Var. **pulchellum** Lange, Meddel. Nat. Floren. 2, Aart. 7 : 169 (1865); Aschers. & Graebn., Syn. Mitteleur. Fl. 6(2) : 524 (1908); Hossain, Not. Roy. Bot. Gard. Edinb. 23 : 448 (1961); Zoh. in Davis, Fl. Turkey 3 : 409 (1970). *Typus* : La Coruna, in rupibus maritimis 1851–52, *Lange* (P lecto.).

T. fragiferum L. var. *alicola* Gib. & Belli, *loc. cit.* 22.
T. bonanni Presl var. *aragonense* Willk. & Lange, Prodr. Fl. Hisp. 3 : 361 (1877).

Selected specimens : PORTUGAL : Arrenodes de Fara, 1882, *d'A. Guimarares* 1684 (P). SPAIN : in saxosis inter calles Puerte de Daroca et Vanta del Puerte pr. Molina in Aragonia, 1850, *Willkomm* 458 (P). CORSICA : Etang d'Urbino, 1913, *Briquet* (G). CRETE : Distr. Malerizi pr. Gazi, 1942, *Rechinger* 14030 (W). TURKEY : pr. Niğde, Aksaray-Sultanhani, 900–950 m, 1958, *Davis* & *Hedge* 032820 (E). CYPRUS : Mts. Trodos-Platros, 1,300 m, 1939, *Feinbrun* (HUJ). SYRIA : Mt. Hermon, Wadi Shibbah, 1,200 m, 1929, *Eig* (HUJ). LEBANON : mountains above Ehden, ca. 1,600 m, 1931, *Eig* & *Zohary* (HUJ). ISRAEL : Upper Galilee, Tarshiha, 1954, *Waisel* 16793 (HUJ). IRAQ : N. of Baghdad, 1946, *Robertson* 4 (E). IRAN : prov. Kerman, inter Kerman et Niris, 2,000 m, 1892, *Bornmüller* 3638 (W). AFGHANISTAN : Panjas, 2,600 m, 1948, *Rechinger* 2710 (E); bei Kabul 1,780 m, 1950, *Gilli* 1668 (W). TRANSCASPIA : ad fines Persiae, Aschabad, Soluklu, 1,900 m, *Sintenis* 1016 (W). Also seen from other Mediterranean and European countries.

111d. T. fragiferum L. Var. **modestum** (Boiss.) Gib. & Belli, *loc. cit.* 22; Post, Fl. Syr. Pal. Sin. 240 (1883–1896). *Typus* : Lebanon : In declivibus Libani supra Cedros, 1846, *Boissier* (G lecto.).

T. modestum Boiss., Diagn. ser. 1, 9 : 27 (1849) et Fl. 2 : 137 (1872); Dinsmore in Post, Fl. Syr. Pal. Sin. ed. 2, 1 : 344 (1932).

Selected specimens : LEBANON : ad Bsherre et circa Cedretum in jugo montis Makmel, 2,400 m, 1885, *Kotschy* 296 (W). We have also collected this variety in other parts of the Lebanon mountains, e.g., Djebel Matrafe, Ras el Barkawiyeh, 2,320 m; between Ehden and Talie, 2,015–2,100 m (all in HUJ).

111e. T. fragiferum L. Var. **orthodon** Zoh. in Zoh. & Heller, Israel J. Bot. 19 : 320 (1970). *Typus*: Afghanistan, 2,600 m, 1948, *M. Köie* 2710 (E holo.).

112. T. physodes Stev. ex M. B., Fl. Taur.-Cauc. 2 : 217 (1808).

Perennial with thick, fusiform roots, glabrous or glabrescent, (10–)20–50 cm. Stems few to many, decumbent to erect, not rooting from nodes, leafy; lower leaves long-petioled, upper ones short-petioled to almost subsessile; petioles often hairy; stipules often crowded on lower part of stems, lanceolate, herbaceous or membranous between nerves, with free portion long-cuspidate; leaflets (0.5–)1–2(–3) × 0.5–1.5 cm, very short-petiolulate, varying in same plant from obovate-cuneate to elliptical and ovate and rarely lanceolate, rounded or retuse at apex, mucronate, acutely dentate or somewhat spinulose-dentate, glabrous or hairy only beneath. Peduncles shorter or somewhat longer than leaves. Heads few, terminal and axillary, 1.5–2.5 cm across, ovoid to globular, not involucrate. Bracteoles membranous, shorter than pedicels, the latter sometimes as long as calyx tube. Calyx tubular, bilabiate, white or reddish, hairy all over or only on upper face or altogether glabrous; tube 3–4 mm, strongly and unilaterally inflating after flowering, about as long as slightly unequal, lanceolate-subulate teeth; upper teeth often somewhat longer than others, soon recurving. Corolla pink; standard 1–1.4 cm, somewhat deflexed, free, entire or somewhat retuse or denticulate at apex, much longer than wings and keel. Fruiting calyx obovoid, inflated on upper adaxial side, villous or glabrous, reticulately veined. Pod oblong, 1–2-seeded. Seeds brown, 1–1.2 mm. Dispersal both by single calyces and entire heads. Fl. April–August. 2n = 16.

112a. T. physodes Stev. ex M. B. Var. **physodes**. Boiss., Fl. 2 : 136 (1872); Bobrov, Acta Inst. Bot. Acad. Sci. URSS ser. 1, 6 : 257 (1947). *Typus*: (Georgia) Iberia, *Steven* (H lecto. LE photo).

T. alatum Biv., Stirp. Rar. Sic. 4 : 14 (1816).
T. cupani Tin., Pl. Rar. Sic. Pug. 1 : 16 (1817).
Galearia cupani (Tin.) Presl, Symb. Bot. 1 : 50 (1831).
T. anomalum Bory & Chaub., Exped. Morée, Bot. 3 : t. 26 (1823) et Nouv. Fl. Pelop. 51 (1838).
T. ovatifolium Bory & Chaub., Nouv. Fl. Pelop. 51, t. 28, f. 1 (1838).
T. durandoi Pomel, Nouv. Mat. Fl. Atl. 101 (1874).
T. clausonis Pomel, *loc. cit.*
T. raddeanum Trautv., Acta Hort. Petrop. 10 : 105 (1887).
T. fragiferum L. ssp. *physodes* (Stev.) Gib. & Belli var. *sericocalyx*, var. *balansae* et var. *durandoi* (Pomel) Gib. & Belli, Mem. Accad. Sci. Torino ser. 2, 41 : 30 et 31 (1890).
T. amani Post & Beauverd ex Dinsm., Publ. Amer. Univ. Beirut Nat. Sci. ser. 2 : 5 (1932).
T. rechingeri Rothm., Bot. Jahrb. 73 : 438 (1944).

Icon.: Gib. & Belli, *loc. cit.* t. 1, f. 4.

Calyx tube densely villous all over or on upper face (side) only.

Hab.: Open ground, clearings in forests and maquis, grassy places; s.l.–1,800 m.

Gen. Distr.: Portugal, S. Europe, Greece, Crete, Cyprus, Caucasus, Turkey, W. Iran, W. Syria, N. Iraq, Palestine, N. Africa.

Selected specimens: PORTUGAL: Lisboa pr. Estremadura, pr. Bellas, 1938, *Rothmaler* 13097 (G). ITALY: Calabria orientalis pr. urb. Geroie, 1877, *Porta* & *Rigo* 94 (P). SICILY: Prota Mts., 1822, *Presl* (W). YUGOSLAVIA: Dalmatia, Cattaro, 1926, *Korb* (W). GREECE: Attica, Mt. Parnes, *Guil.* 111 (HUJ). CRETE: 8 mi. E. of Ag. Barbara, 480 m, 1964, *Orshan* 07503-4 (HUJ); Distr. Siria prop. Chandros, 600 m, 1942, *Rechinger* 12676 (G, holotype of *T. rechingeri*). TURKEY: in Monte Amanus, 1903, *Shepard* 209 (HUJ, holotype of *T. amani*); Central Anatolia, Kizilcahamam, 1963, *Orshan* 51331 (HUJ). CYPRUS: ad rivulos in reg. Prodromus, 1862, *Kotschy* 704 (W). ALGERIA: in pratis montis Dira, 1,200 m, 1936, *Maire* (HUJ); *ibid.*: Mont Mouzaia, 1858, *Clauson* (P, holotype of *T. clausonis* Pomel). MOROCCO: Atlas Centr., env. d'Ifrane, 1964, *Sauvage* 12351 (HUJ, as var. *durandoi*). IRAN: pr. Azerbaidjan, Mt. Astara pr. Asbine, *Mirdamadi* 557 (W). Also recorded by Hossain (*loc.cit.*) from Palestine; we did not find a single specimen in Palestine among the hundreds observed.

112b. T. physodes Stev. ex M. B. Var. **psilocalyx** Boiss., Fl. 2: 136 (1872); Hossain, Not. Roy. Bot. Gard. Edinb. 23: 450 (1961). *Typus*: Antilebanon pr. Rascheya, 1846, *Boissier* (G !, iso. K).

T. sclerorrhizum Boiss., Diagn. ser. 1, 9: 28 (1849).
T. germaniciae Post ex Gib. & Belli, *loc. cit.* 30 nom. nud.; Post, Fl. Syr. Pal. Sin. 240 (1883–1896) et ed. 2, 1: 344 (1932).
T. physodes Stev. ex M. B. var. *sclerorrhizum* (Boiss.) Dinsm. in Post, Fl. Syr. Pal. Sin. ed. 2, 1: 344 (1932).
T. amani Post & Beauverd var. *glabrescens* Thiéb., Fl. Lib.-Syr. 2: 28 (1940).

Icon.: [Plate 109].

Calyx totally glabrous, rarely with few scattered hairs.

Selected specimens: TURKEY: Amanus, Bailan, 1862, *Kotschy* 299 (W); prov. Gaziantep, Duluk Baba about 7 km N. of Gaziantep, 1,100 m, 1957, *Davis* & *Hedge* D. 28060 (E). SYRIA: 74 km S.W. of Damascus, 1933, *Eig* & *Zohary* (HUJ). PALESTINE: Upper Galilee, betw. Sassa and Hurfesh, 1925, *Naftolsky* 606 (HUJ) (also found in the Judean Mts. and Gilead). IRAQ: near Amadia, 1932, *Majed Mustafa* 3599 (K).

113. T. tumens Stev. ex M. B., Fl. Taur.-Cauc. 2: 217 (1808).

Perennial, rhizomatous, glabrous, somewhat woody at base, 15–40 cm. Stems many, ascending or procumbent. Lower leaves long-petioled, upper ones almost sessile; stipules ovate-lanceolate, prominently nerved, semi-membranous, free portion subulate; leaflets 0.8–2 × 0.6–1.5 cm, obovate with cuneate base, retuse or rounded at apex, prominently nerved, acutely dentate-serrate. Heads hemispherical to spherical, (13–)15–18 mm across, not involucrate, axillary and terminal on slender, hairy or glabrous peduncles longer than leaves. Pedicels very short to as much as half the length of calyx tube, hairy or glabrous, deflexed after flowering. Bracteoles white,

Plate 109. *T. physodes* Stev. ex M.B. Var. *psilocalyx* Boiss.
(Palestine : *Naftolsky* 16578, HUJ).
Flowering plant; flower; fruiting calyx with persistent corolla.

membranous, minute to obsolete. Flowers (6–)7–9 mm. Calyx bilabiate; tube cylindrical, later inflating on upper side, densely hairy; teeth lanceolate to subulate, the upper ones somewhat spreading, the lower ones erect, all shorter than to as long as tube. Corolla pink to pale purple, not resupinate, almost twice as long as calyx, marcescent-persistent. Fruiting calyx 2(–3) mm across, transversely wrinkled. Pod stipitate, ovate, glabrous to hairy above, 1(–2)-seeded, with erect style longer than pod. Seeds brown, ovoid, 1–1.2 mm. Fl. May–August. 2n = 16.

Hab. : Forests, forest clearings, damp meadows.

Gen. Distr. : Turkey, Iran, Caucasus.

113a. T. tumens Stev. ex M. B. Var. **tumens**. Boiss., Fl. 2 : 136 (1872); Bobrov, Acta Inst. Bot. Acad. Sci. URSS ser. 1, 6 : 258 (1947). *Typus* : In Iberia ad fluvium Iberum et Alazanium, *Steven* (H lecto.).

Galearia tumens (Stev. ex M. B.) Presl, Symb. Bot. 1 : 50 (1831).
T. tumens Stev. ex M. B. var. *majus* Boiss., *loc. cit.* 137.
T. fragiferum L. ssp. *tumens* (Stev. ex M. B.) Gib. & Belli, *loc. cit.* 34.

Icon. : [Plate 110].

Heads 1–1.3 cm across. Flowers 6–7(–8) mm long. Calyx teeth shorter than tube.

Selected specimens : TURKEY : N. E. Anatolia, above Artvin, 940 m, 1964, *Zohary* & *Plitmann* 2561-5 (HUJ). CAUCASUS : Azerbaydzhan, in graminosis prope Lenkoran, 1836, *Hohenacker* (HUJ). IRAN : in Valle Sheheristanek montium Elburz, ca. 2,200 m, 1902, *Bornmüller* 6599 (K).

113b. T. tumens Stev. ex M. B. Var. **talyschense** (Chalilov) Zoh. in Zoh. & Heller, Israel J. Bot. 19 : 322 (1970). *Typus* : (Azerbaydzhan) Lerik, angustiae fl. Orand-czai, 6.6.1946, *Chalilov* (holo.).

T. talyschense Chalilov, Not. Syst. Inst. Bot. Acad. Sci. URSS 12 : 118 (1950).

Heads 1.8–2 (–2.2)cm. Flowers longer than in var. *tumens*. Calyx teeth as long as tube.

114. T. resupinatum L., Sp. Pl. 771 (1753).

Annual, glabrous or glabrescent, 20–60 cm. Stems mostly furrowed and hollow, ascending or erect, branching mainly from lower part. Leaves with hairy or glabrous petioles; stipules membranous, ovate-oblong, with longitudinally striated nerves, united below, free portion subulate, as long as or longer than united part, those of upper leaves much shorter; leaflets 1–2.5 cm long, rhomboidal or ovate-oblong and tapering at base, spinulose-dentate with alternately larger and smaller, upwardly oriented teeth, apex rounded-tapering (never truncate or emarginate). Peduncles usually much longer than leaf. Head axillary, many-flowered, 1–1.5 cm across. Pedicels thick, glabrous, one-quarter to one-half the length of white, glabrous calyx

Plate 110. *T. tumens* Stev. ex M.B. Var. *tumens*
(Iran : *Bot. Exp.* 2048, K).
Plant in flower and fruit; flower; fruiting calyx with persistent corolla.

tube. Flowering calyx lobes linear-lanceolate, much shorter than tube, unequal, the upper ones much longer. Corolla resupinate, deep pink to purple, fragrant; standard oblong, notched at apex and apiculate in sinus, (5–)6–8 mm long; wings longer than keel and about one-half the length of standard. Fruiting head stellately globular. Fruiting calyces inflated, ovoid or elliptical, up to 8 mm long, with prominent reticulate venation, diverging from one another; upper lobes long, divergent. Pod membranous, lenticular, dehiscing at thickened sutures, 1-seeded. Seed ovoid, brown, about 1.2 mm long. Dispersal by single calyces or by entire heads. Fl. March–May. 2n = 14, 16, 32.

Hab. : Mostly heavy soils; fields, riverbanks, roadsides, waste places.

Gen. Distr. : C. and S. Europe, all Mediterranean countries, S. W. Asia.

1. Stems hollow, up to 80 cm long and up to 5 mm thick. Flowering heads dense, 1–1.5 cm across. Flowers 0.8–1 cm. Leaflets up to 3 cm long. Plants mostly cultivated. **114c.** var. **majus.**
– Stems mostly solid, 20–40(–60) cm long, slenderer than above. Flowering heads loose, less than 1 cm across. Leaflets 1–1.5(–2) cm long. Plants not cultivated. 2
2. Fruiting heads up to 1.3–2 cm across. **114a** var. **resupinatum**
– Fruiting heads 0.8–1 cm. **114b.** var. **microcephalum**

114a. T. resupinatum L. Var. **resupinatum**. Gib. & Belli, Mem. Accad. Sci. Torino ser. 2, 41 : 10 (1890); Bobrov, Acta Inst. Bot. Acad. Sci. URSS ser. 1, 6 : 259 (1947); Hossain, Not. Roy. Bot. Gard. Edinb. 23 : 451 (1961) p. p. *Typus* : Described from Europe, Herb. Linn. 930/52 (BM).

T. bicorne Forssk., Fl. Aeg.-Arab. 139 (1775).
T. formosum Curt. ex DC., Prodr. 2 : 200 (1825) non D'Urv. (1822).
Galearia resupinata (L.) Presl, Symb. Bot. 1 : 50 (1831).
T. resupinatum L. var. *robustum* Rouy, Fl. Fr. 5 : 92 (1899).
T. resupinatum L. var. *gracile* Rouy, *loc. cit.* 93.

Icon. : [Plate 111]. Reichenb., Ic. Fl. Germ. Helv. 22 : t. 107 (1903); Gib. & Belli, *loc. cit.* t. 1, f. 1; Zohary, Fl. Pal. t. 244.

Selected specimens : SWITZERLAND : Genève, bord de lac, 1936, *Becherer* (G). AUSTRIA : Vienna, 1879, *Braun* (W). HUNGARIA AUSTRALIS : in pratis inundatis ad Bazias, etc., 1822, *Borbas* (W). ROMANIA : inter pag. Gradiste et Comona, 1957, *Molea* 91 (HUJ). PORTUGAL : Paro, 1936, *Kostermans* et al. 716 (HUJ). SPAIN : Algecinar, 1873, *Winkler* (W); env. de Carveneros, 1880, *Martinez* (W). MALTA : 1835, *Brenner* (HUJ). FRANCE : Montpellier, bois de Gramant, 1838, *Sender* (W). ITALY : pr. Santa Severa in Latis, 1895, *Pappi* (HUJ); SICILY : Messina, 1881, *Borzi* (W). YUGOSLAVIA : Suva Planina, *Ilic* (W); Pravosa bei Ragusa, 1886, *Pichler* (W). TURKEY : Smyrna, 1827, *Fleischer* 109 (W); pr. Adana, distr. Feke, Sencan Dere, betw. Gürümze and Süphandere, 1,000 m, 1952, *Davis* et al. 19620 (W). CYPRUS : pr. Monast. Cantara, 1880, *Sintenis* 409 (G). SYRIA : ad canalem Mahmudie, 1855, *Kotschy* 900 (W); Homs, 1,000 ft, 1910, *Haradjian* 3311 (G). LEBANON : Beit Meri, 1878, *Peyron* 64 (G). PALESTINE : Sharon Plain, Benyamina-Caesarea, 1932, *Davis* 4431 (HUJ) (common elsewhere). EGYPT : prope Abu Zabel, 1835,

Plate 111. *T. resupinatum* L. Var. *resupinatum*
(Palestine : *Plitmann*, HUJ).
Plant in flower and fruit; flower; fruiting calyx.

Schimper 10 (W); in Egypt inf., 1835, *Wiest* 127 (K). LIBYA: Medmet Fajun, 1876, *Ascherson* (P). ALGERIA: Env. d'Algèr, 1851, *Jamin* (W); Biskra in hortis, 1904, *Chevalier* (K). MOROCCO: Tanger, 1803, *Schousbee* (W). CANARY ISLANDS: Gomera, Baranco de Aqua Hilba, 1905, *Pitard* 130 (G). IRAQ: Kallat Shergat, 170–900 m, *Maresh* 71 (W); Amara, S. of Ali Gharbi, 1958, *Haddad* 125743 (K). CAUCASUS: Azerbaydzhan inter Lenkoran et Astara ad viam, 1946, *Chalilov* (HUJ). IRAN: S. of Astara, 1962, *Furse* 2560 (W); Cha Bosan, 500 m, *Köie* 321 (W). AFGHANISTAN: Kurum Valley, 1879, *Aitchison* 468 (P). PAKISTAN: Kachemir Nuzafarabad, 1953, *Steward* 183 (G). Also seen specimens from other countries of the Mediterranean region and from Crimea, Dagestan, England (introduced), etc.

114b. T. resupinatum L. Var. **microcephalum** Zoh. in Zoh. & Heller, Israel J. Bot. 19 : 324 (1970) et Zoh., Fl. Pal. 2 : 456 (1972). *Typus*: Israel, Hula Plain, W. of Gonen, 1963, *Zohary* et al. 624534 (HUJ holo.).

Selected specimens: FRANCE: Port Juvenal à Montpellier, 1835 (G). CORSICA: Bonifacia, 1880, *Reverchon* (W). GREECE: Peloponnesus, pr. Corinthou Naupliae, 1857, *Orphanides* 603 (W). TURKEY: Bulgardagh ad ripas fluvii Sari, 1853, *Kotschy* 361a (G); Constantinopole, *Clementi* (P). LEBANON: Beirut, 1931, *Zohary* (HUJ). IRAQ: Semedscha, inter Baghdad et Mosul, 1910, *Handel-Mazzetti* (W). IRAN: Khavon, 1,800 m, 1936, *Rechinger* (W).

114c. T. resupinatum L. Var. **majus** Boiss., Fl. 2 : 137 (1872). *Typus*: Iran: Persia borealis maritima, *Buhse* (P lecto.).

T. suaveolens Willd., Hort. Berol. 1 : 108 (1816).
T. resupinatum L. var. *suaveolens* (Willd.) Dinsm. in Post, Fl. Syr. Pal. Sin. ed. 2, 1 : 345 (1932).

Selected specimens: IRAN: Perse, cultivé (P); pr. Sistan, Zabol-Nasratabad in incultis, ca. 600 m, 1948, *Rechinger* f. 4113 (W); Luristan, Kalvar, 1940, leg. (?) 15986 (W); in camp. ad radices montium Elburz frequens, *Kotschy* (W); Hammadan, foot of Mt. Alvand, 2,210 m, 1965, *Danin, Baum* & *Plitmann* 65-874 (HUJ). AFGHANISTAN: (Hazarajat) Besud, Kajao Valley, 1954, *Thesiger* (W); Kabul Umgeb. Dar-ul-Aman, 1,700 m, 1935, *Scheibe* (W).

Note: This is the cultivated variety widespread in Iran, Afghanistan and adjacent countries. As in the case of other cultivated plants, no specimens of this variety have yet been collected in primary habitats. It is no doubt a constant form and its taxonomic state (whether specific or varietal) still requires re-examination. In Herbarium Boissier true var. *majus* is placed together with other varieties. Hossain (1961) has misinterpreted var. *majus* and intermixed it with var. *resupinatum*.

115. T. clusii Godr. & Gren. in Gren. & Godr., Fl. Fr. 1 : 414 (1849).

Annual, sparsely patulous-hairy or glabrous, 10–20 cm. Stems many, slightly furrowed or striate, erect or ascending, diffuse, branching from base. Petioles decreasing in length towards apex of stem; stipules submembranous, obovate-oblong, green-nerved, united for over half their length, free portion triangular-lanceolate, acuminate, as long as or longer than united part; leaflets obovate with rounded (not truncate) apex, often with bright stripe across, about 0.7–1 × 0.5–0.7 cm, remotely spinulose-denticulate, mucronate at apex, the uppermost ones narrower, cuneate. Peduncles slender, longer than leaves, somewhat thickening and arcuate in fruit. Heads many-flowered, about 1 cm or less across. Pedicels about one-fourth the length of tubular calyx. Lower calyx lobes white-margined, lanceolate-linear, much shorter than whitish tube. Corolla pink, 6–7 mm long; standard oblong, deeply notched, two-thirds the length of wings which are somewhat longer than keel. Fruiting head globular, 0.6–0.8(–1) cm in diam., white to pinkish, deflexed. Fruiting calyces compact, without interspaces but accompanied by projecting parts or withered corollas, unilaterally inflated, obliquely ovoid or globular, 3–4 mm long, glabrous or hairy, reticulately veined, with 2 long, glabrous, horizontal, filiform lobes at conical apex. Pod membranous, ovoid, 2 mm long, 1(–2)-seeded. Seed oblong, yellowish-brown. Dispersal by single calyces, rarely by entire heads. Fl. March–April. 2n = 16.

Hab. : Batha, garigue, fields and roadsides.

Gen. Distr. : France, Sicily, Turkey, Syria, Lebanon, Palestine, Egypt, Iraq, Iran, Morocco. Also recorded from Spain and Tunisia.

1. Fruiting heads smoothly globular, densely covered with white wool that entirely conceals nervation of fruiting calyces; tips of fruiting calyces and teeth not protruding from surface of head. **115b.** var. **gossypinum**
– Fruiting heads with protruding calyx tips; fruiting calyces hairy or glabrous with visible nervation 2
2. Heads about 0.8–1 cm. Peduncles 2 cm or longer. **115a.** var. **clusii**
– Heads 6–7 mm in diam. Peduncles 0 to about 1 cm. **115c.** var. **kahiricum**

115a. T. clusii Godr. & Gren. Var. **clusii**. Opphr., Bull. Soc. Bot. Fr. 108 : 56 (1961). *Typus* : France, Montpellier, 1828, *Delile* (n.v.).

T. resupinatum Guss., Fl. Sic. Syn. 2 : 344 (1843) non L. (1753).
T. resupinatum L. var. *minus* Boiss., Fl. 2 : 137 (1872).
T. clusii Godr. & Gren. var. *glabrum* Evenari in Opphr. & Evenari, Bull. Soc. Bot. Genève ser. 2, 31 : 292 (1941).

Icon. : [Plate 112].

Selected specimens : TURKEY : Bei Brussa, 1873, *Pichler* 34 (G). LEBANON : env. d'Ehden, 1856, *Blanche* 3067 (G). PALESTINE : Upper Jordan Valley, env. of Wadi Bira, 1963, *Plitmann* 12/19 (HUJ).

115b. T. clusii Godr. & Gren. Var. **gossypinum** Zoh. in Zoh. & Heller, Israel J. Bot. 19 : 325 (1970) et Zoh., Fl. Pal. 2 : 457 (1972). *Typus* : Israel, Hula Plain, env. of Gonen, rocky basalt hill, wadi bed, 17.4.1962, *Plitmann* 1610 (HUJ !).

?*T. tomentosum* L. var. *longipedunculatum* Hossain, Not. Roy. Bot. Gard. Edinb. 23 : 454 (1961).

Icon. : [Plate 112].

Selected specimens : PALESTINE : Lower Galilee, between Arbel and Mrar, W. of Wadi Rubaidiya, 1963, *Plitmann* 18/13 (HUJ); Samaria, Bat Shlomo road, 10 km from Haifa road, 1963, *Zohary* & *Plitmann* 1341 (HUJ); also common elsewhere.

115c. T. clusii Godr. & Gren. Var. **kahiricum** Zoh. in Zoh. & Heller, Israel J. Bot. 19 : 326 (1970). *Typus* : (Egypt) bords de Nile à Kasr el Aini, 1908, *Burdet* 213 (G !).

Selected specimens : EGYPT : infer., pr. Abu Zabel, 1835, leg. (?) 19 (W); Cairo, 1836, *Kotschy* 590 (W); pyramids de Gizeh, 1890, *Barbey* 284 (G). MOROCCO : Forêt de Marmora, 1961, *de Wilde* et al. 165b (WAG).

Note : Flowering specimens of *T. clusii* not separated into the above varieties (lacking fruiting heads) were observed also in the following localities : IRAQ: distr. Amara, ca. 70 km N. of Amasia, 1957, *Rechinger* 14171 (W). IRAN : distr. Luristan, Dorud, 1941, leg. (?) 17430 (W); 50 mi. from Khuramabad, 1951, *Stutz* 1912 (W). Also seen from Samos and Rhodes.
T. clusii is very similar to *T. resupinatum*, but is easily distinguished from the latter by the structure of the fruiting head and fruiting calyx. In *T. resupinatum* the fruiting head is clearly stellate, with the fruiting calyces projecting from the surface and diverging from one another in their upper part, while in *T. clusii* the fruiting calyces are congested and adhere to one another all along, so that the fruiting head is either evenly globular or very slightly lobuled by the protruding necks and teeth of the individual fruiting calyces. Other distinguishing characteristics are the globular shape of the fruiting calyx, the marcescent corollas remaining on the fruiting head, and the obovate, not rhombic leaflets of the upper leaves. *T. clusii* also differs from *T. resupinatum* by being a plant of much drier habitats. The flowers of *T. clusii* are not fragrant.
The three varieties of *T. clusii* appear to us to be heterophyletic in their origin : while var. *clusii* and var. *kahiricum* are clear derivatives of *T. resupinatum*, var. *gossypinum* points more to the "Formenkreis" of *T. tomentosum* as its origin. Future studies will no doubt have to reconsider the taxonomic value of var. *gossypinum*.

116. T. tomentosum L., Sp. Pl. 771 (1753).

Annual, glabrous or sparingly hairy, 10–20 cm. Stems many, terete or furrowed, erect, ascending or decumbent, mostly branching from base. Petioles long, glabrous; stipules 0.8–1.2 cm, ovate, green or green-striped, connate for about half their length, free portion shorter than united part, triangular to triangular-lanceolate; leaflets 0.4–1.2(–1.5) cm, those of lower leaves obovate, those of upper ones oblong-cuneate to obtriangular-cuneate, acutely dentate, glabrous or slightly hairy beneath. Flowering heads 6–7(–8) mm across, sessile and mostly involucrate by upper leaves or pedunculate, many-flowered. Pedicels minute. Bracteoles cup-shaped. Calyx tubular, hairy above; teeth linear to lanceolate, much shorter than tube. Corolla 3–6 mm, pink; standard ovate, notched, less than twice as long as ovate wings, the latter somewhat longer than keel. Fruiting head 0.8–1 cm, white,

Plate 112. *T. clusii* Godr. & Gren. Var. *gossypinum* Zoh.
(Palestine : *Plitmann* 12/8, HUJ).
Plant in flower and fruit; fruiting calyx (left below); flowers of Var. *clusii* (left);
fruiting branch and fruiting calyx of same variety (right).

yellowish-white, sordid white or pinkish. Fruiting calyx 3–6 mm, inflated, ovoid, glabrescent or pilose-wooly or felty, tapering above to or terminating abruptly in neck-like column crowned with subulate, hairy, speading or curved upper teeth. Pod ovoid to globular, 1–2-seeded. Seeds scarcely 1 mm, yellowish or brown or mottled with brown. Fl. February–April. 2n = 16.

Hab. : Grassy places, fields, batha, steppes and roadsides, also sandy flats.

Gen. Distr. : All the Mediterranean countries and the adjacent Irano-Turanian ones. The East Mediterranean province is probably the main center of varietal differentiation of this species.

Note : *T. tomentosum* is very polymorphous. The following varieties so far distinguished proved constant :

1. Fruiting heads rather loose and lobed (individual fruiting calyces somewhat separated and divergent from one another in their upper part); upper calyx teeth setaceous, strongly recurved and projecting from periphery of head together with conical or cylindrical calyx neck; fruiting heads partly sessile and partly borne on mostly 1–1.5 cm long peduncles. Upper leaflets long-obtriangular, cuneate. Desert plants. **116c.** var. **curvisepalum**
– Fruiting heads evenly and smoothly globular; calyx teeth not as above 2
2. Fruiting heads all crowded together at base of plant and overtopped by long-petioled leaves. **116f.** var. **chthonocephalum**
– Fruiting heads along or on top of stems and branches 3
3. Fruiting heads pilose; reticulate nerves of calyces visible; upper calyx teeth horizontal, often appressed to tube. Peduncles conspicuous, deflexed. **116a.** var. **tomentosum**
– Fruiting heads cottony or tomentose; reticulate nerves of calyces concealed by hairy cover; neck of fruiting calyx and teeth visible or hidden among fleece. Peduncles not deflexed 4
4. Flowering heads 7–8(–9) mm across. Lower calyx teeth linear. Fruiting heads grey or pinkish or sordid-white 5
– Flowering heads 5–6 mm across. Lower calyx teeth lanceolate or triangular. Fruiting heads cottony-white or yellowish-white (rarely almost glabrescent). **116b .** var. **lanatum**
5. Heads all sessile, upper ones often in pairs. **116d.** var. **orientale**
– Heads, at least part of them, short- or long-peduncled. **116e.** var. **philistaeum**

116a. T. tomentosum L. Var. **tomentosum**; Boiss., Fl. 2 : 138 (1872); Aschers. & Graebn., Syn. Mitteleur. Fl. 6, 2 : 523 (1908); Bobrov, Acta Inst. Bot. Acad. Sci. URSS ser. 1, 6 : 263 (1947). *Lectotypus* : Magn. monsp. t. 264 (1686).

Galearia tomentosa (L.) Presl, Symb. Bot. 1 : 50 (1831).
T. resupinatum L. ssp. *tomentosum* (L.) Gib. & Belli, Mem. Accad. Sci. Torino ser. 2, 41 : 17 (1890).
T. tomentosum L. var. *pedunculatum* Náb., Publ. Fac. Sci. Univ. Masaryk Brno 35 : 71 (1923).

Icon. : [Plate 113].

Selected specimens : SWITZERLAND : San d'Anabida, *Chodat* 594 (G). PORTUGAL : Aveiro : Xina, 1954, *J. Matos* et al. 4765 (HUJ). SPAIN : Granada, 1864, *Del Campo* (G).

MALTA : 1939, *Brenner* (HUJ). FRANCE : Montpellier, 1930 *Eig* (HUJ); Herault, rettains sablonneux, vignes à Roquehaute, entre Agde et Beziers, 1881, *Neura* 43 (W). ITALY : Cosanza (Calabria), 1893, *St. Lager* (G). SICILY : in campis sterilibus Palermo, *Todaro* 1190 (HUJ). YUGOSLAVIA : Sebenico, 1902, *Ronniger* (W); Spalato, 1926, *Korb* (W). GREECE : Attica, 1841, *Spruner* (G); Izis Santorin, Phira-Pyrgos, 1933, *Ronniger* (W). CRETE : 1896, *Heldreich* (G). TURKEY : Lycia, Adalia in graminosis, 1860, *Bourgeau* (P); N. Anatolia, Buyuk Ada, 1939, *Post* (G); Trapezunt, 1853, *Huet du Pavillon* (G). SYRIA : in arvis lapidosis collinis pr. Aleppum, 1841, *Kotschy* 75 (G); Mt. Hermon, Rachaya, 1926, *Berton* (HUJ). LEBANON : Saida, bord du chemin, 1853, *Gaillardot* (P). PALESTINE : Arabia Petraea, 1846, *Boissier* (G); foot of Mt. Gilboa, 1930, *Naftolsky* (HUJ); Upper Galilee, Ain Zaytun to Kerem ben Zimra, 1956, *Feinbrun* (HUJ). EGYPT : Cairo-Alexandria desert road, S. of Amria, 1961, *Täckholm* (G). LIBYA : El Mechili, Uadi Ramla, 1933, *Pampanini* 4106 (G). TUNISIA : Sidi Athman, 1909, *Cvenod* (G). ALGERIA : ca. Mustapha, 1893, *Chevalier* (HUJ); env. d'Alger, 1838, *Bové* 26 (G); Constantine, 1888, *Gerod* 21 (G). MOROCCO : Massif Beni-Snassen, Martingorey du-Kiss, moissons, 1929, *Faure* (HUJ). CANARY ISLANDS : in vicinit. pag. Santa Ursula, La Quinta, 1933, *Asplund* 185 (G); Teneriffa, 1820, *Courant* (G). IRAQ : Mosul-Zakho, ca. 250 m, 1954, *Guest* 13282 (K). IRAN : Luristan, Canguleh, 1948, *Bellard* (W). Also known from the Caucasus (Bobrov, 1947).

116b. T. tomentosum L. Var. **lanatum** Zoh. in Zoh. & Heller, Israel J. Bot. 19 : 328 (1970) et Zoh., Fl. Pal. 2 : 457 (1972). *Typus* : Palestine, Samaria, betw. Wadi Ara and Um el Fahm, 1963, *Zohary* (HUJ holo.).

Icon. : [Plate 113].

Selected specimens : SPAIN : Madrid, 1934, *Albo* (G). TURKEY : Prov. Mersin, distr. Anamur, Anamure-Gilindere, 1956, *Davis* & *Polunin* D. 26015 (HUJ). SYRIA : ca. Aleppo, 1931, *Zohary* (HUJ). In Palestine not rare; we have seen specimens from Acco Plain, Sharon Plain, Philistean Plain, Upper Galilee, Mt. Carmel, Mt. Gilboa, Judean Mts., Dan Valley, Upper Jordan Valley.

116c. T. tomentosum L. Var. **curvisepalum** (V. Täckh.) Thiéb., Fl., Lib.-Syr. 2 : 28 (1940). *Typus* : Egypt : Sinai, Rafa, Bir el Meleha at the coast, *G. Täckholm* (S holo.).

Icon. : [Plate 113].

Selected specimens : PALESTINE : Negev, 2 km S.W. of Be'er Sheva, 1965, *Heller* 119 (HUJ). [Also found in other parts of Palestine, e.g., Philistean Plain, env. of Nitsanim, 1954, *Jaffa* 110 (HUJ)]. IRAN : in coll. apricis ad pagum Dalechi inter Abuschir et Shiraz, 1842, *Kotschy* 202 (G).

116d. T. tomentosum L. Var. **orientale** Bornm., Verh. Zool.-Bot. Ges. Wien 48 : 581 (1898). *Typus* : (Palestine) Jaffa, *Bornmüller* 448a (B).

T. hebraeum Bobrov, Acta Inst. Bot. Acad. Sci. URSS ser. 1, 6 : 259 (1947).

Icon.: [Plate 113].

Selected specimens: FRANCE: Bois des Doseares (Montpellier), 30 m, *Kümmel* 119 (WAG). TURKEY: Alexandrette, 50–100 m, 1932, *Delbès* (HUJ). CYPRUS: Agia Irini (Morphou), 1941, *Davis* 2553 (E) (det. by Hossain as var. *curvisepalum*). ISRAEL: Sharon Plain, Atlith near the castle, 1956, *Orshan* 17392 (HUJ). SINAI: El-'Arish, 1925, *Eig* (HUJ). MOROCCO: Forêt de Marmora, triangle Sali-Monod-Kantera, 1961, *de Wilde* et al. 165b (WAG).

116e. T. tomentosum L. Var. **philistaeum** Zoh. in Zoh. & Heller, Israel J. Bot. 19: 329 (1970). *Typus*: Palestine: Philistean Plain, Tel Nof, road W. of Ramle, fallow field, 1963, *Zohary* & *Plitmann* 5412 (HUJ holo.).

Note: In Palestine the last variety has also been collected in the Sharon Plain, Dan Valley and Judean Mts. Not observed outside Palestine. All of the above varieties are fairly readily recognizable even at a first glance. All of them have been tested for constancy and proved genetically fixed. There are, no doubt, certain other morphological and ecological characteristics which are linked to those pointed out above but have not been adequately examined as yet.

116f. T. tomentosum L. Var. **chthonocephalum** Bornm., Beih. Bot. Centralbl. 31, 2: 204 (1914). *Typus*: Lebanon: Ad Antilibani radices occidentales in declivibus supra Baalbek, 1,130–1,150 m, 1910, *Bornmüller* 11671 (E lecto.).

Selected specimens: SYRIA: Kheilan pr. Aleppo, 1865, *Haussknecht* (G). CYPRUS: Nicosia Aerodrome, 1943, *Evenari* (HUJ). PALESTINE: Upper Galilee, env. of Rama, 1927, *Smoly* (HUJ); Ammon: Ain Suella, 1929, *Eig* & *Zohary* 892 (HUJ) (also collected in Upper Galilee, Judean Mts., Judean Desert, N. and C. Negev). EGYPT: ad Mariout pr. Alexandriam inter segetes, 1878, *Letourneux* 187 (G). LIBYA: Derna, 1887, *Taubert* 174 (K).

Note: This variety is a basicarpic form in which al the fruiting heads are crowded at the base of the plant. Similar varieties occur in *Astragalus*, *Trigonella* and others.

117. T. bullatum Boiss. & Hausskn. in Boiss., Fl. 2: 138 (1872).

A tiny annual, glabrous or very sparingly hairy, 5–15 cm. Stems solitary or few, erect, branching from base. Petioles somewhat canaliculate, those of lower leaves very long; leaflets not over 8 × 4 mm, obovate-cuneate, retuse, rather remotely denticulate, glabrous; stipules spreading, ovate, acuminate, green-nerved, united for more than half their length, free portion lanceolate. Peduncles about 1 cm, erect in flowers, bowed in fruit. Flowering heads minute, 4–9 mm across, 10–15-flowered. Pedicels minute. Calyx 2–3 mm, tubular; teeth 1 mm or less, triangular. Corolla 2 mm long, pink or whitish; standard broad-ovate, somewhat longer than wings. Fruiting head 5–8 mm; fruiting calyces relatively few, well discernible, glabrous or hairy, reticulately nerved; upper calyx teeth minute, triangular, spinulose, often re-

Plate 113. *T. tomentosum* L. 1–3. Var. *orientale* Bornm. : 1. Plant in flower and fruit; 2. flower; 3. fruiting calyx; 4–5. Var. *lanatum* Zoh. : 4. fruiting branch; 5. fruiting calyx; 6. Var. *tomentosum* : fruiting head; fruiting calyx; 7. Var. *curvisepalum* (V. Täckh.) Thiéb. : fruiting branch; fruiting calyx.

duced to callosities. Pod about 1.2 mm, lenticular, follicularly dehiscent, 1-seeded. Seed ovoid, yellowish, 0.8 mm. Fl. March–April. 2n = 16.

Hab. : Fields and roadsides.

Gen. Distr. : Turkey, Syria, Iraq, Iran, Israel, Jordan, Egypt.

117a. T. bullatum Boiss. & Hausskn. Var. **bullatum**. *Typus* : (Syria) In deserto fl. Chabur, 1867, *Haussknecht* (BM lecto., iso. G).

T. bullatum Boiss. & Hausskn. var. *glabrescens* Post, Fl. Syr. Pal. Sin. 241 (1883–1896).
T. resupinatum L. ssp. *tomentosum* (L.) Gib. & Belli var. *bullatum* (Boiss. & Hausskn.) Gib. & Belli, Mem. Accad. Sci. Torino ser. 2, 41 : 17 (1890).
T. tomentosum L. var. *glabrescens* (Post) Hausskn. & Bornm, Mitt. Thüring. Bot. Ver. 24. : 39 (1908).
T. tomentosum L. ssp. *bullatum* (Boiss. & Hausskn.) Opphr., Bull. Soc. Bot. Fr. 108 : 56 (1961).

Icon. : [Plate 114].

Fruiting heads 5–6 mm. Upper calyx teeth minute, 1 mm or less.

Selected specimens : TURKEY : inter Aintab et Aleppo, 1865, *Haussknecht* (G); env. d'Antioche, 1935, *Delbès* (HUJ). SYRIA : Telejin-Abudhur (Aleppo distr.), 1931, *Zohary* (HUJ). PALESTINE : Sharon Plain, Magdiel, 1927, *Naftolsky* 191 (HUJ) (also collected in Philistean Plain, Upper and Lower Galilee, Esdraelon Plain, Samaria, Judean Mts., Dan Valley, Hula Plain, Upper Jordan Valley). IRAQ : Garaghan, 1932, *Guest* 1876 (HUJ). IRAN : N.W. of Laristan-Sheshom, 700–750 m, 1963, *Jacobs* 6389 (W).

117b. T. bullatum Boiss. & Hausskn. Var. **macrosphaerum** Zoh. in Zoh. & Heller, Israel J. Bot. 19 : 331 (1970). *Typus* : Hula Plain, Hatsbani River banks, S.E. of Mayan Barukh, alluvial soil, 1963, *Zohary* & *Plitmann* 524558 (HUJ !).

Fruiting heads up to about 8–9 mm in diam. Calyx teeth considerably longer than in typical variety.

Selected specimens : SYRIA : S.W. of Damascus, 950 m, 1933, *Eig* & *Zohary* (HUJ). PALESTINE : Sharon Plain, Petah Tiqwah, 1927, *Naftolsky* (HUJ). IRAN : Khuzistan, betw. Harmuz and Bagh e Malik, 700 m, 1959, leg. (?) 3758 (K); Laristan, 3 km W. of Murabad, S. of Kashghan, 1960, *Wright* & *Bent* 433-201 (K).

Note : *T. bullatum* is, without doubt, an E. Mediterranean counterpart of the mainly W. Mediterranean *T. tomentosum* var. *tomentosum*. It is distinguishable from the latter by several characters, of which the following should be mentioned : (1) the smaller flowering and fruiting head; (2) the triangular lower calyx teeth; (3) the very short upper teeth of the fruiting calyx, which are sometimes abortive and transformed into callosities; (4) the very slender and symmetrically bowed peduncles that bear all the heads of the plant. These characters proved to be constant. *T. bullatum* occurs in a form with glabrous and pilose heads but there are many intermediate transitions so that distinction of two varieties, as done by Evenari (*loc. cit.*), is not feasible.

Plate 114. *T. bullatum* Boiss. & Hausskn. Var. *bullatum*
(Palestine : *Plitmann* 10, HUJ).
Plant in flower and fruit; flowering head; fruiting calyx.

Section Five. CHRONOSEMIUM Ser.

in DC., Prodr. 2 : 204 (1825)

Genus *Chrysaspis* Desv., Fl. Anjou 338 (1827).
Section *Lotophyllum* Reichenb., Icon. exot. 1 : 7 (1827).
Genus *Amarenus* Presl, Symb. Bot. 1 : 46 (1830).
Subgenus *Chronosemium* (Ser.) Hossain, Not. Roy. Bot. Gard. Edinb. 23 : 470 (1961).

Lectotypus : *T. badium* Schreb. (Hossain, *loc. cit.*).

Annuals. Inflorescence many-flowered and then capitate-globular to ovoid, rarely few-flowered and -racemed. Flowers shortly pedicellate, reflexed in fruit. Bracts minute to 0. Calyx often campanulate, not growing in fruit, 5-nerved; throat glabrous, open; calyx teeth unequally long, the 2 posterior ones distinctly shorter than the 3 anterior ones. Petals yellow, purple or pink (never white), persistent, turning scarious in fruit with spoon- or boat-shaped standard. Pod stipitate, hidden in fruiting corolla, 1–2-seeded.
Seventeen species, primarily limited to European and Mediterranean countries. Adventive in American and African countries.

Synopsis of Species in Section Five

Section Five. CHRONOSEMIUM Ser.
- Series i. STIPITATA
  - 118. T. sintenisii Freyn
- Series ii. COMOSA
  - 119. T. billardieri Spreng.
  - 120. T. philistaeum Zoh.
  - 121. T. grandiflorum Schreb.
  - 122. T. boissieri Guss.
  - 123. T. brutium Ten.
- Series iii. BADIA
  - 124. T. badium Schreb.
  - 125. T. spadiceum L.
  - 126. T. erubescens Fenzl
  - 127. T. sebastianii Savi
- Series iv. AGRARIA
  - 128. T. aureum Poll.
  - 129. T. velenovskyi Vandas
  - 130. T. campestre Schreb.
  - 131. T. patens Schreb.
- Series v. FILIFORMIA
  - 132. T. dubium Sibth.

Section Five: Chronosemium

133. T. micranthum Viv.
134. T. dolopium Heldr. & Hausskn.

Key to Species

1. Flowers about 2mm long. Pod as long as corolla, regularly dehiscing by valves. Pedicels 1.5–2 mm long. Terminal leaflet of leaves sessile or subsessile. Low, decumbent, much branched annuals with axillary 2–6(–20)-flowered heads. Heads about 4 mm across. **133. T. micranthum**
– Plant not as above 2
2. Standard boat-shaped in fruit. Flowers about 4 mm long, yellow. Heads 5–7 mm in diameter 3
– Not as above 4
3. Terminal leaflet petiolate. Pedicels about 1 mm long. Calyx one-third to one-fourth the length of corolla. Leaflets obovate. Standard obsoletely or not at all ribbed. **132. T. dubium**
– Terminal leaflet sessile. Leaflets oblong-elliptical. Pedicels of uppermost flowers up to 3(–4) mm long. Calyx almost as long as corolla. Standard prominently ribbed. **127. T. sebastinanii**
4. (2). Heads 2–8?flowered. Flowers purple, 1.2 cm or more in length. Fruiting calyx purple or purple-tipped. **118. T. sintenisii**
– Not as above 5
5. Terminal leaflet of leaves sessile 6
– Terminal leaflet (at least of upper leaves) distinctly petiolulate 9
6. Standard spoon-shaped, i.e., with flat or concave limb narrowing more or less abruptly into canaliculate claw which is nerveless but has 3–4 furrows 7
– Standard spatulate, folded or boat-shaped without claw as above; nerves continuing to base 8
7. Fruiting heads cone-like with dense, imbricate flowers. Stipules oblong. Pedicels less than half the length of shorter part of calyx tube. Standard notched. **128. T. aureum**
– Fruiting heads loose, spherical, with spreading flowers. Stipules ovate-oblong with somewhat auriculate base. Pedicels half the length of shorter part of calyx tube or longer. Standard not notched. **129. T. velenovskyi**
8. Fruiting heads mostly cylindrical, spicate, 8–10 mm broad, blackish-brown. Flowers 4–6 mm. **125. T. spadiceum**
– Fruiting heads globular or ovoid, 1.2–2 cm broad, light brown. Flowers 7–10 mm. **124. T. badium**
9 (5). Corolla violet or pink at anthesis 10
– Corolla yellow or orange at anthesis, sometimes becoming brownish 14
10. Standard spoon-shaped, with flat or concave limb narrowing abruptly into canaliculate claw with 3–4 furrows 11
– Standard boat-shaped or folded, without distinct claw 13
11. Fruiting head cone-like, globular, with dense, imbricate flowers. Flowers 5–6 mm. Limb of standard orbicular. **130b. T. campestre** var. **lagrangei**
– Not as above 12
12. Fruiting head spicate, up to 2–3 × 0.8–1 cm. Fruiting corolla 7 mm. Leaflets oblong-linear. **119. T. billardieri**
– Fruiting head ovoid. Fruiting corolla 10 mm. Leaflets obovate to broad-elliptical. **121. T. grandiflorum**

13 (10). Flowers up to 12 mm long. Heads spicate, up to 2–3 cm. Leaflets several times as long as broad. **120. T. philistaeum**
– Flowers 4–6(–7) mm long. Heads ovoid or globular, scarcely 1 cm long. Leaflet scarcely twice as long as broad. **126. T. erubescens**
14 (9). Flowering and fruiting heads hemispherical or conical or obconical. Fruiting standard not more than 3 mm broad. Flowers never imbricated. **131. T. patens**
– Not as above 15
15. Peduncles densely patulous-pubescent. **122. T. boissieri**
– Peduncles more or less adpressed-hairy or glabrous 16
16. Leaflets several times as long as broad, linear-lanceolate to narrow-elliptical. **134. T. dolopium**
– Leaflets scarcely 1–2 times as long as broad, obovate to obtriangular 17
17. Fruiting head less than 1 cm in diam., cone-like, with dense, imbricate, 6–8 mm long flowers. Lower pedicels 0.5 mm, upper ones somewhat longer. Calyx 3 mm long. Leaflets oblong-elliptical. **130. T. campestre**
– Fruiting head not cone-like, with more or less lax, up to 10 mm long flowers. Calyx up to 4 mm long. Leaflets mostly obovate. **123. T. brutium**

Series i. STIPITATA Zoh. (ser. nov.)*

Annual or biennial. Flowers few, large (0.8–1 cm), loose, pink to purple. Standard conduplicate, boat-shaped in fruit. Pod very long-stipitate.

118. T. sintenisii Freyn, Oesterr. Bot. Zeitschr. 41 : 404 (1891). *Typus* : Pontus australis : Sumila in pratis alpinis uliginosis ad Karakapan, 3 Aug., 1889, *Sintenis* 1641 (W ! holo., iso. P !).

T. stipitatum Boiss. & Bal. in Boiss., Fl. 2 : 149 (1872) non Clos (1847).

Icon : [Plate 115].

Annual or biennial, glabrous to glabrescent, 5–15(–20) cm. Stems single or few, erect or ascending, poorly branching. Leaves sparse with petioles decreasing in length towards apex, the uppermost ones almost opposite; stipules ovate to ovate-oblong, rounded at base, acute; leaflets 0.8–1.2 × 0.6–0.8 cm, all sessile, obovate with cuneate base, apex emarginate, margin obsoletely dentate. Peduncles rather short, terminal, erect. Heads 1.5–2.5 × 1.5 cm, ovate to obconical, few(4–15)-flowered. Pedicels slightly hairy to glabrous, shorter than calyx tube. Calyx glabrous, 5-nerved; teeth very unequal, the upper ones minute, about one-fourth the length of tube, the 2 lateral ones as long as tube, the lowermost one longer than tube, all purple-tipped. Corolla 1–1.2 cm, pink to purple in flower, becoming lurid brown in fruit, erect before and deflexed after flowering, about 3–4 times as long as calyx; standard obovate-oblong, conduplicate, somewhat longer than long-clawed wings; keel as long as wings. Ovary borne on 3–4 mm long stipe, ellipsoidal; style curved,

* See Appendix.

Plate 115. *T. sintenisii* Freyn
(Turkey : *Davis* & *Hedge* D. 32153, E).
Plant in flower and fruit; flowers.

much longer than ovary. Pod 2 mm long, rather leathery, one-half to one-third the length of stipe and much shorter than style, ovoid. Seed up to 2 mm, smooth, long, brown. Fl. July–August.

Hab.: Damp subalpine meadows; 2,000–2,800 m.

Gen. Distr.: N.E. Turkey, Transcaucasia.

Selected specimens: TURKEY: Rize: "Vallée sous-alpine de Djimil" (Lazistan), 2,300 m, 1866, *Balansa* 1404 (holo. of *T. stipitatum* G !); Prov. Giresun: Balabandağlari above Tamdere at Avsar, 2,400 m, 1952, *Davis* D. 20568, *Dodds, Cetik* (E); Prov. Trabzon: N. side of Soğanli Dag above Çaykara, 2,000–2,200 m, 1957, *Davis* & *Hedge* D. 32153 (E); Prov. Gümüsane: Karagoelldagh, 1894, Sintenis 7157 (FI). Recorded by Bobrov, Fl. URSS, from Transcaucasia.

Series ii. COMOSA Gib. & Belli
Malpighia 3: 205 (1889)

Series *Speciosa* Bobr., Acta Inst. Bot. Acad. Sci. URSS ser. 1, 6: 243 (1947).

Annuals. Heads many-flowered. Standard with distinct, carinate claw and flat-orbicular or obovate or fan-shaped, broad limb, not folded in fruit.

119. T. billardieri Spreng., Syst. Veg. 3: 211 (1826; "billarderii"). *Typus*: Juxta Beruthum Syriae a 1786–87, *Labillardiere* (FI, holo., iso. K).

T. comosum Labill., Ic. Pl. Syr. Dec. 5: 15 (1812) non L., Sp. Pl. 767 (1753); Boiss., Fl. 2: 150 (1872).

T. speciosum Willd. sensu Ser. in DC., Prodr. 2: 205 (1825).

Icon.: [Plate 116]. Labill., *loc. cit.* t. 10; Zoh., Fl. Pal. t. 233.

Annual, glabrous except for patulous-hairy peduncles, 10–15 cm. Stems erect or ascending, simple or branching from base. Leaves petiolate; stipules long, oblong-lanceolate, acute, acuminate; leaflets 0.8–1.8 cm, obovate-oblong to oblong-elliptical, cuneate at base, obtuse or truncate to emarginate at tip, denticulate mainly in upper half, the terminal one long-petiolulate. Peduncles stout, rather short. Head 1.5–3 cm, spicate, cylindrical. Pedicels very short, deflexed in fruit. Calyx patulous-hirsute; tube membranous, very short; upper teeth minute, triangular, the lower ones linear-lanceolate, obtuse, prominently nerved, with ciliate hairs arising from tuberculate base. Corolla 6–8 mm, erect in flower, pale pink to lilac and flesh-coloured; standard with indurated, canaliculate claw and orbicular limb, entire, denticulate at margin, prominently nerved and scarious in fruit; wings divergent, long-clawed, much shorter than standard and somewhat longer than keel. Pod ellipsoidal, 1.5 mm thick and long-stipitate, 1-seeded, membranous, somewhat hirsute; style thick, short. Seed ovoid-oblong, yellowish. Fl. March–April. 2n = 16

Hab.: Sandy clay.

Gen. Distr.: Coasts of Lebanon and Israel.

Plate 116. *T. billardieri* Spreng.
(Palestine : *T. Segal* 16668, HUJ).
Plant in flower and fruit; fruiting calyces with persistent corolla (anterior and posterior views).

Selected specimens: LEBANON: Sables ferrugineux du cap de Bayrouth, 1869, *Blanche* 1006 (G). ISRAEL: Sharon Plain: Pardesiya, 1950, *Segal* 16669 (HUJ); *ibid.*: Karkur to Pardes-Hanna, 1962, *Plitmann* 28319 (HUJ); Bir Salem (n. Jaffa), 1922, *J. E. Dinsmore* (HUJ).

120. T. philistaeum Zoh. in Gruenberg-Fertig & Zoh., Israel J. Bot. 19 : 296 (1970); Zoh., Fl. Pal. 2 : 164 (1972).

Annual, glabrous or appressed-hairy, 10–25 cm. Stems single or few, ascending, terete, dichotomously branching above. Lower leaves long-petioled, upper ones short-petioled; stipules herbaceous, ovate to oblong, entire, sparingly patulous-hirsute, many-nerved, lower half adnate to petioles, upper free portion lanceolate; leaflets 1–2(–2.5) cm, oblong to elliptical to linear and filiform, long-dentate in upper half or entire, glabrous to appressed- or patulous-hairy, the terminal one with long petiolule. Heads spicate, 1–6 × 1.5–2 cm, long-peduncled, many-flowered, ovoid, elongating in fruit. Pedicels short, initially erect then deflexed. Calyx small, membranous; tube obliquely truncate; the 2 upper teeth short, glabrous, triangular, the 3 lower ones patulous-hairy, lanceolate-subulate, almost equal, 2–4 times as long as upper ones. Corolla about 1 cm, white to pinkish-white, turning pink or violet or flesh-coloured; standard 2–3(–4) times as long as calyx, longer than divergent wings, with broadly elliptical, many-nerved, denticulate limb, and canaliculate, prominently carinate claw; keel erect, much shorter than wings. Pod obliquely ellipsoidal, 1.5 mm long, 1-seeded, stipitate, membranous, carinate; style short, strongly coiled. Seed 1.2 mm, ovoid, yellowish-brown. Fl. March–May.

Hab.: Sandy soil, along coast.

Gen. Distr.: Israel, N. Sinai.

1. Leaflets all toothed, the upper ones oblong to linear. **120a.** var. **philistaeum**
– Leaflets (all or only those of the uppermost leaves) filiform, entire. **120b.** var. **filifolium**

120a. T. philistaeum Zoh. Var. **philistaeum.** *Typus*: In arenosis Palaestinae circa Gaza, April–May 1846, *Boissier* (G holo., iso.: K, P).

T. stenophyllum Boiss., Diagn. ser. 1, 9 : 30 (1849) et Fl. 2 : 151 (1872) non Nutt., Journ. Acad. Philad. N. S. 1 : 151 (1847).

Icon.: [Plate 117]. Zoh., *loc. cit.* t. 234.

Selected specimens: ISRAEL: Coastal Galilee: N. of Nahariya, 1955, *Zohary* 17368 (HUJ); Sharon Plain: Kefar Ganim, 1927, *Naftolsky* 17365 (HUJ); Philistean Plain: Gaza, 1927, *Eig, Zohary* & *Feinbrun* 124192 (HUJ); Negev: Hatserim, 1950, *Zohary* & *D'Angelis* 19502 (HUJ); Coastal Negev: Rafiah, env. of railway station, 1925, *Eig* 19253 (HUJ).

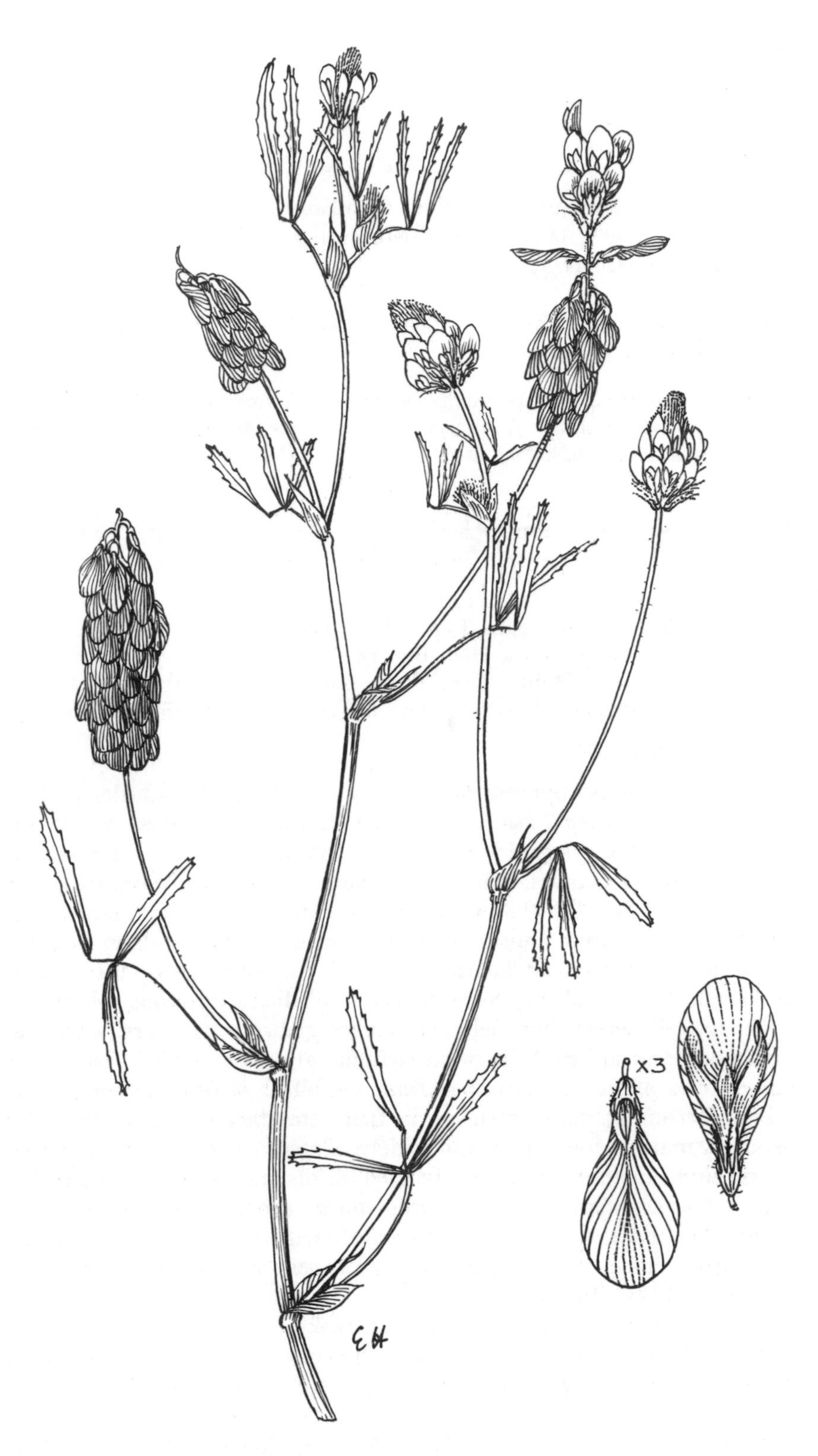

Plate 117. *T. philistaeum* Zoh. Var. *philistaeum*
(Palestine : *D. Zohary*, HUJ).
Flowering and fruiting branch; fruiting calyces
with persistent corolla (anterior and posterior views).

120b. T. philistaeum Zoh. Var. **filifolium** Zoh., Fl. Pal. 2 : 456 (1972). *Typus* : Israel : N. Negev, between Yad-Mordechai and Erez, sandy-limestone hill, 1962, *Plitmann* 11101 (HUJ holo.).

Selected specimens : ISRAEL : S. Shefela : Ashdoth, 1954, *Jaffe* 19541 (HUJ); Coastal Negev : Rafiah, 1925, *Naftolsky* 17367 (HUJ).

Note : Among the many specimens seen there is a form with a 2-coloured head, i.e., flowers changing colour with age. This feature is also characteristic of *T. dichroanthum* which similarly grows in the coastal plain of Israel.

121. T. grandiflorum Schreb., Nov. Act. Nat. Cur. 3 : 477 (1767). *Typus* : Habitat in Crete : Vieweg (lecto. : Herb. Willd. Cat. no.14240, B).

T. speciosum Willd., Sp. Pl. 3 : 1382 (1802).
T. gussoni Ten., Pl. Rar. Sicil. Pugill. 1 : 17 (1817).
T. plicatum Presl ex Sweet, Hort. Brit. ed. 2 : 137 (1830).
Amarenus speciosus (Willd.) Presl, Symb. Bot. 1 : 46 (1830).
T. violaceum Davidoff, Trav. Soc. Bulg. Sci. Nat. 8, 69 (1915).

Icon. : [Plate 118].

Annual, glabrous, appressed-pubescent to sparingly patulous-villous, 10–30 cm. Stems single or many, ascending, rarely decumbent or erect, poorly branching, leafy. Leaves petioled; stipules ovate to broad-lanceolate, herbaceous, auriculate, short-adnate to petioles, acute, acuminate to short-cuspidate, irregularly denticulate; leaflets 0.8–1.5 × 0.4–0.8 cm, oblong to elliptical, rarely obovate, truncate to deeply emarginate at apex, upper ones conspicuously petiolulate, remotely denticulate. Peduncles terminal and lateral, 3–7 cm long, appressed-pubescent. Heads globular or ovoid, 1.3 × 1.2 cm, 8–20-flowered. Pedicels glabrous, about as long as calyx tube, initially erect then deflexed. Calyx glabrous or hairy with few bristles especially on teeth apices; tube broad-cylindrical; lower teeth 2–3 times as long as tube, upper ones usually shorter than tube. Corolla 3–4 times as long as calyx, violet, the colour becoming more intensive in fruit; standard spoon-shaped, 7–10 mm, somewhat carinate in lower part and widely dilated into orbicular or fan-shaped, 5–6 mm broad limb, toothed and partly fringed all around; wings divergent, long-clawed, oblong, shorter than standard; keel much shorter than wings. Pod 1.5–2 mm, ovoid-ellipsoidal, membranous, 1-seeded, with style as long as or longer than pod and stipe, almost twice as long as pod proper. Seed about 1.2 mm, yellowish-brown. Fl. April–May. 2n = 16.

Hab. : Forest, among scrub and on rocks.

Gen. Distr. : Italy, Sicily, Albania, Yugoslavia, Greece, Aegean Islands, Cyclades, Rhodes, Crete, Bulgaria, Turkey, Syria, Lebanon, Iraq, N. W. Iran.

Selected specimens : ITALY : Calabria, Cosenza 900 m, 1912, *Lopez* 1865 (K). SICILY : Nebrodim Pomieri, Castellobuobo, 1855, *Huet du Pavillon* s. n. (G). ALBANIA : Gjinokastore (Argirokastron), 1,500 ft, 1933, *Alston* & *Sandwith* 1144 (K). GREECE : Südhänge des Parnassos östl. von Arachowa, 900 m, 1967, *I. Denil* (HUJ). CYCLADES : Amrigos at Langadha, 150 m, 1940, *Davis* 1526 (E). RHODES : in monte Prophet Elias, pr. Salakos, ca.

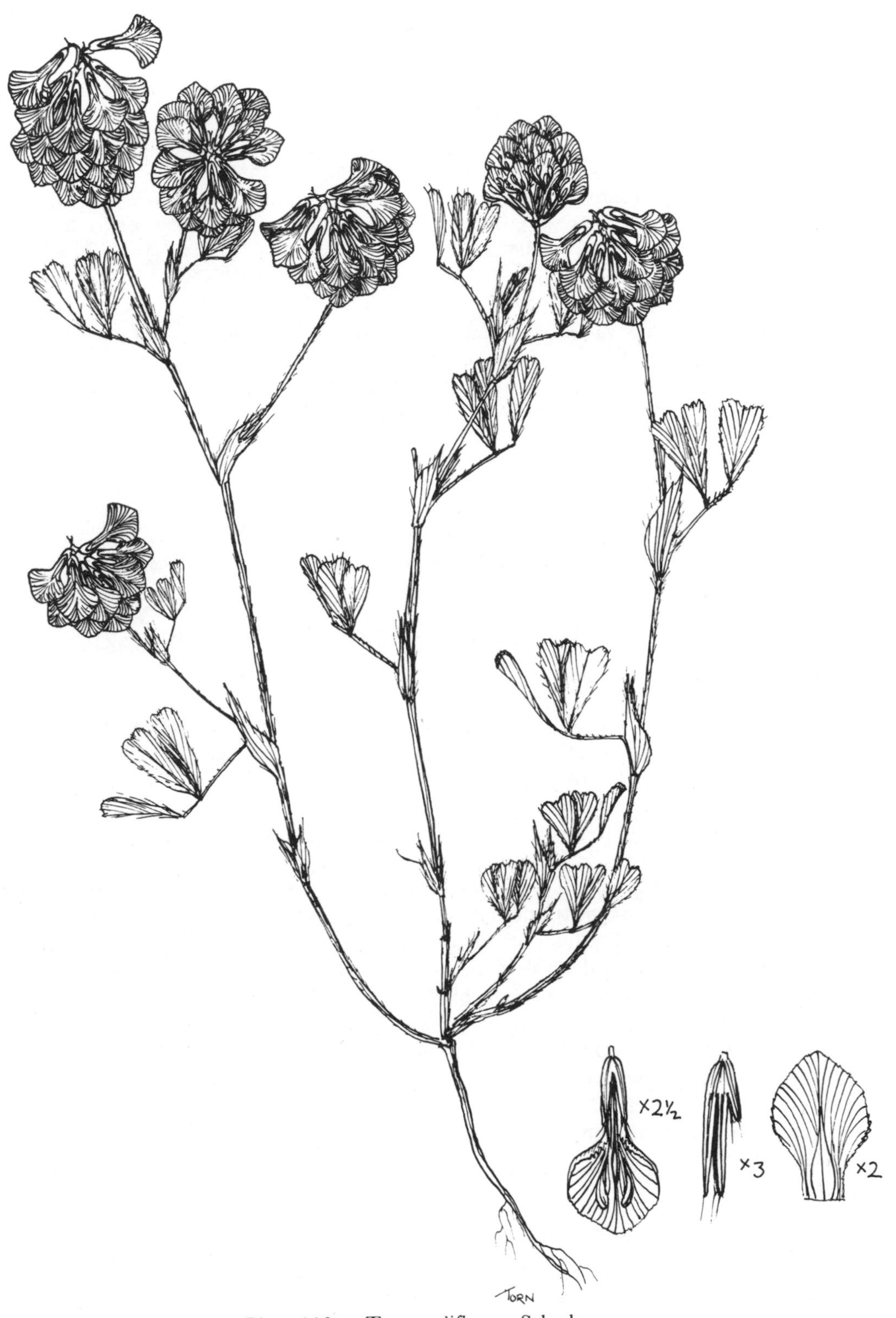

Plate 118. *T. grandiflorum* Schreb.
(Greece : *F. Guiol* 137, HUJ).
Plant in flower and fruit; fruiting calyx with persistent corolla; fruiting calyx; standard.

600 m, 1935, *K. H. & F. Rechinger* 7085 (BM). SYRIA: Djebel Seman, 600 m, 1908, *Haradjian* 2119 (E); Mt. Hermon n. Majdal Shams, 1968, *Danin & Peri* (HUJ). LEBANON: ex jugo montis Libani Djebel Baruk dicto, supra Ain Zehalteh, 1,600–2,200 m, 1877, *Bull* 2104 (K). IRAQ: Kuh-Sefin, Shaklava, 900 m, 1893, *Bornmüller* 1151 (K); Jebel Sindjar (above Beled Sindjar), 1,060 m, 1933, *Eig & Zohary* 3274 (HUJ). N.W. IRAN: Khossamabad, 1,320 m, 1929, *Cowan & Darlington* 1018 (K).

Note: *T. grandiflorum* varies considerably in the size of the head (especially when in fruit), the indumentum and the size of the individual flower. It differs from *T. boissieri*, among others, by its standard which terminates abruptly in a claw; the standard is flat in the fruit and its margin is less conspicuously toothed than that in *T. boissieri*.

122. T. boissieri Guss. ex Soyer-Willemet & Godr. in Mem. Soc. Roy. Sci. Nancy 1846: 220 (1847). *Typus*: Rupibus Calidis Argolidis, IV 1842, *Boissier* (as *T. speciosum* Willd., NSP, holo., iso. K).

T. speciosum sensu Boiss., Diagn. ser. 1, 2: 33 (1843) non Willd. (1802).

Icon.: [Plate 119]. Zoh., Fl. Pal. 2: t. 236 (1972).

Annual, patulous-villous above, 15–40 cm. Stems few or many, terete, erect, branching mainly in upper parts. Leaves long-petioled; stipules oblong-lanceolate, acuminate; leaflets 1–2 cm, ovate to elliptical to obovate, obtuse, truncate or slightly emarginate at apex, dentate, the terminal one long-petiolulate. Peduncles rather thick, often longer than leaves, rarely as long or shorter, patulous-villous. Heads ovoid, 10–30-flowered. Pedicels as long as or longer than calyx tube, initially erect or spreading, then deflexed. Calyx membranous, white, 5-nerved, glabrous or partly hairy; upper 2 calyx teeth triangular, the others lanceolate-subulate, at least twice as long as upper ones, all ciliate or terminating in long, scattered bristles. Corolla 7–8 mm, yellow or cream-coloured in flower, chestnut-coloured in fruit; standard with broad-elliptical to strongly obovate blade, deeply denticulate, gradually tapering to folded, canaliculate claw, 2–3 times as long as calyx, much longer than wings, the latter slightly longer than keel. Pod about 1.2–1.5 mm, as long as style, 1-seeded, with thick stipe almost twice as long as pod. Seed 1.2 mm, oblong, yellowish-brown. Fl. April–May. 2n = 16.

Hab.: Shady places in batha, maquis and forests.

Gen. Distr.: Greece, Aegean Islands, Rhodes, Crete, Turkey, Cyprus, Syria, Israel, Iraq.

Selected specimens: GREECE: Nauplia, 1885, *Haussknecht* (B). TURKEY: Lydia, Montis Sipyli in regione inferiore, 1906, *Bornmüller* 9382 (E). RHODES: Phileremos ad Trianda, 1935, *Rechinger* 7007 (BM). CRETE: Central, 28 km E. of the Lashite Plateau, on the way to Ag Nikolaos, 280 m, 1964, *Zohary & Orshan* 29406-14 (HUJ). CYPRUS: Akantho, 600 ft, *Crofirton* 768 (K). ISRAEL: E. Upper Galilee, Jiftah, Canyon near Nebi-Yusha, 1954, *Shenkar* 16576 (HUJ). SYRIA: Mt. Hermon, Wadi Afre, 1968, *Shmida* (HUJ). IRAQ: Acra, 1,500–3,000 ft, 1932, *Qaimagom* 3082 (K).

Plate 119. *T. boissieri* Guss.
(Palestine : *D. Jaffe* 16433, HUJ).
Plant in flower; part of fruiting branch; fruiting calyx with persistent corolla.

123. T. brutium Ten., Fl. Neap. Prodr. App. 5 : 24 (1826) et Viagg. Calabr. 126 (1827). *Typus* : Calabria, Monte Cocuzzo, 1826 (NAP).

T. speciosum Bory & Chaub., Nuov. Fl. Pelop. 51 (1838) non Willd. (1802), nec Boiss. (1843) et Griseb. (1842).
T. procumbens L. var. *pauciflorum* Griseb., Spicil. Fl. Bith. Rum. 1 : 36 (1843).
T. mesogitanum Boiss., Diagn. ser. 1, 2 : 34 (1843) et Fl. Or. 2 : 152 (1872).
T. aurantiacum Boiss. & Sprun. in Boiss., Diagn. ser. 1, 2 : 33 (1843) et Fl. Or. 2 : 182 (1872).

Icon. : [Plate 120].

Annual, hairy, 5–20 cm. Stems many, ascending, branching. Leaves with rather long to very short petioles; stipules ovate to semi-cordate at base, often auriculate; leaflets 0.5–1 × 0.3–0.6 cm, obovate-cuneate or obovate, truncate or emarginate, margins serrulate only from above middle, terminal one in upper leaves with conspicuous petiolule up to 2.5 mm or subsessile. Inflorescences 1–2 cm in diam., much longer than subtending leaves. Pedicels 0.5–1 mm long, deflexed after anthesis. Calyx 2–3 mm long; tube obconical, 5-nerved; teeth unequal, the upper ones triangular-linear, somewhat shorter than to about as long as tube, the lower ones much longer than tube. Corolla 8–10 mm long, yellow, remaining so or becoming dark brownish after anthesis; standard obovate, denticulate in upper part or all around. Pod about 1.5–2 mm, long-stipitate, about as long as or shorter than style, 1-seeded, obovoid. Seed 1.2 mm, oblong, brown. Fl. June–July.

Hab. : Mountain slopes and open bush.

Gen. Distr. : S. Italy, Albania, Greece, Crete, European Turkey, W. Anatolia.

Selected specimens : ITALY : Calabria, Prov. Cosanza, 500–1,000 m, 1898, *Rigo* 4045 (E); Basilicata : M. Coccovello, 1,800 m, 1899, *Fiori* (FI). ALBANIA : Olycerea Olpochari distr. Jaucina, 1896, *Baldacci* 68 (BM). GREECE : Laconia bor. Megali Anastosova, 1897, *Zahn* 1417 (B); auf Bergen in Morca, 1841, *Spruner* (G, type of *T. aurantiacum*). TURKEY : in regione supra Mesogis, supra Tralles, 1842, *Boissier* (G, type of *T. mesogitanum*).

Note : *T. aurantiacum* and *T. mesogitanum* were synonymized with *T. brutium* after meticulous examination of many specimens from the localities mentioned above. The distinguishing markers given by Boissier in the above-cited Diagnoses are of minor importance and not adequate even for retaining these forms as varieties.

Series iii. BADIA Bobr.

Acta Inst. Bot. Acad. Sci. URSS ser. 1, 6 : 239 (1947).
Incl. Ser. *Sebastiana* Bobr., *loc. cit.* 246

Standard without claw, conduplicate, carinate at back, upper part somewhat dilated, spoon-shaped.

124. T. badium Schreb. in Sturm., Deutschl. Fl. Abt. 1, Bd. 4, Heft 16, t. 104 (1804). *Typus* : Illustration t. 104 in Sturm *loc. cit.* (lecto.).

Plate 120. *T. brutium* Ten.
Plant in flower and fruit; fruiting calyx with persistent corolla; fruiting calyx; standard; wing; keel.

T. spadiceum Vill., Hist. Pl. Dauph. 3 : 491 (1788) non L.
T. spadiceum L. var. *badium* Lapyer., Hist. Abr. Pyr. 439 (1813).
T. aureum Geners., Elench. Scepus. n. 69, ex Wahlenb., Fl. Carpat. 222 (1814) non Polich.
Amarenus badius (Schreb.) Presl, Symb. Bot. 1 : 46 (1830).
T. rytidosemium Boiss. & Hoh. in Boiss., Diagn. ser. 1, 9 : 29 (1841) et Fl. Or. 2 : 149 (1872).
T. badium Ledeb., Fl. Ross. 1 : 556 (1843).
T. rivulare Boiss. & Bal. in Boiss., Diagn. ser. 2, 6 : 49 (1859)
T. pseudo-badium Velen. in Abhandl. Böhm. Ges. Wiss. 1889, 2 : 33 (1890) et Fl. Bulg. 141 (1891).
T. ponticum Alb. in Soc. Imp. Hortic. Odessa 7 (1892).
T. badium Schreb. ssp. *rytidosemium* (Boiss. & Hoh.) Hossain, Not. Roy. Bot. Gard. Edinb. 23 : 473 (1961).
T. rytidosemium Boiss. & Hoh. var. *rivulare* (Boiss. & Bal.) Zoh. in Davis, Fl. Turkey 3 : 401 (1970).

Icon. : [Plate 121].

Perennial with thick taproot, appressed-hairy to glabrescent, 10–40 cm, slender, erect, sparingly branching; branches ascending. Leaves rather long-petioled, the uppermost ones more or less opposite; stipules 1–2 cm, adnate to petioles for about two-thirds of their length, with free portion lanceolate, acute to acuminate; leaflets 1–3 × 0.6–1.5 cm, sessile or with short petiolules, rhombic or broad-elliptical, rounded, truncate or notched at apex, margins denticulate especially along upper half. Peduncles erect, up to 8 cm, sometimes in pairs. Heads 1–1.5 cm in flower and up to 2.5 cm in fruit. Pedicels about as long as calyx tube, deflexed after flowering. Calyx 3–4 mm long; tube broadly cylindrical, 5-nerved; teeth linear, often terminating in 1–2 bristles, the upper ones shorter than the lower ones, 2–3 times longer than tube. Corolla 0.7–1 cm, pale yellow to golden yellow, often becoming chestnut-brown in fruit; standard broadly oblong-elliptical, spoon-shaped, conduplicate, usually entire, longitudinally nerved, gradually tapering at base, much longer than wings and keel. Pod about 2 mm with 0.5 mm long style, ellipsoidal, tapering at both ends, 1–2-seeded. Seeds 1.4 mm, ovoid, brown. Fl. July–August. 2n = 14.

Hab. : Moist calcareous rocks and in alluvial gravelly soil in alpine pastures.

Gen. Distr. : Spain, C. France, Australia, Germany, Switzerland, Hungary, N. Italy, Poland, Czechoslovakia, Yugoslavia, Albania, Romania, Bulgaria, Turkey, Iran.

Selected specimens : SPAIN : Seo d'Urgel Mont-Cady, 1847, *Bourgeau* 717 (K). FRANCE : Saint Sorlin-d'Arves (Savoie), 2,200 m, 1855, *Didier* 1885 (B). AUSTRIA : Linz, Kerschbaumer Alpe, 1905, *Richter* 17852 (B). GERMANY : Bavaria, Berchtesgaden, 1849, *Einsele* 467 (G). SWITZERLAND : Murren, 1904, *Gregor* (K). ITALY : Karerpass, 1,800 m, 1926, *Engelhardt* (B). YUGOSLAVIA : Hajla Massif, 20 km W.N.W. of Pec, 7,100 ft, 1959, *Cook* et al. 308 (K). ROMANIA : Transilvania, distr. Hunedoara, Mtibus Ratezat, 1,900–2,100 m, 1929, *Nyárády* 1989 (HUJ). BULGARIA : Vidosch-Gebirge nächst Sofia. 1890, *Pichler* & *Barbey* 1145 (K isotype of *T. pseudobadium* Velen.). TURKEY : Lazistan : Vallé sous-alpine de Djimil, 2,300 m, 1871, *Balansa* 1405 (G). IRAN : prov. Teheran, in monte Totschal, 1843, *Kotschy* 704 (G). TRANSCAUCASIA : Mont Dagh, 1894, *Alboff* 77 (P).

Plate 121. *T. badium* Schreb.
(Romania : *E. I. Nyárády* 1989, HUJ).
Plant in flower and fruit; fruiting calyx with persistent corolla; standard; pod; fruiting calyx.

Note : After much hesitation and detailed examination of the many specimens from Turkey and Iran, we arrived at the conclusion, already reached by Gibelli & Belli, that the two species described by Boissier (*T. rytidosemium* and *T. rivulare*) should be synonymized with *T. badium* Schreb. without accrediting them even with the rank of varieties.

125. T. spadiceum L., Fl. Suec. ed. 2 : 261 (1755); Boiss., Fl. 2 : 150 (1872). *Typus* : Described from Sweden (Hb. Linn. 930/62).

T. montanum L., Sp. Pl. 772 (1753) non *loc. cit.* 770.
T. litigiosum Desv., Ann. Sci. Nat. ser. 1, 13 : 329 (1828).
Amarenus spadiceus (L.) Presl, Symb. Bot. 1 : 46 (1830).

Icon. : [Plate 122]

Annual or biennial, appressed-pubescent to glabrescent, 10–40 cm. Stems few or single, erect, poorly branching, somewhat flexuous, leafy. Leaves, except for lowest ones, short-petioled, upper ones almost opposite; stipules herbaceous, adnate to petioles for almost two-thirds of their length, free portion oblong-ovate, acute; leaflets 0.8–2 × 0.6–1 cm, all sessile, those of lowest leaves obovate, the rest elliptical or oblong-elliptical with cuneate base and obtuse to truncate, sometimes notched apex, margin denticulate along upper part. Peduncles terminal, up to 5 cm long, erect. Heads up to 1.5–2(–2.5) × 1–1.2 cm, 50–70-flowered, ovoid in flower, cylindrical in fruit. Pedicels much shorter than tube of calyx, initially erect and later deflexed. Calyx about 2–3 mm, 5-nerved; tube obconical, glabrous; teeth long-villous, the upper ones as long as tube, the lower ones 2–3 times as long. Corolla 4–6 mm, yellowish, soon turning light brown, later blackish-brown; standard obovate, not folded, later cuculate in upper part, grooved lengthwise, almost twice as long as very short-clawed wings; keel much shorter than wings. Pod 1.5 mm long, elliptical, membranous, with style as long as pod and stipe one-third of this length, 1-seeded. Seed 1–1.5 mm. Fl. June–July. 2n = 14.

Hab. : Damp places, near streams, meadows.

Gen. Distr. : Almost the whole of Europe, Turkey, Iran.

Selected specimens : SWEDEN : Dalarna, Paroecia Hademora, Vikmanslytten, 1915, *Samuelsson* 1062 (K). FINLAND : Satakunta par. Karkku Jarvenlaka, 1878, *Hjett* 275 (K). LATVIA : Prov. Latgale Volkenberg, 1923, *Mednis* 2310 (K). GERMANY : Bavaria, Oberfranken Stadsteinach, 350 m, 1906, *Puchtler* 1018 (G). CZECHOSLOVAKIA : M. Velosa Zlatni-Mostove, 1,400 m, 1952, *Vihodzeissky* 161 (K). ROMANIA : Transilvania, distr. Ciuc., 850 m, 1925, *Nyarady* 553 (K). SPAIN : Puerto de Leitariegos, 1835, *Durieu* 355 (K). FRANCE : Le Balgoux, ca. 1,475 m, 1950, *Drummond* & *Sandwith* 411 (K). EUROP. RUSSIA : Tambov., Krasnaja Slobodka, 1903, *Schiraevsky* 3427 (BM). TURKEY : Prov. Ankara, distr. Kizilçahamam, Işik Dag near summit, 1960, *Khan*, *Franco* & *Ratcliffe* 729 (E). AZERBAIJAN : Nachitschaven, Batabad, 2,360–2,400 m, 1934, *Prilipko* & *Sultanov* (HUJ). ARMENIA : prope Cubuckly, 1927, *Schelkownikow* & *Kara-Mursa* (HUJ). IRAN : Koschadara, *Szovitz* 487 (K).

Plate 122. *T. spadiceum* L.
(Czechoslovakia : *G. Sirjaev* 234, HUJ).
Plant in flower and fruit; fruiting calyx; fruiting calyx with persistent corolla; standard.

Note: Except for the varying length of the fruiting heads, all the other characters are more or less the same in all the specimens examined, so that there is no room for subdivision of this species. *T. spadiceum* is instantly distinguished from *T. badium* by the size of the flowers, the shape of the standard, the campanulate calyx and the dark brown fruiting heads.

126. T. erubescens Fenzl, Pugill. 5 (1842); Boiss., Fl. 2 : 151 (1872); Zoh., Fl. Pal. 2 : 166, t. 237 (1972); Hossain, Not. Roy. Bot. Gard. Edinb. 23 : 476 (1961). *Typus*: Turkey: Hatay prope Suedia (Samandagi) ad Ostia Orontis, *Kotschy* 141 (W).

T. speciosum Willd. var. *erubescens* (Fenzl) Fenzl, Ill. Pl. Syr. 14 (1843).

Icon.: [Plate 123].

Annual, appressed- or patulous-pubescent, 10–40 cm. Stems slender, few or many, terete, erect or ascending, rarely decumbent, branching above. Leaves rather long-petioled, shorter-petioled in upper part; stipules submembranous at margin, free portion oblong, acute; leaflets 0.6–1 × 0.3–0.6 cm, hairy or ciliate, obovate, cuneate at base, rounded or almost truncate or emarginate at apex, denticulate in upper half, all with very short petiolules or lacking petiolules. Peduncles axillary, almost filiform, 2–5 times as long as leaves, the terminal ones shorter. Heads 0.6–1 cm across, many-flowered, loose, broad-ovoid to globular. Pedicels shorter than calyx tube. Calyx about 2 mm, glabrous; tube membranous, white, with 5 obscure nerves; 2 upper teeth triangular, almost as long as tube, 3 lower ones linear-lanceolate to subulate, often with a few hairs at apex, mostly 1.5–2 times as long as tube. Corolla 4–5 mm, pink or lilac; standard with oblong-spatulate limb, abruptly narrowing in claw, carinate at back, conduplicate in fruit, flesh-coloured to dark brown, longer than wings and violet-tipped keel. Pod long- and thick-stipitate, as long as or somewhat longer than coiled style, 1-seeded. Seed 1 mm, ovoid, brown. Fl. March–May.

Hab.: Mainly in shady places. Batha, maquis and scree.

Gen. Distr.: Turkey, W. Syria, Lebanon, Israel, Ethiopia.

Selected specimens: TURKEY: Cassius (Akra Da.), 900 m, *Meinertzhagen* (E). SYRIA MEDIA: about Riha, 1931, *Zohary* 19317 (HUJ). LEBANON: Nahr-el-Kalb prope Beirut, 1931, *Zohary* 19311 (HUJ). ISRAEL: Upper Galilee, Beit Jann to Har Meiron, 1951, *Haran* 16742 (HUJ); Mt. Carmel: Wadi Shaanan, 1927, *Naftolsky* 16739 (HUJ). ETHIOPIA: Lölho, 8,000 ft, 1854, *Schimper* 821 (G).

Note: *T. erubescens* is readily distinguishable from *T. patens* by the following characters: 1) leaflets – even uppermost ones – obovate, long-denticulate and hairy or ciliate; 2) flowers lilac with purplish spot at tip of keel; 3) stipe and style often longer than pod; 4) lower calyx teeth at most twice as long as tube, often shorter; 5) stipules submembranous.

127. T. sebastianii Savi, Lett. al Sebast. 2 (1815) et Diar. Med. Flajani 2 (1815); Sebast., Rom. Pl. fasc. 2 : 14 (1815); Boiss., Fl. 2 : 155 (1872). *Typus*: Described from Italy.

Amarenus sebastianii (Savi) Presl, Symb. Bot. 1 : 46 (1830).

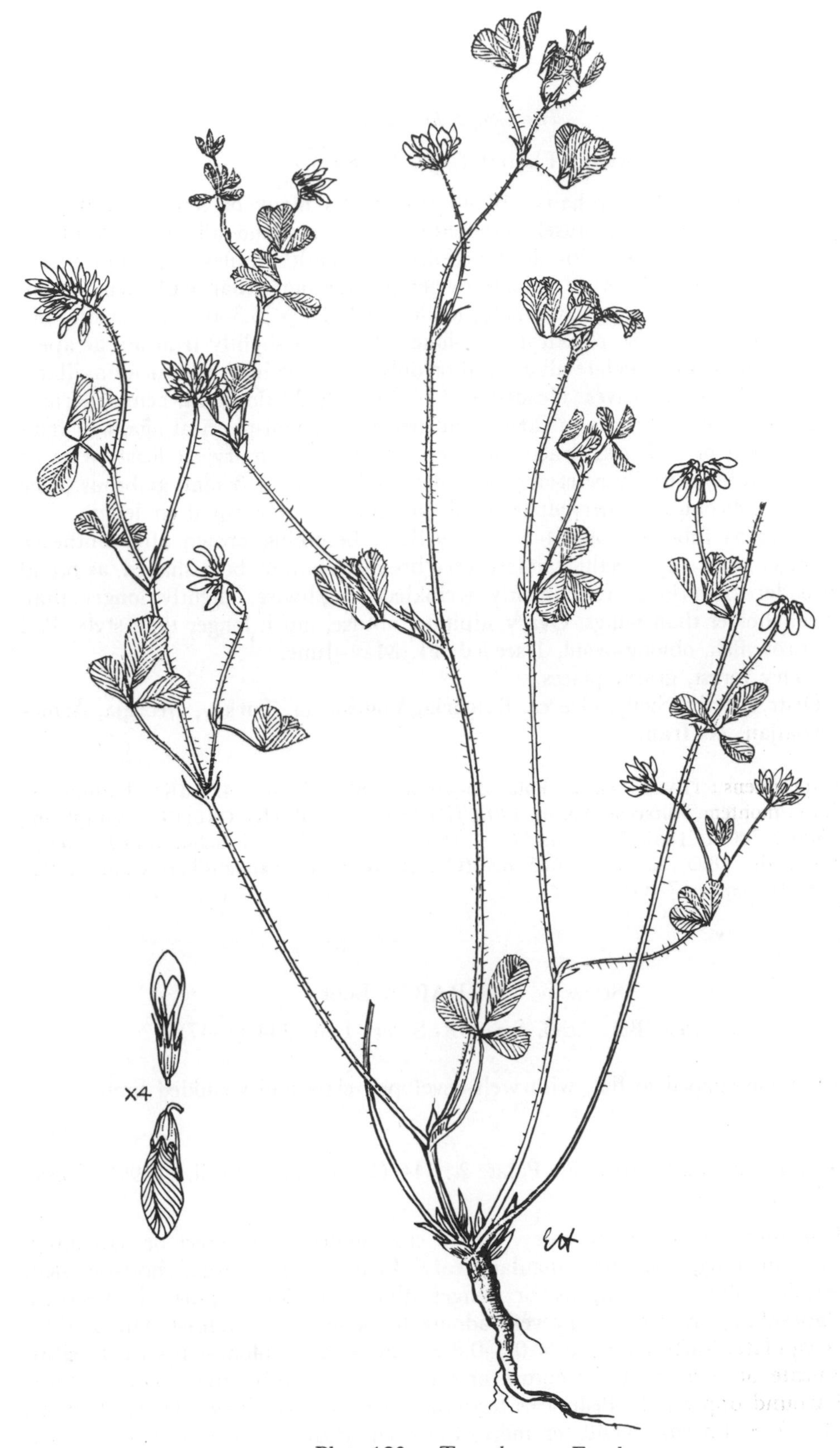

Plate 123 *T. erubescens* Fenzl
(Palestine : *Amdursky* & *Bosnier* 249, HUJ).
Plant in flower and fruit; flower; fruiting calyx with persistent corolla.

Icon.: [Plate 124]. Fiori, Ic. Fl. Ital. f. 2022 (1899).

Annual, glabrous or slightly hairy, 10–40 cm. Stems many or few, ascending or procumbent, rarely erect, profusely branching from base and all along; branches spreading, leafy. All leaves with short petioles (1–2 cm); stipules submembranous, lanceolate, longitudinally nerved, adnate to petioles for more than half their length, free portion ovate-triangular, acuminate; leaflets 0.4–0.8 × 0.3–0.7 cm, sessile, all, except the lowermost ones, elliptical to oblong, obtuse to slightly truncate at apex, margin denticulate and ciliolate all around or only in upper half. Peduncles axillary, shorter than subtending leaves. Heads 0.7–1 × 1 cm, 8–20-flowered, hemispherical, turning globular or umbellate in flower and becoming hemispherical again in fruit due to recurving of pedicels. Bracts rudimentary. Pedicels hairy, at least twice as long as calyx tube, initially erect and later deflexed. Calyx 2–3 mm, glabrous; tube about 1 mm, obconical, 5-nerved; teeth all subulate, nearly equal in length, 2–4 times as long as tube. Corolla 4–5 mm, yellow, becoming brown after anthesis, slightly longer than calyx teeth; standard very broadly obovate, boat-shaped, as broad as or broader than long, prominently wrinkled lengthwise, slightly longer than wings; keel shorter than wings. Ovary stipitate, obtuse, much longer than style. Pod leathery, brownish, oblong-ovoid, 1-seeded. Fl. May–June.

Hab.: Dry forest, grassy places.

Gen. Distr.: Italy, Sicily, Greece, Bulgaria, Yugoslavia, Turkey, Georgia, Armenia, Azerbaijan, N. Iran.

Selected specimens: ITALY: Istria, Pola, 35–70 m, 1884, *Pichler* 422 (K); Latium, S. Marinella et montem Tolfaccia, 360 m, 1902, *Chiovenda* 452 bis (K). GREECE: Agrapham, Pindi, 3,500–3,700 ft., 1885, *Haussknecht* s. n. (K). TURKEY: Prov. Istanbul, près Lagkeuy, entre Abindagh, 1893, *Aznavour* 663 bis (G). DAGESTAN: Kaitagh-Taleassran, 1902, *Alexenko* & *Woronow* 67 (G).

Series iv. AGRARIA Bobr.

Acta Inst. Bot. Acad. Sci. URSS ser. 1, 6: 244 (1947)

Standard spoon-shaped to flat, with well-developed claw and wrinkled limb.

128. T. aureum Poll., Hist. Pl. Palat. 2: 344 (1777) non Thuill. (1799); Boiss., Fl. 2: 153 (1872).

Annual or biennial, appressed-hairy, 20–60 cm. Stems many, erect or ascending, profusely branching, obsoletely angular, leafy. Leaves rather large, short-petioled, with petioles often as long as or longer than stipules; stipules herbaceous, oblong-lanceolate, prominently nerved, adnate to petioles for at least half of their length, cuspidate; leaflets 1.5–2.5 × 0.6–0.8 cm, all sessile, oblong-elliptical to elliptical, cuneate at base, acute or obtuse or truncate to slightly emarginate at apex, dentate around upper half. Peduncles terminal, often 2–5 cm long. Heads globular, mostly 1.2–2 × 1.3 cm, 20–40 (or more)-flowered. Pedicels about 1 mm. Calyx 2

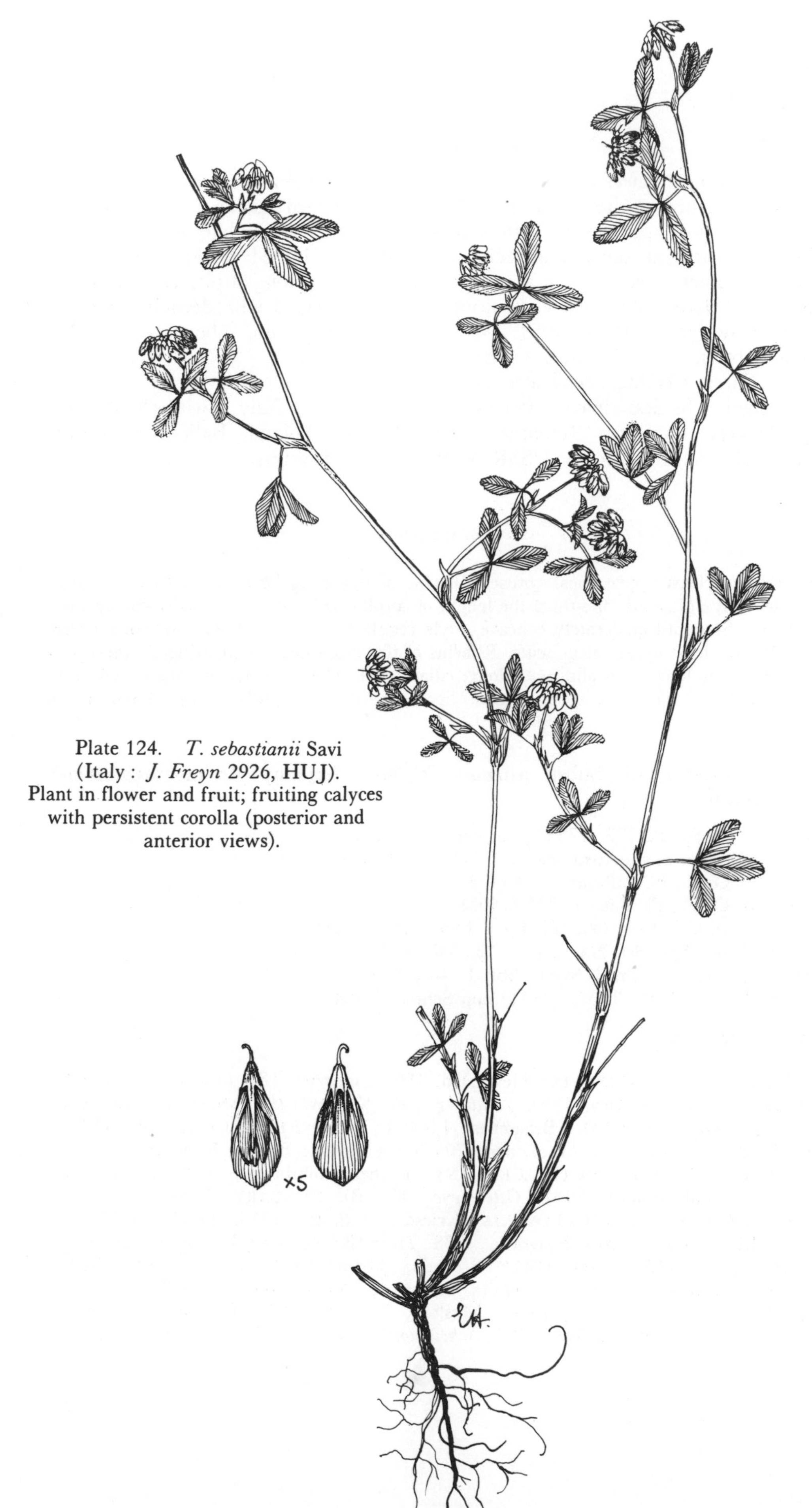

Plate 124. *T. sebastianii* Savi (Italy : *J. Freyn* 2926, HUJ). Plant in flower and fruit; fruiting calyces with persistent corolla (posterior and anterior views).

mm; tube obconical; lower teeth 2–2.5 times as long as upper ones. Corolla 7–8 mm, yellow, turning brown after anthesis and in fruit; standard carinate, limb shallowly or deeply spoon-shaped above with canaliculate claw, denticulate in lower part, deeply emarginate at apex, longitudinally furrowed-wrinkled; wings shorter than standard, divergent; keel as long as wings. Ovary oblong, long-stipitate; style terminal or lateral, longer than ovary. Fruiting head cone-shaped with densely imbricate flowers. Pod longer than style, 1-seeded. Seed 1.2 mm, obovoid-ellipsoidal, yellowish-brown. Fl. July–August. 2n = 14.

Hab.: Forest clearings, roadsides, etc.

Gen. Distr.: S. Scandinavia, Portugal, Spain, France, Italy, Switzerland, Germany, Austria, Hungary, Romania, Czechoslovakia, Poland, Balkan Peninsula, Turkey, Lebanon, N. Iran, S. USSR, Caucasus, Transcaucasia.

Key to Subspecies

1. Leaflets (at least upper ones) obtuse. Rhachis of flowering heads with short, cup-like bracts. Calyx glabrous, one-third the length of corolla which turns yellowish-brown after anthesis. Standard moderately concave. Style regular. **128a.** subsp. **aureum**
– Leaflets (at least upper ones) acute. Rhachis of flowering heads with fringed bracts. Calyx nearly as long as corolla, with long, ciliate teeth. Corolla turning dark brown after anthesis. Standard deeply spoon-shaped. Style lateral. **128b.** subsp. **barbulatum**

128a. T. aureum Poll. Subsp. **aureum**. *Typus*: Germany: In montosis silvosis circa Steinbach.

T. agrarium L., Sp. Pl. 772 (1753) p. p. nom. ambiguum.
T. strepens Crantz, Stirp. Austr. ed. 2, 5 : 411 (1769) nom. illeg.
T. tumescens Gilib., Fl. Lithuan. 2 : 4 (1782).
T. campestre Gmel., Fl. Bad. 3 : 237 (1808).
Chrysaspis candollei Desv., Obs. Pl. Env. Angers 165 (1818).
T. fuscum Desv., Ann. Sci. Nat. ser. 1, 13 : 330 (1826).
Amarenus agrarius (L.) Presl, Symb. Bot. 1 : 46 (1830).
T. badium Puccin., Syn. Pl. 371 (1841) non Schreb. (1804).

Icon.: [Plate 125].

Selected specimens. SCANDINAVIA: Hevgeslad, 1907, *Anderson* (B). FINLAND: Nylandia par. Kyrkslätt, Pag. Österley, 1907, *Lindberg* 764 (K). SWITZERLAND: A. Lunsbruch, 1904, *Gregor* (K). PORTUGAL: Bragomea, 1,500 ft, 1959, *Epsom College* 367 (BM). SPAIN: Cerdagne, 1,450 m, 1926, *Sennen* 5705 (G). FRANCE: Bois de Randamne, Pay-de Dâme, 1854, *Pecoq* et al. 689 (B). GERMANY: Lychen, Rohrbruch, 1886, *Heitaud* (B). AUSTRIA: bei Palmäurdorf, 1931, *Ostermeyer* 12 (B). HUNGARY: Potoscsaba, 1916, *Filarszky* 854 (BM). CZECHOSLOVAKIA: Kriesdorfer Sattel, 1910, *Fiedler* 6863 (B). POLAND: Monte Zorek, prope Kalwaria, 1938, *Trela* (B). RUSSIA: Leningrad, Dacznoje, 1920, *Roshevitz* 3428b (BM). ITALY: Venetia, 30 m, 1904, *Beguinot* 299 (BM). ROMANIA: Autonoma Maghiara, Toplita, Lemca Bradului, 600 m, 1958, *Chirilà* 92 (HUJ). BULGARIA: M. Vitosa, Zlatni Moslove, 1,400 m, 1951, *Hinkova* 45 (HUJ). ARMENIA: Wargavar et Bugakiar, 1929, *Schelkovnikov* & *Kara Murza* (HUJ). AZERBAI-

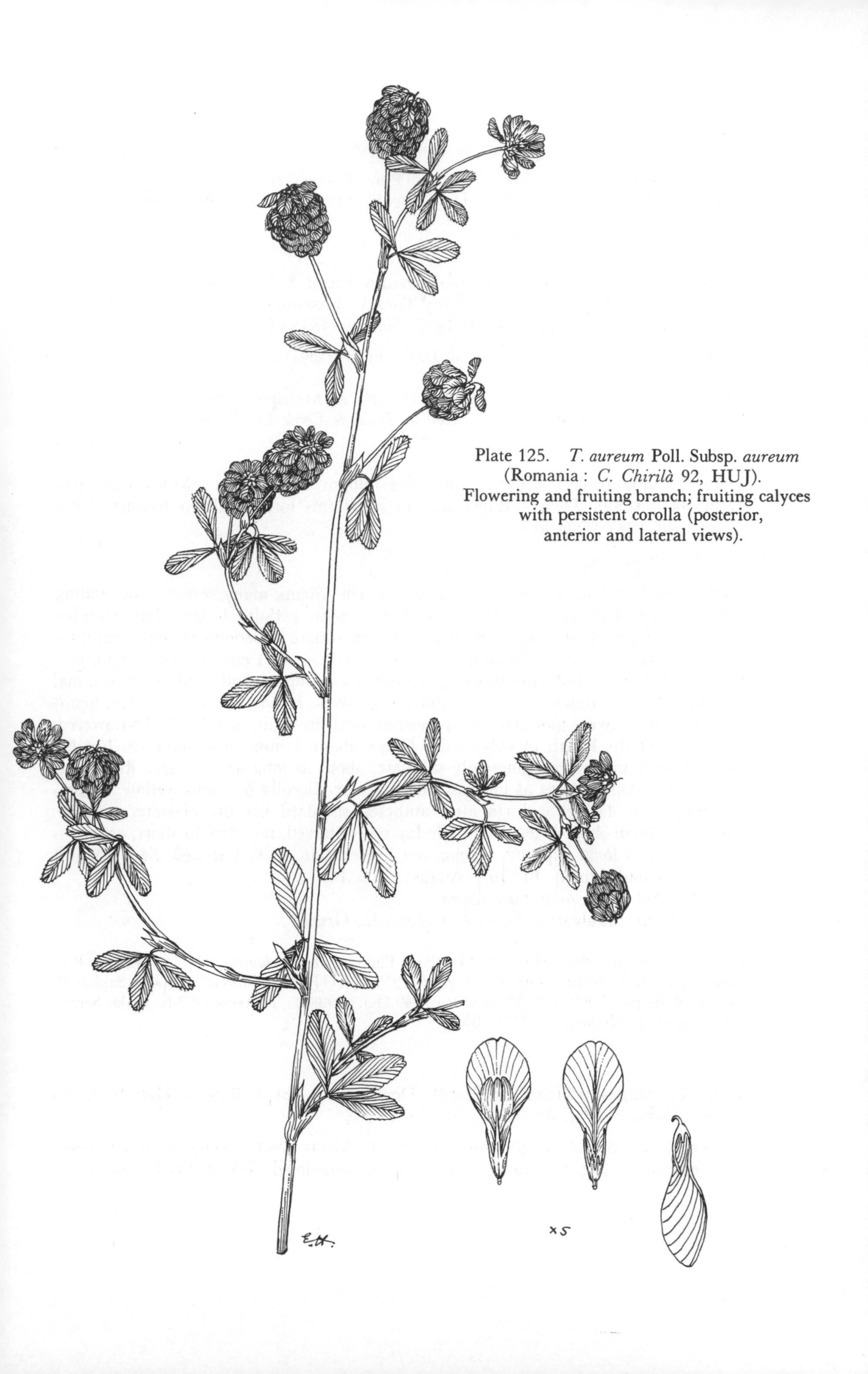

Plate 125. *T. aureum* Poll. Subsp. *aureum* (Romania : *C. Chirilà* 92, HUJ).
Flowering and fruiting branch; fruiting calyces with persistent corolla (posterior, anterior and lateral views).

JAN : Nijal-dagh, *Chalilov* 15 (B). TURKEY : Prov. Trabzon, N. slope of Soganli Dag above Çaykara, 1,300 m, *Davis* & *Hedge* D. 32076 (E). LEBANON : Bir Hassem, Beirut, 1945, *Gilbert* 12 (BM).

128b. T. aureum Poll. Subsp. **barbulatum** Freyn & Sint. ex Freyn in Oesterr. Bot. Zeitschr. 43 : 337 (1893). *Typus* : Turkey (Kastamonu) : Tossia in pratis subalpinis montes Giaurdagh, 29.VII.1892, *Sintenis* 4835 (P).

T. barbulatum (Freyn & Sint.) Zoh. in Davis, Fl. Turkey 3 : 404 (1970).

Selected specimens : TURKEY : Tavchan-dagh, près de Merzifaun, 1903, *Manissadjian* 15 (G); Prov. Bolu : Mudurnu to Göl, 1,100 m, *Davis* & *Coode* D. 37124 (E).

129. T. velenovskyi Vandas, Sitz. Ber. Böhm. Ges. Wiss. (Math.-Nat. Kl.) 1888 : 441 (1889). *Typus* : Bulgaria : "in subalpinis totius Balcani frequens" leg. *Velenovsky* 1887 (E lecto.).

Icon. : [Plate 126].

Annual or biennial, appressed-hairy, 20–40 cm. Stems many, erect or ascending, leafy, branched. Leaves rather short-petioled, with petioles longer than stipules; stipules herbaceous, oblong, prominently nerved, adnate to petioles for only one-third of their length, not auriculate at base; leaflets 0.6–2 × 0.3–1 cm, obovate to elliptical, cuneate at base, rounded or truncate, dentate around upper half. Peduncles terminal and axillary, 2–3 times as long as subtending leaves. Heads 1–2 × 1–1.5 cm, hemispherical in flower, globular or somewhat oval in fruit, loose, 30–40-flowered. Pedicels half the length of calyx tube. Calyx about 4 mm; tube glabrous, broadly campanulate; upper teeth narrowly subulate, about as long as or longer than tube, lower ones about 3 times as long as tube, acicular. Corolla 6–7 mm, yellow, becoming somewhat darker or lurid after anthesis; standard broadly obovate, carinate, shallowly spoon-shaped, longitudinally furrowed-nerved, tapering to short, canaliculate claw. Pod long, stipitate, oblong, not longer than style, 1-seeded. Seed about 1 mm, ellipsoidal, brown. Fl. July–August. 2n = 16.

Hab. : Subalpine mountain slopes.

Gen. Distr. : Bulgaria, Albania, Yugoslavia, Greece.

Selected specimens : BULGARIA : Serbia, Stara Planina, 1894 *Adamović* (HUJ); Montes Rila Planina, 1,100–1,600 m, 1930, *Rechinger* f. 1939a (HUJ). YUGOSLAVIA : Popova Sapka, 30 km W. of Skopje, 5,500 ft, 1959, *Stainton* 7869 (K). GREECE : Macedonia, Mt. Leila, Serrai, 1,400 m, 1936, *Rechinger* 10734 (BM).

130. T. campestre Schreb. in Sturm., Deutschl. Fl. Abt. 1, Band 4, Heft 16, t. 253 (1804); Zoh., Fl. Pal. 2 : 165, t. 235 (1972).

Annual, hairy or almost glabrous, 10–30 cm. Stems erect, ascending or prostrate, simple or branching. Leaves short- to rather long-petioled; stipules herbaceous, ovate

Plate 126. *T. velenovskyi* Vandas
(Bulgaria : *Adamović*, HUJ).
Plant in flower; flower.

to oblong, long-acuminate, adnate to petioles for about half their length; leaflets 0.8–1.6 × 0.4–0.8 cm, rhombic to oblong-elliptical with cuneate base, truncate or retuse at apex, denticulate in upper half, terminal one long-petiolulate. Peduncles as long as or longer or shorter than leaves. Heads 0.8–1.5 × 0.7–1 cm, many-flowered, often globular. Pedicels shorter than calyx, soon deflexing. Flowers numerous, rather dense, later becoming densely imbricate, often forming cone-shaped capitulum. Calyx about 3 mm, white; tube 5-nerved, glabrous or rarely slightly hairy, membranous; 2 upper teeth very short, triangular or lanceolate, others at least twice as long as tube, long-subulate. Corolla (4–)5–6(–7) mm, pale to bright yellow, rarely pink to purple, turning brown in fruit; standard with orbicular limb, flat or spoon-shaped, denticulate at margin. Pod longer than style, stipitate. Seed solitary, 1 mm, ovoid-lenticular. Fl. February–April(–October). 2n = 14.

Hab. : Fields, batha and roadsides.

Gen. Distr. : Whole of Europe, all the countries of the Mediterranean region, Caucasus, Iran, Iraq (also introduced into several countries of the New World and some tropical regions of Afroasia).

1. Flowers yellow. Stems and leaves green. **130a.** var. **campestre**
– Flowers violet or purplish 2
2. Leaves green. Peduncles as long as or shorter than leaves. **130b.** var. **lagrangei**
– Leaves and stem stained purplish. Peduncles much longer than leaves. **130c.** var. **paphium**

130a. T. campestre Schreb. Var. **campestre**. *Typus* : Illustration in Sturm, *loc. cit.* t. 253 (lecto.).

T. agrarium L., Sp. Pl. 772 (1753) p. p.
T. procumbens L., Fl. Suec. ed. 2 : 261 (1755) p. p. nomen ambiguum, non Sp. Pl. 772 (1753).
T. spadiceum Thuill., Fl. Par. ed. 2 : 385 (1799) non L. (1755).
T. ciliatum Poir., Encycl. 8 : 28 (1808).
T. pseudo-procumbens C. C. Gmel., Fl. Bad. 3 : 240 (1808).
Chrysaspis campestre Desv., Obs. Pl. Env. Angers. 164 (1818).
Chrysaspis procumbens Desv., *loc. cit.*
T. minimum Barton, Comp. Fl. Philad. 2 : 74 (1824).
T. procumbens-nanum Ser. in DC., Prodr. 2 : 205 (1825).
Amarenus procumbens (L.) Presl, Symb. Bot. 1 : 46 (1830).
T. lupulinum Gualdenst. ex Ledeb., Fl. Ross. 1 : 557 (1843).
T. schreberi Jord. ex Reut., Cat. Pl. Genev. ed. 2 : 49 (1861).
T. thionanthum Hausskn. in Mitt. Thür. Bot. Ver. 5 : 71 (1885)
T. glaucescens Hausskn. in Mitt. Thür. Bot. Ver. N. F. 5 : 79 (1893).
T. erythranthum Halácsy in Oesterr. Bot. Zeitschr. 56 : 208 (1906).
T. pumilum Hossain, Not. Roy. Bot. Gard. Edinb. 23 : 479 (1961).

Icon : [Plate 127].

Selected specimens : DENMARK : Iylland, Varde, 1890. *Raunkiaer* (HUJ). POLAND : Tarnow, Sciernisko, 1952, *Tacik* 434 (HUJ). CZECHOSLOVAKIA : Moravia, Krumlov, 350 m,

Plate 127. *T. campestre* Schreb. Var. *campestre* (Palestine : *Zohary*, HUJ). Plant in flower and fruit; flower.

1926, *Podpera* & *Sirjaev* 232 I (HUJ). ENGLAND : Oxfordshire, Bladon Heath, 1943, *Turrill* (K). FRANCE : S. of Banyuls. 50–100 m, 1956, *Segal* 195 (HUJ). PORTUGAL : Arravido monastery, 1933, *Atchley* 290 (K). SPAIN : S. Roque, 1844–6, *Willkomm* 677 (BM). ITALY : Apulien : Mt. Gargano, Manfredonia, 1964, *Podlech* 9489 (HUJ). YUGOSLAVIA : Jurjevo, Krasno, 180 m, 1966, *Petermann* (HUJ). ROMANIA : Transsilvania : Cluj, 350 m, 1937, *Bujorean* 1990a (HUJ). BULGARIA : Varna, 1907, *Schneider* 222 (HUJ). GREECE : Insula Thasos, ad Limenas, 1891, *Bornmüller* & *Sintenis* 282 (B); Attika : Kap Sunion, 1967, *Pfadenhauer* (HUJ); Attica, Athenae, Mt. Pentelikon, *Haussknecht* 1885 (K, type of *T. thionanthum*). ALGERIA : Oran à Santa-Cruz, 1928, *Taura* (HUJ). MOROCCO : Marsa Saguira, 50 m, 1927, *Font Quer* 304 (B). TURKEY : Istanbul-Sariyer, Belgrad Ormaine, Balabandere Yamaçlari, 1959, *Yaltirik* 2483 (E). SYRIA : Arsouz, 20–50 m, 1933, *Delbes* 353 (HUJ). LEBANON : Env. of Waley Nabi Ayoub, 1,300 m, 1934, *Bot. Dept.* 19347 (HUJ). ISRAEL : Ein Harod, *Naftolsky* 16420 (HUJ); Shefela, ca. 10 km E. of Ramle, 1952, *Zohary* 16594 (HUJ). CRETE : N. Crete, 15 km W. of Rothymnon on the way to Khania, 150 m, 1964, *Zohary* & *Orshan* 01502-10 (HUJ). CYPRUS : Nicosia airfield, 1944, *Evenari* 19446 (HUJ). IRAN : Lorestan, Ilam, 700 m, 1963, *Jacobs* 6300 (K). IRAQ : Jabal Atshau, 500 m, 1933, *Eig* & *Zohary* 19336 (HUJ). SAUDI ARABIA : Gebel Soda, 9,000 ft, 1952, *Polhill* 158 (BM). AFGHANISTAN : Hindu Kush, Salang Pass, 600 ft, 1966, *Furse* 8060 (K).

130b. T. campestre Schreb. Var. **lagrangei** (Boiss.) Zoh. (comb. nov.). *Typus* Environs de Syria (Archipelago), 5.IV.1865, *M. Legrange* s. n. (G lecto.).

T. procumbens L. var. *erythranthum* Griseb., Spicil. 1 : 36 (1843).
T. lagrangei Boiss., Fl. Or. 2 : 154 (1872).
T. procumbens L. var. *lagrangei* (Boiss.) Heldr. exs. ex Attica (1878).
T. glaucum Hausskn. ex Nym., Consp. 180 (1878).

Hab. : Rocky places.
Gen. Distr. : Greece and Islands, Crete, Lebanon, Syria.

Selected specimens : GREECE : Attica, m. Pentebici, 1,000–2,500 ft, 1887, *Heldreich* 916 (G). RHODES : Phileremos ad Triandra, 1935, *Rechinger* 2009 (BM). CRETE : Kissamos, 1884, *Reverchon* 235 (E).

130c. T. campestre Schreb. Var. **paphium** (Meikle) Zoh. (comb. nov.). *Typus* : Cyprus, summit of Khorteri above Pavrostis Psokas, 4,000 ft, *Meikle* (K).

T. campestre Schreb. ssp. *paphium* Meikle in Hooker's Icon. Plant. 37, t. 3652 (1969); Meikle, Fl. Cyprus 1 : 470 (1978).

Hab. : Dry grassy mountain summit; 4,000 ft.
Gen. Distr. : Cyprus (probably endemic). (Only specimen cited as type.)

Note : Four forms can be distinguished within var. *campestre* : (1) with long-peduncled heads and appressed-hirsute stems; (2) with short-peduncled to subsessile heads and more or less patulous-pubescent stems; (3) with short-peduncled heads and appressed-pubescent stems, and (4) with medium-sized peduncles and appressed hairs. One can name the first *T. campestre* Schreb. var. *campestre*, and the second *T. campestre* Schreb. var. *subsessile* (Boiss.)

Hayek, but the last two forms are undoubtedly transitional. The same holds true for the degree of glaucescence which may appear in all the four forms mentioned. Thus, there is no room for a further subdivision of the typical variety.

131. T. patens Schreb. in Sturm., Deutschl. Fl. Abt. 1, Band 4, 16, t. 256 (1804).
Typus: Illustration in Sturm, *loc. cit.* t. 256 (lecto.).

T. aureum Thuill., Fl. Cor. 2 : 385 (1799) non Poll. (1777).
T. spadiceum Dubois, Fl. Orleans no. 1683 (1803) non L. (1755).
T. agrarium Mérat, Nouv. Fl. Env. Par. 202 (1812).
T. parisiense DC., Fl. Fr. Suppl. 562 (1815).
T. procumbens Loisel., Fl. Gall. ed. 2 : 127 (1828) non L. (1753).
Amarenus patens (Schreb.) Presl, Symb. Bot. 1 : 46 (1830).
T. speciosum Marg. & Reut., Fl. Zanti 41 (1841) non Willd. (1802).

Icon.: [Plate 128].

Annual, glabrous to slightly appressed-hairy, 10–50 cm. Stems many, poorly branched, erect or ascending, very slender. Leaves sparse, short-petioled; stipules herbaceous, ovate to oblong, auriculate at base, free portion ovate-triangular, acute to acuminate, shorter than petioles; leaflets 0.5–1.8 × 0.3–0.4 cm, narrow-obovate to oblong, the uppermost ones elliptical, cuneate at base, denticulate in upper part, acute or rounded, truncate or notched at apex, the terminal one petiolulate. Peduncles very slender, longer than subtending leaves, erect. Heads 0.8–1(–1.5) cm, axillary, hemispherical to globular, 10–20-flowered; rhachis usually hairy. Bracts minute to almost 0. Pedicels 2–3 mm, initially erect or spreading, then deflexed, strongly curved. Calyx glabrous, one-third to one-half the length of corolla; teeth lanceolate-subulate, unequal, the lower ones up to twice as long as tube, the upper ones shorter than tube. Corolla about 6 mm, golden-yellow, becoming dark brown and deflexed in fruit; standard conduplicate, somewhat spoon-shaped and carinate in upper part; wings long, clawed, shorter than standard; keel acute, almost as long as wings. Ovary long-stipitate, as long as style. Pod 1-seeded. Seed yellow to brown, ellipsoidal, shiny. Fl. June–August. $2n = 16$.

Hab.: Meadows.

Gen. Distr.: France, Spain, Corsica, Italy, Switzerland, Belgium, Austria, Serbia, Romania, Poland, Hungary, Bulgaria, Greece, Turkey, Crete, Syria, Israel.

Selected specimens: FRANCE: Sisteron, Basses Alpes, 1957, *Simonet* (G). SPAIN: Guipuzcoa, 1895, *Gandoger* 222 (K). ITALY: Pisa (Toscana), 1857, *Savi* 2241 (G). SWITZERLAND: Bai Lugano, 1843, *Thomas* 2218 (K). BELGIUM: Blankenberghe, 1923, *Lambert* (K). AUSTRIA: Styria, Hochenegg, 320 m, 1898, *Hayek* (K). ROMANIA: Banatus distr., Caras-Severin, 150 m, 1923, *Nyarady* 554 (HUJ). POLAND: Jazdowiczki prope Proszowice, 1959, *Tacik* 546 (HUJ). HUNGARY: Kapornak, im Zalaer Kunutato, 1874, *Ernst* (K). BULGARIA: Gorni Losen, 1954, *Gancev* 458 (HUJ). GREECE: Messenia, vallis Nedontis, Kalamata, 1897, *Zahn* 1418 (K). TURKEY: ex agro Byzantino secus Bosphorum, 1877, *Ball* 2537 (E). SYRIA: Soukluk, 1931, *Zohary* 10519 (HUJ). ISRAEL: Coastal Plain, n. Ein Hay, *Eig* (HUJ).

Plate 128. *T. patens* Schreb.
(Bulgaria : *Gančev* & *Kočev* 458, HUJ).
Flowering and fruiting branch.

Series v. FILIFORMIA Gib. & Belli
Malpighia 3 : 250 (1889) pro stirpe.

Heads 3–20-flowered. Corolla 2–4 mm, light yellow; standard boat-shaped.

132. T. dubium Sibth., Fl. Oxon. 231 (1794). *Typus* : Described from England; the Illustration in Curtis, Fl. Lond. t. 307.

T. procumbens L., Sp. Pl. 772 (1753) p. p.; Boiss., Fl. Or. 2 : 154 (1872).
T. filiforme L., Fl. Suec. ed. 2 : 261 (1755) non Sp. Pl. 773 (1753).
T. minus Smith in Relham, Fl. Cantabr. ed. 2 : 290 (1802).
T. flavum Presl, Fl. Sic. p. XXI (1826).
Amarenus flavus (Presl) Presl, Symb. Bot. 1 : 46 (1830).
T. luteolum Schur. in Verh. Naturf. Ver. Bruem. 15, 2 : 179 (1877).
T. parviflorum Bunge ex Nym., Consp. 180 (1878).

Icon. : [Plate 129].

Annual, glabrous or sparingly hairy, 20–40 cm. Stems many, often brownish, slightly furrowed, somewhat flexuous, erect to ascending, poorly branching. Leaves very short-petioled; stipules herbaceous, ovate, acute, short-adnate to petioles, 3–5 mm long; leaflets 0.8–1 × 0.4–0.7 cm, obovate, cuneate at base, rounded or sligthly notched at apex, dentate around upper part, bluish-green, terminal ones long-petiolulate. Peduncles axillary, filiform, much longer than subtending leaves. Heads 8–9 × 6–7 mm, rather dense, 3–20-flowered, hemispherical. Pedicels 1 mm or less, erect, later recurved. Calyx 1.5–2 mm; tube campanulate, 5-nerved, glabrous; lower teeth almost twice as long as tube, upper ones shorter than tube. Corolla about 4 mm, yellow, becoming brownish in fruit; standard ovate, smooth, conduplicate with funnel-shaped bundle of nerves in each half, entire or obscurely denticulate; wings clawed, shorter than standard. Ovary long-stipitate, longer than style. Pod 1-seeded, with style one-third to one-fourth the length of pod. Seed ellipsoidal, 1.3 mm long, light brown. Fl. May–October. 2n = 16, 28.

Hab. : Sandy places, edges of pastures.

Gen. Distr. : Holland, Scandinavia, Belgium, British Isles, France, Portugal, Spain, Italy, Hungary, Poland, Romania, Czechoslovakia, Balkan Peninsula, Turkey, Cyprus, Israel, S. and C. Russia, Caucasus, USA (introduced).

Selected specimens : HOLLAND : between Oegstgeest and Sassenheim, 1948, *Van Ooststroom* 11093 (K). BELGIUM : Prov. Limburg, Munsterbilzen, 1960, *Gadella* et. al. (HUJ). ENGLAND : Claphell, 1945, *Taylor* 155/21 (K). SWITZERLAND : Shane, Svalöv, 1924, *Tedin* 1059 (BM). PORTUGAL : Almeida : Portas da Cruz, 1884, *Ricardo da Cunha* 1507 (HUJ). SPAIN : Guepuecoa, Andoain, 1895, *Gandoger* 121 (K). ITALY : Pisa (Toscana), 1860, *Savi* 347 bis (BM). HUNGARY : Obesse, comit, Bacs-Bodrog, 1915, *Kovacs* 852 (K). POLAND : Dublany, 1907, *Raciborski* 189 (K). CZECHOSLOVAKIA : Montis Bi'le' Karpaty, 1934, *Weber* (HUJ). ROMANIA : Banatus distr. Timiș-Torontal, 90 m, 1941, *Bujorean* 2442 (HUJ). TURKEY : Tekirdağ : 3 mi. W. of Tekirdağ, 100 m, 1965, *Coode* & *Jones* with *Yusuf Dönmez* 2824 (E). CYPRUS : Khandria, 4,000 ft, 1962, *Meikle* 2837 (K). ISRAEL : Sharon Plain, env. of Ain Hai, 1929, *Eig* (HUJ); Acco Plain, heavy soils, Jidro, 1926,

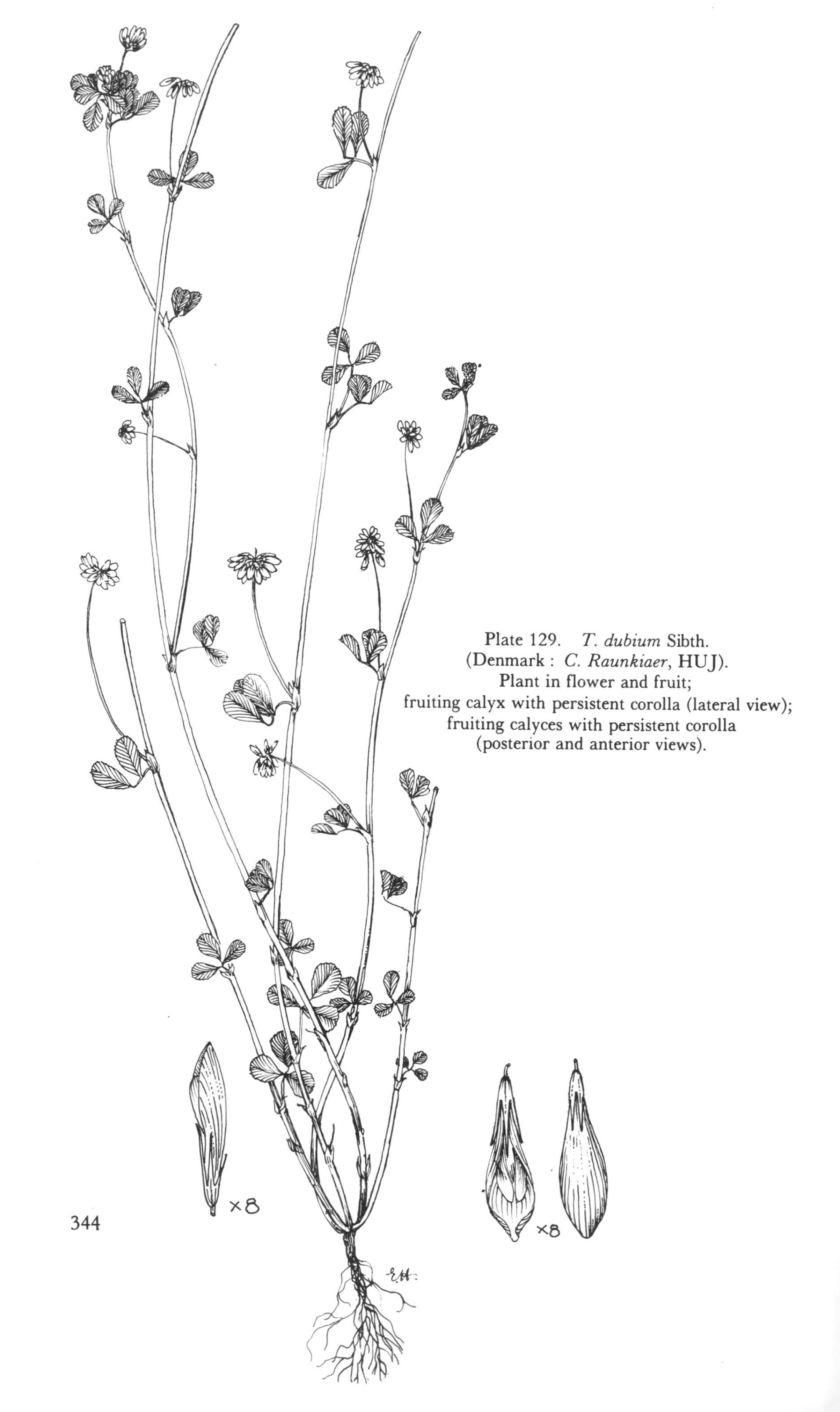

Plate 129. *T. dubium* Sibth.
(Denmark : *C. Raunkiaer*, HUJ).
Plant in flower and fruit;
fruiting calyx with persistent corolla (lateral view);
fruiting calyces with persistent corolla
(posterior and anterior views).

Zohary (HUJ). CRIMEA: Distr. Yalta, Nikita, 100 m, 1959, *Davis* 33106 (E).

Note: The differences between *T. dubium* and *T. micranthum* are very obvious. Here are a few outstanding ones:

	T. dubium	*T. micranthum*
Leaflets	Obscurely dentate	Coarsely dentate
Upper leaflet	Petiolulate	Sessile or subsessile
Stipules	3–5 mm	1.5–2 mm
Peduncles	Ascending, rigid	Curved, filiform
Heads	Dense, 5–20-flowered	Loose, 2–6(–10)-flowered
Pedicels	Scarcely 1 mm long	1.5 mm long
Standard	With prominent nerves in two bundles	With obsolete nerves in one bundle
Pod	Much shorter than corolla, irregularly torn	As long as corolla, dehiscent by valves
Seed	Ellipsoidal	Obovoid

133. T. micranthum Viv., Fl. Lyb. Spec. 45, t. 19, f. 3 (1824). *Typus*: In montibus Cyrenaicae (GE, not seen).

T. filiforme L., Sp. Pl. 773 (1753) p. p. nomen ambiguum.
T. procumbens Poll., Hist. Pl. Palat. 2: 345 (1777) non L. (1753).
T. luteum Lam., Fl. Fr. 2: 604 (1778) nomen ambiguum.
T. controversum Jan ex DC., Prodr. 2: 206 (1825).
Amarenus filiformis (L.) Presl, Symb. Bot. 1: 46 (1830).
T. capiliforme Del. ex Nym., Consp. 181 (1878).
T. delilii Dalb. ex Gib. & Belli in Malpighia 3: 230 (1889).

Icon.: [Plate 130].

Annual, glabrous except in upper parts, 5–30 cm. Stems few or many, filiform, erect or decumbent. Lowest leaves petiolate, others subsessile; stipules oblong, acute, free almost from base; leaflets 5–8 × 3–6 mm, obovate to oblong, cuneate at base, truncate or retuse at apex, coarsely denticulate in upper part, terminal ones subsessile. Peduncles capillary, longer than leaves, later recurved. Heads about 4 mm in diam., very loose, 2–6(–10)-flowered. Pedicels shorter than or almost as long as calyx, becoming deflexed. Calyx about 1.5 mm; tube membranous, white, 5-nerved, glabrous; teeth linear, the 2 upper ones as long as or shorter than tube, the 3 lower ones longer than tube, often terminating in two bristles. Corolla 1.5–3(–4) mm, yellow, membranous in fruit, persistent; standard oblong, boat-shaped, somewhat longer than keel and wings. Pod stipitate, free, slightly exserting from calyx, regularly dehiscent, lenticular to orbicular, 1–2-seeded. Seeds ovoid-reniform, somewhat flattened, dark brown, shiny. Fl. April–May. 2n = 14, 16, 32.

Hab.: Seasonally inundated ground, fields, sands, etc.

Gen. Distr.: S. Norway, Denmark, British Isles, France, Iberian Peninsula, Italy, Hungary, Romania, Balkan Peninsula, Greece, Turkey, Syria, Lebanon, Israel, Iran, Caucasus, N. Africa, Canary Islands, Madagascar.

Plate 130. *T. micranthum* Viv.
(Palestine : *Naftolsky* 16748, HUJ).
Plant in flower and fruit; fruiting calyx with pod and persistent corolla.

Selected specimens: FRANCE: d'Angers, Maine et Loire, 1851, *Guepin* 33 bis (K). PORTUGAL: Oliveira do Hospital, Ribeira, 1954, *Matos* et al. 5014 (HUJ). SPAIN: Cerdagne, Luvia, Estavar, 1,250–1,400 m, 1928, *Sennen* 6557 (BM as *T. praticolum* Sennen). CORSICA: Bastelica, 1878, *Reverchon* 150 (HUJ). HUNGARY: Obesse, comit, Bacs-Bodrog, 1916, *Kovacs* 684 (K). ROMANIA: Banatus distr., Timis-Torontal, 90 m, 1941, *Bujorean* 2443 (HUJ). GREECE: N.E. Macedonia, Hogiatrias, 4,600 ft, 1932, *Alston* & *Sandwith* 91 (K). TURKEY: Prov. Istanbul, Chichli, face cimetière grec., 1895, *Aznavour* 670 (G). ISRAEL: Sharon Plain: Birket el Battich, 1927, *Naftolsky* 16752 (HUJ). IRAN: Ardebil-Astara, 4,000 ft, 1963, *Bowels Bot. Exped.* 2308 (K). MOROCCO: Asib de Ktama, 1,400 m, 1929, *Font Quer* 254 (G).

134. T. dolopium Heldr. & Hausskn. ex Halac., Consp. Fl. Gr. 1 : 405 (1901); Gib. & Belli, Malpighia 3 : 228 (1889). *Typus*: [Greece] Agrapha (Dolopia Veterum) in regione inferiore nr. Pindi, circa monasterium Korona in nemorosis quercinis. Alt. 3,500–3,700 ft, 20–28 June, 1885, *Heldreich* (B holo., iso. E).

T. patens Schreb. var. *koronense* Hausskn., Mitt. Thür. Bot. Ver. N. F. 5 : 78 (1895).

Icon.: [Plate 131].

Annual, appressed-hairy, 10–30 cm. Stems few, erect, leafy, grooved, branched all over. Leaves rather short-petioled; stipules 0.8–1.2 cm long, adnate to petioles for only one-third of their length, ovate-oblong, the free portion lanceolate, acute, prominently nerved, auriculate at base; leaflets 0.6–2 × 0.2–0.7 cm, the lower ones oblong-ovate, cuneate at base, rounded at apex, denticulate in upper half, the upper ones oblong-lanceolate to elliptical, narrowly lanceolate, acute, denticulate, apiculate, the terminal one rather long-petioled. Heads 1.2–1.8 cm in diam., loose, many-flowered, ovoid to globular. Peduncles at least twice as long as subtending leaves. Pedicels 1.5–2 mm, deflexed after anthesis. Calyx about 3 mm long; tube membranous, sparsely hairy; upper teeth shorter than tube, lower ones about twice as long as tube, all narrowly lanceolate-subulate, hairy at margin especially when young. Corolla about 8 mm long, yellow, turning brown after anthesis; standard obovate-oblong, longitudinally nerved, carinate and tapering to claw. Pod about as long as style or longer, 1-seeded. Fl. June.

Hab.: Mountain meadows.

Gen. Distr.: N. Greece (endemic).

Specimens seen: GREECE: (type specimen cited above). Another specimen from S. Macedonia above Chilandari, leg. *Hill, Sandwich* & *Turril* (K), seems to be intermediate (in the length of the pedicels and the flower colour) between *T. dolopium* and *T. brutium*.

Note: *T. dolopium* is believed by some to be a hybrid between *T. brutium* and *T. patens*.

Plate 131. *T. dolopium* Heldr. & Hausskn.
Flowering and fruiting branches; flowers.

Section Six. TRIFOLIUM

Lagopoda L., Sp. Pl. 767 (1753) p. p.
Lagopus Ser. in DC., Prodr. 2 : 189 (1825) p. p.; Bernh., Syst. Verz. Pfl. Erfurt 220 (1890) pro gen.; Lojac., Nuov. Giorn. Bot. Ital. 15 : 228 (1883b); Gib. & Belli, Mem. Accad. Sci. Torino ser. 2, 39 : 1 (1889); Aschers. & Graebn., Syn. Mitteleur. Fl. 6/2 : 526 (1908); Herm., Feddes Repert. 43 : 318 (1938); Bobrov, Acta Inst. Bot. Acad. Sci. URSS 1, 6 : 265 (1947) pro subgen. p. p.; Koch, Syn. 1 : 184 (1837) pro sect. p. p.; Boiss., Fl. Or. 2 : 110 (1872) pro sect.
Eulagopus Lojac., *loc. cit.* 232, pro sect.
Trifolium C. Presl, Symb. Bot. 1 : 48 (1832) pro gen.; Hossain, Not. Roy. Bot. Gard. Edinb. 23 : 397 (1961) pro subgen.

Lectotypus : *T. pratense* L. (Hitchcock & Green in Proposals by British Botanists 177, 1929).

Annuals or perennials. Heads falsely terminal or axillary, sessile or pedunculate. Flowers bractless, very rarely with few bracts at base of head. Calyx 10–20-nerved, hairy or rarely glabrous with unequal or equal teeth; throat usually closed by bilabiate callosity, if open then provided with hairy ring or narrowed by protruding ring. Corolla mostly partly united. Pod included in calyx tube, 1-, very rarely 2-seeded. Dispersal by single fruiting calyces or by entire fruiting heads.

Synopsis of Species in Section Six

Section Six. TRIFOLIUM
Subsection I. TRIFOLIUM
135. T. pratense L.
136. T. mazanderanicum Rech. fil.
137. T. noricum Wulf.
138. T. pallidum Waldst. & Kit.
139. T. diffusum Ehrh.
Subsection II. INTERMEDIA
140. T. medium L.
141. T. pignantii Brogn. & Bory
142. T. heldreichianum (Gib. & Belli) Hausskn.
143. T. patulum Tausch
144. T. velebiticum Deg.
145. T. wettsteinii Dörfl. & Hay.
Subsection III. OCHROLEUCA
146. T. longidentatum Náb.
147. T. ochroleucum Huds.
148. T. caucasicum Tausch
149. T. canescens Willd.

150. T. davisii Hossain
151. T. trichocephalum M.B.
152. T. caudatum Boiss.
153. T. pannonicum Jacq.

Subsection IV. ALPESTRIA

154. T. rubens L.
155. T. alpestre L.

Subsection V. STELLATA

156. T. stellatum L.
157. T. incarnatum L.
158. T. sylvaticum Gérard ex Loisel.

Subsection VI. STENOSEMIUM

159. T. striatum L.

Subsection VII. TRICHOPTERA

160. T. bocconei Savi
161. T. trichopterum Panč.

Subsection VIII. SCABROIDEA

162. T. scabrum L.
163. T. lucanicum Gasp. ex Guss.
164. T. dalmaticum Vis.

Subsection IX. PHLEOIDEA

165. T. gemellum Pourr. ex Willd.
166. T. phleoides Pourr. ex Willd.
167. T. ligusticum Balb. ex Loisel.

Subsection X. LAPPACEA

168. T. hirtum All.
169. T. cherleri L.
170. T. lappaceum L.
171. T. barbeyi Gib. & Belli
172. T. congestum Guss.

Subsection XI. ARVENSIA

173. T. arvense L.
174. T. affine C. Presl
175. T. saxatile All.
176. T. stipulaceum Thunb.

Subsection XII. ANGUSTIFOLIA

177. T. angustifolium L.
178 T. purpureum Loisel.
179. T. blancheanum Boiss.
180. T. roussaeanum Boiss.
181. T. dichroanthum Boiss.
182. T. palaestinum Boiss.
183. T. dichroanthoides Rech. fil.
184. T. haussknechtii Boiss.
185. T. prophetarum Hossain
186. T. dasyurum C. Presl

Subsection XIII. ALEXANDRINA

187. T. salmoneum Mout.
188. T. apertum Bobrov
189. T. berytheum Boiss. & Bl.

190. T. meironense Zoh. & Lern.
191. T. vavilovii Eig
192. T. alexandrinum L.

Subsection XIV. SQUAMOSA

193. T. squamosum L.
194. T. cinctum DC.

Subsection XV. URCEOLATA

195. T. juliani Batt.
196. T. daveauanum Thell.
197. T. miegeanum Maire
198. T. squarrosum L.
199. T. obscurum Savi
200. T. constantinopolitanum Ser.
201. T. leucanthum M.B.

Subsection XVI. ECHINATA

202. T. latinum Seb.
203. T. echinatum M.B.

Subsection XVII. CLYPEATA

204. T. clypeatum L.
205. T. scutatum Boiss.
206. T. plebeium Boiss.

Key to Species

1. Calyx tube 15–20-nerved 2
– Calyx tube 10-nerved 9
2. Fruiting heads depressed, disk-like, not disarticulating into fruiting calyces, but separating as whole, together with stipular involucre. Flowers white or cream-coloured. Stipules not cuspidate at apex. **169. T. cherleri**
– Fruiting heads and other characters not as above 3
3. Heads few, 3–7-flowered, all axillary, surrounded by broadened leaf stipules and overtopped by relatively long-petioled leaves. **172. T. congestum**
– Heads terminal, not as above 4
4. Leaflets obovate to obcordate, scarcely twice as long as broad. Annuals 5
– Leaflets oblong-lanceolate to elliptical, 3–6(or more) times as long as broad. Perennials 7
5. Calyx teeth shorter than tube. Corolla pink, twice as long as calyx. Dwarf caespitose plants. **171. T. barbeyi**
– Calyx teeth much longer than tube 6
6. Calyx tube long-haired. Heads sessile, involucrate, disarticulating at maturity into fruiting calyces. **168. T. hirtum**
– Calyx tube glabrous or glabrescent. Heads pedunculate, not disarticulating at maturity. **170. T. lappaceum**
7. Calyx teeth sharp-pointed. Leaflets entire, rounded-obtuse. Stems usually glabrous. **140c. T. medium** var. **sarosiense**
– Calyx teeth blunt. Leaflets denticulate, acute or mucronate 8
8. Calyx tube mostly glabrous. Stems and stipules glabrous or almost so. Free portion of stipules triangular-lanceolate. **154. T. rubens**

– Calyx tube, stems and stipules hairy. Free portion of stipules subulate-cuspidate. **155. T. alpestre**

9 (2). Flowers small, 0.3–1.3(–1.4) cm 10

– Flowers large, 1.5–3 cm 56

10. Leaflets obovate or ovate, rarely broadly elliptical, 1–1.5(–2) times as long as broad 11

– Leaflets all or mostly (especially those of middle and upper leaves) elliptical, lanceolate or linear, usually (2.5)3–7 times as long as broad 25

11. Free portion of stipules ovate to ovate-lanceolate, obtuse or acute, not provided with subulate cusp or awn or long mucro 12

– Free portion of stipules with long-aristate or subulate or mucronate tip 13

12. Calyx teeth sharp-pointed, equal. **157. T. incarnatum**

– Calyx teeth blunt, unequal. **158. T. sylvaticum**

13. Leaflets with lateral nerves thickened and arcuate-recurved at margin. Annuals with sessile, axillary or axillary and terminal heads about 1(–2) cm long, not or tardily disarticulating at maturity 14

– Not as above 16

14. Teeth of fruiting calyx linear-subulate, triangular at base, almost twice as long as tube, membranous at margin, stellately spreading. Corolla up to twice as long as calyx. **164. T. dalmaticum**

– Teeth of fruiting calyx broader. Corolla as long as or slightly longer than calyx 15

15. Teeth of flowering calyx as long as or shorter than tube, broadly lanceolate, not subulate and not membranous, those of fruiting calyx as long or slightly longer than tube. Fruiting heads cuneate at base. **162. T. scabrum**

– Teeth of flowering calyx lanceolate-subulate or subulate, much longer than tube, those of fruiting calyx densely ciliate-hirsute. Fruiting heads broad at base, not cuneate. **163. T. lucanicum**

16. Corolla generally shorter than calyx (rarely as long as calyx but then calyx tube urceolate) 17

– Corolla longer than calyx 18

17. Heads globular or ovate, frequently subsessile or short-peduncled, often in pairs. Stems appressed-hairy. **165. T. gemellum**

– Heads oblong or cylindrical, more or less long-peduncled, solitary. Stems patulous-hairy. **167. T. ligusticum**

18. Throat of fruiting calyx closed by bilabiate callosity 19

– Throat of fruiting calyx open, glabrous or hairy 22

19. Fruiting calyx teeth (all or part) narrower at base than in upper part; fruiting calyx 1 cm or more in length. **199. T. obscurum**

– Fruiting calyx teeth not as above 20

20. Heads provided at base with 6–7-lobed, membranous involucre. Calyx tube not urceolate. **194. T. cinctum**

– Heads not as above 21

21. Lower calyx tooth ovate, triangular to broadly lanceolate; fruiting calyx tube glabrous. **205. T. scutatum**

– Lower calyx tooth lanceolate to subulate; calyx tube hairy. **206. T. plebeium**

22. Wings of corolla appressed-hairy throughout or near auricles only. Calyx teeth connivent in fruit, markedly unequal. **161. T. trichopterum**

– Wings of corolla glabrous. Teeth of calyx equal 23

23. Flowers about 1.3 cm long. **138. T. pallidum**

– Flower 6–9 mm 24

24. Fruiting calyx cylindrical or campanulate. ***T. pulchellum**
– Fruiting calyx urceolate, globular or fusiform. **158. T. striatum**
25 (10). Corolla shorter than or as long as calyx 26
– Corolla considerably longer than calyx 35
26. Teeth of fruiting calyx lanceolate, all or part considerably narrower at base than in middle; fruiting calyx about 1 cm long. **199. T. obscurum**
– Teeth of fruiting calyx not as above 27
27. Wings of corolla hairy, at least near auricle. Teeth of calyx plumose 28
– Wings of corolla glabrous. Teeth of calyx plumose or not 29
28. Teeth of calyx equal, about twice as long as tube, erect or spreading. **174. T. affine**
– Teeth of calyx as long as or shorter than tube, connivent. **160. T. bocconei**
29. Throat of calyx closed by bilabiate callosity 30
– Throat of calyx open, glabrous or hairy 31
30. Calyx much shorter than 1 cm. **165. T. gemellum**
– Calyx 1 cm or longer. **177. T. angustifolium**
31. Fruiting calyx with stellately spreading teeth. Leaflets mostly truncate 32
– Fruiting calyx with erect or connivent teeth 33
32. Teeth of fruiting calyx united at base by broad, membranous, collar-like seam. Stems appressed-hairy. Heads mostly cylindrical. **166. T. phleoides**
– Teeth of fruiting calyx without broad, membranous, collar-like seam at base. Stems patulous-hairy. Heads globular or ovoid. **165. T. gemellum**
33. Teeth of fruiting calyx mostly plumose and 2 or 3 times as long as calyx tube. Corolla not or slightly exserted from calyx tube and much shorter than teeth. Fruiting heads white or reddish. **173. T. arvense**
– Not as above 34
34. Throat of calyx glabrous. Prostrate alpine plants. **175. T. saxatile**
– Throat of calyx hairy. Erect, South African plants. **176. T. stipulaceum**
35 (25). Teeth ot calyx blunt, usually terminating in one or more rough bristles 36
– Teeth of calyx sharp-pointed, usually terminating in glabrous tip 40
36. Teeth of calyx equal or very nearly so 37
– Teeth of calyx distinctly unequal, the lower one considerably longer than others 38
37. Teeth of fruiting calyx about twice as long as tube, finely plumose. Ovary stipitate. **185. T. prophetarum**
– Teeth of fruiting calyx about as long as tube, not plumose. **178c. T. purpureum** var. **pamphylicum**
38. Corolla white or pinkish. Stems repeatedly and dichotomously branched, appressed-pubescent. Calyx covered with white, silky hairs. Leaflets narrowly oblanceolate, oblong-obovate or obovate; upper leaves opposite. **184. T. haussknechtii**
– Not as above 39
39. Calyx covered with coarse, almost patulous bristles. Upper leaflets almost elliptical, about twice as long as broad. Littoral, procumbent plants with almost globular, about 1 cm long flowering heads. **179. T. blancheanum**
– Calyx covered with shiny hairs (often of two kinds : appressed and spreading). Stems patulous- or antrorsely pubescent. Leaflets oblong-obovate, the lower ones sometimes emarginate. **180. T. roussaeanum**

* Doubtful species, cf. M. Zohary (1972). A revision of the species of *Trifolium* sect. *Trifolium* (Leguminosae). III. Taxonomic treatment (continuation). Candollea 27/2 : 263.

40. Throat of fruiting calyx entirely open or slightly narrowed by inwardly intruding, epidermal or callous, hairy or glabrous ring 41
– Throat of fruiting calyx entirely closed by bilabiate callosity, sometimes leaving only narrow slit between lips 46
41. Fruiting heads not disarticulating at maturity, 1–1.3 cm broad. Throat of calyx with ring of short hairs. Seeds about 2 mm long. Calyx teeth not purple-tipped; lower one of lower flowers often somewhat oblique or irregularly bifid. Cultivated and subspontaneous. **192. T. alexandrinum**
– Fruiting heads readily disarticulating at maturity, 0.8–1 cm broad. Seeds usually smaller. Calyx teeth not as above 42
42. Throat of calyx provided with ring of long, spreading hairs 43
– Throat of calyx glabrous or sparingly appressed-hairy 45
43. Flowers pink. **190. T. meironense**
– Flowers white or cream-coloured 44
44. Calyx teeth triangular at base, all 3-nerved with purple tips; tube of fruiting calyx white between nerves. **188. T. apertum**
– Calyx teeth narrow, usually 1-nerved, not purple-tipped; fruiting calyx tube covered with tuft of long grey or brownish hairs. **189. T. berytheum**
45. Calyx teeth purple-tipped, lower one twice as long as others; throat of calyx widely open. **191. T. vavilovii**
– Calyx teeth not purple-tipped, lower one shorter than above; throat of calyx with callous ring, half open. **187. T. salmoneum**
46. Annuals. Fruiting heads globular or somewhat ovoid, 1–1.5 cm. Fruiting calyces very dense, adnate to one another at base and to rhachis, and not separating from rhachis at maturity. Flowers pink or white to cream-coloured and then often with keel pinkish at tip; teeth of fruiting calyx erect, divergent or rarely somewhat stellately spreading. Uppermost leaves opposite 47
– Heads and calyx not as above 49
47. Fruiting calyx with 1-nerved or nerveless teeth and glabrous or almost glabrous, white tube. **203. T. echinatum**
– Fruiting calyx with 3-nerved teeth and hairy, green tube. **202. T. latinum**
48. Fruiting calyx tube obconical or campanulate, not constricted at throat, white, usually glabrous; all or shorter calyx teeth shorter than tube. Wings of corolla narrower and mostly shorter than keel 49
– Calyx not as above; fruiting calyx tube urceolate and markedly constricted at throat 50
49. Heads provided with involucre of membranous bracts. **194. T. cinctum**
– Heads without involucre of bracts. **193. T. squamosum**
50. Teeth of fruiting calyx equal or very nearly so 51
– Teeth of fruiting calyx distinctly unequal 55
51. Fruiting calyx teeth (all or part of them) narrower at base than in middle; fruiting calyx 1 cm or more in length. **199. T. obscurum**
– Fruiting calyx teeth broader at base than in middle 52
52. Fruiting calyx teeth triangular, about half the length of tube. Fruiting heads about 0.8–1 cm broad. Peduncles appressed-hairy. **196. T. daveauanum**
– Fruiting calyx teeth triangular-lanceolate, considerably longer than above 53
53. Calyx teeth almost as long as tube, narrowing abruptly into subulate tip from 2 mm broad base; tube appressed-hairy to almost glabrous. Corolla as long as or shorter than calyx. **197. T. miegeanum**
– Calyx teeth tapering gradually into subulate tip; tube patulous-hairy 54
54. Fruiting heads globular or semi-globular, rather dense. Calyx tube almost globular in fruit, shorter than teeth. **201. T. leucanthum**

– Fruiting heads spicate; calyces rather remote. Fruiting calyx tube elongated, much longer than teeth. **195. T. juliani**

55. Fruiting heads about 2 cm broad. Corolla only slightly longer than longest calyx tooth. Fruiting calyx teeth 3–5-nerved; tube and teeth covered with rough, tuberculate hairs. **198. T. squarrosum**

– Fruiting heads about 1 cm broad. Corolla much longer than longest calyx tooth. Fruiting calyx teeth 1–3-nerved; tube not covered with tuberculate hairs at base. **200. T. constantinopolitanum**

56 (9). Leaflets (all or most of them) obovate or very short-elliptical, usually about as long as broad, rarely up to about twice as long as broad 57

– Leaflets linear-elliptical or oblong-oblanceolate to lanceolate, usually more than twice as long as broad 70

57. Stipules ovate or obovate, rounded or somewhat tapering above, but never terminating in mucro or awn or cusp-like appendage 58

– Stipules terminating in long-mucronate or aristate or caudate or cuspidate, free upper portion 61

58. Corolla not or scarcely exserting from calyx. Throat of calyx closed by thick tuft of hairs; teeth equal. Annuals. **156. T. stellatum**

– Corolla exserting considerably from calyx. Throat of calyx open or closed by bilabiate callosity. Annuals or perennials 59

59. Calyx teeth ovate-triangular, the lower one about 5 mm broad in fruit; throat of calyx closed by bilabiate callosity. Corolla pink to white, 3 times as long as calyx. Annuals. **204. T. clypeatum**

– Calyx teeth subulate or lanceolate-subulate, not exceeding 2 mm in width; throat of calyx not closed by bilabiate callosity 60

60. Alpine, rhizomatous, appressed-pubescent perennials. Nerves of leaflets thickened and recurved. Heads sessile. **145. T. wettsteinii**

– Lowland or montane, patulous-pubescent annuals. Nerves not thickened, directed upwards. Heads pedunculate. **157. T. incarnatum**

61. Calyx or calyx tube usually glabrous 62

– Calyx, or at least calyx tube, usually hairy at least in upper part 63

62. Calyx teeth longer than tube. **140. T. medium**

– Calyx teeth shorter than tube. **144. T. velebiticum**

63. Calyx teeth blunt 64

– Calyx teeth sharp-pointed 67

64. Free portion of stipules abruptly narrowed into setaceous-aristate tip. Flowers usually crimson or pink (rarely white). **135. T. pratense**

– Free portion of stipules gradually narrowed into lanceolate-subulate cusp 65

65. Calyx large, 1.5 cm or more in length. **141. T. pignantii**

– Calyx much smaller 66

66. Annuals. Leaves all along stems and branches. Stems appressed-hairy. **138. T. pallidum**

– Perennials. Leaves (all or most of them) arising from base of stem, which is covered with dead stipules. **137. T. noricum**

67. Calyx teeth glabrous or glabrescent, all 1-nerved, all or part of them with membranous margins at base. Stems sparingly appressed-pubescent. Leaflets obovate, 1–1.5 times as long as broad. Corolla twice as long as calyx. **142. T. heldreichianum**

– Not as above 68

68. Leaflets elliptical to ovate, all or almost all retuse and less than twice as long as broad. Flowers white or yellow. Calyx half the length of corolla. **148. T. canescens**

– Plants differing from above by one or more characters 69

69. Corolla cream-coloured, not longer or only slightly longer than calyx teeth or longest calyx tooth. Leaves all crowded in rosette; leaflets short-elliptical or ovate, scarcely twice as long as broad; free portion of stipules lanceolate-subulate. Heads up to 2 cm. Subalpine plants. **150. T. davisii**
– Corolla purple, 3 times as long as calyx. Free portion of stipules foliaceous, lanceolate. **152. T. caudatum**

70 (56). Calyx teeth sharp-pointed 71
– Calyx teeth blunt 76

71. Free portion of stipules leaf-like, oblong-lanceolate, not subulate, 0.6–2 cm broad; leaflets ovate to rhombic. Lower calyx tooth 5-nerved. Heads large, about 3 × 2 cm, often subtended at base by series of small, membranous or green but well-discernible bracts. Flowers up to 3 cm long. **146. T. longidentatum**
– Free portion of stipules not as above. Heads and calyx not as above 72

72. Calyx tube usually glabrous. **140. T. medium**
– Calyx tube hairy 73

73. All or upper teeth of fruiting calyx triangular-ovate, many-nerved. Upper leaves opposite. Annual. **198. T. squarrosum**
– Teeth of fruiting calyx linear, subulate or filiform 74

74. Lower tooth of fruiting calyx up to 1.5 cm, deflexed. Heads sessile or subsessile, involucrate, up to 3 cm across. Leaflets of middle and upper leaves elliptical, mucronate or acute. Flowers 1.5–2 cm, pink or white-yellowish, turning reddish. **148. T. caucasicum**
– Not as above 75

75. Flowers about 1.5 cm. Heads mostly ovoid, sessile, 1.5–3 cm long. Leaflets oblong to elliptical, usually up to 3 cm long, all except uppermost ones obtuse and often retuse. **147. T. ochroleucum**
– Flowers up to 2.5 cm. Heads ovoid-oblong, elongating in fruit, up to 4–6 cm, usually long-peduncled. Leaflets of middle and upper leaves oblong to linear. **153. T. pannonicum**

76. Corolla about as long as calyx, i.e., not longer or very slightly longer than longest calyx tooth 77
– Corolla distinctly longer than calyx 82

77. Perennials, mostly montane or subalpine. Leaflets 1 cm or more in breadth. Calyx teeth unequal. Stems appressed-hairy. Head 2 cm or more in breadth 78
– Not as above 79

78. Flowers purple. Stems leafy. **136. T. mazanderanicum**
– Flowers yellow or cream-coloured. Most leaves radical. **150. T. davisii**

79. Stems patulous-pubescent. Calyx teeth much longer than tube. **139. T. diffusum**
– Stems appressed-hairy 80

80. Calyx teeth equal or almost so. Uppermost leaves opposite 81
– Calyx teeth distinctly unequal. All leaves alternate. **177. T. angustifolium**

81. Calyx up to 1.5 cm long. **186. T. dasyurum**
– Calyx 6–7 mm long. **185. T. prophetarum**

82. Perennials. Stems appressed-hairy. Petioles shorter than stipules, often as long as adnate parts of stipules; leaflets linear-oblong, sparingly hairy at margins. Corolla purple. **143. T. patulum**
– Annuals or perennials. Not as above 83

83. Throat of fruiting calyx open or with hairy ring only. Subalpine caespitose perennials with patulous-hairy stems. Heads about 3 cm across Corolla white or cream-coloured or tinted with purple, 2–3 times as long as calyx 84

– Throat of fruiting calyx closed by bilabiate callosity. Plants not as above 85
84. Lower flowers of head subtended by small bracts inside involucre formed by upper leaves. Leaflets oblong or elliptical. **137. T. noricum**
– Flowers without bracts. Leaflets ovate. Calyx tube often glabrous. **141 T. pignantii**
85. Perennials. Heads ovoid, up to 5 cm long and 3–4 cm broad. Leaflets elliptical to oblong. **151. T. trichocephalum**
– Annuals. Heads narrower or smaller 86
86. Calyx covered with appressed, silky, shiny hairs. Corolla white or pink. Leaflets linear-elliptical or oblanceolate. **184. T. haussknechtii**
– Hairs of calyx not appressed-silky and not shiny 87
87. Stems appressed-pubescent 88
– Stems patulous- or subpatulous-pubescent 89
88. Leaflets elliptical, oblong to linear, 2–5 × 0.3–1 cm. Corolla purple or purple with pink. Fruiting heads ellipsoidal or cylindrical. **178. T. purpureum**
– Leaflets oblong-cuneate, 1–1.5 × 0.2–0.5 cm. Heads ovoid or obovoid. Flowers pink or flesh-coloured. **183. T. dichroanthoides**
89. Flowers white, slightly or not at all exserting from calyx. Heads, including fruiting ones, not exceeding 2.5 cm. Flowering heads 1-coloured. **182. T. palaestinum**
Flowers lilac to pale purple, much longer than above, at least in fruiting stage. Flowering heads 2-coloured. **181. T. dichroanthum**

Subsection I. TRIFOLIUM

Pratensia Gib. & Belli, Mem. Accad. Sci Torino ser. 2, 39 : 300 (1889) pro stirpe

Perennials or annuals. Calyx 10-nerved; teeth as long as or shorter than tube, the lower one often longer; throat open, provided with epidermal protuberance and or hairy ring. Corolla marcescent, usually much longer than calyx.

135. T. pratense L., Sp. Pl. 768 (1753).

Perennial with fusiform root and short stock but without runners, patulous- to appressed-pubescent or glabrescent (with hairs whitish, often arising from tubercles), 20–60 cm. Stems many, arising from basal leaves, simple or branched, erect or ascending or decumbent, furrowed to angular. Lower leaves long-petioled, middle and upper ones short-petioled, uppermost ones subsessile; stipules ovate-lanceolate, adnate to petioles, membranous, with green or red nerves, hairy or glabrescent, free portion abruptly mucronulate or cuspidate, usually much smaller than lower part; leaflets very short, petiolulate, mostly 1.5–3(–5) × 0.7–1.5 cm, obovate to obovate-oblong or broadly elliptical, rarely (and in upper leaves only) oblong-lanceolate, obtuse or acutish or retuse, almost toothless, appressed-hairy on both faces or only beneath, often spotted. Heads terminal, solitary or in pairs, globular or ovoid, mostly involucrate by the stipules of reduced leaves, rarely pedunculate. Flowers many, usually 1.5–1.8 cm long, dense. Calyx tubular-campanulate,

10-nerved, whitish-green, sometimes with reddish tint, often appressed- or patulous-hairy (rarely tube glabrous); throat with slight annular (epidermal not callous), hairy thickening; teeth unequal, erect, subulate, blunt, often ciliate or patulous-hairy, the lower one longer than tube and others. Corolla reddish-purple to pink, rarely yellowish-white or white; standard longer than wings and keel, notched. Ovary 2(1)-ovuled. Pod ovoid, membranous with cartilaginous, shiny upper part. Seed 1, oblong-ovoid, tuberculate, yellow, brownish or violet. Fl. May–September. 2n = 14, 28, 56.

Hab. : Meadows, grassy plots, roadsides, near water, glades, etc.

Gen. Distr. : All Europe (except extreme North), Western Asia, the Mediterranean region (except its S.E. part). Very widespread and mostly cultivated throughout the entire Northern Hemisphere and elsewhere.

Note : *T. pratense* is extremely polymorphic and no less than 40 binomials have been published from the "Formenkreis" of this species. There are, however, very few experimental data that testify to the constancy of these "species" and those which have been reduced to subspecies or varieties. Although there is a vast amount of information on this widely grown clover, which is perhaps the oldest one under cultivation, there is yet a need for an authoritative taxonomist-breeder to decide upon the taxonomy of this species, which undoubtedly calls for special treatment.

The present authors were unable to undertake such a critical examination of the numerous forms, so that the following subdivision into 6 varieties is only a tentative one, and does not embrace all the forms recorded in the literature, which very often overlap and intergrade. In future it will be necessary to seek more reliable characters for assessing the systematic rank and value of the infraspecific taxa.

1. Calyx altogether (incl. teeth) glabrous. **135f**. var. **rhodopeum**
– Calyx (at least along teeth) hairy 2
2. Stems 20–40 cm, not fistulous, decumbent or ascending, with densely appressed, white hairs, rarely almost glabrescent. **135a**. var. **pratense**
– Stems glabrous above or all over, or densely covered with patulous hairs 3
3. Stems 40–100 cm, mostly erect, thick and fistulous, sparingly hairy to glabrous. Heads large, often in pairs. Calyx teeth usually reddish. Leaflets up to 3–5 cm long, often spotted. **135c**. var. **sativum**
– Plants not as above 4
4. Stems short, 10–30 cm, decumbent or ascending, arcuate, more or less densely patulous-villous or hirsute, rarely appressed-hairy. Stipules hairy all over. Heads many-flowered, up to 3 cm across. Calyx densely hairy, often reddish. Corolla white to dirty white, sometimes yellowish or reddish. Plants often growing in alpine and subalpine meadows. **135e**. var. **villosum**
– Plants not as above 5
5. Stems much branched, patulous-hairy. Leaflets 3–4 × 1–1.5 cm, those of lower leaves oblong-ovate, not retuse, those of upper ones lanceolate, acute; stipules herbaceous, long, glabrous. Heads large and dense, involucrate. Corolla carmine-red. **135b**. var. **americanum**
– Stems mostly appressed-hairy above and patulous-hairy below. Leaflets much smaller than above, those of uppermost leaves lanceolate, denticulate; stipules with long ciliate cusp. Heads rather small, often pedunculate. All calyx teeth usually longer than tube. Corolla yellowish-white or pink. **135d**. var. **maritimum**

135a. T. pratense L. Var. **pratense**. *Typus* : Described from Europe, Herb. Cliff. 375 (BM).

T. pratense L. var. *spontaneum* Willk., Führer 535 (1863).
T. pratense L. var. *sylvestre* Ducomm., Taschanb. 169 (1869).
T. borysthenicum Gruner, Bull. Soc. Nat. Mosc. 41, 2 : 140 (1869).
T. pratense L. var. *genuinum* Rouy, Fl. Fr. 5 : 119 (1889).
T. pratense L. var. *collinum* Gib. & Belli, Mem. Accad. Sci. Torino ser. 2, 39 : 302 (1889).
T. pratense L. var. *anatolicum* Freyn, Bull. Herb. Boiss. 3 : 177 (1895).
T. pratense L. subsp. *eupratense* Aschers. & Graebn., Syn. Mitteleur. Fl. 6/2 : 548 (1908).
T. fontanum Bobrov in Fl. URSS 11 : 251 (1945).

Icon. : [Plate 132].

The number of forms or races that have been distinguished among populations of this variety is very large. Part of them is recorded by Ascherson & Graebner (1908), Hegi (1923–1924), Savulescu & Rayss (1952) and others. They are based mainly on vegetative characters. Neither the descriptions of these taxa nor the cumbersome nomenclature can be critically treated or even fully recorded here.

Selected specimens : SWEDEN : Göteborg Landvetter s. n. Buaras, 1930, *Borgvall* (HUJ). CZECHOSLOVAKIA : Olomouc in pratis inter Repcin et Horka, 220 m, 1933, *Otruba* 831 (HUJ). SWITZERLAND : between Buffalo and Ofen Pass, 1950, *Feinbrun* (HUJ) FRANCE : Montpellier, Castelnau Cres, 1931, *Eig* (HUJ). ITALY : in cultis, alt. 5 m, pr. Lussinpiccolo, 1923, *Lusina* (HUJ). TURKEY : Prov. Zonguldak, Keltepe above Karabük Pastures, 1,300 m, 1963, *Davis* & *Coode* 560 (E). IRAN : Takestan to Sandjan about 20 km E. of Sandjan, 1,740 m, 1965, *Danin* & *Plitmann* 65929 (HUJ). TRANSCAUCASIA : Azerbaijan, distr. Vank-Disal, prope p. Gadrut in horto, 1935, *Heideman* (HUJ). AFGHANISTAN : Prov. Kabul, between the Unai and Hajigak Passes, wet meadow, 2,950 m, 1962, *Hedge* & *Wendelbo* 4588 (HUJ).

135b. T. pratense L. Var. **americanum** Harz, Bot. Centralbl. 45 : 106 (1891).

T. expansum Waldst. & Kit., Pl. Rar. Hung. 3 : 263, t. 237 (1807).
T. diffusum sensu Baumg., Enum. 2 : 372 (1816) non Ehrh. (1792).
T. pratense L. var. *majus* Boiss., Fl. 2 : 115 (1872).
T. pratense L. var. *expansum* (Waldst. & Kit.) Hausskn., Mitt. Bot. Ver. Thür. N. F. 5 : 72 (1893); Aschers. & Graebn., *loc. cit.* 555.

Selected specimens : PORTUGAL : Porto de Sfos Sladeira de Leima, 1952, *Fernandes* et al. 4264 (HUJ). FRANCE : Boisserau, Montpellier, 1930, *Eig* (HUJ). ITALY : in pratis humidiusculis pr. Romani, 1922, *Lusina* (HUJ). GREECE : Cephalonia, Mt. Menos, 1934, *Guiol* 287 (HUJ). S. RUSSIA : Aktubinskaia Gub., env. of Telmir, bank of R. Telmir, 1926, *Iljin* et al. (HUJ). TURKEY : Distr. Bingöl, 20 km E. of Bingöl, oak forest, 1,800 m, 1963, *Zohary* 473124 (HUJ). IRAN : Elburz Mts., about 100 km N. of Karaj, *M.* & *D. Zohary* 734 (HUJ). TRANSCAUCASIA : pr. Nachitschevan inter segetes, 1934, *Prilipko* (HUJ).

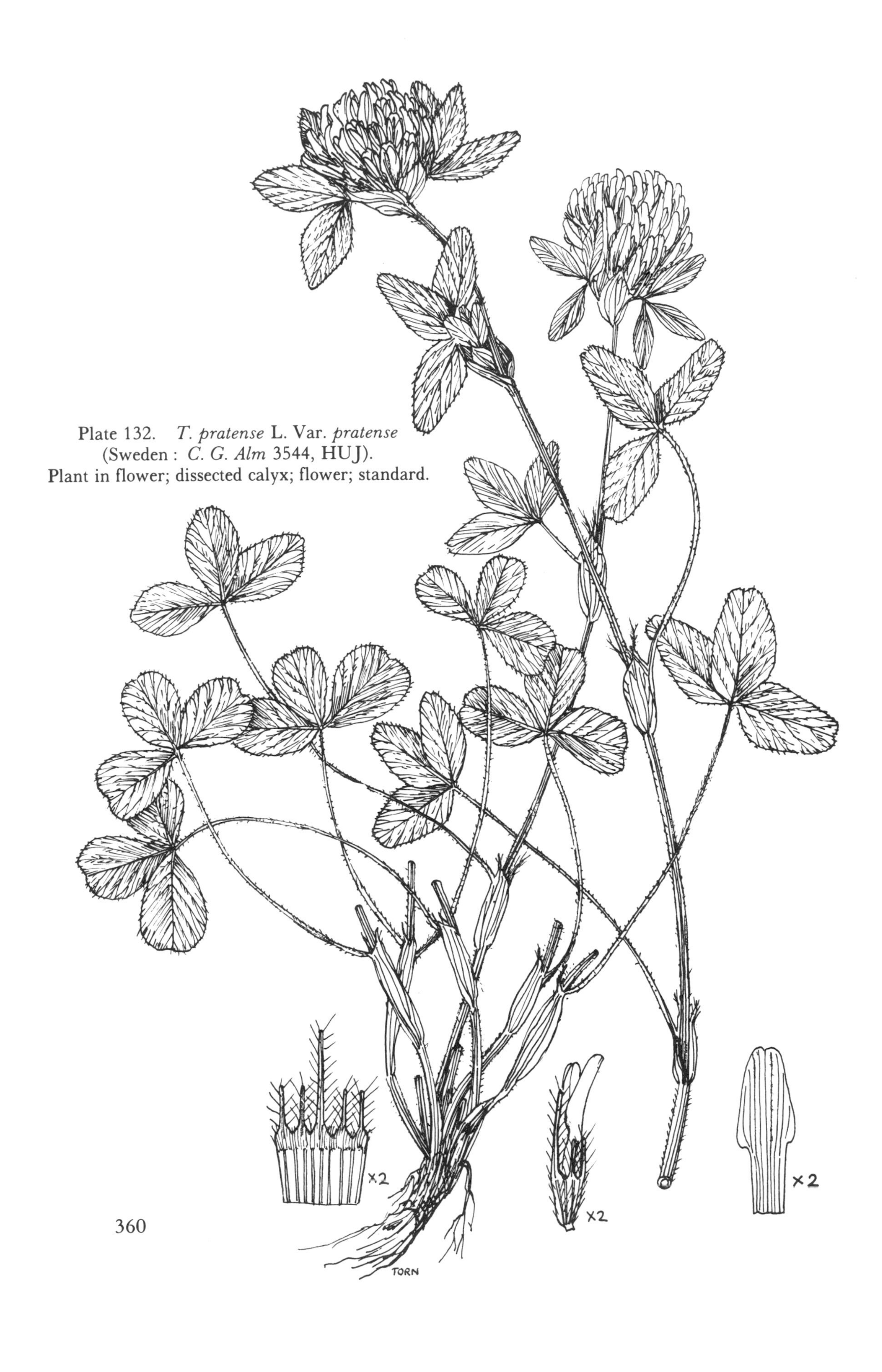

Plate 132. *T. pratense* L. Var. *pratense*
(Sweden : *C. G. Alm* 3544, HUJ).
Plant in flower; dissected calyx; flower; standard.

135c. T. pratense L. Var. **sativum** Schreb. in Sturm., Deutschl. Fl. 4 : 15, t. 236 (1804); Gib. & Belli, *loc. cit.* 302; Aschers. & Graebn., *loc. cit.* 552.

T. pensylvanicum Willd., Enum. Hort. Berol. 793 (1809).
T. bracteatum Schousb. in Willd., *loc. cit.* 792.
T. sativum (Sturm.) Crome ex Boenn., Prodr. Fl. Monast. 222 (1824).
T. baeticum Boiss., Voy. Midi Esp. 2 : 726 (1845).
T. pratense L. var. *fistulosum* Schur ex Schur, Enum. Pl. Trans. 154 (1866) pro syn.

Selected specimens : SCOTLAND : Canty Bay, N. Berwick, 1956, *Currie* 0748 (HUJ). FINLAND : Karelia austr. par Vehkalahti, Pyhältö in area ad domum, 1960, *Fagerström* (HUJ). SPAIN : in montibus supra Algesiras, 1849, *Boiss.* & *Reut.* (K, isotype of *T. baeticum*). TURKEY : Bosphorus, *Beykos* to *Adampol*, 1905 (G); S. W. of Trabzon, grassy western slopes, 1,600 m, *Hennipman* 1805 (G). IRAN : Hamadan, foot of Mt. Alvand, irrigated plots, by runnels, 2,100 m, 1965, *Danin, Baum* & *Plitmann* 65-876 (HUJ). AFGHANISTAN : Kuram Valley, *Harsukh* 14867 (K). TUNISIA : Makter in humidis, 1903, *Murbeck* (G). MOROCCO : entre Khasbet et Hadjab et Dar Kaid, 1918, *Benoist* (G).

Note : Also seen from Iraq, Canada and elsewhere. Recorded from S.W. Spain, Sicily, etc.

135d. T. pratense L. Var. **maritimum** Zabel, Arch. Ver. Freunde Naturg. Meklenburg 14 : 31 (1859).

T. pratense L. var. *villosum* Wahlenb., Fl. Gothob. 73 (1824) non DC. (1805).
T. pratense L. var. *australe* Freyn, Verh. Zool.-Bot. Ges. Wien 27 : 309 (1878).
T. pratense L. var. *hirsutum* Pahnsch., Arch. Naturk. Liv.-Est. -u. Kurl. ser. 2, 9 : 287 (1881) non var. *hirsutum* Boiss. (1840).
T. seravschanicum Ovcz. ex Bobrov in Fl. URSS 11 : 252 (1945).

Selected specimens : PORTUGAL : Ponte do Ratiço entre Murça e Palheiros, 1955, *Fernandes* et al. 5430 (HUJ). SPAIN : Tardad, Villalba (Lugo), 1951, *Orosa* (HUJ). TRANSCAUCASIA : Distr. Norachen, village Achura, 1947, *Chalilov* (HUJ).

135e. T. pratense L. Var. **villosum** DC. in Lam. & DC., Fl. Fr. ed. 3, 4 : 526 (1805).

T. pratense L. var. *alpinum* Hoppe in Sturm., Deutschl. Fl. 8, 32 : t.512 (1812).
T. pratense L. var. *frigidum* Gaudin, Fl. Helv. 4 : 582 (1829) subs. illegit.
T. pratense L. var. *nivale* Koch, Syn. Fl. Germ. 168 (1837) subs. illegit.
T. nivale Sieb. ex Koch, *loc. cit.* pro syn.
T. pratense L. subsp. *nivale* (Sieb. ex Koch) Aschers. & Graebn., Syn. Mitteleur. Fl. 6/2 : 557 (1908).

Specimens seen : SWITZERLAND : Grödner Joch, 2,100 m, 1928, *Bojko* (HUJ).

135f. T. pratense L. Var. **rhodopeum** Velen., Sitzb. Böhm. Ges. Wiss. 29 : 8 (1894).

Specimens seen : BULGARIA : in pratis alpinis ad Levoroso, 1895, *Stribrny* (HUJ). Also recorded from Tyrol, Yugoslavia, etc.

136. T. mazanderanicum Rech. fil., Ann. Naturhist. Mus. Wien 51 : 401 (1941). *Typus* : Iran : distr. Mazanderan, Lichtungen der Laubwälder im Talar-Tal ca. Abbasabad und Cahi, ca. 400 m, 1937, *Rechinger* 2031 (W holo.).

Icon. : [Plate 133].

Perennial, crisp-patulous-hairy, 70–110 cm. Stems thick (3–4 mm), fistulous, flexuous, procumbent or ascending, slightly striate, branching from base, brown, purplish or violet, with elongated internodes. Stipules membranous, brownish, prominently nerved, patulous-hairy, adnate to petioles for at least two-thirds of their length, free portion triangular-lanceolate terminating in 5–6 mm long seta; petioles of lower leaves up to twice as long as leaflets, those of upper leaves half the length of leaflets, long, patulous-villous; leaflets of lower leaves up to 5 × 2 cm, oblong-lanceolate with almost entire margin, cuneate at base, usually rather obtuse and mucronulate. Heads solitary, terminal, many-flowered, globular to subovoid, 2–2.5 cm in diam., subtended by upper leaves. Flowers about 1.7 cm. Calyx tube 4.5–6 mm long, membranous, brown, 10-nerved; teeth all setaceous-acuminate from triangular base, sparingly patulous-setose with bristles arising from tubercle, truncate at apex; throat open, provided with ring-like, hairy protuberance. Corolla brown-purplish (when dry), shorter to slightly longer than lower calyx teeth. Fl. August.

Hab. : Forest clearings.

Gen. Distr. : Iran (endemic).

Note : *T. mazanderanicum* recalls *T. pratense* at a first glance and differs unpronouncedly from it by the relatively short corolla (not longer or only slightly longer than calyx teeth), the leaves, the calyx, etc. However, in the light of the very great polymorphism of *T. pratense*, the independence of this binomial should be re-examined.

137. T. noricum Wulf. in Roem., Arch. Bot. 3 : 387 (1805).

Perennial with woody stock, usually densely patulous-hairy, 5–15 cm. Stems many, ascending or decumbent, not branching, leafy. Leaves mostly radical, long-petioled, the uppermost ones subsessile; stipules of middle leaves oblong, connate and adnate to petioles up to about one-half of their length, membranous, hairy throughout, green-nerved, free portion triangular, long-acuminate; leaflets (0.3 –)1.5 – 2.5 × (0.2–)0.8–1.2 cm, ovate-elliptical to oblong-lanceolate to obovate-cuneate, obtuse, rarely acutish or mucronulate, sparingly long-haired beneath. Heads solitary, terminal, often involucrate by 1 or 2 leaves, globular, 2.5–3.5 cm across. Flowers numerous, about 1.5 cm long. Calyx densely patulous-hirsute; tube cylindrical-campanulate; throat of calyx with ring of long hairs; teeth slightly unequal, about as long as calyx, linear-subulate, blunt, with wide sinuses between them. Corolla white or

Plate 133. *T. mazanderanicum* Rech. f.
(Iran : *Rechinger* 2031, W).
Flowering branch; fruiting calyx; dissected calyx; standard; flower.

cream-coloured, marcescent, 2–3 times as long as calyx; standard longer than wings and keel. Pod membranous. Seed 1, brown. Fl. July–August. 2n = 16.

Hab. : Rock fissures, on alpine and subalpine slopes; alpine meadows.

Gen. Distr. : Switzerland, Austria, Italy, Yugoslavia, Albania, Greece.

1. Wings of corolla glabrous. Leaflets of upper leaves 2–3 × 1–1.2 cm, elliptical to ovate-lanceolate. Head up to 4 cm in diameter. **137a.** var. **noricum**
– Wings of corolla pilose. Leaflets smaller. Heads smaller. **137b.** var. **ottonis**

137a. T. noricum Wulf. Var. **noricum**; Koch, Syn. Fl. Germ. ed. 2 : 186 (1843); Boiss., Fl. 2 : 116 (1867). *Typus* : In alpibus Carinthiae superioris, *Wulf.* & *Hoppe* (n. v.).

T. praetutianum Guss. ex Ser. in DC., Prodr. 2 : 202 (1825; err. typogr. *"prutetianum")*.
T. noricum Wulf. var. *praetutianum* (Guss. ex Ser.) Gib. & Belli, Mem. Accad. Sci. Torino ser. 2, 39 : 315 (1889).
T. noricum Wulf. var. *hirsuta* Wettst., Alban. 38 (1892).
T. noricum Wulf. f. *biceps* Beck, Ann. Naturh. Hofmus. Wien 11 : 72 (1896).
T. praetutianum Guss. var. *brevitrichum* Bald., Riv. Coll. Bot. Alb. 48 (1898–1899).

Icon. : [Plate 134a]. Reichenb., Icon. 22 : t. 2136, f. 1–3 (1903).

Selected specimens : SWITZERLAND : Glacier du Rhône, 1886, *Thomas* (K). AUSTRIA : Kärnthen, 1907, *Korb* (W). ITALY : Loc. Venetia, Prov. di Udine in pascuis montis Lodma, alt. 1,900–2,000 m, 1903, *Porta* (K). YUGOSLAVIA : Montenegro (G). ALBANIA : Korob, 1918, *Kümmerle* 1918 (K). GREECE : Epirus orientalis, in rupibus calcareis cacuminis Mt. Tsumerka, 1893, *Halácsy* (W).

137b. T. noricum Wulf. Var. **ottonis** (Sprun. ex Boiss.) Zoh., Candollea 27 : 116 (1972).

T. ottonis Sprun. ex Boiss., Diagn. ser. 1, 2 : 28 (1843) et Fl. 2 : 116 (1872).

Icon. : [Plate 134b].

Selected specimens : GREECE : in summis m. Veluchi, 7,000 ft, 1887, *Heldreich* 3291 (G); in m. Velugo, 7,200 ft, *Spruner* (G holo.).

138. T. pallidum Waldst. & Kit., Pl. Rar. Hung. 1 : 35 (1800–1801); Boiss., Fl. 2 : 125 (1872). *Typus* : (Hungary) in pratis et arvis planioris Banatus et Comitatus Bihariensis inter Magno Varadinum & Szent Jobb.

T. flavescens Tin., Pugill. 1 : 15 (1817).
T. villosum J. & C. Presl, Delic. Prag. 48 (1882).
T. pratense L. subsp. *pallidum* (Waldst. & Kit.) Gib. & Belli, Mem. Accad. Sci. Torino ser. 2, 39 : 309, 414 (1889).

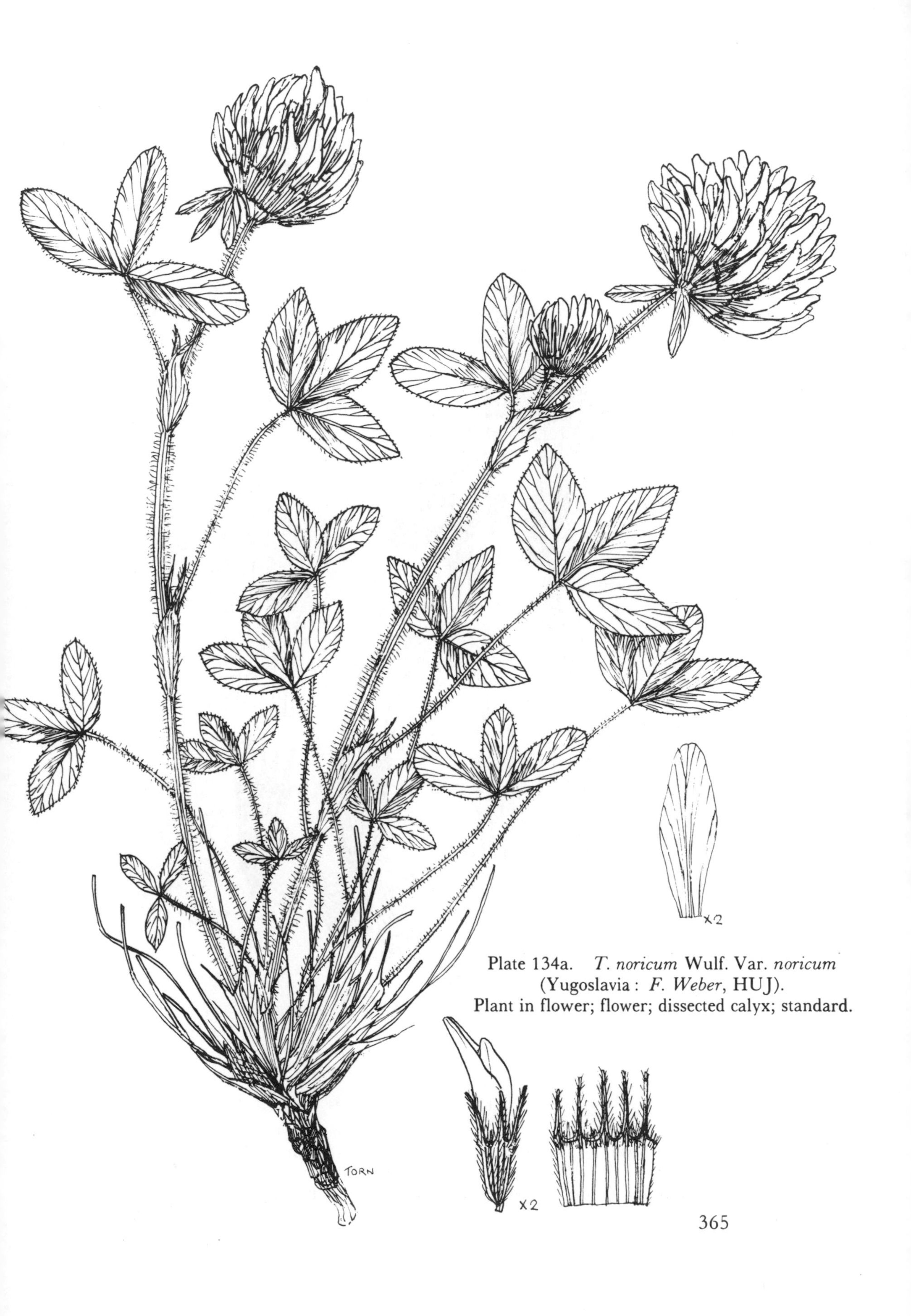

Plate 134a. *T. noricum* Wulf. Var. *noricum*
(Yugoslavia : *F. Weber*, HUJ).
Plant in flower; flower; dissected calyx; standard.

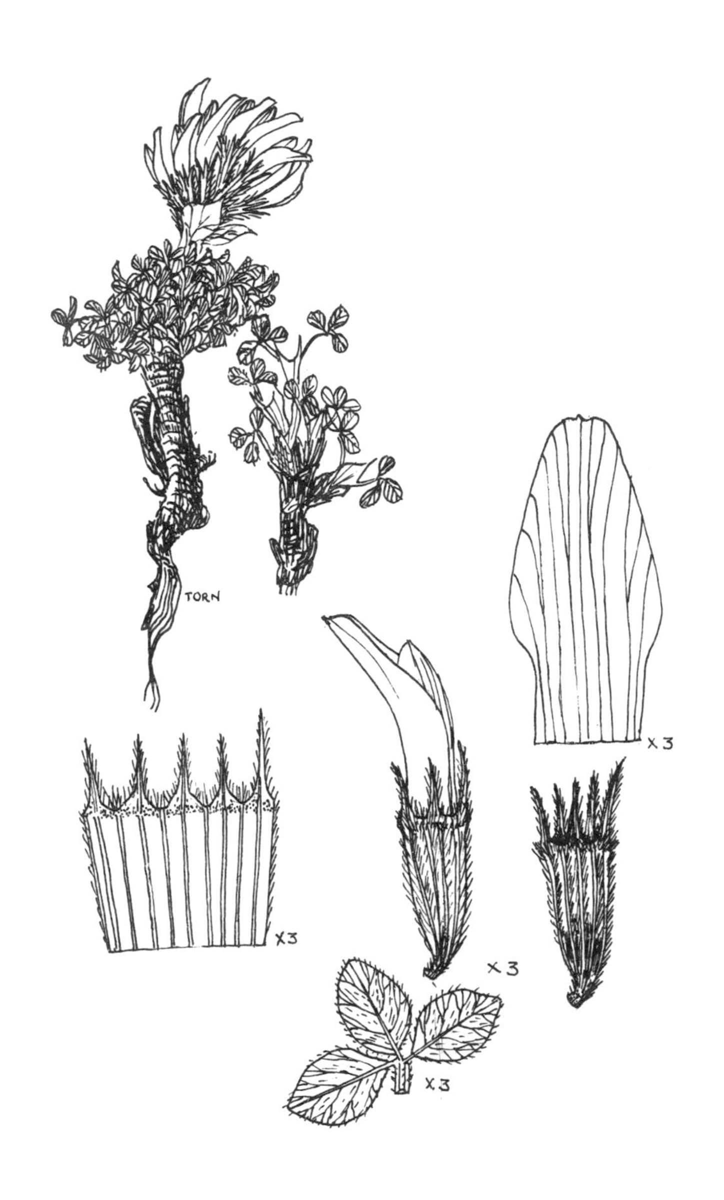

Plate 134b. *T. noricum* Wulf. Var. *ottonis* (Sprun. ex Boiss.) Zoh.
(Italy : *Porta* 584, P).
Plant in flower; sterile branch; dissected calyx; leaf; flower; calyx; standard.

Icon.: [Plate 135]. Reichenb., Icon. 22 : t. 2133, f. 2, 3 (1903); Waldst. & Kit., *loc. cit.* t. 36; Rech., Pl. Ban. t. 14, f. 30 (1828, as *T. procerum*).

Annual or biennial, patulous-villous, usually 20–50 cm. Stems numerous, ascending or erect, mostly caespitose, branching from base, leafy. Leaves rather small, with petioles decreasing in length towards apex; stipules ovate-oblong, the upper ones often ovate, all membranous, with darker nerves, connate and adnate to petioles for up to three-quarters of their length, free portion triangular, abruptly aristate-acuminate, green, ciliate; leaflets 1–2 × 0.5–1.5 cm, obovate with cuneate base to almost rhombic, 1.5 cm, truncate or retuse at apex, often mucronulate and denticulate, with appressed, silky hairs, ciliate at margins. Heads (1–)1.5–2 cm, solitary, terminal, involucrate by upper leaves, globular to ovoid. Flowers dense, about 1.3 cm. Calyx tube obconical, patulous-hairy, 10-nerved; throat with protruding hairy ring but open; teeth almost equal or the lower one longer, 5-nerved, subulate, blunt, longer than tube. Corolla white or cream with pinkish apex, twice as long as calyx. Pod membranous, with cartilaginous operculum. Seed 1, brown. Fl. May–July. 2n = 16.

Hab.: Grassy places, rocks and boulders.

Gen. Distr.: Belgium, Hungary, Romania, France, Italy, Balkan Peninsula, Turkey, S.W. Russia, N. Africa.

Selected specimens: ITALY: Pâturages des terrains argileux près de Pisa (Toscane), 1857, *Savi* 241 (HUJ). YUGOSLAVIA: zwischen Fiume und Volosca längs der Strasse bei Vragna, *Noë* 1362 (HUJ). GREECE: Hauteur de Pylos Messenia, 1844, *Heldreich* 167 (G). CRETE: Platania, 1883, *Reverchon* (G). RHODES: Bord des chemins, près Salakos, 1870, *Bourgeau* 44 (P). TURKEY: Prov. Aydin, Samsun Dag 10 km S. of Davutlar, 280 m, *Pinus nigra* forest, S. slope, 1962, *Dudley* D.34964 (HUJ). ALGERIA: El Afroum pr. Alger, 1875, *Cosson* (G). MOROCCO: Forêt entre Ozron et Toumhadit, 1918, *Benoist* 647 (G).

Note: *T. pallidum* varies in the size of its vegetative parts (e.g., leaves and stems and density of indumentum), the flower colour, the length of the calyx teeth, etc.

139. T. diffusum Ehrh., Beitr. Naturk. 7 : 165 (1792); Boiss., Fl. 2 : 125 (1872). *Typus*: Described from material cultivated in Ehrhardt's Garden, Göttingen.

T. purpurascens Roth, Catal. 1 : 91 (1797).
T. ciliosum Thuill., Fl. Env. Par. ed. 2 : 380 (1799).
T. pratense L. subsp. *diffusum* (Ehrh.) Gib. & Belli, Mem. Accad. Sci. Torino ser. 2, 39 : 313 (1889).

Icon.: [Plate 136].

Annual, with long, patulous hairs, 20–50 cm. Stems few or many, ascending or erect, branching. Leaves varying in size and in length of petioles, alternate, the uppermost pair opposite; stipules of lower leaves oblong, lanceolate, membranous, green-nerved, connate for over half their length, free portion lanceolate, subulate, green, patulous-hairy; leaflets 2–3 × 0.8–1.5 cm, those of lower leaves obovate, those of

Plate 135. *T. pallidum* Waldst. & Kit.
(Algeria : *H. d'Alleizette* 7, HUJ).
Plant in flower; flower.

Plate 136. *T. diffusum* Ehrh.
(Turkey: *Zohary* 57012, HUJ).
Plant in flower; flower; fruiting calyx; pod.

upper ones broad-elliptical to oblong-cuneate, hairy, denticulate in upper part. Heads up to 3 cm across, globular, solitary, terminal, short-peduncled or sessile, the fruiting ones ovate, elongating. Flowers 0.8–1.4 cm. Calyx tubular-campanulate; tube 10-nerved, long-pilose; throat open with ciliate ring; teeth subulate, 2–3 times as long as tube, setaceous, with triangular, 3-nerved base, blunt at apex, with long, patulous hairs, lower one slightly longer than others. Corolla usually not longer or only very slightly longer than calyx, purplish-pink; standard with ovate, somewhat auriculate limb; wings acute. Ovary short-stipitate, 1–2-ovuled. Pod membranous with cartilaginous operculum. Seeds 1–2, globular. Fl. June–August. 2n = 16.

Hab.: Forests and forest clearings, meadows.

Gen. Distr.: British Isles, Germany, Czechoslovakia, Hungary, Romania, Portugal, Spain, France and Corsica, Italy and islands, Balkan Peninsula, S. Russia, Turkey and Transcaucasia.

Selected specimens: HUNGARY: Hadju, Debreczen, 1914, *Rapaico* 661 (FI). ROMANIA: Dobrudscha, 1872, *Sintenis* 701 (K). PORTUGAL: Castello Branco, Monte de Massana, 1882, *Ricardo da Cunha* 1506 (HUJ). SPAIN: Sommet de la Sierra de Segura, 1850, *Bourgeau* 637 (G). FRANCE: Près à Sorede (Pyrénées), 1903, *Conill* (G). GREECE: Agrapha in reg. inf. m. Pindi, 1885, *Heldreich* (G). BULGARIA: in collibus Philippopolium, 1,886 m, *Stribrny* (P). TURKEY: Distr. Muş betw. Muş and Tatvan, deciduous oak forest, 1,300 m, 1964, *Zohary* & *Plitmann* 2160-50 (HUJ). TRANSCAUCASIA: Armenia, distr. Migri 3–4 km ad N.W. pr. Kartschevan, in silvis, 1,400–1,500 m, 1934, *Karjagin* (HUJ).

Subsection II. INTERMEDIA (Gib. & Belli) Bobrov

Acta Inst. Bot. Acad. Sci. URSS, ser. 1, 6: 283 (1947)

Perennials. Calyx tube 10(rarely 12–20)-nerved, usually glabrous, rarely hairy; lower tooth of calyx longer than upper ones, all more or less spreading in fruit; throat of calyx closed by callosity. Corolla marcescent, much longer than calyx.

140. T. medium L., Amoen. Acad. 4: 105 (1759).

Perennial, rhizomatous, appressed-hairy to glabrescent, 20–50 cm. Stems flexuous, ascending, rarely decumbent, mostly branching. Leaves basal and cauline, the lower ones long-petioled, the upper ones short-petioled; stipules linear, adnate to petioles up to one-half of their length, membranous to herbaceous, green- or red-nerved, free portion subulate, lanceolate and often foliaceous, ciliate; leaflets 1.5–5 × 0.8–2 cm, acutish, remotely denticulate to entire, with arcuate, many-forked nerves, hairy throughout to glabrous. Heads globular to ovoid, 1.5–2 cm in diam., often terminal and solitary, rarely in pairs, sessile or short-peduncled. Flowers 1.5–1.8 cm long. Calyx 10(–12–20)-nerved; tube cylindrical-campanulate, glabrous, rarely hairy, whitish-green; throat closed by bilabiate callosity; teeth subulate-setaceous, sharp-pointed, ciliate, unequal, the shorter ones considerably longer than tube.

Corolla about twice as long as calyx, purple, rarely white, later turning brownish; limb of standard ovate, rounded at apex, longer than wings and keel. Pod ovoid-globular, dehiscent. Seed 1, triangular, pale brown. Fl. May–August. 2n = 48, 78, 80, 84, 96–98, 126.

Hab. : Forests and among scrub; on cultivated hillsides; also in meadows and pastures up to 2,000 m; also cultivated.

Gen. Distr. : British Isles, Norway, Sweden, Finland, Germany, Switzerland, Austria, Poland, Czechoslovakia, Hungary, Romania, Portugal, Spain, France, Italy, Balkan Peninsula, W. Russia, Turkey, Iran, Transcaucasia and elsewhere.

Note : *T. medium* is extremely polymorphic. The following varieties include the bulk (but probably not all) of the forms encountered in this species:

1. Calyx 13(20)-nerved, with teeth much longer than tube. Lower flowers of head often accompanied by bracts up to 3 mm long. **140c.** var. **sarosiense**
– Calyx 10(14)-nerved 2
2. Stems with spreading hairs above. **140d.** var. **pseudomedium**
– Stems appressed-hairy or glabrescent above 3
3. Upper calyx teeth as long as or shorter than tube. **140a.** var. **medium**
– Upper 4 calyx teeth longer than tube. **140b.** var. **banaticum**

140a. T. medium L. Var. **medium**. *Typus* : Described from England, Hb. Linn. 930/27.

T. alpestre Poll., Hist. Pl. Palat. 2 : 335 (1777) non L. (1763).
T. bithynicum Boiss., Diagn. ser. 1, 9 : 31 (1849).
T. medium L. var. *majus* Boiss., Fl. 2 : 114 (1872).
T. medium L. var. *eriocalycinum* Hausskn., Mitt. Thür. Bot. Ver. 5 : 2 (1887).
T. medium L. subsp. *skorpili* Velen., Sitzb. Böhm. Ges. Wiss. 29 : 7 (1894).
T. medium L. subsp. *flexuosum* (Jacq.) Aschers. & Graebn. var. *typicum* Aschers. & Graebn., Syn. Mitteleur. Fl. 6/2 : 567 (1908).

Icon. : [Plate 137]. Reichenb., Icon. 22 : t. 2135, f. 1–2 : 1–6 (1903).

Selected specimens : FINLAND : Nylandia, par Esbo, Gammelgard Jura in margine viae, 1960, *Nordström* (HUJ). POLAND : Ojców pov. Olkusz, in prato silvatico, 1952, *Wróblówna* et al. 435b (HUJ). SWITZERLAND : Davos-Platz Graubunden, 1,600 m, 1918, *Schibler* (HUJ). SPAIN : Catalogne : Pyrénées à Ripoll, 680 m, 1913, *Sennen* (HUJ). FRANCE : Haute-Savoie, Forêt des Senbuis, 1,000–1,200 m, 1931, *Eig* (HUJ). ITALY : prope Torre Alfina in Latio, 1900, *Pappi* (HUJ). BULGARIA : pr. Nova Mahala, 1895, *Stribrny* (HUJ). RUSSIA : Distr. of Moskva, Ostankino, 1946, *Ewtuchova* 3690 (HUJ). TURKEY : above Artvin, N. slope, forest and forest remnants, 940 m, 1964, *Zohary* & *Plitmann* 2561-37 (HUJ). TRANSCAUCASIA : Azerbaijan, distr. Shemacha pr. pag. Marjevka ad marginem sup. silvae, 1934, *Heideman* & *Beideman* (HUJ); pr. Baku, 1930, *Grossheim* (HUJ).

140b. T. medium L. Var. **banaticum** Heuff., Enum. Pl. Banat., in Verh. Zool.-Bot. Ges. Wien 8 : Abh. 89 (1858).

T. medium L. subsp. *banaticum* (Heuff.) Hendryck, Preslia 28 : 405 (1956).

Plate 137. *T. medium* L. Var. *medium*
(Russia : *Ewtuchova* 3690, HUJ).
Plant in flower; flower.

Selected specimens: SWITZERLAND: Canton Vaud: Prairie au Sentier, Vallée de Joux, 1931, *Eig* (HUJ). ROMANIA: Banatus, distr. Severin in herbosis montis Allion pr. oppidum Orsovo, alt. ca. 300 m, 1942, *Borzo* et al. 2446 (HUJ). PORTUGAL: Carris de Famalicao num lameiro, 1951, *Fernandes* et al. 3764 (HUJ). TURKEY: env. of Istanbul, Jeni Schiftleck (près de Alemdagh), 1900, *Aznavour* (G).

140c. T. medium L. Var. **sarosiense** (Hazsl.) Savul. & Rayss, Fl. Rum. 5 : 208 (1952).

T. sarosiense Hazsl., Ejsz. Magyar. 76 (1864); Neilr., Diagn. Ung. Slav. 35 (1867); Janka, Trif.-Lot. Europ., in Termész. Füzet. 8/3 : 159 (1884).
T. medium L. subsp. *sarosiense* (Hazsl.) Simonkai, Enum. Fl. Transs. 180 (1887).
T. flexuosum Jacq. subsp. *sarosiense* (Hazsl.) Gib. & Belli, Mem. Accad. Sci. Torino ser. 2, 39 : 333 (1889).

Icon.: Reichenb., Icon. 22 : t. 2158 (1903). 2n = 48.

Selected specimens: ROMANIA: Transilvania: ad margines silvarum frondosarum pr. opp. Rodna, solo trachitico, 800–1,000 m, *Porcius* 3203 (P).

140d. T. medium L. Var. **pseudomedium** (Hausskn.) Halác., Consp. Fl. Gr. 1 : 378 (1900). 2n = 64, 65.

T. pseudo-medium Hausskn., Mitt. Thür. Bot. Ver. 5 : 70 (1887).
T. medium L. subsp. *balcanicum* Velen., Fl. Bulg. 135 (1891).

Selected specimens: GREECE: Mt. Korax, pr. Musinitza, 1,800 m, *Leonis* 375 (K).

Note: For a more extensive treatment of the subdivision of *T. medium* the reader is referred to the various European floras. There is, however, such a divergence of opinions on the amount, naming and rank of the infraspecific taxa of this species that only a meticulous experimental study can cast some light on this question.

141. T. pignantii Brogn. & Bory in Fauché & Chaub., Expéd. Sci. Morée 3, 2 (Bot.): 219, n° 1029 (1832); Vis., Fl. Dalm. 3 : 295 (1850–52); Boiss., Fl. 2 : 115 (1872). *Typus*: In montibus supra Patras (Pignant) (n. v., prob. in FI).

T. pallidum Bory & Chaub., Fl. Pelop. 50, t. 28, f. 3 (1838) non Waldst. & Kit. (1800–1801).
T. fulcratum Griseb., Spicil. Fl. Rumel 1 : 26 (1843).
T. pichleri Vis. ex Pichler (ed. P. Honi), Oesterr. Bot. Z. 19 : 156 (1869) nom. nud.
T. flexuosum Jacq. var. *pignantii* (Fauché & Chaub.) Gib. & Belli, Mem. Accad. Sci. Torino ser. 2, 39 : 330 (1889).

Icon.: [Plate 138]. Fauché & Chaub., *loc. cit.* Atlas t. 26, f. 2 (1835–38).

Perennial with woody stock and horizontal stolons, sparsely patulous-villous, 20–30 cm. Stems flexuous, ascending, slightly branching. Leaves, except for uppermost

Plate 138. *T. pignantii* Brogn. & Bory (Yugoslavia : *F. Weber*, HUJ).
Plant in flower; dissected calyx, flower; standard.

ones, long-petioled; stipules lanceolate, lower part remotely denticulate, those of lower leaves adnate to petioles for up to one-third or one-half of their length, membranous, pale green, nerved, those of upper leaves almost free, with free portion broad-lanceolate, caudate, green, foliaceous; leaflets 1–2 × 0.8–1.2 cm, obovate to elliptical with cuneate base, obtuse or usually notched at apex or mucronulate, remotely denticulate, glabrous or sparingly appressed-hairy and ciliate. Heads 2–3 cm across, spherical to ovoid, often involucrate by upper leaves, rarely pedunculate. Flowers pedicellate, especially the lower ones, about 2–2.3 cm. Calyx tube cylindrical-campanulate, 10-nerved, glabrous or glabrescent except for upper part; throat closed by bilabiate callosity; teeth almost equal in length, blunt, setaceous, with very broad sinuses at base, long-ciliate, much longer than tube. Corolla cream-coloured, 1.5 times as long as calyx; standard oblong-lanceolate, conduplicate and falcate, much longer than wings and keel. Pod 1-seeded. Seed ovoid-spherical, reddish or brown. Fl. May–June. 2n = 16.

Hab. : Forest; up to and above 1,000 m.

Gen. Distr. : Yugoslavia, Albania, Greece, Bulgaria.

Selected specimens : YUGOSLAVIA : Dalmatien, im Kastanienhaine zw. Castelnuovo und Meligere in der Bucht von Cattaro, 1928, *Vetter* (HUJ). GREECE : Bistrica Gorna Dzumaga, 1,100 m, 1931, *Stojanoff* et al. s. n. (B).

142. T. heldreichianum (Gib. & Belli) Hausskn., Mitt. Thür. Bot. Ver. N. F. 5 : 72 (1893) (cf. Gruenberg-Fertig, Israel J. Bot. 21 : 2 1972). *Typus* : (Greece) In memorosis quercinis Pindi Dolopici reg. infer. in ditione mon. Korona, 1885, *Heldreich* (W lecto.).

T. heldreichianum Hausskn. ex Gib. & Belli, Mem. Accad. Sci. Torino ser. 2, 39 : 333 (1889) pro syn. *T. heldreichii* Hausskn. in Sched. & in litt.; ex Nyman, Consp. Fl. Europ. Suppl. 2, 1 : 90 (1889) nom. nud.

T. flexuosum Jacq. subsp. *heldreichianum* Gib. & Belli., *loc. cit.*

"*T. heldreichianum* Hausskn." Hayek, Prodr. Fl. Penins. Balc. 1 : 870 (1926).

Icon. : [Plate 139]. Gib. & Belli, *loc. cit.* t. 5, f. 3 & 3 bis.

Perennial, caespitose, sparingly hairy to glabrescent, 20–50 cm. Stems slender, arcuate, ascending, sparingly branched, leafy all along. Cauline leaves short-petiolate; stipules membranous, green-nerved, appressed-hairy to glabrescent, those of upper leaves with connate lower parts, adnate to petioles for 1 cm of their length, free portion lanceolate-subulate, herbaceous; leaflets 1–2 × 0.5–1 cm, pale green, denticulate all around margin, obtuse to retuse, those of lower leaves obovate, those of upper ones elliptical, prominently nerved beneath, lateral nerves arcuate, deflexed near margin. Heads globular, 2 cm across, short-pedunculate, many-flowered. Flowers about 2 cm. Calyx appressed-pubescent, 10-nerved; tube obconical, about 4 mm long; teeth short-pointed, unequal, 1–3-nerved, linear-subulate, mostly glabrous or glabrescent with membranous margin at base, the upper ones usually somewhat longer than tube; throat of calyx closed by callosity. Corolla pink, about twice as long as calyx; standard oblong. Pod ovoid, membranous. Seed 1, brown. 2n = 16.

Plate 139. *T. heldreichianum* (Gib. & Belli) Hausskn.
(Greece : *K. H. & F. Rechinger* 9538, K).
Plant in flower; leaf; flower; standard.

Hab.: Fields and mountain slopes; up to 1,700 m.
Gen. Distr.: Albania, N. Greece, Bulgaria, Turkey.

Selected specimens: ALBANIA: in campis ad Sesce sub. m. Bali prope Andrijevica, *Baldacci* 260 (W). GREECE: Agrapha (Dolopia veterum), in reg. sup. Pindi mon. Korona, alt. 5,500 ped., substratu schistoso, 1885, *Heldreich* (W.K). BULGARIA: in prolis alpinis Zovorova, 1875, *Stribrny* (K); Mt. Rhodopa supra pagum Iasmos (Jasi-Köj), ca. 500 m. 1936, *K.H.* & *F. Rechinger* 9538 (K). TURKEY: Prov. Bursa, in reg. inf. mont. Keschish dagh (Ula dagh), 200–300 m, 1889, *Bornmüller* 4316 (K).

143. T. patulum Tausch, Syll. Soc. Bot. Ratisb. 2 : 245 (1826–28); Vis., Fl. Dalm. 3 : 294 (1850–52); Boiss., Fl. 2 : 114 (1872); Janka, Trif.-Lot. Europ., in Terméza Füzet. 8/3 : 158 (1884). *Typus*: Probably in LE (n. v.).

T. medium Griseb., Spicil. Fl. Rumel. 1 : 25 (1843) non L. (1759).
T. longestipulatum Ebel, Zwelf Tage Monten. Dalm. 2 : 84 (1844) non Loisel.
T. flexuosum Jacq. subsp. *patulum* (Tausch) Gib. & Belli, Mem. Accad. Sci. Torino ser. 2, 39 : 333 (1889).

Icon. [Plate 140]. Ebel, *loc. cit.* t. 4, f. 2; Reichenb., Icon. 22 : t. 2132, f. 1 : 1 (1903).

Perennial, appressed-hairy, 20–35(–60) cm. Stems many, erect or ascending, somewhat flexuous, divaricately branching. Leaves all short-petioled, the upper ones with petioles shorter than stipules; stipules linear, adnate part as long as petiole, free portion lanceolate-subulate, acuminate, hairy, green-nerved; leaflets 1–2.5 × 0.3–0.6 cm, oblong to linear, obtuse, mucronulate, entire or obsoletely denticulate, mostly sparingly hairy to glabrescent, with prominent midrib and many arcuate, forked, lateral nerves. Heads ovoid to oblong, 2–3 × 2–3 cm, often involucrate by 2 uppermost leaves, rarely short-pedunculate, many-flowered. Flowers up to 16 mm long, rather dense. Calyx tube cylindrical, tapering at base, with densely appressed or antrorse hairs, 10-nerved; throat entirely closed at maturity by bilabiate callosity; teeth subulate-setaceous with broad sinuses at base, rather blunt, plumose, very unequal, somewhat longer than tube, the lower one longer, all divergent or spreading in fruit. Corolla mostly longer than calyx, purple; standard oblong, obtuse, as long as or somewhat longer than wings. Fruit ovoid, membranous. Seed 1, ovoid, brownish. Fl. May–June.
Hab.: Woodland and scrub.
Gen. Distr.: Italy, Yugoslavia, Albania, Greece.

Selected specimens: YUGOSLAVIA: Dalmatia, bei Cattaro, 1872, *Pichler* (G). ALBANIA: in umbrosis sylvarum Rumia ad V Mikulia ditio Primorise, 1898, *Baldacci* (K). GREECE: in monte Malevo, Laconiae prope Hagios Petros, 3,000 ft, 1857, *Orphanides* (G).

144. T. velebiticum Deg., Mag. Bot. Lap. 10 : 113 (1911) incl. var. *gracile* Deg. *Typus*: Dalmatia in humosis inter saxa aridissima montis Velebit, alt. 900 m, 1905 (B).

Plate 140. *T. patulum* Tausch
(Yugoslavia : *Pichler* 1214, HUJ).
Plant in flower; flower; fruiting calyx from above; dissected calyx.

Perennial, appressed-pubescent. Stems simple or branched. Stipules membranous, ovate-lanceolate, reticulately veined, free portion ovate-lanceolate, ciliate; leaflets ovate, cuneate. Heads globular. Calyx glabrous except for upper part of teeth; teeth shorter than tube, unequal. Corolla pink.
Hab. : Rocks of mountains.
Gen. Distr. : N.W. Yugoslavia, Albania.

Specimens seen : ALBANIA : in castaneis ad monasterium Visoki Decani, 600–800 m, 1933, *Rechinger fil.* et al. 1416 (S).

145. T. wettsteinii Dörfl. & Hay., Oesterr. Bot. Zeitschr. 70 : 16 (1921). *Typus* : N.O. Albania, Gipfelregion des Pashtrick, 1918, *Zerny* (W).

Icon. : [Plate 141].

Perennial, caespitose, woody at base, with appressed, white hairs, 10–15 cm. Stems ascending. Leaves small, delicate, with filiform petioles, mainly crowded on lower part of stem and branches; stipules oblong, membranous, prominently nerved below, united and adnate to petiole up to about half their length, upper portion ovate-oblong, obtuse (neither cuspidate nor acuminate), often obsoletely denticulate; leaflets 0.5–1 × 0.3–0.6 cm, obovate-cuneate with rounded or retuse apex, entire. Head about 2 cm across, sessile, broad-obovoid or hemispherical, subtended by very broad stipules and leaflets, 12–20-flowered. Flowers about 2 cm. Calyx tube sparingly appressed-hairy, pale at base, becoming purplish at apex, cylindrical-campanulate, 10-nerved; throat open with hairy ring within; teeth sharp-pointed, subulate, 1–3-nerved, unequal, the shorter ones somewhat longer than or as long as tube. Corolla pink, about twice as long as calyx; standard with elliptical limb, slightly longer than wings. Pod unknown. Fl. July.
Hab. : Alpine rocks.
Gen. Distr. : Yugoslavia, Albania.

Specimen seen : N. ALBANIA : Alpentriften in der Gipfelregion des Pashtrick, ca. 1,800–1,900 m, 1918, *Zerny* (W).

Subsection III. OCHROLEUCA Gib. & Belli

Mem. Accad. Sci. Torino ser. 2, 39 : 351 (1889) pro stirpe

Perennials. Calyx 10-nerved; teeth linear-subulate, often indurated in fruit, with triangular base, 1–3-nerved, unequal, stellately spreading or otherwise diverging in fruit; throat of calyx closed by bilabiate or annular callosity. Corolla rather large (1.5–3 cm), mostly yellowish-white.

146. T. longidentatum Nábělek, Publ. Fac. Sci. Univ. Masaryk Brno 35 : 69, 76, f. 7/6, t. 3, f. 4 (1923). *Typus* : In Kurdistaniae Turcicae distr. Hakkari : mons Maidanoke, supra pagum Hasithe dit. Gulamerik, 1,800 m, 20.6.1910, *Nábělek* 2848 (n. v.).

Plate 141. *T. wettsteinii* Dörfl. & Hay.
(Albania : *J. Dörfler* 931, HUJ).
Plant in flower; flower; standard.

Icon. : [Plate 142].

Perennial with woody stock, appressed-pubescent, 24–40 cm. Stems simple or slightly branching, erect or ascending. Leaves large, the lower ones very long-petioled, those on middle and upper part of stems short-petioled; stipules of lower and middle leaves 2–3 cm long, lower part connate and adnate to petioles for four-fifths of their length, slightly membranous, free portion ovate-triangular to lanceolate, green, strongly nerved, glabrous, those of upper leaves up to 2 cm or more in length, foliaceous; leaflets oblong-elliptical, 1.5–4 × 1–2 cm, obtuse to slightly emarginate. Heads pedunculate, 2–4 × 1–2 cm, ovate, elongating in fruit. Flowers dense, numerous, up to 2.5 cm. Calyx antrorsely villous; tube white, cylindrical, prominently nerved; teeth unequal, lanceolate-subulate, 3-nerved below, all longer than tube, plumose up to top, sharp-pointed. Corolla pale pink, almost twice as long as calyx. Fruiting calyx with deflexed lower tooth; throat closed by callosity. Pod membranous, with cartilaginous apex and strongly bent beak. Seed 1, brown, oblong, 2 mm. Fl. June–August.

Hab. : Subalpine meadows.

Gen. Distr. : Turkey (endemic).

Selected specimens : TURKEY : Prov. Van, distr. Şatak Kavuşşahap dag, 9,000 ft, slopes 1954, *Davis* 23070 (E); Prov. Bitlis/Van, 10 km S.E. of Pelli, 9,000 ft, 1958, *Davis* 22533 (E).

147. T. ochroleucum Huds., Fl. Angl. 283 (1762; *"ochroleucon"*); L., Syst. Nat. ed. 12, 3 : 233 (1768); Ledeb., Fl. Ross. 1 : 544 (1843) p. p.; Boiss., Fl. 2 : 116 (1872); sensu Gib. & Belli, Mem. Accad. Sci. Torino ser. 2, 39 : 352 (1889) non L. (1768). *Typus* : (England) "Habitat in pratis et pascuis siccioribus, in comitatibus Essexiensi, Cantabrigiensi, frequens" (BM).

T. roseum J. & C. Presl, Delic. Prag. 50 (1822).

T. ochroleucum Huds. var. *roseum* (J. & C. Presl) Guss., Fl. Sic. Prodr. 2 : 498 (1828); Gib. & Belli, *loc. cit.*

T. cinarescens Kit. (ed. Kanitz), Linnaea 32 : 619 (1863) subs. illegit.

T. dipsaceum Camus, Cat. Pl. Fr. 64 (1888) non Thuill. (1789–90).

Icon. : [Plate 143]. Schlecht., Lang. & Schenk, Fl. Deutschl. 23 : t. 2373 (1885); Jacquin, Fl. Austr. 1 : t. 40 (1773); Ross-Craig, Draw. Brit. Pl. 7 : 26 (1954).

Perennial with creeping stock, forming tufts, with appressed or antrorse hairs, lower parts often patulous-hairy, 20–50 cm. Stems many, slender, ascending to erect, often branching, densely leafy at base and with scattered leaves above. Lower and middle leaves long-petioled, upper ones almost sessile; stipules 2–5 cm, adnate to petioles in their lower half, membranous between green nerves, free portion subulate, green, plumose-hairy; leaflets of lower leaves obovate to oblong-cuneate, those of middle and upper leaves oblong-elliptical to lanceolate, 2–5 × 0.5–1 cm, entire or minutely denticulate in upper part, hairy on both faces and often punctulate on lower surface, with obtuse, rounded, often somewhat retuse apex, the uppermost ones mucronulate. Heads mostly solitary, globular or ovoid, later elongating, 1–2(–3) cm across, sessile

Plate 142. *T. longidentatum* Náb. (Turkey : *Davis* 45675, HUJ). Flowering and fruiting branch; flower; standard; wing; keel; dissected calyx.

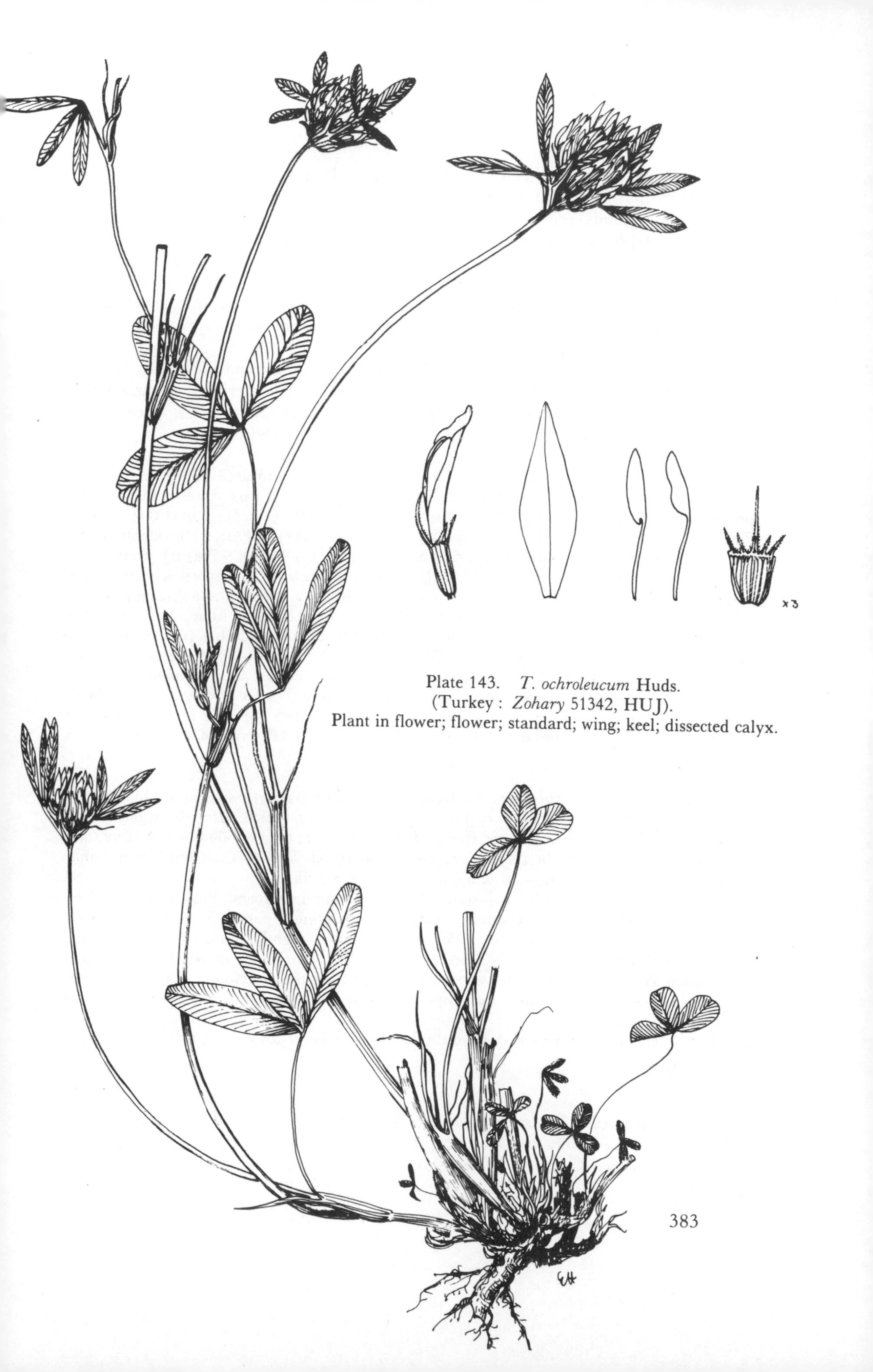

Plate 143. *T. ochroleucum* Huds.
(Turkey : *Zohary* 51342, HUJ).
Plant in flower; flower; standard; wing; keel; dissected calyx.

or on short peduncle subtended by opposite leaves. Flowers many, 15–18 mm. Calyx obconical-cylindrical, up to 8 mm long, hirsute or villous, prominently nerved, often with minute glands between hairs; teeth lanceolate-subulate, with 3 prominent nerves, ciliate-plumose, sharp-pointed, the upper ones less than half the length of tube, the lower one about twice as long as the upper ones, all stellately spreading when ripe; throat of calyx closed by bilabiate callosity. Corolla cream-coloured, turning brown when dry (rarely corolla pink); standard oblong-lanceolate, much longer than wings and keel. Pod ovoid, dehiscent, with leathery operculum. Seeds ovoid, brownish. Fl. June–July. 2n = 16.

Hab. : Grassy patches in forest, among shrubs and along roadsides; 400–2,000 m.

Gen. Distr. : British Isles, Germany, Czechoslovakia, Switzerland, Austria, Hungary, Romania, Poland, Portugal, E. Spain, France, Italy, Yugoslavia, Greece, Bulgaria, S. Russia, Turkey, Iran, Algeria, Morocco and elsewhere.

Selected specimens : CZECHOSLOVAKIA : Montes Bile Karpaty, in declivibus graminosis, pr. Velka, 1934, *Weber* (HUJ). ROMANIA : Transilvania, distr. Turda, in foenatis subsalsis prope oppid. Turda, 357 m, 1940, *Ghiuta* & *Todor* 2445 (HUJ). PORTUGAL : Villar Formoso : Alto de Raza, 1884, *Ricardo de Cunha* 1508 (HUJ). SPAIN : Barcelona, massif du Tibidabo, Riesa de Vallpidiera, 1926, *Sennen* (HUJ). FRANCE : Prairies argileuses à Rohrbach-les-Bitche, 1837, *Schultz* 51 (HUJ). ITALY : Eurusia in nemorosis collium agri Florentini, 1875, *Levier* (G); SICILY, in montosis sylvaticis, Palermo alla Pizzuta, *Todaro* 391 [HUJ, as var. *roseum* (J. & C. Presl) Guss.]. YUGOSLAVIA : Fiume, *Noë* 1365 (G); Illyria, in nemorosis Tergesti, *Tommasini* (G). TURKEY : Prov. Çoruh (Artvin), Borcka to Hopa, 450 m, 1957, *Davis* & *Hedge* D. 29866 (K). IRAN : 12 km W. of Astara, 3,000 ft, 1929, *Cowan* & *Darlington* 2503 (K). CAUCASUS : bei Gudaur auf Alpenwiesen, 1873, *Rehmann* (G). ALGERIA : Oran, Djebel Nador, 1918, *Faure* (G). MOROCCO : Moyen Atlas, Daiët Ochlef, côteaux calcaires boisés, 1,800 m, 1923, *Jahandiez* 588 (G); prairies sèches au Djebel, localité Zanda (Grand Atlas), 1927, *Jahandiez* 366 (P, type of var. *abbreviatum* Maire, inedit.).

Note : Rouy refers to *T. ochroleucum* a form named by him (var. α) *pallidulum* (Jord.) Rouy, Fl. Fr. 5 : 124 (1899; = *T. pallidum* Jord., Pugill. 56, 1852), which differs from the typical form by the lesser constriction of the throat of the calyx tube, the very long lower calyx tooth which is twice as long as the calyx tube, the shorter corolla which is not much longer than the calyx, and the light yellow to white colour of the corolla.

Lindberg describes *T. ochroleucum* subsp. *lamprotrichum* in Ofvers. Finska Vetensk Soc. Förh. 48 : 55 (1906) : stems glabrous except for uppermost part which has appressed indumentum; leaflets of larger leaves 3.5–5 × 1.2–1.6 cm, slightly crenulate in upper part; free portion of stipules subulate, 2–3 cm long; calyx dark coloured with obsolete nerves and shiny hairs. We have not seen the type specimen, but from the description it seems that this taxon should be included within the scope of this very polymorphic species which merits experimental treatment.

The East Mediterranean material generally has much shorter leaves and in this regard approaches *T. caudatum*.

148. T. caucasicum Tausch, Syll. Soc. Bot. Ratisb. 2 : 245 (1826–28); Boiss., Fl. 2 : 156 (1872) p. p.; Zoh. in Davis, Fl. Turk. 3 : 420 (1970). *Typus* : (Georgia) Ex agro Georgipolitani Caucasi (LE photo).

T. pannonicum sensu M. B., Fl. Taur.-Cauc. 2 : 212 (1808) non L. (1771) nec Jacq. (1767).
T. ochroleucum sensu M. B., *loc. cit.* 211, non Huds. (1762) nec L. (1768).
T. squarrosum sensu M. B., *loc. cit.* 214, non L. (1753); Boiss., *loc. cit.* 117.
T. cassium Boiss., Diagn. ser. 1, 9 : 23 (1849) et Fl. 2 : 117 (1872).
T. marschallii Rouy, Fl. Fr. 5 : 114 (1889) non Savi (1808–10).
T. inaequale sensu Gorssh., Fl. Kavk. 2 : 279 (1930) non Lojac. (1878).

Icon. : [Plate 144]. Bobrov, Acta Inst. Bot. Acad. Sci. URSS ser. 1/6 : 267 (1947).

Perennial, appressed-canescent, 20–50 cm. Stems many, simple, ascending or erect, terete, striate, arising from slender, woody stock. Leaves long-petiolate except for uppermost ones; stipules lanceolate, membranous between nerves, lower third adnate to petioles, free portion gradually tapering into long-subulate, green, plumose cusp; leaflets elliptical-oblong to elliptical-lanceolate, entire or slightly denticulate in upper part, with appressed long hairs and minute glands between them (on lower surface), those of lower leaves or all of those of upper leaves only acutish-mucronate. Heads ovoid in flower, oblong to cylindrical in fruit, sessile or subsessile, often involucrate by 2 opposite leaves. Flowers up to 2 cm long. Flowering calyx tubular, up to 1.8 mm long (incl. teeth), appressed- or antrorsely villous, with 10 prominent nerves; teeth unequal, setaceous-subulate, 1–3-nerved, plumose, sharp-pointed, the lower one twice as long as the upper ones, deflexed. Corolla longer than lower calyx tooth, cream-coloured, becoming dark reddish when dried; standard much longer than wings, with tapering limb. Fruiting calyx growing in width, somewhat urceolate and restricted beneath throat; throat closed by bilabiate callosity; teeth often stellately spreading in fruit, 3-nerved. Pod 1.5–2 mm long. Seed 1, brown, reniform. Fl. May–August.

Hab. : Grassy places in mountains; woods and scrub.

Gen. Distr. : Turkey, Syria, Caucasus.

Selected specimens : TURKEY : Amanus Mts., western slopes between Karagouz and Bagajak, 200–800 m, Pinetum, 1932, *Eig* & *Zohary* 105 (HUJ); *ibid.*, descent de Bakladja à Karagouz, 1,000–1,400 m, Gabbres de serpentines, 1932, *Delbès* 133 (HUJ); Taurus Mts., env. of Gözne (N.W. of Mersin), 1,000–1,100 m, maquis, 1931, *Eig* & *Zohary* 233 (HUJ). SYRIA : "In silvaticis regionis inferioris montis Cassii", 1846, *Boissier* (lectotype of *T. cassium* G, isotype K); about 45 km N. of Lattakie, deep shady wadi 1932, *Eig* & *Zohary* 472 (HUJ).

Note : *T. caucasicum* Tausch is very close to *T. ochroleucum* Huds. or *T. pannonicum* Jacq. and therefore has been largely confused with them. Our specimens from Turkey correspond fully to the Russian specimens of which we have seen the phototype.
T. cassium Boiss. is undoubtedly *T. caucasicum* Tausch, not seen by Boissier, who in his flora records *T. cassium* Boiss., *T. squarrosum* M. B. and *T. caucasicum* Tausch as separate species. Gibelli & Belli seem to be incorrect in interpreting *T. caucasicum* (*T. cassium*) as a hybrid between *T. ochroleucum* and *T. pannonicum*.

149. T. canescens Willd., Sp. Pl. 3 : 1369 (1802); Boiss., Fl. 2 : 117 (1872). *Typus* : (Turkey) Habitat in Cappadocia, Via seg. 14204 (B holo.).

Plate 144. *T. caucasicum* Tausch (Turkey: *Delbès*, HUJ).
Plant in flower; fruiting calyx from above; dissected calyx; flower; standard.

T. hohenackeri Jaub. & Sp., Ill. Pl. Or. 2 : 55 (1845).

Icon. : [Plate 145]. Bot. Mag. 19 : t. 1168 (1809); Jaub. & Sp., *loc. cit.* t. 140.

Perennial herb, patulous- or antrorsely pubescent, 10–35 cm. Stems many or few, simple, arising from woody base, densely leafy below, sparingly above. Lower leaves very long-petioled, uppermost ones sessile, opposite; stipules 1.5–3 cm, hairy, adnate part membranous, free portion herbaceous, subulate, with broad base; leaflets thin, obovate to ovate to broad-elliptical, 1–3 × 0.5–1.5 cm, with brighter midrib and delicate, lateral, ascending nerves, hairy on both faces, truncate and more or less deeply notched at apex. Head 2–4 × 1.5–3 cm, terminal, sessile and involucrate by upper leaves, mostly ovoid, becoming cylindrical in fruit. Flowers many, up to 3 cm long. Calyx tubular, about 1 cm, hairy; tube 10-nerved; teeth lanceolate-subulate, 1–3-nerved, sharp-pointed at tip, unequal, longer than tube. Corolla pale yellow; standard with ovate, tapering-ligulate limb, obtuse and slightly notched at apex, much longer than wings; keel shorter than wings, often with violet tip. Fruiting calyx closed by bilabiate callosity, with stellately spreading teeth. Pod about 1.5 mm, obovoid, leathery at apex. Seed 1. Fl. May–August. 2n = 18, 48.

Hab. : Subalpine meadows, riverbanks; up to 2,600 m.

Gen. Distr. : Turkey, Iran, Transcaucasia.

Selected specimens : TURKEY : Prov. Rize : vallée sous-alpine de Djimil vers 2,000 m, 1866, *Balansa* 1403 (G); distr. Amasya, 6 km W. of Çakiralan, ca. 700 m, 1963, *Zohary* & *Orshan* 10761 (HUJ). IRAN : Prov. Mazandaran : in valle fluvii Chalus, 2,000–2,600 m, Pal-e Zanguleh, 1948, *K. H.* & *F. Rechinger* 6335 (K, P). TRANSCAUCASIA : in montosis et proclivibus Georgiae Caucasicae, 1834 (type of *T. hohenackeri* Jaub. & Sp., G.); Georgia, pr. Tiflis, distr. Borzhom pr. pagum Bakuriani, 1,500 m in pratis, 1925, *Kozlovsky* 338 (G).

150. T. davisii Hossain, Not. Roy. Bot. Gard. Edinb. 23 : 403 (1961). *Typus* : Turkey, d. Bahçe, Dumanli dag above Haruniye, 700–900 m, *Davis* & *Hedge* D. 26861 (E ! holo., iso. : BM, K).

Icon. : [Plate 146].

Perennial, rhizomatous, somewhat caespitose herbs, with antrorse to appressed hairs, 10–15 cm. Stems poorly branching from base. Leaves mostly crowded at base of plant, all, including 2 opposite upper ones, long-petioled; stipules linear-lanceolate, purplish, membranous, 2–2.5 cm, adnate to petioles to about middle, prominently nerved, patulous-hairy, free portion subulate-setaceous; leaflets 0.8–2 × 0.6–1.5 cm, ovate, rhombic to short-elliptical, entire, those of lower leaves retuse, those of upper ones obtuse, mucronulate, appressed-hairy, with reddish spots. Head 2.5(–3) cm across, broad-ovoid to globular, solitary, terminal, on 4–7 cm long peduncles. Flowers many, 2.5 cm long. Calyx 2.2–2.5 cm, almost as long as corolla; tube cylindrical-campanulate, with 10 prominent nerves, hairy along nerves and minutely glandular-punctate; teeth linear-subulate to setose, sharp-pointed, unequal, considerably longer than tube, the lower one slightly longer than others, all 1-nerved; throat of young calyx slightly thickened (that of fruiting one probably closed). Corolla

Plate 145. *T. canescens* Willd.
(Armenia : *Schelkovnikov* & *Kara-Murza*, HUJ)
Plant in flower; dissected calyx;
fruiting calyx from above; flower.

Plate 146. *T. davisii* Hossain
Plant in flower; leaf (somewhat enlarged); standard; dissected calyx; flower.

white, turning cream-coloured when dried; standard 2–2.5 cm, oblong, truncate at apex by about 5 mm, longer than wings and keel. Ovary 2-ovuled. Pod unknown. Fl. April.

Hab.: Open woodland and amongst scrub; up to 1,000 m.
Gen. Distr.: Turkey (endemic).

Specimens seen: TURKEY: Hatay, Dörtyol, Kuzuculo to Bülke, 1,000 m, *Coode & Jones* 411 (E).

Note: The differences between *T. davisii* and the closely related *T. canescens* are as follows:

T. davisii	*T. canescens*
Leaflets of radical leaves ovate, retuse.	Leaflets of radical leaves usually broadly elliptical and deeply notched.
Floral leaves distinctly petiolate.	Petioles of floral leaves very short, much shorter than stipules.
Calyx teeth 9–12 mm long.	Calyx teeth 5–7 mm long.
Corolla as long as or very slightly longer than calyx.	Corolla nearly twice as long as calyx.

The above differentiating characters are almost all quantitative and can scarcely satisfy the requirements for specific rank. The independence of *T. davisii* thus remains questionable until experimental studies will prove its constancy.

151. T. trichocephalum M. B., Fl. Taur.-Cauc. 2 : 212 (1808) et 3 : 508 (1819) excl. syn. Steven (*T. armenium* Willd.); Boiss., Fl. 2 : 118 (1872); Gib. & Belli, Mem. Accad. Sci. Torino ser. 2, 39 : 361 (1889) pro hybrido; Bobrov, Acta Inst. Bot. Acad. Sci. URSS ser. 1, 6 : 266 (1947). *Typus*: (Georgia) In Caucasi iberici alpestribus, 1,800 m, *C. Mussin-Puschkin* (LE photo).

T. trichocephalum M. B. var. *macrophyllum* Hossain et var. *lonchophyllum* Hossain, Not. Roy. Bot. Gard. Edinb. 23 : 407 (1961).

Icon.: [Plate 147]. Gib. & Belli, *loc. cit.* t. 9, f. 3; Bobrov, *loc. cit.* 267, f. 3.

Perennial, patulous-pubescent, 30–70 cm. Stems many, erect or ascending, simple or slightly branching. Leaves large, the lower ones up to 20 cm (incl. petioles); stipules ovate-oblong, up to 4 cm, prominently nerved and long-villous, adnate to petioles for up to half their length, free portion lanceolate, gradually tapering into subulate cusp; lower leaflets ovate to elliptical, 3–4 × 1.5–2 cm, with appressed, long hairs on either side, entire, retuse, upper ones elliptical to lanceolate, acute, mucronate. Heads large (largest in Section), up to 4–6(–8) × 3–4 cm, many-flowered, terminal, ovoid, sessile or subsessile, subtended by shorter upper leaves. Flowers dense, about 2(–2.5) cm. Calyx patulous-villous; tube cylindrical-campanulate, with 10 prominent nerves, 6–8 mm; teeth unequal, lanceolate-subulate, blunt, as long as or slightly longer than tube, the lower one slightly longer than the others, later elongating, curved; throat closed by bilabiate callosity. Corolla cream-coloured, almost twice as long as calyx; standard oblong, somewhat lingulate in upper part, longer than wings and keel, the latter often with purplish tip. Pod membranous with cartilaginous

Plate 147. *T. trichocephalum* M. B.
(Armenia : *Schelkovnikov* & *Kara-Murza* 281, HUJ).
Plant in flower; dissected calyx; flowers.

operculum. Seed 1. Fl. May–July. 2n = 14, ~ 48.
Hab. : Meadows, woods and swampy soil at higher altitudes.
Gen. Distr. : Turkey, N. Iran, Caucasus, Transcaucasia.

Selected specimens : TURKEY : Prov. Erzerum, 14 mi. from Horasan towards Karaurgan, 2,000 m, 1957, *Davis* & *Hedge* D. 29507 (E); prov. Agri : 2 km S. W. of Hamur (Murat Valley), 1,670 m, sloping meadows, 1966, *Davis* 44021 (E); Prov. Van, 30 km from Başkale to Hoşap, S. of pass, 2,600 m, meadows of stream, 1966, *Davis* 45922 (E); Grand Ararat, about 8,500 ft, 1910, *B. V. D. Post* 3120 (G). IRAN : betw. Teheran and Tabriz, 1859, *Bunge* (P). TRANSCAUCASIA : Armenia, Tiflis, *Wittman* 232 (E).

152. T. caudatum Boiss., Diagn. ser. 1, 9 : 22 (1849) et Fl. 2 : 116 (1872). *Typus* : (Turkey) "In herbidis pratis superioris montis Mesogis supra Tralles in Lydia", June 1842, *Boissier* (G holo.).

T. caudatum Boiss. var. *grosseanum* Aschers. & Sint. ex Boiss., Fl. Suppl. 166 (1888).

Icon. : [Plate 148]. Gib. & Belli, Mem. Accad. Sci. Torino ser. 2, 39 : t. 5 (1889).

Perennial, with patulous or retrorse hairs, 10–30 cm. Stems few, ascending or erect, unbranched or sparingly branching. Leaves long-petioled; stipules (at least upper ones) ovate-triangular, often terminating in broad, foliaceous, green free portion; leaflets 0.8–1.5 × 0.7–1 cm, ovate to ovate-elliptical, slightly retuse, subsessile, entire or almost so, sparingly appressed-hirsute. Heads 2–2.5 × 1.5–2 cm, solitary, terminal, ovoid, sessile and subtended by 2 uppermost leaves or by their stipules. Flowers dense, many, up to 1.8 cm long. Calyx obconical, about 1 cm (incl. longest tooth); tube obconical, with 10 prominent nerves, about 3 mm long, sparingly villous, especially above; teeth subulate, 1-nerved, sharp-pointed, unequal, the upper ones about as long as tube, the lower one about twice as long. Corolla pink or purple or pinkish-cream; standard with oblong, somewhat retuse limb, much longer than wings. Fruiting calyx with white, obconical tube and green, stellately spreading teeth; throat closed by bilabiate callosity. Pod ovoid, 1–2 mm. Seed 1, brown. Fl. May–August.
Hab. : Woods and forest clearings, calcareous soils.
Gen. Distr. : Turkey (endemic).

Selected specimens : TURKEY : Lydia in montis Güme-Dagh (Mesogis) supra oppidum Tere (Tyrrha), 600–700 m, 1906, *Bornmüller* 9284b (E); Bithynia, Brussa in reg. subalpina montis Olympi, 900–1,200 m, 1899, *Bornmüller* 4318 (E); Taurus, Giosna, 1,100 m, Kalkboden, 1895, *Siehe* 114 (E); Prov. Troja, Mt. Ida, in sylvis prope Kareikos, 1883, *Sintenis* 631 (E, isotype of var. *grosseanum* Aschers. & Sint. ex Boiss., Fl. Suppl. *loc. cit.*).

Note : Boissier's descriptions, both in the Diagnoses (*loc. cit.*) and in the Flora (*loc. cit.*), do not fit the entire variational range of this species which is very close to *T. ochroleucum* (and certainly not to *T. pratense* as stated by Boissier); the latter has an open calyx throat and blunt teeth. *T. ochroleucum* differs at a first glance from *T. caudatum* by its appressed indumentum, longer leaflets, smaller heads and flowers, flower colour, etc. The foliaceous shape of the stipules is not a constant character and therefore can not serve as a distinction

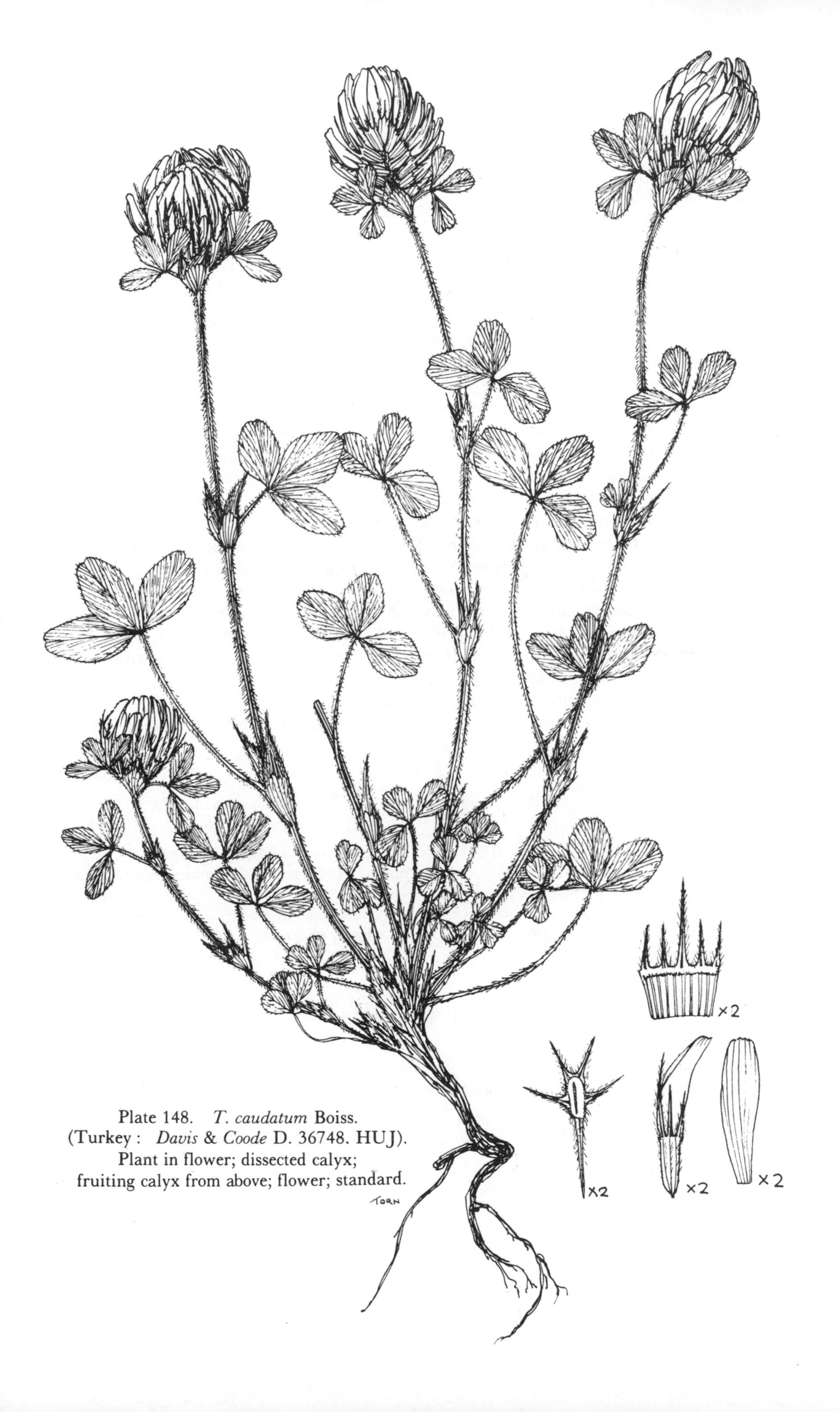

Plate 148. *T. caudatum* Boiss.
(Turkey : *Davis* & *Coode* D. 36748. HUJ).
Plant in flower; dissected calyx;
fruiting calyx from above; flower; standard.

between var. *caudatum* and var. *grosseanum* Aschers. & Sint. ex Boiss. Notwithstanding the above characters, it is worthwhile examining the relations between *T. caudatum* and *T. ochroleucum* experimentally.

153. T. pannonicum Jacq., Obs. Bot. 2 : 21 (1767); L., Mant. Alt. 2 : 276 (1771).

Perennials, patulous or appressed-pubescent, 20–40(–60) cm. Stems erect or ascending, striate, sparsely branched. Leaves long-petioled in lower part, short-petioled to subsessile in upper part; stipules linear-oblong, somewhat membranous, with green nerves, hairy, lower and middle ones adnate to petioles up to half of their length, free portion subulate-linear, green, those of upper leaves longer than petioles; leaflets of lower leaves obovate to elliptical, often not exceeding 1 cm, those of middle and upper leaves oblong-lanceolate, 2–3(–4.5) × 0.5–1.2 cm, obtuse, slightly retuse, the uppermost ones acute, mucronate. Heads solitary, ovoid to ovoid-oblong, later elongating, many-flowered, 2–4(–5) × 2–3(–5) cm, short- or long-peduncled, often subtended by 2 opposite, short-mucronate leaves. Flowers up to 2.5 cm long. Calyx cylindrical, later campanulate; tube 10-nerved, subpatulous-hairy, about 4–5 mm long; teeth subulate, sharp-pointed, ciliate, the upper ones about as long as the lower one, approximately 1.5–2 times as long as tube. Corolla white, almost twice as long as calyx; standard oblong, tapering, much longer than wings and keel. Fruiting calyx with throat closed by bilabiate callosity and stellately spreading or deflexed teeth. Pod membranous with cartilaginous apex, 1-seeded. Seed brown, elliptical, about 2 mm long. Fl. June–August. 2n = 96, 98, ~ 126, 128, 130, ~ 180.

Hab. : Meadows, forest clearings and steppes.

Gen. Distr. : Germany, Poland, Czechoslovakia, Hungary, Romania, France, Italy, Balkan Peninsula, USSR, Turkey.

Key to Subspecies

1. Leaves usually large, 6–7 × 0.8–1.2 cm, the lower ones elliptical to lanceolate, the uppermost ones lanceolate, acute or mucronate. Head not involucrate, long-peduncled. Teeth of calyx not stellately divergent. **153a.** subsp. **pannonicum**
– Leaves much shorter and narrower, oblong-lanceolate to linear, 1.5–3 × 0.2–0.5 cm. Heads mostly involucrate. Teeth of calyx stellately divergent or deflexed. **153b.** subsp. **elongatum**

153a. T. pannonicum Jacq. Subsp. **pannonicum.** *Typus* : Hungary, circa pagum Siglisberg (BM ?).

T. armenium auct. nonnul. non Willd. (1809).

T. pannonicum "L." var. *rubrocalycinum* Podpera, Verh. Zool.-Bot. Ges. Wien 52 : 646 (1902).

Icon. : [Plate 149]. Jacq., *loc. cit.* t. 42; Reichenb., Icon. 22 : t. 2138 (1903).

Selected specimens : POLAND : ad pedem Carpatorum orientalium in pratis, 1929, *Panov* 42 (HUJ). CZECHOSLOVAKIA : Rossia subcarpatica montes Svidovec, in pratis submontanis supra Kevelov, 800 m, 1929, *Domin* & *Deyl* 67 (HUJ). HUNGARY: Monora, *Barth* (K). ROMANIA: Transilvania, distr. Cojocna in locis graminosis ad oppidum Cluj, 1921, *Prodan* 555b (HUJ). YUGOSLAVIA: unter Krain Gerganzberg, *Freyer* (K). ITALY: Prov. Torino, 1908, *Ferrari* et al. 1075 (K).

153b. T. pannonicum Jacq. Subsp. **elongatum** (Willd.) Zoh. in Davis, Fl. Turk. 3 : 422 (1970). *Typus*: (Turkey) Habitat in Galatia, *Sestini* 14203 (B holo., iso. G).

T. elongatum Willd., Sp. Pl. 3, 2 : 1369 (1802).
T. armenium Willd., Enum. Pl. Hort. Berol. 2 : 793 (1809).
T. olympicum Hornem. ex Hook., Bot. Mag. 54 : t. 2790 (1827).
T. sulphureum C. Koch, Linnaea 19 : 63 (1846).

Icon.: [Plate 149]. Bot. Mag., *loc. cit.*; Jaub. & Sp., Ill. Pl. Or. 2 : t. 139 (1845).

Selected specimens: TURKEY: Bithynia, in reg. inf. montis Olympi supra Brussa (Bursa), 300–500 m, 1899, *Bornmüller* 4320 (K, P, G); Sandshak Gumüschkane, Moaddas dagh in pratis silvat., 1894, *Freyn* 5963 (G); Habitat in Armenia (B, holotype of *T. armenium* Willd.).

Note: Distinction between the above two varieties is not always possible. There are transitions, especially in the size and shape of the leaves, the length of the peduncles, etc.; see also Hossain (1961). The following specimens are intermediates between the two recorded varieties: Turkey, prov. Kars, Sarikamis, 2,100 m, D. 30802 (E); Bithynia, *Bornmüller* 4320 (E).

Subsection IV. ALPESTRIA Gib. & Belli

Mem. Accad. Sci. Torino ser. 2, 39 : 334 (1889) pro stirpe;
Aschers. & Graebn., Syn. Mitteleur. Fl. 6/2 : 574 (1908)

Perennials. Calyx 20(very rarely 10 or 12–15)-nerved; lower tooth of calyx longer than tube; throat of calyx narrowed by hairy ring but not closed. Standard almost as long as wings. Pod altogether membranous.

154. T. rubens L., Sp. Pl. 768 (1753).

Perennial with creeping rootstock, glabrous or rarely hairy, (20–)30–80 cm. Stems erect, usually thick, striate to furrowed, mostly simple. Stipules very large, up to 8 × 1 cm (especially the upper ones), adnate to petioles for up to three-quarters of their length, linear to lanceolate, herbaceous, green-nerved, free portion lanceolate, dentate, acuminate; leaflets oblong-lanceolate, with many arcuate, lateral nerves each terminating in a curved tooth, those of lower leaves mostly obtuse to retuse, those of upper ones tapering, acute. Heads mostly twin, oblong-cylindrical, 3–7 × 2–3 cm, usually involucrate at base. Flowers dense, up to 1.6 cm long. Calyx tube cylindrical,

Plate 149. *T. pannonicum* Jacq. Subsp. *pannonicum* (Romania : *Ciocirlan* 89, HUJ).
Plant in flower; standard; flower; dissected calyx; fruiting calyx from above.
Subsp. *elongatum* (Willd.) Zoh. (Turkey : *Manisadjan*, HUJ).
Flowering branch (right).

glabrous, rarely hairy, (12–15–)20-nerved; teeth filiform, mostly long-ciliate, very unequal, blunt, the upper ones much shorter and the lower one much longer than tube; throat of calyx open, hairy. Corolla purple, rarely white, much longer than calyx; standard elliptical, reflexed above. Pod globular, dehiscing by valves. Seed ovoid. Fl. June–August. 2n = 16.

Hab.: Mountain forests, forest edges, stony slopes, among bush.

Gen. Distr.: Germany, Poland, Czechoslovakia, Switzerland, Austria, Hungary, Romania, Spain, France, Italy, Yugoslavia, Albania, C. and S. Russia, Turkey.

1. Calyx tube and often also stems and leaves glabrous. **154a.** var. **rubens**
– Calyx tube, stems and leaves hairy. **154b.** var. **villosum**

154a. T. rubens L. Var. **rubens**; Boiss., Fl. 2 : 113 (1872); Gib. & Belli, Mem. Accad. Sci. Torino ser. 2, 39 : 337 (1889); Aschers. & Graebn., Syn. Mitteleur. Fl. 6/2 : 575 (1908). *Typus*: Described from Switzerland, Hb. Cliff. 375.

Icon.: [Plate 150]. Jacq., Fl. Austr. 4 : t. 385 (1790); Reichenb., Icon. 22 : t. 2137, f. 1–11 (1903); Gib. & Belli, *loc. cit.* t. 5, f. 6.

Selected specimens: GERMANY: Kalbergen in Thüringen, *Fest* 267 (HUJ). CZECHOSLOVAKIA: Montes Bile Karpaty in declivibus stepposis collis prope Lipov, 1934, *Weber* (HUJ); Moravia: Mer.-Or. Uh. Ostroh. Carpati albi, Velko, ca. 460 m, 1933, *Podpera* & *Uxor* 832 (HUJ). SWITZERLAND: Vaud, Pied du Jura: près du Signal d'Orbe, près sec, 505 m, 1951, *Villaret* 12156 (HUJ). AUSTRIA: Pr. Lienz, 1865, *Ausserdorfer* (K). HUNGARY: in monte Mecsek, 1915, *Maly* (K). ROMANIA: Transilvania, distr. Cajocna in herb. montanis ad stationem viae ferreae Cajocna-Boj, ca. 400 m, 1924, *Nyarady* 557 (HUJ).

154b. T. rubens L. Var. **villosum** Vis. & Sacc., Cat. Pl. Vasc. Veneto 260 (1869); var. β Bertol., Fl. It. 8 : 170 (1851), phras. descript.; "var. *villosum* Bert." Gib. & Belli, *loc. cit.*; Aschers. & Graebn., Syn. Mitteleur. Fl. 6/2 : 575 (1908).

T. rubens L. var. *hirsutum* Loeske ex Spribille, K. Gymn. Inowrazlaw Wiss. Beil. Progr. No. 142 : 14 (1888) nom. prov.

Specimens recorded: ITALY: ex olivetis di Romana, Montinio. TURKEY: in lapidosis sylvaticis siccis circa Byzantium (Sestini ex Sibth.) Boiss., *loc. cit.*

155. T. alpestre L., Sp. Pl. ed. 2 : 1082 (1763).

Perennial with creeping rhizome, patulous- or appressed-hairy, 20–50 cm. Stems many, erect or ascending, mostly simple. Leaves rather short-petioled; stipules lanceolate-subulate, membranous between nerves, sheathing at base, lower half adnate to petiole, free portion subulate, villous; leaflets mostly 2–5 × 0.5–0.8 cm, lanceolate to elliptical or linear-lanceolate, denticulate, rigid, often almost leathery,

Plate 150. *T. rubens* L. Var. *rubens* (Switzerland : *P. Villaret* 12156, HUJ). Flowering branch; flower.

hairy on one or both surfaces, with arcuate lateral nerves. Heads about 2 cm long, solitary or more often in pairs, globular to ovate-oblong, mostly involucrate by uppermost leaves, rarely short-peduncled. Flowers 1–1.3 cm, many. Calyx 5–7 mm; tube short-campanulate, 3–5 mm long, 20(very rarely 10–15)-nerved, densely hirsute; teeth unequal, setose, blunt, the lower one twice as long as the upper ones. Corolla much longer than lower tooth of calyx, purple, rarely pink or white; standard spoon-shaped, deflexed at tip, almost as long as wings and keel. Throat of fruiting calyx open, somewhat narrowed by ciliate ring; teeth not stellately spreading. Pod ovoid-globular, dehiscing by valves. Seed 1, ellipsoidal, reddish-brown. Fl. June–August. 2n = 16, 20.

Hab.: Pastures, meadows, forest clearings in the montane belt.

Gen. Distr.: Denmark, Germany, Poland, Switzerland, Austria, Hungary, Spain, France, Italy, Balkan Peninsula, S. Russia, Iran, Caucasus.

1. Calyx 20-nerved — 2
– Calyx 10(11–12)-nerved. — **155c.** var. **durmitoreum**
2. Indumentum of stem appressed. — **155a.** var. **alpestre**
– Indumentum of stem patulous. — **155b.** var. **lanigerum**

155a. T. alpestre L. Var. **alpestre**; Boiss., Fl. 2 : 113 (1872); Gib. & Belli, Mem. Accad. Sci. Torino ser. 2, 39 : 334 (1889). *Typus*: Described from Europe. Hb. Linn. 930/28, photo.

T. alpestre L. var. *distachyum* Ser. in DC., Prodr. 2 : 194 (1825).
T. alpestre L. var. *glabratum* J. C. von Klinggräff, Fl. Preuss. 98 (1848).
T. alpestre L. var. *sericeum* Hausskn., Mitt. Thür. Bot. Ver. N. F. 8 : 23 (1895).
T. alpestre L. var. *eualpestre* Aschers. & Graebn., Syn. Mitteleur. Fl. 6/2 : 576 (1908).

Icon.: [Plate 151]. Jacq., Obs. Bot. 3 : t. 64 (1768); Reichenb., Icon. 22 : t. 2135, f. 3, 7–12 (1903).

Selected specimens: CZECHOSLOVAKIA: Moravia: Mer. Or. Uh. Ostroh. Carpati albi; ad oppid. Velka sub summo monte Machova, 550 m, 1933, *Podpera* & *Uxor* 829 (HUJ). SWITZERLAND: Forêts sur les grès Vosgiens aux env. de Bitche Jum., *Schultz* 23 (HUJ). BULGARIA: Mt. Vitosa in petrosis supra pagum Dragalevci, ca. 1,000 m, solo silicioso, 1951, *Stojanov* 47, (HUJ). TURKEY: Prov. Kars: Zivaret Dag above Yalnizcam. 2,300 m, rocky igneous slopes, 1957. *Davis* & *Hedge* D. 30320 (HUJ). IRAN: Persia, *Szovits* 556 (G). CAUCASUS: In alpe Kasbek ad fl. Terek, 1881, *Brotherus* 239 (G). TRANSCAUCASIA: Azerbaidzhan, distr. Kedabek inter pag. Kedabek mont. Karadagh, 1939, *Chalilov* (HUJ); Armenia, Pratis subalpinis, Jelidja, 1929, *Schelkovnikov* (HUJ).

155b. T. alpestre L. Var. **lanigerum** Ser. in DC., Prodr. 2 : 194 (1825).

T. alpestre L. var. *villosum* Čelak., Prodr. Fl. Böhm. 667 (1875).

Selected specimens: SWITZERLAND: Vaud, Pied du Jura, route Moiry-Juriens, La Bossenaz sur Ferreyres, bois de chênes coupés, alt. 660 m, 1951, *Villaret* 12147 (HUJ).

Plate 151. *T. alpestre* L. Var. *alpestre*
(Bulgaria: *N. Stojanov* 47, HUJ).
Plant in flower; part of leaflet showing nervation;
dissected calyx; flower.

FRANCE: Prairies au nord de la commune à Sr Sarlin pr. Maupienne (Savoie), 1854, *Didier* 1646 (WAG). GREECE: Epirus, Mt. Olytsika, 1929, *Guiol* 1081 (HUJ).

155c. T. alpestre L. Var. **durmitoreum** Rohlena, Sitzb. Böhm. Ges. Wies. 17 : 24 (1903).

Recorded from Montenegro am Durmitor oberhalb Zabljak (Aschers. & Graebn., *loc. cit.* 578) and also from other places.

Subsection V. STELLATA Gib. & Belli

Mem. Accad. Sci. Torino ser. 2, 39 : 293 (1889) pro stirpe, emend. Zoh Candollea 27 : 129 (1972)

Annuals. Calyx 10-nerved; teeth of fruiting calyx more or less equal in length, spreading at maturity, as long as or longer than tube; throat of calyx closed by hairy ring. Corolla about as long as calyx, rarely somewhat longer. Leaflets obovate to obtriangular; stipules ovate, more or less dentate, not cuspidate.

156. T. stellatum L., Sp. Pl. 769 (1753).

Annual, patulous- or appressed-villous, 10-20 cm. Stems few, mostly erect or ascending, sparingly branching or unbranched. Leaves small, lower ones long-petioled, upper ones short-petioled; stipules membranous, ovate, obtuse, dentate, green at margin; leaflets mostly 5–8 × 4–8 mm, obcordate with cuneate base, sharply dentate in upper part. Heads 1.5–2 cm across in flower, about 3 cm in fruit, long-peduncled, many-flowered, broad-obovoid to globular. Flowers pedicellate, loose, mostly 1.4–1.8 cm. Calyx campanulate, 10-nerved, densely subappressed-hispid; tube about half the length of lanceolate-subulate teeth, the latter connate at base; throat densely fleecy but not closed by callosity; fruiting calyx with long and sharp-pointed base, bristly-hairy, with markedly divergent to spreading, sharp-pointed teeth, broadened at base. Corolla white or pink, rarely purple or yellow, mostly shorter than or as long as, rarely somewhat longer than calyx; standard with ovate to oblong limb, slightly longer than wings. Pod short-stipitate, membranous, obovoid to pyriform. Seed 1, ellipsoidal to ovoid, shiny, yellowish. Fl. February–April. 2n = 14.

Hab.: Roadsides, fields and among dwarf shrubs.

Gen. Distr.: British Isles, Portugal, Spain, France, Italy, Balkan Peninsula, Turkey, Syria, Lebanon, Palestine, Iraq, Iran, Caucasus, Transcaucasia, Egypt, Libya, Tunisia, Algeria, Morocco, Canary Islands and elsewhere.

1. Flowers white or pink ... 2
– Flowers yellow. ... **156c.** var. **xanthinum**
2. Plants patulous-hairy ... 3

– Plants appressed-hairy. **156b.** var. **adpressum**
3. Corolla about as long as calyx. **156a.** var. **stellatum**
– Corolla somewhat longer than calyx. **156d.** var. **longiflorum**

156a. T. stellatum L. Var. **stellatum**; Boiss., Fl. 2 : 121 (1872); Gib. & Belli, Mem. Accad. Sci. Torino ser. 2, 39 : 293 (1889). *Typus* : Syntypes from Sicily and Italy, Hb. Cliff. 375.

Icon. : [Plate 152]. Reichenb., Icon. 22 : t. 2143, f. 1–10 (1903).

Selected specimens : HUNGARY : 1823, *Sadler* (W). PORTUGAL : prope Almodo, 1840, *Welwitsch* 482 (W). SPAIN : prope Cadiz, 1873, *Winkler* (W). FRANCE : Friches autour du bois Serres pr. Carcassonne, 1885, *Baichere* 1128 (W). ITALY : in herbidis agri Neapolitani, 1841, *Heldreich* (W). CRETE : 4 km S. of Mires on the way to Kali-Limenes, *Thymus-Andropogon* batha, 100 m, 1964, *Zohary* & *Orshan* 24/4/3 (HUJ). TURKEY : Constantinopole, 1845, *Noë* (W). SYRIA : Aleppo, 1841, *Kotschy* 56 (G). LEBANON : env. of Eden, 1856, *Blanche* 3060 (G). PALESTINE : Jerusalem, Mt. Scopus, 1931, *Amdursky* 137 (HUJ). IRAQ : Zawita, 1956, *Guest* 15393 (K). TRANSCAUCASIA : Daghestan, pr. Derbent versus pagum Sabnova, 1902, *Alexeenko* 68 (G). IRAN : Karerun plain near Bushire, 1928 (K). EGYPT : betw. Matruh and Barran, 1932, *Shabetai* 41 (K). ALGERIA : Saida, 1893, *Blanche* 90 (W). MOROCCO : Pr. Xanen, *Font Quer* 199 (BM). MADEIRA : in pascuis Cabo Gircio, 1865–66, *Mandon* 64 (G). CANARY ISLANDS : between Tenerra and Taburienta, 1913, *Sprague* & *Hutchinson* 1354 (HUJ).

156b. T. stellatum L. Var. **adpressum** Turrill, Kew Bull. 1924 : 327 (1924). *Typus* : Turkey : Thymbra in collibus, 3.5.1883, *Sintenis* 175 (K lecto., isolecto. E).

Selected specimens : GREECE : Dodecanese : Patmos, rough, stony hillside near Scala, 1962, *Gathorne-Hardy* 150 (E). IRAQ : Rowanduz gorge, 1932, *Guest* 2026 (K). IRAN : Railway gorge nr. Khorramabad, 1,200 m, 1936, *Trott* 394 (K).

Note : Also known from Italy, Sicily, Albania, Aegean Islands, Cyprus, etc.

156c. T. stellatum L. Var. **xanthinum** (Freyn) Bald., Nuov. Giorn. Bot. Ital. n. s. 6 : 155 (1899). *Typus* : Greece, in Aetolien, Berge Arapokephalos, nr. Agrinion, 22.5.1878, *Heldreich*; Klissura, 21.5.1878, Heldreich (syntypes E).

T. xanthinum Freyn, Bot. Centralb. 1 : 308 (1880).

Specimens seen : GREECE : Laconia borealis in districto Alagonia ca. Megali-Anastova, 1889, *Zahn* 1527a (W). W. ANATOLIA : Distr. Aydin; S. side of Bafa Göl., 100 m, D. 40748 (E).

156d. T. stellatum L. Var. **longiflorum** Griseb., Spicil. Fl. Rumel. 1 : 19 (1843). *Typus* : No exact typification could be made.

Plate 152. *T. stellatum* L. Var. *stellatum*
(Palestine : *Zohary*, HUJ).
Flowering and fruiting branch; flower; fruiting calyx from above.

Icon. : Sibth. & Sm., Fl. Graec. 8 : t. 750 (1883).

Selected specimens : FRANCE : 1847, *Kralik* 230 (WAG). CORSICA : 1849, *Kralik* 16550 (WAG). TURKEY : "In regione semperviridi Bithyniae et Thraciae alt. 0–1,200 m, copiose in campis siccis prope Modanis..." etc., fide Hossain, *loc.cit.* 434 (1961). PALESTINE : Upper Galilee, Metula, Eocene hills, terra rossa, 1963, *Zohary* & *Plitmann* 12/21 (HUJ).

157. T. incarnatum L., Sp. Pl. 769 (1753).

Annual (rarely biennial), appressed- to patulous-villous, 20–50 cm. Stems few or solitary, erect or ascending, simple or sparingly branching from base. Leaves rather large, all alternate, the lowest ones long-petioled; stipules membranous, the lower part sheathing below, connate and adnate to petioles for three-fifths of their length, free portion green or purplish, ovate, dentate, nerved, patulous-hairy; leaflets 1–2 × 1–1.5 cm, broadly obovate with short-cuneate base, denticulate margins and rounded or retuse apex. Heads oblong, 2–6 × 1–1.5 cm, later elongating, more or less thickly pedunculate. Flowers rather dense, 1–1.5 cm. Calyx about 1 cm long; tube cylindrical, with appressed long hairs, 10-nerved; teeth linear, almost equal, setaceous, sharp-pointed, with dense, long hairs, 1–2 times as long as tube. Corolla red or pink, rarely cream-coloured, usually longer than calyx; standard oblong-elliptical, acute, much longer than wings and keel. Fruiting calyx with open throat narrowed by thickened ring and spreading teeth. Pod ovoid, membranous with cartilaginous apex. Seed 1, greenish-yellow, ovoid, 2 mm. Fl. May–August. 2n = 14.

Hab. : Fields, pastures, meadows, roadsides.

Gen. Distr. : British Isles, Portugal, Spain, France, Italy and islands, Balkan Peninsula, Turkey.

Note : *T. incarnatum* includes the following 2 varieties:

1. Corolla blood-red, rarely white, as long as calyx. Stems stout. Heads dense. **157a.** var. **incarnatum**
– Corolla usually yellowish-white, rarely pink, much longer than calyx. Heads rather loose. Stems slender. Stems and petioles usually appressed-hairy. **157b.** var. **molinerii**

157a. T. incarnatum L. Var. **incarnatum**. *Typus* : Described from Italy, Hb. Linn. 930/32.

T. incarnatum L. var. *sativum* Ducomm., Taschenb. 169 (1869).
T. stellatum L. subsp. *incarnatum* (L.) Gib. & Belli, Mem. Accad. Sci. Torino ser. 2, 39 : 296 (1889).
T. stellatum L. subsp. *incarnatum* var. *elatius* Gib. & Belli, *loc. cit.*

Icon. : [Plate 153]. Reichenb., Icon. 22 : t. 2145, f. 2, 5–12 (1903).

Widely cultivated and more or less naturalized.

Plate 153. *T. incarnatum* L. Var. *incarnatum* (Portugal : *Matos* & *Marques*, HUJ). Plant in flower; standard; dissected calyx; flower; fruiting calyx from above.

Selected specimens : PORTUGAL : Coimbra, Mainca, nos campos cultivados, 1954, *Matos & Marques* (HUJ). YUGOSLAVIA : Belgrad Weiden, 1888, *Bornmüller* (HUJ).

Note : Also seen and recorded from almost all European countries.

157b. T. incarnatum L. Var. **molinerii** (Balb. ex Hornem.) Ser. in DC., Prodr. 2 : 190 (1825). *Typus* : Described from Italy (n. v.).

T. molinerii Balb., Cat. Hort. Acad. Taurin. App. 1 : 17 (1813) nom. nud.
T. molinerii Balb. ex DC. in Lam. & DC., Fl. Fr. 3, 5 : 556 (1815) pro syn.
T. molinerii Balb. ex Hornem., Hort. Hafn. 2 : 715 (1815).
T. stramineum C. Presl, Fl. Sic. 1 : XX (1826).
T. noëanum Reichenb. ex Mert. & Koch in Röhling, Deutschl. Fl. ed. 3, 5 : 265 (1839) pro syn.
T. stellatum L. subvar. *stramineum* (C. Presl) Gib. & Belli, *loc. cit.* 298.

Icon. : Syme, Engl. Bot. 3 : t. 353 (1864).

Hab. : Meadows and roadsides.
Gen. Distr. : Mainly W., S. and C. Europe.

Selected specimens : ROMANIA : Banatus distr., Caraş-Severin, 300 m, 1923, *Borza* et al. 560 (HUJ). FRANCE : env. of Montpellier, Castelnau-Crès roadside, 1931, *Eig* (HUJ). ITALY : Alpe Apuane, Prov. Lucca, Weiden am Weg von Stazzema zum Monte Procinto, ca. 600 m, 1948, *Koch* 48/229 A (HUJ). YUGOSLAVIA : Croatien, Vrebae, *Schlosser* (W); Fiume auf Wiesen, *Noë* 1366 (G, syntype of *T. noëanum* Reichenb.). TURKEY : Distr. Constantinopole (Istanbul) entre Tachamlidja et Dodourlou, 1891, *Aznavour* 643 (G).

Note : Also seen from other Mediterranean countries.

158. T. sylvaticum Gérard ex Loisel., Desv. Journ. Bot. 2 : 367 (1809) et Fl. Gall. ed. 2, 2 : 124, n. 40 (1828). *Typus* : Described from S. France (n. v.).

T. lagopus Pourr. ex Willd., Sp. Pl. 3 : 1365 (1802) non Garsault, Fig. Pl. Anim. Med. t. 602 (1764) nom. inval. nec Gouan, Fl. Monsp. 105 n. 20 (1765) subs. illegit; Gib. & Belli, Mem. Accad. Sci. Torino ser. 2, 39 : 349 (1889).
T. smyrnaeum Boiss., Diagn. ser. 1, 2 : 25 (1843) et Fl. 2 : 121 (1872; "*smyrneum*"); Hossain, Not. Roy. Bot. Gard. Edinb. 23 : 418 (1961).
T. hervieri Freyn, Bull. Herb. Boiss. 1 : 543 (1893).
T. lagopus Pourr. var. *smyrnaeum* (Boiss.) Bornm., Bot. Jahrb. 59 : 475 (1925) comb. inval.
T. bonnevillei Mout., Fl. Djebel Druze 127 (1953).

Icon. : [Plate 154]. Cusin, Herb. Fl. Fr. 6 : t. 1100 (1870); Gib. & Belli, *loc. cit.* t. 6, f. 3 (1889); Mout., *loc. cit.* t. 8, f. 1.

Annual, patulous-hairy, 10–15 cm. Stems many, decumbent or ascending, poorly branching. Leaves small, lower ones long-petioled, upper ones short-petioled to sessile; stipules leaf-like, connate and adnate to petioles for up to one-fourth or one-third of their length, hairy, free portion ovate to ovate-lanceolate, acute or ob-

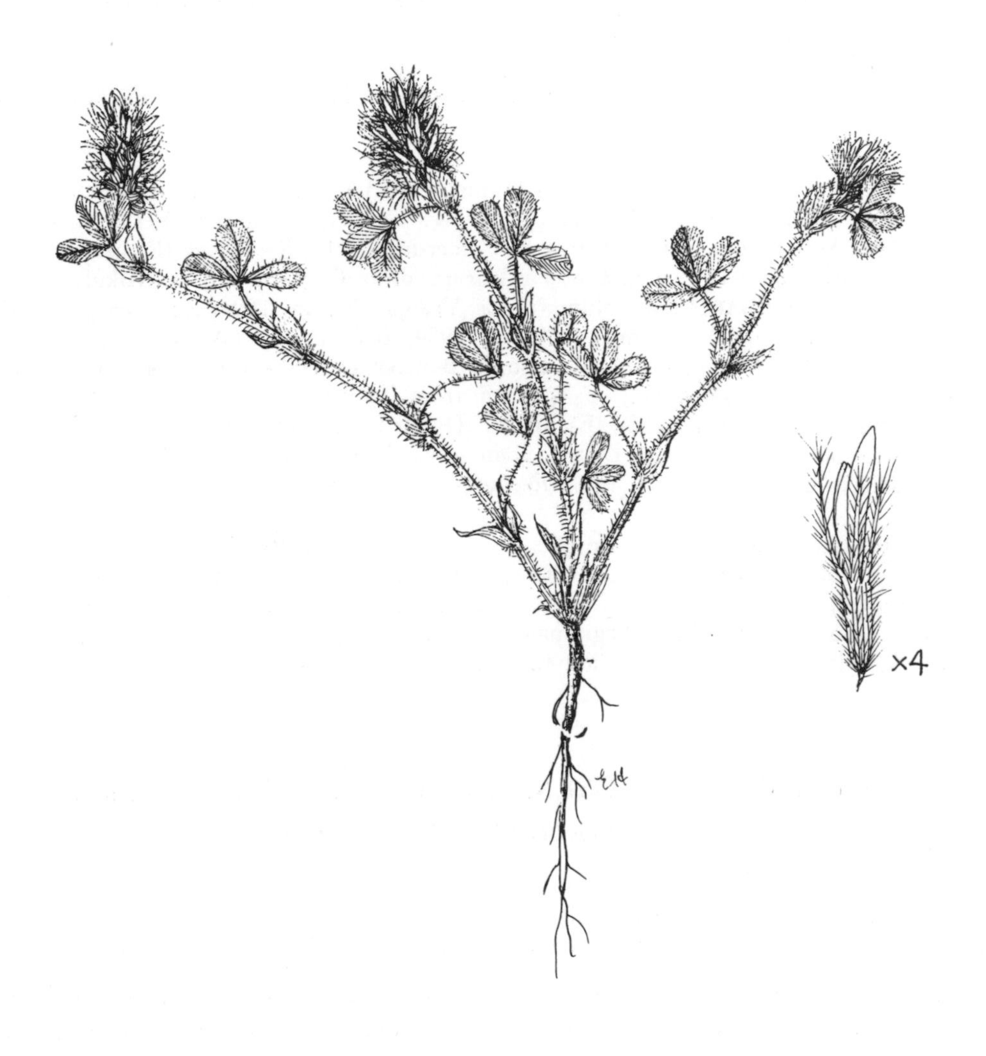

Plate 154. *T. sylvaticum* Gérard ex Loisel.
(Bulgaria : *B. Ahtarov* 459, HUJ).
Plant in flower; flower.

tuse; leaflets (0.6–)0.8–1 × 0.8 mm, small, cuneate, obovate, obtuse or retuse at base, denticulate at apex and mostly mucronulate. Heads terminal, solitary or twin, 1–1.5 × 1 cm, globular to ovate-oblong, many-flowered, often involucrate by upper leaves. Flowers 0.8–1 cm long. Calyx tube cylindrical, with 10 prominent nerves, hispid, becoming globular or urceolate in fruit; throat closed by bilabiate callosity; teeth subulate, rigid, truncate, plumose-ciliate, unequal, the lower one longer, about as long as tube, all stellately spreading at maturity. Corolla pink to purple, about as long as or longer than calyx; standard linear-oblong, obtuse, apiculate, considerably longer than wings and keel. Pod ovoid, membranous. Seed 1, globular, yellowish-brown. Fl. April–May. 2n = 14,16.

Hab. : Hillsides and montane valleys.

Gen. Distr. : Belgium, Portugal, Spain, France, Balkan Peninsula, Turkey, Syria, Iraq, etc.

Selected specimens : PORTUGAL : Base do Monte de S. Bartolomeu, pr. Bragança, terras em pousio, 1958, *Fernandes* et al. 6345 (HUJ). SPAIN : Catalogne, Vilarnadal et S. Clemente, friches, 1907, *Sennen* 256 (HUJ); Sierra de Comeruna, 1892, *Reverchon* (K, isotype of *T. hervieri* Freyn). ALBANIA : in umbrosis sylvarum ditio Primarije ad v. Mikulic, 1898, *Baldacci* 61 (P). BULGARIA : in collinis prope Vranja, 700 m, 1894, *Adamović* s. n. (P). TURKEY : Smyrne, dans les champs en friche, 1954, *Balansa* 159 (W); in montibus circa Smyrnam, mons Corax, jugum Sipyleum, supra Bournabat, Maio–Junio 1842, *Boissier* (G lectotype of *T. smyrnaeum* Boiss.); Uşak, 5 mi. from Uşak on road to Güre, open dry pasture, c. 800 m, *Coode* & *Jones* 2539 (E). SYRIA : Djebel Druze, Tel Chihane, 1943, *Mouterde* (HUJ, as *T. bonnevillei*). IRAQ : Distr. Sulaimaniya, Kurdistan, Montes Avronam ad confines Persiae in dit pagi Tawillo, 1957, *Rechinger* 12432 (W).

Note : The reasons cited by Hossain (*loc. cit.*) for not identifying *T. smyrnaeum* with *T. sylvaticum* are unconvincing. We have seen specimens from Portugal and Spain which are in good agreement with *T. smyrnaeum* Boiss. Rouy (Fl. Fr., 1899), Briquet (Fl. Corse, 1913), and others also synonymize the illegitimate name of *T. lagopus* Pourr. with *T. sylvaticum* Gérard ex Loisel., which is the earliest valid name for this species.

Subsection VI. STENOSEMIUM (Čelak.) Aschers. & Graebn.

Syn. Mitteleur. Fl. 6/2 : 527 (1908)

Section *Stenosemium* Čelak., Oesterr. Bot. Zeitschr. 24 : 75 (1874).
Eleuterosemium Gib. & Belli, Mem. Accad. Sci. Torino ser. 2, 39 : 261 (1889) pro stirpe.

Annuals. Calyx tube 10-nerved, almost globular in fruit; teeth not reflexed at maturity; throat open, but somewhat narrowed by epidermal thickening. Corolla persistent; standard free from other petals and from stamens.

159. T. striatum L., Sp. Pl. 770 (1753); Godr. in Gren. & Godr., Fl. Fr. 1 : 412 (1848–49); Boiss., Fl. 2 : 130 (1872); Gib. & Belli, Mem. Accad. Sci. Torino ser. 2,

39 : 261 (1889). *Typus*: Syntypes from Germany, France and Spain, Hb. Linn. 930/45.

T. conicum Pers. fide Savi, Obs. Trif. 41 (1808–10) non Lag. (1816) nec Kit. ex Nyman (1878).
T. tenuiflorum Ten., Prodr. Fl. Nap. 1 : XLIV (1811–15) et Fl. Nap. 5 : 141, no. 2610, t. 177, f. 1 (1835–38).
T. incanum auct. non J. & C. Presl (1822).
T. kitaibelianum Ser. in DC., Prodr. 2 : 194 (1825).
T. cylindricum Wallr., Beitr. Fl. Hercyn., in Linnaea 14 : 619 (1840).
T. striatum L. var. *spinescens* Lange, Linnaea 12 : 361 (1857) et in Willk. & Lange, Prodr. Fl. Hisp. 3 : 362 (1877).
T. striatum L. var. *kitaibelianum* (Ser.) Heuff., Verh. Zool.-Bot. Ges. Wien 8 : Abh. 89 (1858).
T. striatum L. var. *prostratum* Lange, Bot. Tidskr. 3 : 124 (1860).
T. striatum L. var. *brevidens* Lange, Naturhist. Forenz. Vidensk. Meddel. 2, Aart.VII : 363 (1860–65).
T. striatum L. var. *macrodontum* Boiss., *loc. cit.*
T. striatum L. var. *elatum* Lojac., Tent. Monogr. Trif. Sic. 124 (1878) et Fl. Sic. 1, 2 : 92 (1891).
T. striatum L. var. *strictum* Dreyer ex Lange, Haandb. Danske Fl. ed. 4 : 832 (1888).
T. striatum L. var. *elongatum* Rouy, Fl. Fr. 5 : 101 (1899).
T. striatum L. var. *nanum* Rouy, *loc. cit.*
T. striatum L. var. *longiflorum* Haláс., Consp. Fl. Gr. 1 : 393 (1901).
T. striatum L. var. *incanum* (J. & C. Presl) Aschers. & Graebn., Syn. Mitteleur. Fl. 6/2 : 528 (1908).

Icon.: [Plate 155]. Reichenb., Icon. 22 : t. 2151, f. 1, 1–10 (1903); Gib. & Belli, *loc. cit.* t. 1 (1889); Ross-Craig, Draw. Brit. Pl. 7 : pl. 31 (1954).

Annual, rarely biennial, densely or sparingly patulous-villous, 10–50 cm. Stems many, erect, ascending or decumbent, slender, branched or simple. Petioles long in lower leaves, decreasing in length towards apex of stem; stipules ovate, adnate to petioles and connate for up to half their length, white-submembranous, appressed-hairy, green-nerved, free portion triangular-ovate, terminating abruptly in subulate, indurate, ciliate cusp; leaflets of lower leaves obovate or oblong, obtuse, retuse or obcordate, cuneate at base, those of upper ones obovate to oblong-lanceolate, those of uppermost ones long-cuneate at base, denticulate at apex, usually mucronulate. Heads 0.6–1.5 cm, solitary or in pairs due to spurious dichotomy, pedunculate or sessile, terminal or axillary, ovoid to oblong, often becoming cylindrical. Flowers 5–7 mm long, dense. Calyx tube ellipsoidal, becoming globular or urceolate in fruit, usually appressed-hairy, with 10 prominent nerves; teeth lanceolate-subulate, erect, sharp-pointed, equal or slightly unequal, usually much shorter than tube, rarely the lower one longer than others and as long as tube, all erect or slightly divergent in fruit; throat of calyx open. Corolla pink, marcescent, as long as or shorter or longer than calyx; standard free from other petals and from stamens, oblong, retuse. Pod membranous, obovoid. Seed 1, ovoid, reddish. Fl. May–August. 2n = 14.

Hab.: Roadsides, wastes, pastures, etc.

Gen. Distr.: British Isles, Denmark, Scandinavia, Germany, Switzerland, Hun-

Plate 155. *T. striatum* L.
(Romania: *Borza* & *Nyárády* 559, HUJ)
Plant in flower; flower.

gary, Romania, Portugal, Spain, France, Italy, Sicily, Balkan Peninsula, S. Russia, Crimea, Turkey, Iraq, Iran, Transcaucasia, Tunisia, Algeria, Morocco, Madeira, Canary Islands and elsewhere.

Selected specimens: DENMARK: Jylland, Bjørnsknude, 1890, *Raundiaer* (HUJ). GERMANY: Dresden, *Hübner* 266 (G). SWITZERLAND: Basch, *Lerasche* (G). HUNGARY: ca. Budam, *Rainer-Haerbad* (G). ROMANIA: Reg. Constanta, raion Istria in foenetis ad marg. merid. oppid. Istria (Babadag), 1956, *Todor* et al. 93 (HUJ). SPAIN: Sierra de Sigura, 1890, *Bourgeau* (G). FRANCE: Roches et champs secs des collines exposés au soleil près de Bitche (Moselle), 1853, *Schultz* 34 (HUJ). SICILY: in montosis, *Gasparrini* (W); in herbosis Ficuzza, 1856, *Huet du Pavillon* 46 (G). BULGARIA: in graminosis ad agrorum margines, prope Kalofer (Thrace), 1871, *Janka* (G). TURKEY: near Istanbul, Belgrad forest, Fagetum, 1963, *Zohary* 5663 (HUJ); Sultandagh, Akshehir, 1899, *Bornmüller* 4342 (G). CYPRUS: Prodhromus, 1,320 m, 1855, *Merton* 2393 (K). IRAN: Lahijan, dry banks, 1937, Miss *N. Lindsay* 822 (K). TRANSCAUCASIA: Azerbaidzhan, Lenkoran, 1946, *Chalilov* (HUJ); Abchasia, Suchumkale, 1908, *Markovicz* 363 (G). ALGERIA: Blida, pentes de l'Atlas, au-dessus de la ville, 1881, *Meyer* 1169 bis (G). CANARY ISLANDS: Teneriffae ad caenobium St Diego del Monte (G).

Note: *T. striatum* is an exceedingly polymorphic species and its variability is discussed at length by Gibelli & Belli, (*loc. cit.*) and by Ascherson & Graebner (*loc. cit.*)
The present authors strictly follow Gibelli & Belli in regarding most of the taxa established on the basis of this variability as habitat forms not worthy of being ranked as varieties, especially when one takes into account the many transitions between the forms. At any rate, only a critical (experimental) study of the wealth of forms included in the discussed species will finally determine their constancy and taxonomic value.

Subsection VII. TRICHOPTERA Gib. & Belli

Mem. Accad. Sci. Torino ser. 2, 39 : 274 (1889) pro stirpe;
Aschers. & Graebn., Syn. Mitteleur. Fl. 6/2 : 535 (1908).

Annuals. Calyx 10-nerved; teeth as long as tube or shorter; throat open but hairy. Corolla persistent, shorter or longer than calyx; wings finely pilose, especially near auricles.

160. T. bocconei Savi, Atti Accad. It. 1 : 191, f. 1, 201 (1808–10; "*boccone*").

Annual, appressed- to patulous-pubescent, 10–25 cm. Stems few or many, erect or ascending, diffuse, branching from base. Lower leaves long-petioled, uppermost ones sessile; stipules oblong-lanceolate, adnate along their lower half to petioles, herbaceous to membranous between green to purple nerves, free portion subulate; leaflets of lower leaves less than 1 cm, obovate, deeply notched, those of upper ones oblong with cuneate base, denticulate in upper part, often notched at apex, those of uppermost ones cuneate. Heads sessile, 1–2 × 0.8–1 cm, involucrate by 1 or 2 leaves, ovoid to cylindrical. Flowers up to 6 mm long. Calyx obconical, with densely appressed white hairs and 10 prominent nerves; teeth subulate, unequal, considerably longer to somewhat shorter than tube, erect, often connivent after flowering. Corolla pink to reddish, rarely yellowish, about as long as calyx, rarely longer than longest calyx tooth; standard ovate-oblong, truncate or rounded at apex; wings shorter, hairy. Pod ovoid, 1–2 mm. Seeds small, ovoid, yellowish. Fl. May–October. 2n = 12, 14.

Hab.: Fields, roadsides, open maquis and dwarf-shrub formations.

Gen. Distr.: British Isles, Portugal, Spain, France, Italy, Balkan Peninsula, Turkey, Syria, Tunisia, Algeria, Morocco, Canary Islands.

1. Upper leaflets cuneate-oblong. Corolla not longer or only slightly longer than calyx teeth. **160a.** var. **bocconei**
– Upper leaflets linear or narrowly oblong. Corolla twice as long as calyx. **160b.** var. **tenuifolium**

160a. T. bocconei Savi Var. **bocconei**; Godr. in Gren. & Godr., Fl. Fr. 1 : 411 (1848–49); Boiss., Fl. 2 : 132 (1872); Gib. & Belli, Mem. Accad. Sci. Torino ser. 2, 39 : 274 (1889). *Typus*: Probably Florence (G, one of the syntypes).

T. collinum Bast., Suppl. Fl. Maine-et-Loire 5 (1812).
T. gemellum sensu Lapeyr., Hist. Abr. Pyrén. 437 (1813) non Pourr. ex Willd. (1802).
T. semiglabrum Brot., Phytogr. Lusit. 1 : 155, t. 63, f. 2 (1816).
T. bocconei Savi var. *gracile* Rouy, Fl. Fr. 5 : 102 (1899).
T. bocconei Savi var. *cylindricum* Rouy, *loc. cit.*
T. bocconei Savi var. *macedonicum* Adamović, Denkschr. Akad. Wiss. 74 : 130 (1904).
T. bocconei Savi var. *pomelii* Maire in Sched. Herb. Paris.

Icon.: [Plate 156a]. Boccone, Mus. Piant. 1 : 104 (1697); Gib. & Belli, *loc. cit.* t.1, f. 4; Reichenb., Icon. 22 : t. 2149, f. 1, 1–8(1903); Ross-Craig, Draw. Brit. Pl. 7 : 30 (1954).

Selected specimens: PORTUGAL: Coimbra, Penedo da Meditaçao, 1950, *Matos & Matos* (HUJ). SPAIN: Montes de Toledo, dry ground, *Quercus ilex* scrub, 930 m, 1959, *M. Y. Sandwith* 5602 (K). FRANCE: Prov. Var, Bord des champs et des bois, etc., pres de Pierrefeu, 1865, *Chamberon* 1048 (HUJ). SARDINIA: Arroundiss. de Tempis, 1882, *Reverchon* 314 (W). ITALY: Toscana, Monte Pisano, 1862, *Savi* 1164 ter ? (G). SICILY: in vineis et campis aridis agri Panormitani, 1840, *Heldreich* (G). YUGOSLAVIA: Polo, Mt. Toibom, 1901, *Untchy* (W). GREECE: Athos, *Frivaldszky* (W). BULGARIA: Mazedonia in herbidis saxosis Mt. Kortiac, 1908, *Adamović* (W). TURKEY: Prov. Istanbul, Calendie près de Yinikeny, 1895, *Aznavour* (G). ALGERIA: env. d'Alger, Corteaux du Hamma, lieux herbeux, 800 m, 1953, *d'Alleizette* (HUJ). MOROCCO: Tanger, Djebel Kebir, *Schousboe* 33 (K)

Note: The two forms distinguished by Rouy (*loc. cit.*), which intergrade with one another and with the typical form *T. bocconei* var. *macedonicum* Adamović, need further study.

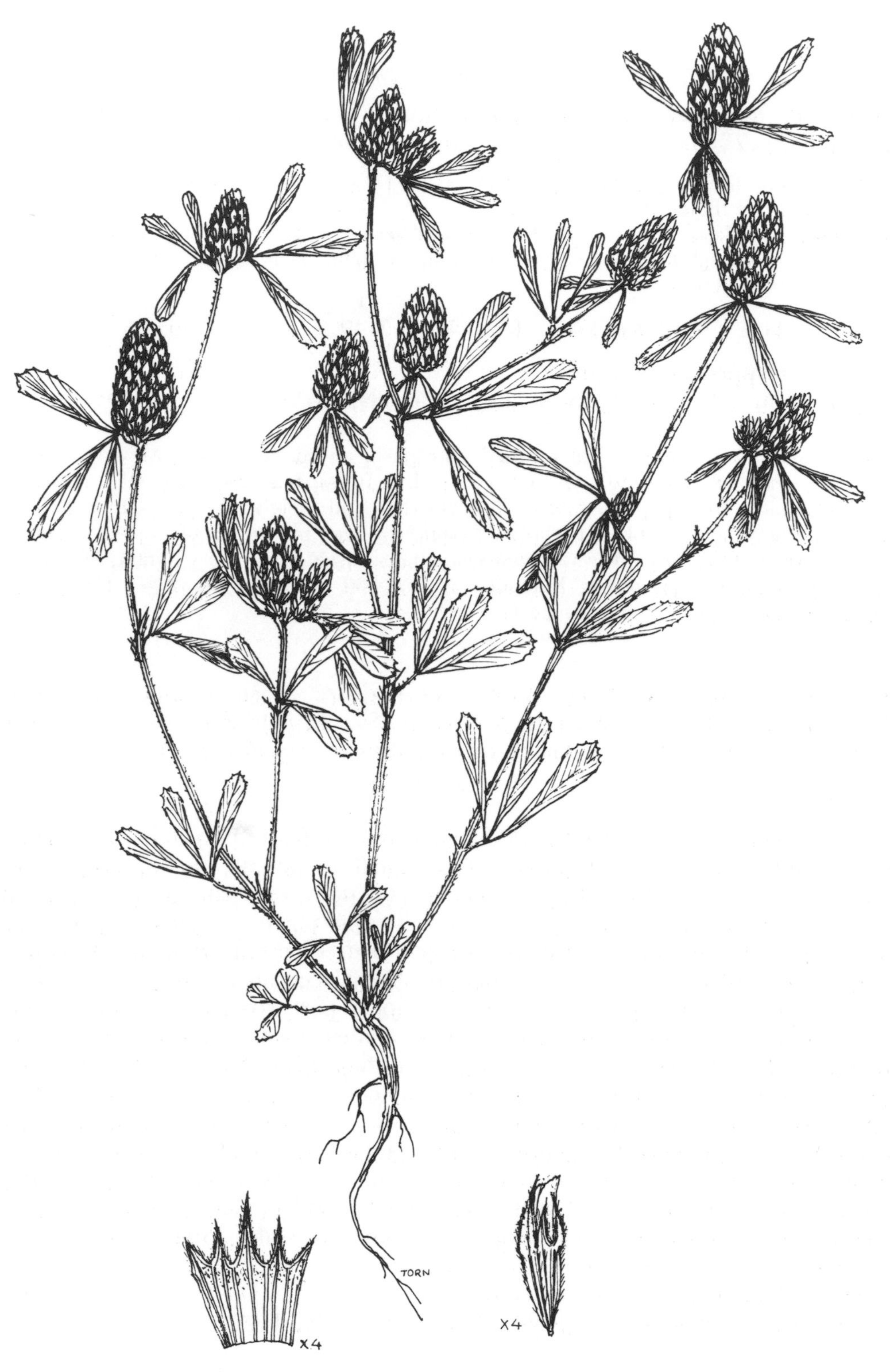

Plate 156a. *T. bocconei* Savi Var. *bocconei*
Plant in flower; dissected calyx; flower.

160b. T. bocconei Savi Var. **tenuifolium** (Ten.) Griseb., Spicil. Fl. Rumel. 1 : 23(1843). *Typus*: Described from Italy (probably in NAP).

T. tenuifolium Ten., Prodr. Fl. Nap. XLIV (1811–15) et Fl. Nap. 5 : 145, no. 2622, t. 177, f. 3(1835–38); Boiss., Fl. 2 : 132 (1872).
T. tenoreanum Boiss. & Sprun. in Boiss., Diagn. ser. 1, 2 : 26 (1843) subs. illegit.
T. bocconei Savi subsp. *tenuifolium* (Ten.) Gib. & Belli, Mem. Accad. Sci. Torino ser. 2, 39 : 277, 410 (1889).

Icon.: [Plate 156b]. Reichenb., Icon. 22 : t. 2149, f. 2, 9–16(1903).

Hab.: Steppes, dry hillsides.
Gen. Distr.: S. Italy, Sicily, Balkan Peninsula, Crete.

Selected specimens: ITALY: Italia austr., Tapygia in olivetis ca. pagum Alessano, 10–100 ft, 1875, *Porta* & *Rigo* 301 (P, HUJ). SICILY: in sepibus, *Parlatore* (P). YUGOSLAVIA: Dalmatia, Verrai, 1905, *Calloro* (G). ALBANIA: ad Stoja in arenosis, 1898, *Baldacci* 62 (W). GREECE: Agrapha (Dolopia veterum) in reg. inf. m. Pindi ca. monasterium Korona, 1885, *Heldreich* (W); Macedonia orientalis, distr. Kozani in montibus Pierra loco dicta Vlassia inter Velvedos et Kataphygion, 500–800 m, 1956, *Rechinger* 17841 (W). CRETE: Canee, 1894, *H. de Voasieu* (P).

161. T. trichopterum Panč., Verh. Zool.-Bot. Ges. Wien 6 : Abh. 480 (1856); Boiss., Fl. 2 : 131 (1872); Aschers. & Graebn., Syn. Mitteleur. Fl. 6/2 : 538 (1908). *Typus*: Yugoslavia, Boračer Felsen in Kragujevacer, *Sapetza* (G ?).

Icon.: [Plate 157].

Annual, appressed-hairy, 10–30 cm. Stems many, profusely branching, procumbent to ascending, arcuate or flexuose. Leaves small to minute, the lowermost ones long-petioled, the middle and upper ones short-petioled to sessile; stipules of middle leaves membranous, oblong-lanceolate, connate and adnate to petioles up to about half of their length, appressed-hairy, red-nerved, free portion triangular-lanceolate, mucronate; leaflets up to 1–1.6 cm, obovate-oblong, truncate or retuse, long-cuneate at base, prominently nerved, denticulate in upper part, sparingly appressed-hairy. Heads 1–2.5(–3) × 0.8–1 cm, sessile, axillary and terminal, all involucrate by broad stipules of leaves, globular to ovoid, later elongating, many-flowered. Flowers 0.6–1 cm, very dense. Calyx-tube obconical to campanulate, somewhat growing in fruit, 10-nerved, covered with white, rough, appressed and patulous hairs; throat open, hairy, but not thickened by callosity; teeth setaceous, rather blunt, unequal, shorter than to as long as tube, erect or connivent in fruit. Corolla considerably longer than calyx, cream- to flesh-coloured, marcescent; standard oblong, much longer than hairy wings. Pod ovoid, membranous, irregularly rupturing. Seed 1.5 mm, ovoid to globular, brown. Fl. May–September. 2n = 14.
Hab.: Dry stony hillsides and wastes.
Gen. Distr.: Yugoslavia, Albania, Greece, Bulgaria, Turkey.

Specimens seen: GREECE: above Armensko, 1932, *Alston Sandwith* 14 (K). BULGARIA: M. Sredna Gora in declivibus graminosis supra urbem Koprivstica ad ca. 1,050 m, 1951,

Plate 156b. *T. bocconei* Savi Var. *tenuifolium* (Ten.) Griseb.
Plant in flower; flower; standard; keel; wing; flower; dissected calyx.

Plate 157. *T. trichopterum* Panč.
Plant in flower; leaf (enlarged);
dissected calyx; flower; fruiting calyx.

Stojanov & *Achtarov* 46 (HUJ); M. Rila pr. monasterium Rila, 1939, *Lindberg* (HUJ); in asperis graniticis supra Vranja, Serb. merid., 1874, *Pancić* (G). TURKEY: Thrace, 200 ft, 1934, *Ottman-Euren* 1460 (K).

Subsection VIII. SCABROIDEA Gib. & Belli

Mem. Accad. Sci. Torino ser. 2, 39 : 286 (1889) pro stirpe; Aschers. & Graebn., Syn. Mitteleur. Fl. 6/2 : 540 (1908)

Annuals. Calyx 10-nerved, pilose; teeth usually shorter than tube, the lower one longer, somewhat spiny at tip; throat open but narrowed by annular callosity. Leaves with lateral nerves, thickened and recurved. Fruiting heads not disarticulating, falling off as single dispersal unit.

162. T. scabrum L., Sp. Pl. 770 (1753); Boiss., Fl. 2 : 130 (1872); Gib. & Belli, Mem. Accad. Sci. Torino ser. 2, 39 : 286 (1889); Aschers. & Graebn., Syn. Mitteleur. Fl. 6/2 : 540 (1908); Hossain, Not. Roy. Bot. Gard. Edinb. 23 : 430 (1961). *Typus*: Syntypes from the British Isles, France and Italy, Hb. Cliff.

Icon.: [Plate 158]. Reichenb., Icon. 22 : 2152, f. 2, 9–16 (1903); Hegi, Ill. Fl. Mitteleur. 4, 3 : 1326, f. 1410 (1923); Ross-Craig, Draw. Brit. Pl. 7 : pl. 32 (1954).

Annual, appressed-hairy, 6–20 cm. Stems many to few or solitary, flexuous, striate, ascending or decumbent, sparingly branching or unbranched. Leaves small, the lower ones with petioles longer than leaflets, the upper ones short-petioled or subsessile; stipules membranous, free portion short-triangular, with subulate tip; leaflets about 1 cm or more, thickish, obovate to oblong, with cuneate base, denticulate. Peduncles 0. Heads axillary, few- to many-flowered, ovoid to obovoid, 0.7–1(–2) cm. Calyx 5–7 mm; tube leathery to ligneous, cylindrical, appressed-hairy, prominently nerved; teeth lanceolate, unequal, rigid, with 1 prominent nerve, rather spiny, the lower one somewhat longer. Corolla white or pinkish, shorter than or as long as calyx. Teeth of fruiting calyx erect and usually only the lower one recurved. Pod membranous, ovoid. Seed 1, 1.5 mm long, ovoid-oblong, reddish-brown. Fl. January–June. 2n = 10, 16.

Hab.: Mainly stony ground and rock crevices.

Gen. Distr.: British Isles, Belgium, Netherlands, Germany, Romania, Portugal, Spain, France, Italy, Yugoslavia, Albania, Greece, Crete, Bulgaria, Crimea, Turkey, Cyprus, Syria, Lebanon, Palestine, Iraq, Iran, Egypt, Libya, Tunisia, Algeria, Morocco, Canary Islands.

Selected specimens: PORTUGAL: Insula Berlanga, solo granitico pr. Peniche, 1924, *Mendonca* 2074 (HUJ). FRANCE: Montpelllier, 1905, *Thellung* 517 (HUJ); pelouses agricoles près de Lardy, env. Paris, 1845, *Kralik* 8366 (P). ITALY: Puglia Murgie di S. Elio, 1950 (G); leg. *Scarbimutti* (HUJ). GREECE: Attica: Mt. Parnas, 1933, *Guiol* 41

Plate 158. *T. scabrum* L.
(Greece : *F. Guiol* 41, HUJ).
Plant in flower and fruit; flower; fruiting branch
(Romania : *Lusing*, HUJ).

(HUJ). CRETE : 28 km E. of the Lashiti Plateau on the way to Ag. Nikolaos, 280 m, 1964, *Zohary* & *Orshan* 29406 (HUJ). RHODES : Callithaca nr. spring, rocks, slope facing the sea, 1961, *Feinbrun* 85 (HUJ). TURKEY : ad vias Smyrnae, 1827, *Fleischer* (K); pr. Adapazari, 45 km S. of Akyazi, 720 m, *Orshan* 54222 (HUJ). CYPRUS : Scarinou (distr. Larnaca), shingle in river bed, 1941, *Davis* 1691 (E). SYRIA : Région inf. du Jebel Seman, 360 m, 1908, *Haradjian* 2044 (K); Mt. Hermon, 1 hour from Shibba, ca. 1,100 m, 1924, *Eig* (HUJ). LEBANON : Cimetière musulman aux Tell, bords du chemin Tripoli, 1854, *Blanche* 1188 (G). PALESTINE : Acco Plain near Kishon River, sandy soil, 1962, *Zohary* (HUJ). IRAN : Lahijan, Ghilan, 1937, *Lindsay* 817 (K). ALGERIA : Alger, 1840, *Fauché* 595 (G). MOROCCO : Jebel Hadid pres de Mogador, 1867, *Balansa* (G). CANARY ISLANDS : Palma, Caldera betw. Tenerra and Taburiente, 1913, *Sprague* & *Hutchinson*, 440 (HUJ).

Note : Only slight variations have been observed in *T. scabrum*. However, the dimensions of the head vary from 0.5–1.5(–2) cm in fruit. Larger heads are found together with small ones in the same specimen. One of the most striking differences between *T. scabrum* and its related species is the lanceolate, relatively broad and lignified calyx teeth that remain erect or spread slightly sidewards; only the anterior one is often deflexed.

163. T. lucanicum Gasp. ex Guss., Fl. Sic. Prodr. 2 : 494 (1828). *Typus* : Sicily, Monte delle piuna e della Rosa, *Gasparrini* (n. v.).

T. dalmaticum sensu Boiss., Fl. 2 : 131 (1872) et sensu Godr. in Gren. & Godr., Fl. Fr. 1 : 411 (1848–49) non Vis. (1829).
T. compactum Post, Fl. Syr. Pal. Sin. 239 (1883–96) et ed. 2, 1 : 341 (1932); Hossain, Not. Roy. Bot. Gard. Edinb. 23 : 431 (1961) p. p.
T. scabrum L. var. *majus* Gib. & Belli, Mem. Accad. Sci. Torino ser. 2, 39 : 287 (1889).
T. scabrum L. subsp. *lucanicum* (Gasp. ex Guss.) Rouy, Fl. Fr. 5 : 109 (1899).
T. scabrum L. var. *lucanicum* (Gasp. ex Guss.) Halác., Consp. Fl. Gr. 1 : 391 (1901).

Icon. : [Plate 159].

Annual, appressed-hairy, patulous-pubescent below, 10–50 cm. Stems many, erect or ascending, furrowed, sparingly branching. Lower leaves long-petioled, upper ones short-petioled to subsessile; stipules ovate-oblong, adnate to petioles for up to half of their length, membranous between prominent nerves, free portion triangular-subulate, green, hirsute; leaflets varying from base to apex, those of lower leaves broad-obovate to almost orbicular, those of upper ones oblong-cuneate to elliptical, all with thick nerves, recurved near margin, denticulate or mucronulate. Heads 1–2 × 1 cm, somewhat elongating in fruit, axillary and terminal, the latter often in pairs, unequal in size, each head involucrate by upper, broadly stipulate leaf. Flowers many, dense, 7–9 mm. Calyx tube cylindrical, with prominent nerves which thicken considerably in fruit, more or less patulous-hairy between nerves; teeth subulate, unequal, the lower one longer, the 2 upper ones shortest, all densely patulous-hirsute up to blunt tip, which terminates in tuft of hairs (in *T. scabrum* teeth are sharp-pointed, lanceolate, appressed-hairy and nerves of tube are less prominent); throat of fruiting calyx narrowed (but not closed) by hairy, annular callosity. Corolla flesh-coloured (when dry), usually longer than calyx (in *T. scabrum* often shorter or as long as calyx); standard much longer than wings and keel. Fruiting calyx gibbous or thickened at throat; teeth much broadened, deflexed or stellately spreading. Pod

Plate 159. *T. lucanicum* Gasp. ex Guss.
(Italy : *Adr. Fiori* 2891, HUJ).
Plant in flower; flower.

membranous, indehiscent or irregularly rupturing. Seeds about 1.5 mm, oblong, brown. Dispersal by separation of entire heads. Fl. April–June.

Hab.: Dry places, rocky ground, etc.

Gen. Distr.: France, Italy, Balkan Peninsula, Turkey.

Selected specimens: ITALY: Puglia, 1841, *Gussone* (G). YUGOSLAVIA: Dalmatia, 1840, *Pichler* (G) (this specimen has been compared with an authentic specimen of this species by Gibelli & Belli). TURKEY: Distr. Constantinopole, env. of Pendik, 1906, *Aznavour* (G); Elmali, sur les collines, 1860, *Bourgeau* (G); distr. Malatya, 20–30 km W. of Malatya, 1,600 m, 1963, *Zohary* et al. 27521 (HUJ).

Note: *T. lucanicum* Gasp. ex Guss. differs from *T. scabrum* in the following characters: heads larger with broader base (not cuneate); generally many more and somewhat longer flowers per head; flowering calyx with cylindrical tube and subulate teeth, hirsute and ciliate all along and terminating in tuft of stiff hairs (not lanceolate and usually terminating in naked point as in *T. scabrum*); fruiting calyx tube prominently nerved, hairy between nerves (and not obscurely nerved and hairy throughout or rarely glabrous as in *T. scabrum*); teeth less indurated, all or part of them reflexed or spreading; corolla generally somewhat longer than calyx. Geographically this is a N.E. Mediterranean species and does not reach the southern border of Turkey.
Despite these differences, there is a series of forms linking *T. lucanicum* with *T. scabrum*, so that the specific rank of the former is retained here only tentatively.

164. T. dalmaticum Vis., Pl. Rar. Dalm., in Flora 12, Ergänzungsbl. 1 : 21, no. 31 (1829) et Fl. Dalm. 3 : 293 (1850–52) non sensu Ten. (1835) nec sensu Boiss. (1872). *Typus*: (Dalmatia) leg. *Alschinger* & *Freyn* (fide Gib. & Belli, *loc. cit.* 291, PAD).

T. maculatum Host, Fl. Austr. 2 : 373 (1831).
T. rotundifolium Bory & Chaub., Nouv. Fl. Pelop. 50 (1838) non Sm. in Sibth. & Sm. (1833).
T. filicaule Boiss. & Heldr. in Boiss., Diagn. ser. 1, 9 : 24 (1849).
T. scabrum L. subsp. *dalmaticum* (Vis.) Gib. & Bellli, Mem. Accad. Sci. Torino ser. 2, 39 : 292 (1889).

Icon.: [Plate 160]. Vis., Fl. Dalm., *loc. cit.* t. 45 (very poor drawing).

Annual, sparingly patulous-pubescent in lower part and appressed-hairy in upper part, 10–25 cm. Stems furrowed, mostly procumbent, with ascending branches. Lower leaves long-petioled, those at middle and apex of stem sessile or subsessile; stipules oblong-linear, membranous, with prominent nerves, adnate to petioles for over half of their length, abruptly terminating in subulate, green, plumose free portion; leaflets up to 1–1.3 × 0.6–0.8 cm, cuneate, denticulate, apiculate with thick nerves recurved near margin, those of lower leaves obovate, those of upper ones obovate-oblong. Heads all or mostly terminal, 1–2(–3) × 0.8–1 cm, involucrate by upper leaves. Flowers many, 1–1.2 cm. Calyx tube with 10 prominent nerves, hairy or with tuft of hairs near base; calyx teeth sparsely hirsute, lanceolate-subulate to narrow-linear, rigid, unequal, the upper 2 almost as long as tube, the lower one

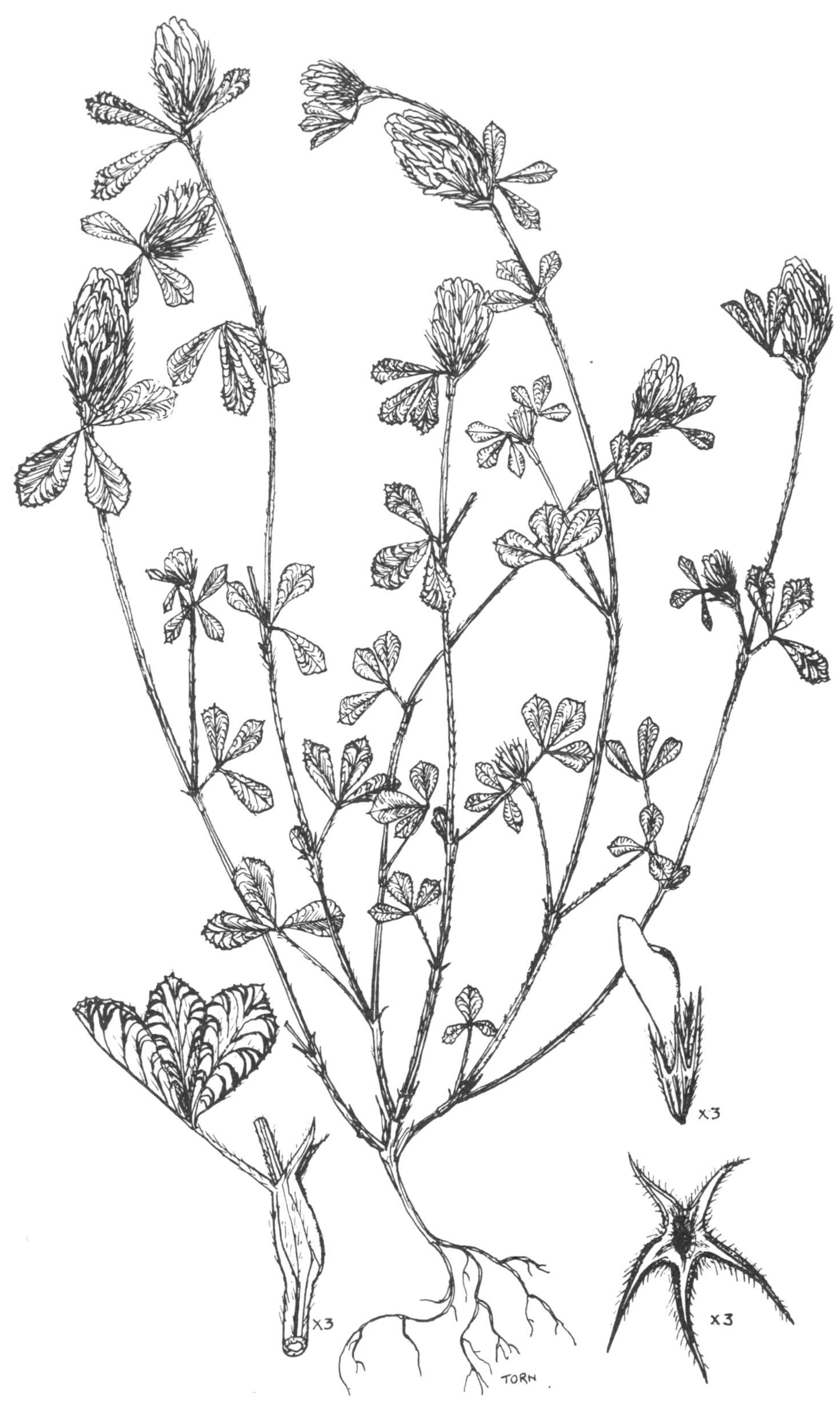

Plate 160. *T. dalmaticum* Vis.
Plant in flower; leaf (enlarged); flower; fruiting calyx from above.

much longer; fruiting calyx increasing slightly in size; teeth indurating, stellately spreading; throat narrowed by callous ring (not closed). Corolla pink or pale flesh-coloured, 1.5–2 times as long as calyx; standard longer than wings. Pod ellipsoidal, somewhat compressed, membranous, irregularly torn, with thickened apex. Seed 1–1.5 mm long, solitary, obovoid. Fl. April–June. 2n = 10.

The above description was made from a specimen which was compared by Gibelli & Belli with the authentic specimens of Visiani's collection.

Hab.: Grassy places.

Gen. Distr.: Balkan Peninsula.

Selected specimens: YUGOSLAVIA: Serbia in pascuis m. Pljackovica, 800 m, 1896, *Adamović* (HUJ); Bosnia, Sehovica pr. Aphoda-Sarajevo, 570 m, 1904, *Maly* (HUJ); Dalmatia, Cattaro, *Pichler* (G). GREECE: Thessalia, Malakasi in vineis, 1896, *Freyn* 625 (HUJ); Taygeti, au dessous du village Malta, 1894, *Heldreich* 199 (G, holotype of *T. filicaule* Boiss. & Heldr.); Laconia boreo-occidentalis, 1896, *Zahn* 1318 (HUJ).

Note: The three species of the *Scabroidea* group are very close to one another and probably exchange genes in their overlapping areas. Although it is quite easy to tell them apart at a first glance, distinction according to qualitative characters is not always feasible because of the transitional forms between them. The whole group requires a more thorough experimental study.

Subsection IX. PHLEOIDEA Gib. & Belli

Mem. Accad. Sci. Torino ser. 2, 39: 274 (1889) pro **stirpe**

Annuals. Calyx 10-nerved; teeth broadened at base, spreading in fruit; throat open but narrowed by callose or ciliate ring. Corolla persistent, not longer or only slightly longer than calyx.

165. T. gemellum Pourr. ex Willd., Sp. Pl. 3: 1376 (1802).

Annual with patulous or antrorse hairs, 10–35 cm. Stems diffuse, erect or ascending, sparsely branching. Leaves small, all except uppermost ones long-petioled; stipules membranous, green-nerved, hairy, those of middle leaves up to 1 cm long, adnate to petioles for up to two-thirds of their length, free portion lanceolate, acuminate; leaflets villous, 0.6–1.5 cm, obovate to oblong-cuneate, denticulate at apex, rounded or truncate or setuse at tip. Heads terminal, 0.8–1.8 × 0.6–1 cm, ovoid to oblong, involucrate by upper leaves, sessile or very short-peduncled. Flowers many, dense. Calyx villous, campanulate, prominently nerved, 5–7 mm long; tube much shorter to longer than almost equal, subulate, setaceous, plumose, 1-nerved teeth; throat open but provided with annular, hispid or glabrous thickening. Corolla marcescent, pink, considerably shorter than calyx, rarely somewhat longer. Pod ovoid, compressed, membranous, 1-seeded. Seed almost globular, 1 mm, yellow to brown. Fl. May–June. 2n = 14.

Hab. : Dry hills, fields, meadows, etc.

Gen. Distr. : Portugal, Spain, Algeria, Morocco.

1. Branches terminating in tufts of leaves almost concealing spherical heads. Throat of calyx open, glabrous. **165d.** var. **phyllocephalum**
– Not as above 2
2. Corolla longer than calyx. All calyx teeth shorter than tube. **165c** var. **atlanticum**
– Corolla shorter than calyx. Calyx teeth as long as or longer than tube 3
3. Heads greyish-white. Calyx very densely long-hispid; throat of calyx with narrow annular callosity. **165b.** var. **leucocephalum**
– Heads greenish. Calyx short-haired; throat of calyx with narrow annular callosity. **165a.** var. **gemellum**

165a. T. gemellum Pourr. ex Willd. Var. **gemellum**. *Typus* : Habitat in Hispania, *Pourret* 228 (B holo.).

T. sphaerocephalum Coss. in Schedae non Desf. (1799) nec. *T. phleoides* Pourr. ex Willd. subsp. *gemellum* sensu Gib. & Belli, Mem. Accad. Sci. Torino ser. 2, 39 : 282 (1889) nec. *T. phleoides* Pourr. ex Willd. subsp. *pseudogemellum* Thell. in Aschers. & Graebn., Syn. Mitteleur. Fl. 6/2 : 539 (1908)

Icon. : [Plate 161]. C. Vicioso, Rev. Gen. Trif., in Anal. Inst. Bot. Cavanilles 11 : t. 11 (1953).

Selected specimens : SPAIN : Sierra Morena, La Carolina, 1851–2, *Lange* (P). ALGERIA : Prov. d'Oran, montagnes à Thlemcen, 1842, *Dueieu de Maisonneuve* (K); pelouses du Djebel Ksel, près Geryville (El Biad), sud de la prov. d'Oran, 1856, *Kralik* 218 (W).

Note : The specimens from Thlemcen are erroneously labelled *T. sphaerocephalum* Desf., this name being a synonym of *T. cherleri* L. Although the above specimens differ markedly from typical *T. gemellum* by their spherical heads, they are tentatively included within this variety.

165b. T. gemellum Pourr. ex Willd. Var. **leucocephalum** Zoh., Candollea 27 : 140 (1972). *Typus* : Spain : Madrid, campos de la Monelra, 1954, *Borja* (HUJ holo).

165c. T. gemellum Pourr. ex Willd. Var. **atlanticum** (Ball) Maire in Jah. & Maire, Cat. Pl. Maroc 2 : 391 (1932).

T. atlanticum Ball, Journ. Bot. 11 : 306 (1873).

Specimens seen : MOROCCO : in pascuis Atlantis medii ca. Iphane, solo calcareo, 1,600 m, 1936, *Maire* (K).

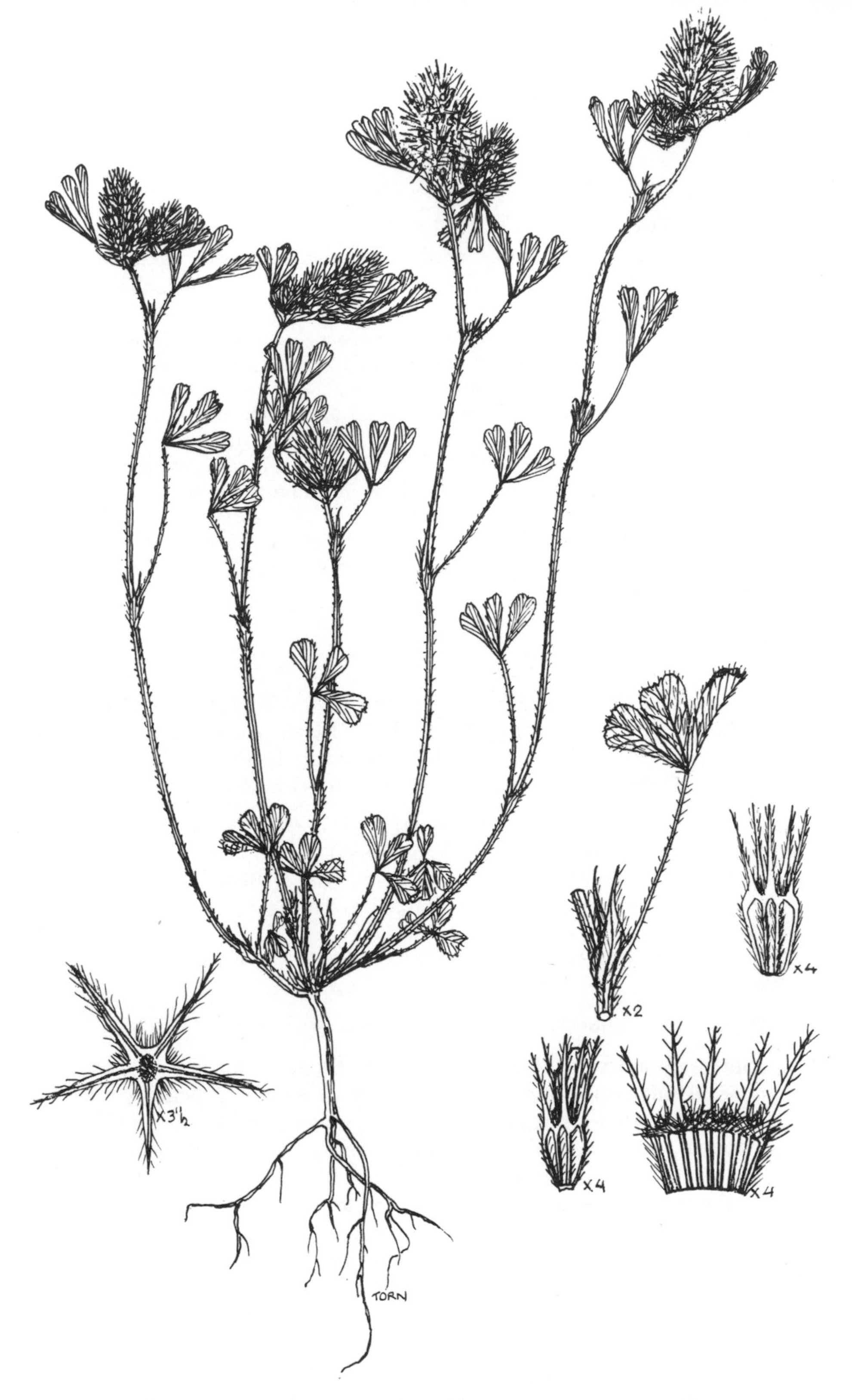

Plate 161. *T. gemellum* Pourr. ex Willd. Var. *gemellum*
Plant in flower; fruiting calyx from above; leaf (enlarged); fruiting calyx; flower; dissected calyx.

165d. T. gemellum Pourr. ex Willd. Var. **phyllocephalum** Zoh., *loc. cit.* *Typus*: Algérie; pelouse à la base nord du Djebel Tongsur, près Balara, 1853, *Balansa* 930 (HUJ holo).

Note: *T. gemellum* var. *gemellum* is a rather heterogeneous taxon. It varies chiefly in the following characters. The heads are spherical to ovoid and oblong. (A form with exclusively spherical heads, represented in several herbaria under *T. sphaerocephalum* Coss., should perhaps be separated as an independent variety.) The indumentum of the calyx in this species varies from hirsute to hispid, the latter giving the head a greyish-white colour. This form has been recorded here as var. *leucocephalum*. The throat of the calyx is constantly provided with a callose ring that varies greatly in thickness and indumentum (from hispid to glabrous). The teeth of the calyx vary considerably in their proportions. The corolla is not longer than the calyx except in var. *atlanticum*. The length of the peduncles varies likewise, but this character is taxonomically unreliable. *T. gemellum* merits much more intensive study. The above division is only tentative.

166. T. phleoides Pourr. ex Willd., Sp. Pl. 3 : 1377 (1802); Willk. & Lange, Prodr. Fl. Hisp. 3 : 370 (1877); Gib. & Belli, Mem. Accad. Sci. Torino ser. 2, 39 : 279 (1889); C. Vicioso, Rev. Gen. Trif., in Anal. Inst. Bot. Cavanilles 11 : 289 (1953). *Typus*: Habitat in Hispania, *Pourret* 230 (B holo.).

T. erinaceum M. B., Fl. Taur.-Cauc. 3 : 510 (1819).
T. minae Lojac., Nuov. Giorn. Bot. It. 15 : 262 (1883).
T. phleoides Pourr. ex Willd. subsp. *gemellum* sensu Gib. & Belli, *loc. cit.* 282.
T. phleoides Pourr. subsp. *audigieri* Fouc., Bull. Soc. Bot. France 47 : 89 (1890).
T. willkommii Chab., Bull. Herb. Boiss. 3 : 145 (1895).
T. phleoides Pourr. subsp. *pseudogemellum* Thell. in Aschers. & Graebn., Syn. Mitteleur. Fl. 6/2 : 539 (1908).

Icon: [Plate 162]. Fouc., *loc. cit.* t. 3; Vicioso, *loc. cit.* t. 12.

Annual, appressed-pubescent to glabrescent, 10–40 cm. Stems few or many, erect or ascending, branching from base. Leaves varying in length of petioles from base to apex, all alternate; stipules membranous, pilose, oblong, adnate to petioles for up to half their length, free portion green, lanceolate-subulate; leaflets of lower leaves obovate, retuse, those of upper ones elliptical to oblong-linear, (0.8–) 2 × 0.3(–0.5) cm, mostly slightly cuneate, denticulate, mucronate. Heads usually terminal, solitary, 1–2 × 0.6–1 cm, ovoid to oblong or cylindrical, mostly long-peduncled, rarely subsessile. Flowers many, dense, small, 4(–5–6) mm. Calyx tube with 10 prominent nerves, campanulate, about 2–3 mm long, green, antrorsely hairy, constricted below throat; teeth almost equal, as long as or somewhat shorter than tube, triangular-lanceolate, broadly membranous at both margins of base, erect or divergent in flower, often stellately spreading in fruit, displaying broad, collar-like seam around base; throat of calyx with glabrous or hairy, very slightly callous ring. Corolla pale pink, as long as to much shorter than calyx. Pod obovoid, membranous, with cartilaginous apex. Seed 1, about 1 mm long, brown. Fl. April–May. 2n = 14.

Hab.: Grassy places, forest clearings, among shrubs.

Plate 162. *T. phleoides* Pourr. ex Willd.
Plant in flower; leaf (enlarged); fruiting calyx from above; dissected calyx; flower; fruiting calyx.

Gen. Distr. : Germany, Portugal, Spain, France, Corsica, Sardinia, S. Italy, Sicily, Greece, Bulgaria, Crimea, Turkey, N. Iran, Caucasus, Algeria, Morocco.

Selected specimens : SPAIN : Monte Lucio in apricis pratis, *Bavo* (W); lieux herbeux du bois de la Sierra de Segura, 1850, *Bourgeau* 899 (P, under *T. gemellum*); El Escorial, 1876, *Hackel* (HUJ). ITALY : Lucania, Potenza in silva Pallareta, loco Monte-Grosso, 900–1,200 m, 1925, *Gavioli* 2892 (HUJ). FRANCE : Blois, les Ponts Chartrains, 1871, *Franchet* (P). SICILY : in apricis pratis montosis, Piana dei Greci, *Todaro* 289 (P). BULGARIA : M. Ljulin in pascuis saxosis loco Gradiste, 900 m, 1951, *Achtarov* et al. 163 (HUJ). TURKEY : N.E. Anatolia: Szandschak Gümüschkhane, Moaldasdagh in pratis silvat., 1894, *Sintenis* 5964 (W); Phrygia, Sultandagh, in subalpinis, 1,600 m, 1899, *Bornmüller* 4332 (G). ALGERIA : Oran, dans les clairières de broussailles, 1852, *Balansa* 611 (W). MOROCCO : Dj. Habibi, 1910–14, *Gandoger* (W); El Hajeb to Ifrane, Forest of Jaba, 1,100 m, 1961, *J. J. F. E.* et al. 2401 (HUJ).

Note : Although *T. phleoides* is clearly distinguished from *T. gemellum*, we found some specimens which could be regarded as intermediates between them. The transitions are mainly manifested in the occurrence of a membranous margin at the base of the calyx teeth or in the thickness of the callosity. Especially variable is the shape of the leaflets, and it is perhaps feasible to divide the plants into forms : those with lanceolate leaflets and those with obovate to elliptical leaflets. The length of the peduncles also varies.

167. T. ligusticum Balb. ex Loisel., Fl. Gall. 731 (1807); Boiss., Fl. 2 : 120 (1872); Gib. & Belli, Mem. Accad. Sci. Torino ser. 2, 39 : 283, t. 1, f. 7 (1889); C. Vicioso, Rev. Gen. Trif., in Anal. Inst. Bot. Cavanilles 11 : 305 (1953). *Typus* : Italy, habitat in Liguria, circa Savonem.

T. gemellum Savi, Atti Accad. It. 1 : 202 (1808) non Pourr. ex Willd. (1802) nec. Ser. (1825).
T. arrectisetum Brot., Phytogr. Lusit. 1 : 152, t. 63, f. 1 (1816).
T. conicum Lag., Gen. Sp. Pl. 23 : 305 (1816) non Pers. (1810).
T. aristatum Willd. ex Link., Enum. Hort. Berol. Alt. 2 : 262 (1822) pro syn.
T. broteri Link, *loc. cit.*, pro syn.
T. lagascanum Ser. in DC., Prodr. 2 : 194 (1825).

Icon. : [Plate 163]. C. Vicioso, *loc. cit.* t. 13; Reichenb., Icon. 22 : t. 2153, f. 1, 1–7 (1903).

Annual, patulous-pubescent, 15–50 cm. Stems few, mostly dark green or brown, erect or ascending or decumbent, poorly branching, leafy. Leaves rather large, most of them petioled, the uppermost ones sessile, all alternate; stipules membranous, pilose, about 1 cm long, oblong-triangular, adnate to petioles up to the middle, the upper portion lanceolate-setaceous; leaflets 0.6–2 × 0.3–1(–1.5) cm, pubescent, all obovate, tapering at base, often emarginate and obscurely denticulate in upper part. Heads (0.8–)1–2(–3) × 0.8–1.4 cm, ellipsoidal or ovoid, mostly long-peduncled, many-flowered, solitary, rarely in pairs due to falsely dichotomous branching. Flowers 5-7 mm long, short-pedicelled. Calyx tube 2 mm, campanulate, with 10 prominent nerves and coarsely hairy all over; teeth 4–5 mm long, equal in length, setaceous, with very broad-triangular base and broad-membranous margins, often

Plate 163. *T. ligusticum* Balb. ex Loisel.
(Algeria : *d'Alleizette*, HUJ).
Plant in flower and fruit; leaf (enlarged); dissected calyx; flower; fruiting calyx from above.

blackish, erect or divergent, but not or only very rarely stellately spreading; throat of calyx narrowed by annular callosity. Corolla pale pink, shorter than calyx. Pod ovoid, pyriform, membranous. Seed 1, spherical, brown. Fl. May–July 2n = 12.

Hab. : Fields, waste plots, stony hillsides, etc.

Gen. Distr. : Portugal, Spain, France, Italy, Greece, Turkey, Lebanon, Tunisia, Algeria, Morocco, Madeira, Canary Islands.

Selected specimens : PORTUGAL : Castello Branco, Carvalhinha, 1885, *Ricardo da Cunha* 830 (HUJ). FRANCE : Toulon (Var), 1859, *Le Jolis* 94 (HUJ). ITALY : Roma, Insuggherata, 1888, 3504 (W). SICILY : Fucizza, in arenosis herbosis submontosis, *Todaro* 286 (HUJ). TURKEY : Paşabahçe Poloneskõy (env. Istanbul), 1939, *Post* (G). LEBANON : infra Beitmeri in Pinetis, 1884, *Peyron* 1512 (K, G). ALGERIA : Depart. Alger, env. d'Alger, forêt de la Réghaia, 150 m, 1953, *d'Alleizette* (HUJ).

Note : *T. ligusticum* differs at first glance from the other two species of Subsect. *Phleoidea* in the following characters : stem slender, sparingly branching; hairs patulous; leaflets, even those of uppermost leaves, broadly obovate; heads mostly ovoid to ellipsoidal; flowers larger (5–7 mm); calyx teeth dark, 2–3 times as long as tube.

The 3 species of Subsect. *Phleoidae* constitute a natural group distinguished by the following : the short corolla (except in *T. gemellum* var. *atlanticum*); the minute calyx tube, becoming urceolate at maturity; the callose ring (hairy or glabrous) of the open calyx throat. Further investigations will have to determine the rank of the recorded species, because they undoubtedly exchange genes in overlapping areas.

Subsection X. LAPPACEA Gib. & Belli

Mem. Accad. Sci. Torino ser. 2, 39 : 318 (1889) pro stirpe

Annuals. Calyx 20-nerved; teeth equal or almost so, with broad 3–5-nerved base or 1-nerved, erect or slightly divergent in fruit; throat of calyx open, hairy or glabrous. Corolla marcescent, slightly to considerably longer than calyx.

168. T. hirtum All., Auct. Fl. Pedem. 20 (1789); Boiss., Fl. 2 : 119 (1872); Gib. & Belli, Mem. Accad. Sci. Torino ser. 2, 39 : 321 (1889). *Typus* : Italy "In Monte serrato secus agros", *Allioni* (TO).

T. hispidum Desf., Fl. Atl. 2 : 200 (1799).
T. pictum Roth, Cat. 2 : 101 (1800).
T. hirsutum Ten., Fl. Nap. 5 : 143 (1836–38) incl. var. *pictum* (Roth) Ten., *loc. cit.*
T. oxypetasum Heldr. & Sart. in Orph., Fl. Graec. exsicc. 320.

Icon. : [Plate 164]. Desf., *loc. cit.* t. 209, f. 1; Reichenb., Icon. 22 : t. 2147, f. 1, 1–7 (1903); Gib. & Belli, *loc. cit.* t. 4, f. 3.

Annual, patulous- or retrorsely villous, 10–20 cm. Stems few, striate, reddish, ascending, arcuate or straight, branching mostly from base, leafy (especially at base).

Plate 164. *T. hirtum* All.
(Greece : *I. Deml*, HUJ).
Plant in flower; dissected calyx; fruiting calyx from above; flower; fruiting calyx.

Leaves petioled, except for uppermost ones; stipules membranous, green-nerved, lanceolate, 1–1.5 cm, adnate to petioles for one-third of their length, free portion long-subulate, ciliate-villous; leaflets 0.8–1.5 × 0.6–1 cm, appressed-hairy, obovate to oblong-cuneate, denticulate in upper part, obtuse to retuse at apex. Heads globular, about 2 cm across, white-villous, solitary, involucrate by very broad, almost orbicular stipules of reduced leaves or also developed ones, with green nerves and margins and acuminate apex. Flowers 1.5–1.7 cm long, dense. Calyx tube obconical, densely covered with appressed, shiny hairs which conceal the 20 nerves (visible in fruiting calyces); teeth setiform, 1-nerved, almost equal, plumose, blunt, 2–3 times as long as tube. Corolla purple, exceeding calyx in length; standard lanceolate, acute. Style adnate up to middle to staminal tube. Fruiting calyx increasing in width and separating at maturity from axis (in contrast to *T. cherleri* in which dispersal unit is entire head); throat and inner face of fruiting calyx tube hairy. Pod obovoid, membranous with somewhat leathery apex. Seed 1, 2–2.2 mm long, yellow. Fl. April–June. 2n = 10.

Hab.: Fields, roadsides and dwarf-shrub formations.

Gen. Distr.: Portugal, Spain, France, Corsica, Italy, Balkan Peninsula, Crete, Crimea, Turkey, Cyprus, Syria, Lebanon, Iraq, Iran, Caucasus, Tunisia, Algeria, Morocco.

Selected specimens: PORTUGAL: Monte, pr. Monte du S. Bartolomew nos Arredores de Bragança, 1958, *Fernandes* et al. 6328 (HUJ). SPAIN: Morena, Puerto Despenaperas, 1876, *Hackel* (W). FRANCE: Bords des chemins des terrains schisteux au Vigau (Gard), 1862, *Tuezkiewicz* 3358 (G). CORSICA: Niolo, 1914, *Ronniger* (E). ITALY: Pavia, 1843, *Pinard* (W). GREECE: Agrapha (Dolopia veterum): in regione inferiori m. Pindi ca. monast. Korona in nemorosis quercinis, 3,500–3,700 ft, 1885, *Heldreich* (W). BULGARIA: in desertis ad Nova Mahala, 1833, *Stribrny* (G); Kalofer auf Wiesen, 1874, *Pichler* 159 (W). CRIMEA: Tauria, leg. *Parreulis* 147 (W); Yalta, 1898, *Golde* 313 (K). TURKEY: Izmir: inter arboreta Smyrnae, 1827, *Fleischer* (K, as *T. hispidum* Desf.); prov. Ankara: Hacikadin valley nr. Keçioren, 1952, *Davis* & *Dodds* 18729 (E). SYRIA: Nussairy Mts., Ain Halakim, 2,500–3,000 ft, 1910, *Haradjian* 3504 (W). LEBANON: Beitmeri, 1879, *Peyron* 39 (G); Beirut, 1900, Delessert (G). IRAQ: Kurdistan, in montis Kuh Safin reg. pag. Schaklava (dit. Erbil), 800 m, 1893, *Bornmüller* 1125 (K). CAUCASUS: prope Helenendorf, 1836, *Hohenacker* (W). MOROCCO: Tanger (G).

169. T. cherleri L., Demonstr. Pl. 21 (1753) et Amoen. Acad. 3 : 418 (1756); Boiss., Fl. 2 : 119 (1872). *Typus*: Described from a plant grown at Uppsala.

T. sphaerocephalum Desf., Fl. Atl. 2 : 201, t. 209, f. 2 (1799).
T. phlebocalyx Fenzl in Tchihat., As. Min. Bot. 1 : 29 (1860).
T. hirtum All. subsp. *cherleri* (L.) Gib. & Belli, Mem. Accad. Sci. Torino ser. 2, 39 : 324 (1889).

Icon.: [Plate 165]. Moris, Fl. Sard. 1 : t. 61 (1837); Reichenb., Icon. 22 : t. 2142, f. 2, 7–15 (1903); Gib. & Belli, *loc. cit.* t. 4, f. 4; Sibth. & Sm., Fl. Gr. 8 : t. 745 (1833).

Plate 165. *T. cherleri* L.
(Palestine : *U. Plitmann* 37, HUJ).
Plant in flower; fruiting heads from below and above; fruiting calyx; flower.

Annual, patulous-hirsute, 5–25 cm. Stems few to many, furrowed, rather thick, ascending or decumbent, poorly branching. Leaves long-petioled except upper ones which are often subsessile; stipules membranous, green-nerved, ovate-oblong, united for more than half their length, hairy, free portion short, triangular-lanceolate; leaflets obovate-cuneate, 0.6–1.2 × 0.4–0.8 cm, retuse to obcordate, denticulate in upper part. Heads sessile, many-flowered, ovoid-globular, 0.6–1.2 cm across, involucrate by 3 flat, orbicular, membranous or foliaceous, often coloured and densely hairy stipules. Flowers (0.8–)1–1.2 cm. Calyx densely hirsute; tube campanulate, 20-nerved, whitish; teeth green, more or less equal, usually blunt with triangular base, considerably longer than tube. Corolla as long as or slightly (sometimes considerably) longer (rarely shorter) than calyx, white to cream-coloured, rarely reddish; standard oblong, acute, somewhat longer than wings. Fruiting calyces adnate to rhachis at their bases and not separating at maturity; teeth erect, densely plumose; tube densely villous inside; throat without callosity. Pod membranous, ovoid, cartilaginous at apex. Seeds about 2 mm, brown. Dispersal synaptospermic, i.e., by separation of entire flattened fruiting head with its bracts from peduncle. Fl. March–May. 2n = 10.

Hab.: Damp soils, grassy places and roadsides.

Gen. Distr.: Portugal, Spain, France, Corsica, Italy, Sicily, Balkan Peninsula, Crete, Turkey, Cyprus, Syria, Lebanon, Palestine, Iraq, Iran, Egypt, Libya, Tunisia, Algeria, Morocco, Canary Islands.

Selected specimens: PORTUGAL: prope Alcantara, 1828, *Hall* (W). SPAIN: Catalogne, Vilarnadal, friches, 1907, *Sennen* 257 (HUJ). FRANCE: Var, env. d'Hyères, 1904, *d'Alleizette* (HUJ). CORSICA: Campo di Loro, 1911, *Thellung* (HUJ). ITALY: Env. de Rome, 1855, *Pittoni* (W). SICILY: Palermo, *Todaro* (W). GREECE: Saloniki, *Friedrichtal* 1012 (W); Graecia, 1936, *Kotschy* 713 (W). TURKEY: Env. of Smyrna, 1854, *Balansa* 171 (W). CYPRUS: Platres and hillsides, 1941, *Davis* 3187 (HUJ). SYRIA: in graminosis Djebel Muhassan, Aleppo, 1865, *Haussknecht* (G). LEBANON: Env. de Beyrouth, bord de la mer, 1811, *Gaillardot* 1737 (G). IRAQ: Banks of the Bigger Zab near Kirkuk, 1956, *Crigg* (K). ALGERIA: La Calle (Constantine), Memelon de la Poudrière, 1915, *Duffour* 1645 (HUJ). MOROCCO: Tanger, *Schousboe* 34 (W).

Note: *T. cherleri* varies greatly in habit (from 5 to 30 cm, decumbent to erect), in the size of the leaflets, and also in the length of the corolla which range from much shorter to much longer than the calyx.

170. **T. lappaceum** L., Sp. Pl. 768 (1753).

Annual, sparingly patulous-hirsute or glabrous, 10–30 cm. Stems many or solitary, mostly flexuous, terete, poorly branching. Lower leaves long-petioled, upper ones short-petioled or subsessile; stipules membranous between prominent green nerves, oblong, free portion lanceolate-subulate; leaflets 0.5–1.5 × 0.3–0.8 cm, obovate, cuneate at base, rounded or truncate, rarely emarginate at apex, minutely dentate. Heads initially semi-globular, subsessile, then conspicuously pedunculate, globular, small, later 1–1.4 cm in diam. Calyx 6–7 mm long; tube obconical or campanulate,

20-nerved, white-haired above; teeth equal or almost so, setaceous, with triangular, 5-nerved base, long-ciliate, as long as or longer than tube. Corolla white or pink, as long as or slightly longer than calyx; standard with darker nerves above, almost as long as wings and keel. Fruiting head spinescent, with indurated calyx tube, open, glabrous or ciliate throat and erect or slightly diverging, bristly-ciliate teeth. Pod membranous with leathery operculum, ovoid, long-beaked. Seed 1, ovoid, brownish, shiny, about 1.3 mm long. Fl. March–June. 2n = 16.

Hab. : Grassy places, fields, roadsides and among shrubs.

Gen. Distr. : Azores, Portugal, Spain, France, Italy, Sicily, Yugoslavia, Greece, Crete, Rhodes, Crimea, Turkey, Cyprus, Syria, Lebanon, Palestine, Iraq, Iran, Caucasus, Egypt, Tunisia, Algeria, Morocco, Madeira, Canary Islands.

1. Throat of calyx ciliate. **170a.** var. **lappaceum**
– Throat of calyx naked. **170b.** var. **zoharyi**

170a. T. lappaceum L. Var. **lappaceum**; Boiss., Fl. 2 : 119 (1872); Gib. & Belli, Mem. Accad. Sci. Torino ser. 2, 39 : 319 (1889). *Typus* : Described from S. France, Hb. Linn. 930/28.

T. nervosum C. Presl, Fl. Sic. 1 : XXI (1826).
T. lappaceum L. subsp. *selinuntinum* Tin. ex Nyman, Consp. Fl. Europ. Suppl. 2 : 90 (1889) nom. nud.
T. lappaceum L. var. *brachyodontulum* Hausskn., Mitt. Thür. Bot. Ver. N. F. 5 : 75 (1893).
T. lappaceum L. subsp. *adrianopolitanum* Velen., Fl. Bulg. Suppl. 80 (1898).
T. rhodense Pamp., Boll. Soc. Bot. It. 142 (1925).
T. lappaceum L. var. *rhodense* (Pamp.) Rech. fil., Fl. Aeg. 368 (1942).

Icon. : [Plate 166]. Reichenb., Icon. 22 : t. 2142, f. 1, 1–6 (1903).

Selected specimens : SPAIN : Champs incultes à Lagos, Algarve, 1853, *Bourgeau* 1836 (G). FRANCE : Bassin à Priae, près de Moissac, 1844, *Lagrèze-Fossat* 835 (G). ITALY : Roma, Macchia Mallei, 1888, *Evers* 8006 (G). SICILY : in pratis Ficuzza, 1856, *Huet du Pavillon* 510 (W). YUGOSLAVIA : Istria, bei Polo, 1909, *Kob* (G). CRETE : Knossos, 1915, *Gandoger* (G); Messara Plain, 5 km E. of Mires, 1964, *Zohary* & *Orshan* 25/41 (HUJ). RHODES : Champs incultes, près Bastida, 1870, *Bourgeau* (G). TURKEY : Smyrna, *Boiss.* (G); Smyrna : Sur les collines incultes, 1854, *Balansa* 163 (G); env. de Rize, terrains argileux-calcaires, 1866, *Bal.* 171 (G). CYPRUS : Env. Rizokarpaso, 300–500 ft, 1912, *Haradjian* 205 (G). SYRIA : Mt. Cassius, 1909, *Haradjian* 2970 (G). LEBANON : Dans les cultures à l'embouchure du Sarnok, Saida, 1853, *Blanche* 582 (G). PALESTINE : Judean Mts., W. of Beit Govrin, batha, 1964, *Baum* & *Plitmann* (HUJ). IRAQ : Shaqlava, 900 m, 1947, *Gillet* 8027 (K). IRAN : Mt. Elburz, pr. Derbent, in humidis, 1843, *Kotschy* 243 (G). MOROCCO : Environs de Keira, 1867, *Balansa* (G). MADEIRA : in pascuis Gorgulho, 1865–66, *Maudon* 147 (G).

170b. T. lappaceum L. Var. **zoharyi** Eig, Bull. Inst. Agr. Nat. Hist. 6 : 24 (1927); Dinsmore in Post, Fl. Syr. Pal. Sin. ed. 2, 1 : 334 (1932).

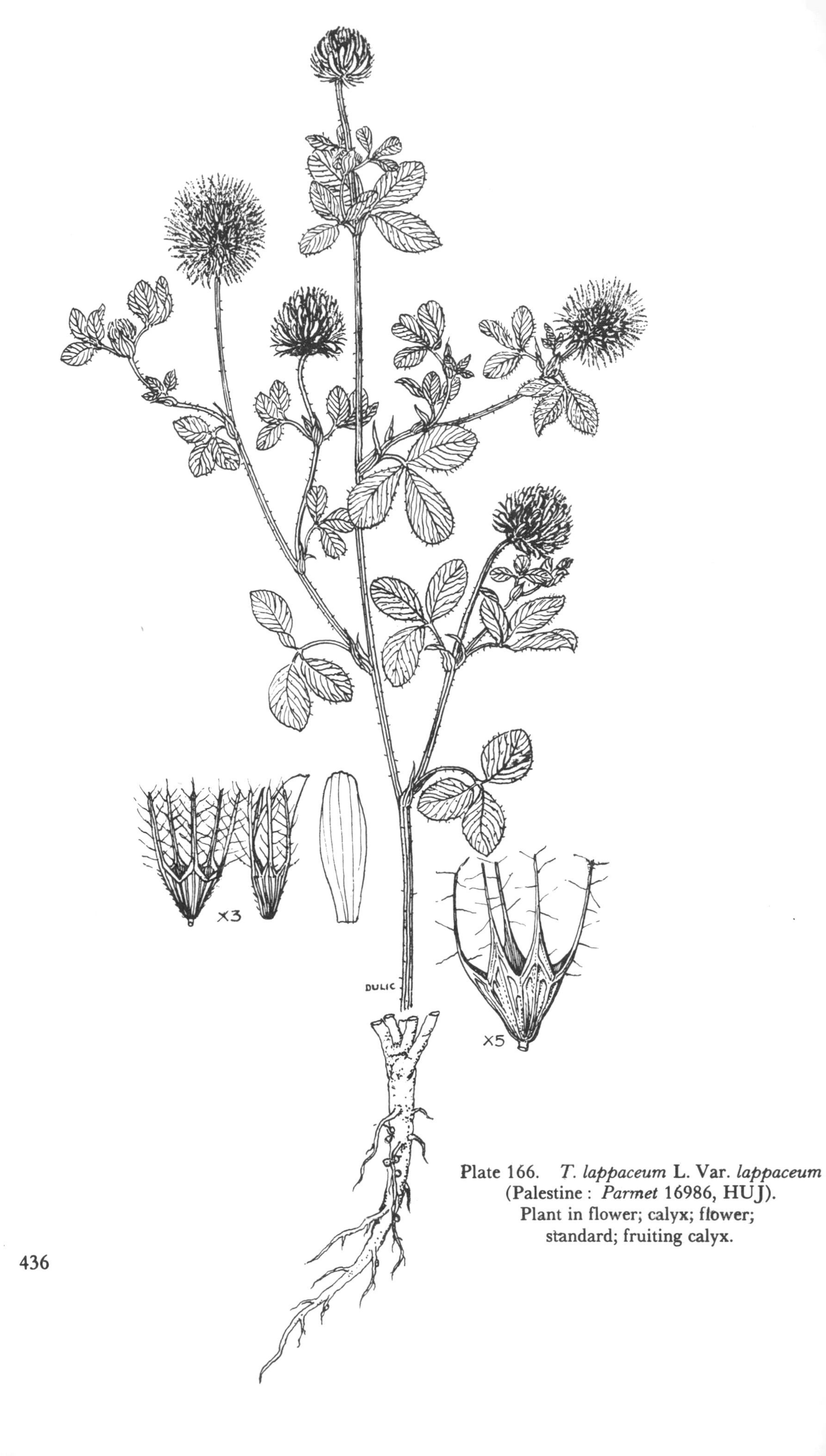

Plate 166. *T. lappaceum* L. Var. *lappaceum* (Palestine : *Parmet* 16986, HUJ). Plant in flower; calyx; flower; standard; fruiting calyx.

Selected specimens: LEBANON: Descent from Wadi Ain Dib towards Qammoua, 1934, *Bot. Dept.* (HUJ). PALESTINE: Esdraelon Plain, Kishon River, east of Mishmar Haemeq, field border, 1963, *Zohary* & *Plitmann* 1522522 (HUJ). IRAN: Derbent i Bassian, 1867, *Haussknecht* 273 (G). EGYPT: Sinai: El Arish, wadi bed, 1925, *Eig* (HUJ).

171. T. barbeyi Gib. & Belli, Atti R. Accad. Sci. Torino 22: 610, t. 8 (1887; *"barbey")*.

Annual, dwarf, hairy, cushion-forming, up to 8 cm. Stems branching, tufted, arising from base, prostrate. Basal and lower leaves crowded, pilose, long-petioled, uppermost ones almost opposite, subtending heads; stipules linear, membranous-scarious, nerved, ciliate at margin, glabrous, with short free portion; leaflets oblong-cuneate to obovate, rounded or scarcely emarginate at apex, hirsute on both faces. Heads subglobular, subtended by stipules of uppermost leaves, short-peduncled. Flowers not numerous, at most 20–25 per head. Flowering calyx tubular-obconical, growing in fruit and turning top-shaped-campanulate; tube villous, 20-nerved; throat with densely hairy ring; teeth slightly unequal, shorter than tube, 5-nerved, broad-triangular, terminating in subulate, hirsute mucro. Corolla pink, twice as long as calyx, marcescent; limb of standard oblong-ovate. Fruiting calyx with broadened, open throat. Pod membranous below, cartilaginous above. Seed 1, subglobular, brown, smooth. Fl. June.

Note: *T. barbeyi* was recorded by Gibelli & Belli from Aegeis, Insula Karpathos, crescit in cultis leg. *Pichler* (1883) and *Forsyth-Major* (1886) (TO).
So far no other specimen of this species has been found.

172. T. congestum Guss., Cat. Pl. Hort. Reg. in Boccadif. 65, 81, n. 17 (1821); Pl. Rar. 311 (1826) et Fl. Sic. Prodr. 2: 489 (1828); Boiss., Fl. 2: 142 (1872); Lojac., Tent. Monogr. Trif. Sic. 142 (1878). *Typus*: Syntypes "in arvis argilosis angustam, Catanam caltanixettam" (FI ?).

Icon.: [Plate 167]. Gib. & Belli, Mem. Accad. Sci. Torino ser. 2, 39: t. 8, f. 1 (1889).

Annual, caespitose, appressed-pubescent to glabrous, 3–7 cm. Stems many, prostrate, rigid, very short, tortuose, overtopped by leaves and covered by withered stipules. Leaves all long-petioled; stipules ovate-triangular, membranous, adnate to petioles for more than half their length, nerved, glabrous, free portion subulate; leaflets 0.8–1.5 × 0.3–0.7 cm, obovate-cuneate, deeply notched, denticulate, villous on both surfaces with some hairs tuberculate at base. Heads semi-globular or ovoid, 1.2–1.5 cm across, short-peduncled or sessile, congested in lower parts of stem, subtended by large, ovate, acute, prominently nerved stipules. Flowers few, 1–1.3 cm long. Calyx tubular, turning campanulate in fruit, leathery, (15–)20-nerved, sparingly hirsute;

Plate 167. *T. congestum* Guss.
(Sicily : *Cetarda* 988, HUJ).
Plant in flower and fruit; fruiting calyx; same dissected.

teeth almost equal, somewhat longer than tube, subulate, with lanceolate-triangular, 3-nerved base; throat of calyx hairy but open (without callosity). Corolla white, as long as calyx, marcescent; standard oblong, notched at apex. Pod membranous, with cartilaginous apex. Seeds solitary, globular, pale brown. Fl. May–June.

Hab.: "Pianta estremamente argilofila" (Gib. & Belli, *loc. cit.*).

Gen. Distr.: S. Italy, Sicily, Malta, Balkan Peninsula, Rhodes.

Specimens seen: SICILY: Gergenti a Maccalulebi, 1882, *Todaro* (P). RHODES: Callavia nella Piano, *Fiori* 230 (FI).

Note: *T. congestum* is a basicarpic plant in which the fruiting heads remain on the plants until the latter vanish. Biologically, it belongs to the category of plants with topochoric dispersal. Some authors refer this species to Sect. *Micranthemum*. However, in view of the lack of bracts and other characteristics, it fits well into Sect. *Trifolium*.

Subsection XI. ARVENSIA Gib. & Belli

Mem. Accad. Sci. Torino ser. 2, 39: 266 (1889) pro stirpe

Calyx tube 10-nerved; teeth densely villous or plumose, not spreading in fruit; throat open (without callosity), hairy. Corolla marcescent, not longer than calyx.

173. T. arvense L., Sp. Pl. 769 (1753).

Annual, with appressed or subpatulous silky hairs, 5–30 cm. Stems solitary, few or many, delicate, flexuous, terete, erect or ascending, mostly branching above. Lower leaves petiolate, upper ones subsessile; stipules membranous between nerves, ovate-oblong, with long-cuspidate tip; leaflets 1–2 × 0.2–0.3(–0.4) cm, linear-oblong to narrow-elliptical with cuneate base, mucronate at apex. Heads axillary and terminal, borne on rather short peduncles, ovoid to cylindrical, 1–2 cm long, elongating in fruit. Calyx subpatulous-hairy, about half the length of equal, setaceous, plumose, often purplish or pink teeth; throat villous, not closed by callosity. Corolla white or pink, persistent, much shorter than calyx; standard obtuse, narrow. Pod membranous, ovoid. Seed 1, 1 mm, globular, yellow. Fl. March–May. 2n = 14, 28.

Hab.: Meadows, grassy places, fields, roadsides, wastes, in open, pioneer plant communities.

Gen. Distr.: British Isles, Denmark, Scandinavian Peninsula, Finland, Siberia, Belgium, Germany, Switzerland, Romania, Portugal, Spain, France, Italy, Balkan Peninsula, S. Russia, Turkey, Cyprus, Syria, Lebanon, Palestine, Iran, Iraq, Egypt, Libya, Tunisia, Algeria, Morocco, Canary Islands, Ethiopia, etc. Also adventive in many other countries.

Note: *T. arvense* is exceedingly polymorphic; most of the presently known forms and varieties are doubtful as to constancy; some grade largely into one another. The following two varieties are tentatively regarded as covering the entire range of morphologic diversity which yet awaits experimental study :

1. Rather robust, mostly erect plants, more or less densely appressed- or patulous-hairy. Calyx 4.5–7 mm, densely hairy; teeth 2–3 times as long as tube, plumose-hairy. **173a.** var. **arvense**
– Delicate, mostly decumbent plants with glabrous or glabrescent stems. Calyx 3–5 mm, glabrous or sparingly hairy; teeth mostly as long as tube (rarely twice as long), often reddish and glabrous or glabrescent. **173b.** var. **gracile**

173a. T. arvense L. Var. **arvense**; Gib. & Belli, Mem. Accad. Sci. Torino ser. 2, 39 : 266 (1889). *Typus* : Described from Europe, Hb. Cliff.

T. arvense L. var. *crassicaule*, var. *perpusillum*, var. *capitatum* Ser. in DC., Prodr. 2 : 191 (1825).
T. brittingeri Weitenw. in Opiz, Naturalientausch 9 : 142 (1825).
T. arvense L. var. *australe* Ten., Fl. Nap. 5 : 141 (1835–38).
T. arvense L. var. *strictior* Mert. & Koch in Röhling, Deutschl. Fl. 5 : 270 (1838).
T. eriocephalum Ledeb., Fl. Ross. 1 : 541 (1843).
T. arvense L. var. *aetnense* Guss., Fl. Sic. Syn. 2 : 236 (1844).
T. lagopinum Jord., Pugill. 57 (1852).
T. agrestinum Jord. ex Boreau, Fl. Centr. Fr. ed. 3, 2 : 153 (1857).
T. sabuletorum Jord. ex Boreau, *loc. cit.*
T. arenivagum Jord. ex Boreau, *loc. cit.*
T. longisetum Boiss. & Bal. in Boiss., Diagn. ser. 2,6 : 47 (1859).
T. arvense L. var. *longisetum* (Boiss. & Bal.) Boiss., Fl. 2 : 120 (1872).
T. arvense L. var. *brachyodon* Čelak., Prodr. Fl. Böhm. 907 (1881).
T. capitulatum Pau in Not. Bot. Fi. Esp. 1 : 9 (1887).
T. arvense L. subsp. "*longisetosum* (Boiss. & Bal.)" Rouy, Fl. Fr. 5 : 104 (1899).
T. arvense L. subsp. *brittingeri* (Weitenw.) Koch var. *maritimum* (Corb.) Rouy, *loc. cit.* 105; Aschers. & Graebn., *loc. cit.* 533.
T. arvense L. subsp. *agrestinum* (Jord. ex Boreau) Rouy var. *alopecuroides* Rouy, *loc. cit.* 106.
T. arvense L. var. *typicum* Beck ex Aschers. & Graebn., Syn. Mitteleur. Fl. 6/2 : 531 (1908).
T. arvense L. var. *latifolium* Pantu, Contr. Fl. Bucur. in Anal. Acad. Rom. Mem. Sect. Sti. ser. 2, 32 : 145 (1910).
T. arvense L. var. *ballii* Murb., Contr. Fl. Maroc. 1 : 62 (1922).
T. arvense L. var. *cyrenaicum* Pamp., Nuov. Giorn. Bot. It. n. s. 31 : 217 (1924).

Icon. : [Plate 168]. Reichenb., Icon. 22 : t. 2146, f. 1–2, 1–11 (1903); Ross-Craig, Draw. Brit. Pl. 7 : pl. 29 (1954).

Distr. : Throughout the area of the species.

Selected specimens : DENMARK : Sjaelland, Orsløv, 1866, *Nielsen* (HUJ). SWEDEN : Uppland Ultuna, vagkant, 1941, *Johansson* (HUJ). FINLAND : Nylandia, par. Teuda Lapvik, 1907, *Ström* 762 (G). CZECHOSLOVAKIA : Bohemia merid. distr. Pelhrimar in pascuis, etc., 600 m, 1962, *Ujcik* 194 (HUJ). ROMANIA : Bucuresti raion Stalin, in foenetis siccis ad marginem septentr. opp. Bucuresti, 1955, *Morlova* et al. 83 (HUJ).

Plate 168. *T. arvense* L. Var. *arvense*
(Palestine : *Naftolsky* 16568, HUJ).
Plant in flower and fruit; flower.

RUSSIA: Uchtomsk nr. Moscow, 1927, *Eig* (HUJ). PORTUGAL: S. Caldeiras, 1905, *Gandoger* (G). SPAIN: Riberas del rio Tajo: Alconetor, 1948, *Rivas-Galiano* (HUJ). ITALY: in incultis pr. Veglia, 70 m, 1920, *Lusina* (HUJ). GREECE: Nisyros, steep uncultivated ground nr. Mandraka, 1963, *Gathorne-Hardy* 365 (E); Attica: Mt. Parnes-Keramindi, 1928, *Guiol* 24 (HUJ). BULGARIA: M. Vitosa in graminosis saxosis apricis, 1,500 m, 1951, *Efremov* 48 (HUJ). CRETE: 47 km S.E. of Iraklion, 610 m, 1964, *Zohary* & *Orshan* 29403 (HUJ). TURKEY: Prov. Çoruh (Artvin) above Borçka, 300 m, shady banks, 1958, *Davis* & *Hedge* D.29864 (E). CYPRUS: Mts. Troodos, Platres, 1,300 m, 1937, *Grizi* (HUJ). SYRIA: Djebel Druz, env. of Souweida, 1,250 m, basalt soil, 1932, *Eig* & *Zohary* (HUJ). LEBANON: Mt. Lebanon, env. of Sir, 1934, *Student* (HUJ). ISRAEL: Coastal Galilee, env. of Tsahal, sandy-clay, fallow fields, 1955, *Feinbrun* 16572 (HUJ). IRAN: Hamadan, Mt. Alvand, steep slopes, granite, rocky soil, 2,250–2,550 m, 1965, *Danin, Baum* & *Plitmann* 65-672 (HUJ). ARMENIA: pr. Eilar in declivibus lapidosis, 1926, *Schelkovnikov* (HUJ). TRANSCASPIA: Kisil Arvat Karakala in valle Joldere, 1901, *Freyn* 1925 (G). LIBYA: Cyrenaica: in dumetis supra Barcam solo calcareo, 400m, 1938, *Maire* & *Weiller* 472 (P, as subsp. *cyrenaicum* Pamp.). ALGERIA: Djurdjura, pelouse vacailleurs, 1,500 m, 1930, *Maire* (P). MOROCCO: in lapidosis arenaceis Anti-Atlantic in ditione Tazerovalt, 500 m, 1934, *Maire* & *Wilczek* (P, type of var. *ballii* Murb.). CANARY ISLANDS: Palma, Al Pass lava stream, 1913, *Sprague* & *Hutchinson* 307 (HUJ).

173b. T. arvense L. Var. **gracile** (Thuill.) DC., Fl. Fr. ed. 3, 4: 530 (1805). *Typus*: Described from near Paris.

T. gracile Thuill., Fl. Env. Par. ed. 2: 383 (1799).
T. arvense L. var. *glabrum* Vis., Fl. Dalm. 3: 292 (1852).
T. rubellum Jord., Pugill. 57 (1852).
T. arvense L. var. *rubellum* (Jord.) Beck, Fl. N. O. 848 (1892).
T. arvense L. subsp. *gracile* (Thuill.) Rouy, Fl. Fr. 5: 107 (1899).

Selected specimens: SWITZERLAND: Tessin, pelouses sèches des terrains granitiques, 1878, *Boullu* 2025 (G). FRANCE: Env. de Lyon, champs sablonneux, 1851, *Marlin* (G). TURKEY: Trabzon, 400 ft, gritty soil, 1934, *Balls* 1570 (E).

174. T. affine C. Presl, Symb. Bot. 1: 54, t. 34 (1832) non Lejeune ex Ser. in DC., Prodr. 2: 195 (1825) pro syn. *Typus*: In Hb. Presl, sine loco (grown from seeds).

T. preslianum Boiss., Diagn. ser. 1, 2: 25 (1843) et Fl. 2: 121 (1872) subs. illegit.

Icon.: [Plate 169].

Annual, with appressed, grey pubescence, 15–25 cm. Stems many, erect or ascending, branching from base. Leaves rather short-petioled, 2–3 cm long; stipules oblong, adnate to petioles up to above middle, subulate, upper ones ovate, acute; leaflets 1–2 × 0.15–0.25 cm, lanceolate-cuneate, entire, obtuse, mucronulate. Heads wooly, pedunculate or subsessile, terminal, ovoid-cylindrical, 1–3 × 1–1.3 cm. Flowers minute, 4–6 mm. Calyx tube campanulate, wooly, 10-nerved; teeth (1–)2–3 times as long as tube, setaceous, almost equal, densely plumose; throat hairy but open. Corolla marcescent, white or pink, about as long as or slightly longer than calyx;

Plate 169. *T. affine* C. Presl
(Turkey : *M.* & *D. Zohary* 69 3, HUJ).
Plant in flower and fruit; flower; standard; wing; keel.

standard oblong, acute or obtuse, notched, longer than wings and keel. Ovary stipitate. Pod ovoid, irregularly dehiscent. Seed 1, yellowish-brown, oblong, 2 mm. Fl. April–May. 2n = 12, 16.

Hab.: In open plant communities, dry hillsides.

Gen. Distr.: Bulgaria, Turkey.

Selected specimens: BULGARIA AUSTR.: in collinis ad Skobelevo, 1898, *Stribrny* 28 (HUJ). TURKEY: Smyrna in montibus supra Bournabat, May, *Boissier* (lecto. of *T. preslianum* Boiss.: G); Constantinopole (Istanbul), in collibus apricis pr. Bostandjik, 1906, *Aznavour* 4861 (G); montis Lydiae et Cariae, 1842, *Boissier* (E).

Note: *T. affine* differs distinctly from *T. arvense* by its entire leaflets, thicker heads, longer corolla (about as long as or rarely slightly longer than calyx) and larger seeds.

175. T. saxatile All., Auct. Method. Stirp. Hort. Taurin 25 (1770–73); Misc. Taurin. 5 : 77 (1774–76) et Fl. Pedem. 1 : 305, 2 : 1168 (1785); Gib. & Belli, Mem. Accad. Sci. Torino ser. 2, 39 : 271 (1889). *Typus*: Hb. Allioni (TO n. v.).

T. thimiflorum Vill., Prosp. 43 (1779); Godr. in Gren. & Godr., Fl. Fr. 1 : 411 (1848–49).

Icon.: [Plate 170]. All., Fl. Pedem. 2: t. 15, f. 3 (1875); Gib. & Belli, *loc. cit.* t. 1, f. 3; Reichenb., Icon. 22: t. 2151, f. 2, 11–20 (1903).

Annual, with greyish pubescence, 5–15 cm. Stems decumbent or ascending, mostly flexuous, branching. Lower leaves long-petioled, upper ones short-petioled; stipules ovate to lanceolate, membranous, reddish with dark nerves, acute to acuminate in lower leaves, broadening in upper ones; leaflets small, obovate-cordate in lower leaves, and oblong with cuneate base and emarginate apex in middle and upper ones. Heads small, solitary or rarely twin, axillary of terminal, depressed-globular, sessile, few-flowered, involucrate by broad stipules of uppermost leaves. Flowers minute, 4–6 mm. Calyx tube cylindrical-ovoid, hairy, becoming urceolate; throat open with hairy ring; teeth lanceolate, sharp-pointed, unequal, initially erect then connivent, all shorter than tube. Corolla whitish to purplish-pink, shorter than calyx; standard oblong, obtuse. Pod ovoid. Seed ovoid, yellowish, shiny. Fl. July–August.

Hab.: Mainly boulders of alpine rivers and foot of glaciers.

Gen. Distr.: Switzerland, Austria, France, Italy.

Selected specimens: SWITZERLAND: Findellen Glacier, 7,500 ft, 1923, leg.(?) (K.). FRANCE: Source d'Arvieran, Chamonix, 1866 (K.).

176. T. stipulaceum Thunb., Prodr. Pl. Cap. 136 (1800). *Typus*: Cape Province, coast from Saldanha Bay to Port Elizabeth, *Thunberg* s. n. (UPS).

T. micropetalum E. Mey., Pl. Afr. Austr. 90 (1836).

Icon.: [Plate 171].

Annual, villous, 8–15 cm. Stems simple or branching from base. Leaves long- and

Plate 170. *T. saxatile* All.
(France : *Grenier*, HUJ).
Plant in flower and fruit; fruiting calyx; standard; wing; keel; leaf (enlarged).

Plate 171. *T. stipulaceum* Thunb.
(Cape : 1845, FI).
Plant in flower and fruit; flower; standard; wing; keel, fruiting calyx.

short-petioled; stipules membranous, nerved, subulate-acuminate, 0.8–1.5 cm long; leaflets 1.2 × 0.4–0.8 cm, obovate-oblong, striate, retuse or emarginate, denticulate. Heads oblong, subsessile or shortly pedunculate. Calyx about 6–8 mm, densely villous, campanulate, 10-nerved; teeth equal or almost so, subulate-setaceous, longer than tube; throat open but hairy inside. Corolla white or red, shorter than calyx. Fruiting calyx with open throat and erect teeth. Pod membranous. Seed reddish-brown.

Hab. : Grassy places, fields, sandy soils near sea.

Gen. Distr. : Cape Province (endemic).

Note : According to Gillett (1952), *T. stipulaceum* is the only species of Sect. *Trifolium* occurring outside the Eastern Holarctis. This discontinuity is very remarkable, but similar ones are known in the case of some other genera.
Its inclusion in Subsect. *Arvensia* is in good accord with the external features of the latter, although it also recalls species of Subsect. *Trichoptera.*

Subsection XII. ANGUSTIFOLIA Gib. & Belli

Mem. Accad. Sci. Torino ser. 2, 39 : 341 (1889) pro stirpe, emend. Zoh.

Annuals. Calyx 10-nerved; teeth unequal or equal, blunt, rarely sharp-pointed, usually spreading in fruit; throat closed by bilabiate callosity. Flowers showy. Corolla usually considerably longer than, rarely as long as calyx, pink or purple or flesh-coloured, rarely white.

177. T. angustifolium L., Sp. Pl. 769 (1753).

Annual, appressed- or subappressed-hirsute, 10–30 cm. Stems solitary or few, ascending or erect, scarcely branched. Lower leaves long-petioled, upper ones short-petioled; stipules partly adnate to petioles, many-nerved, free portion lanceolate, subulate; leaflets 3–5 × 0.2–0.4 cm, narrowly linear-lanceolate, those of upper leaves acute, those of lower ones usually obtuse, appressed-pubescent on both faces, entire or nearly so. Heads 3–8 cm, short-peduncled, spicate, cylindrical or conical. Flowers 1–1.3 cm. Calyx tubular-campanulate, covered with appressed to spreading, stiff hairs arising from tubercle; teeth subulate-setaceous, sharp-pointed (or blunt in var. *intermedium*), the lower one longer than the upper ones. Corolla pale pink to almost purple or white, same length as calyx teeth or very slightly longer; standard notched. Fruiting calyx with entirely closed throat and stellately spreading teeth. Pod ovoid, membranous, with cartilaginous operculum. Seed 1, ovoid, light brown. Fl. March–April. 2n = 14, 16.

Hab. : Fallow fields, steppes, open plant communities, meadows.

Gen. Distr. : Austria, Portugal, Spain, France, Italy, Sicily, Capri, Yugoslavia, Greece, Bulgaria, Crete, Crimea, Turkey, Syria, Lebanon, Israel, Iraq, Iran, Caucasus, Transcaucasia, Algeria, Morocco, Madeira, Canary Islands.

1. Stems erect or ascending, usually 15–50 cm. Leaflets usually longer than petioles. Teeth of calyx sharp-pointed. Corolla usually pink or purple. **177a.** var. **angustifolium**
– Stems mostly decumbent or ascending, usually up to 15 cm. Leaflets usually shorter than petioles. Teeth of calyx blunt. Corolla usually pale pink to whitish. **177b.** var. **intermedium**

177a. T. angustifolium L. Var. **angustifolium**; Boiss., Fl. 2 : 122 (1872); Gib. & Belli, Mem. Accad. Sci. Torino ser. 2, 39 : 341 (1889); Aschers. & Graebn., Syn. Mitteleur. Fl. 6/2 : 579 (1908). *Typus* : described from Italy, Hb. Cliff. 375.

T. angustifolium L. var. *acrogymnum* Maire in Emberg. & Maire, Pl. Maroc. Nov. fasc. 1 : 3 (1902).
T. angustifolium L. var. *acrolophum* Maire in Emberg. & Maire, *loc. cit.*

Icon. : [Plate 172]. Sibth. & Sm., Fl. Graec. 8 : t. 749 (1832); Reichenb., Icon. 22 : t. 2144, f. 1, 1–8 (1903); Gib. & Belli, *loc. cit.* t. 6, f.1.

Selected specimens : AUSTRIA : Albazia, 1900, *Richter* (W). PORTUGAL : Nos taludes, ao longo de via ferrea entre Verride e Maruial, 1949, *Fernandes* & *Sousa* 3137 (HUJ). SPAIN : Felipi, 1882, *de Xativa* (W). FRANCE : Beaumont en Veron, champs calcaires incultes, 1883, *Tourlet* 511 (W). ITALY : Pendici di Aspa, 1881, *Marchesetti* (HUJ). SICILY : Palermo in arvis, 1879, *Lojacono* (W). CAPRI : in loco graminoso, 1908, *Frenckel* (HUJ). YUGOSLAVIA : Lessino, 5 m, Steinboden, 1910, *Marsevic* (W). GREECE : Nisyros in road up outside of crater, 1963, *Gathorne* & *Hardy* 378 (E). BULGARIA : in desertis ad Nova Mahala, 1894, *Stribrny* (W). CRETE : 4 km S. of Mires on the way to Kali Limenes, 1964, *Zohary* & *Orshan* 24/43 (HUJ). CRIMEA : Yalta in nemore quercerto, 1898, *Golde* 212 (W). TURKEY : Prov. Sinop, Promontory of Sinop, 10 m beside the sea, 1962, *Davis* & *Code* 63/1146 (E); Thrace, Belgrad forest, 1963, *Zohary* 5662 (HUJ). SYRIA : Nussairy Mts., Ain Halakim, 2,500–3,000 ft, 1910, *Haradjian* 34956 (G). LEBANON : Beyrouth, *Labillardière* (G). PALESTINE : Mt. Carmel, 1897, *Bornmüller* (K, type of f. *brachystachys* Bornm.). IRAQ : Jarmo, 1955, *Helbaek* 1032 (K). IRAN : N. Iran : Lahijan, Guilan, 1937, *Lindsay* 954 (K). CAUCASUS : prope Soczy ad rupes maritimas, 1926, *Malcev* (K). TRANSCAUCASIA : Azerbajdzhan, distr. Kjurdamir ad meridiem p. Padar, in montanis, 1936, *Grossheim* (HUJ). ALGERIA : Champs sablonneux au sud de la plaine d'Eghris, 1852, *Balansa* 449 (W); tombeau de la Chrétienne, 1929 (P, holotype of var. *acrolophum* Maire). MOROCCO : Forêt de Jaba betw. Al-Hajeb and Ifrane, 18 km N.W. of Ifrane, 1961, *De Wilde* & *Dorgelo* 2551 B (HUJ); in Atlante rifano, Bal Ranina in Quercetis, 1928 (P, holotype of var. *acrogymnum* Maire). MADEIRA : Funchal in arvis, 1900, *Bornmüller* 477 (W). CANARY ISLANDS : Teneriffa : Icad el Alto, 1889, *Brunnen* 1653 (W).

177b. T. angustifolium L. Var. **intermedium** Gib. & Belli, Mem. Accad. Sci. Torino ser. 2, 39 : 342, 420, t. 6 (1889) non var. *intermedium* (Guss.) Gib. & Belli, *loc. cit.* nom. illegit. *Typus* : Sicilia meridionalis Terranova, *Gussone* (NAP? holo).

T. intermedium Guss., Cat. Pl. Boccadif. 65, 82, no. 18 (1821) non Lapeyr. (1818); Boiss., Fl. 2 : 122 (1872).

Icon. : Reichenb., Icon. 22 : t. 2145, f. 1, 1–4 (1903).

Selected specimens : ITALY : Calabria, arenos in Fuimara pr. Statisnem, 1877, *Huter, Porta*

Plate 172. *T. angustifolium* L. Var. *angustifolium*
(Greece : *I. Deml, HUJ*).
Plant in flower and fruit; fruiting calyx from above; flower.

& *Rigo* 101 (W). SICILY: in collibus arenosis supra Scaglietti, 1856, *Huet du Pavillon* 44 (W); in pascuis arenosis Siciliae merid., *Gussone* (G iso.). GREECE: Argolis in collibus siccis inter olivetis prope Poros, 1887, *Heldreich* 914 (W). AEGEAN ISLANDS: Karpathos: Valatha, 1883, *Barbey* 191 (G). CRETE: N. Crete, 17 km W. of Rethymnion on the way to Khania, 1964, *Zohary* & *Orshan* 01502 (HUJ). RHODES: 1870, *Bourgeau* (G). TURKEY: Urfa: 40 km from Viranşehir to Urfa, 700 m, fallow fields, 1966, *Davis* 42412 (E) (doubtful). CYPRUS: ad Livadiam Larnacae, 1862, *Kotschy* (G). ALGERIA: Env. d'Alger, Fort de l Eau, Le Rassoute, alt. 20 m, 1953, *d'Alleizeete* (HUJ).

Note: The taxonomic value of var. *intermedium* is very unclear. According to the structure of the calyx teeth (with blunt points), it should definitely be referred to *T. purpureum* Loisel., and not to *T. angustifolium*. However, by virtue of its small white flowers and narrow leaves it approaches the latter. The issue is further complicated by the occurrence in Algeria of var. *acrolophum* Maire, a form which must be included within typical *T. angustifolium* because of its sharp-pointed calyx teeth. These forms are either hybrids or intermediate forms between *T. angustifolium* and *T. purpureum*. Thus *T. intermedium* can by no means be retained at the rank of a species.

178. T. purpureum Loisel., Fl. Gall. 484 (1807).

Annual, appressed- to subappressed-pubescent, 10–30(–50) cm. Stems few or many, striate, erect, ascending or rarely decumbent, branching above. Leaves with petioles shortening towards apex of stem; stipules oblong-lanceolate, many-nerved, membranous between nerves, 1–1.5 cm, free portion long-subulate; leaflets (1–)2–4(–6) × 0.2–1 cm, oblong-lanceolate to linear (rarely elliptical), acute, mucronulate, obsoletely toothed in upper part, mostly hairy only at margins. Heads many at end of dichotomous or simple branches, ovoid to conical in flower, ovoid-oblong to cylindrical in fruit. Flowers up to 2 cm long, often shorter. Flowering calyx up to 1 cm long, subappressed- to antrorse-hirsute; tube almost tubular, 3 mm or more in length; teeth unequal (rarely equal or almost so), subulate-setaceous, blunt, the lowermost one slightly longer than to twice as long as others. Corolla exserting from calyx by one-third or more, purple above, lilac-whitish below, sometimes totally lilac or whitish; standard with oblong limb one-third longer than wings. Fruiting calyx with slightly obconical, prominently nerved tube, with throat closed by a hairy, bilabiate callosity and teeth divergent or spreading, plumose with hairs arising from tubercle. Pod sessile, ovoid, membranous with cartilaginous apex. Seed 1, ovoid, brown, about 1 mm. Fl. January–June. 2n = 14.

Hab.: Fields, roadsides, open dwarf-shrub communities, stony and sandy sites.

Gen. Distr.: Romania, Azores, Portugal, Spain, France, Italy, Balkan Peninsula, Crimea, Turkey, Syria, Lebanon, Palestine, Egypt, Iraq, Iran, Caucasus, Libya, Tunisia, Algeria, Morocco, Madeira, Canary Islands, and elsewhere.

Note: *T. purpureum* Loisel. is a rather polymorphic species, especially in Lebanon and Palestine. It varies especially in the size of the head and the flowers as well as in the colour of the corolla. The following readily discernible forms are recorded here:

1. Teeth of calyx equal or almost so. **178c.** var. **pamphylicum**

– Teeth of calyx distinctly unequal 2
2. Flowering heads elongated, cylindrical, 2–4 cm or more in length, mostly compact. Flowers very numerous. Fruiting heads 3–6 cm long. Plants usually erect. Leaflets mostly oblong-elliptical. **178a.** var. **purpureum**
– Flowering heads globular or ovoid or obconical, usually not exceeding 2 cm (often only 1 cm). Flowers not very numerous. Fruiting heads 1.5–2 (rarely –3) cm. Plants erect or decumbent. Leaflets narrowly elliptical to linear. **178b.** var. **desvauxii**

178a. T. purpureum Loisel. Var. **purpureum**; Boiss., Fl. 2 : 123 (1872) non Gilib. (1782) nom. illegit. *Typus* : France, ad margines agrorum circa Monspelium, ? *Degland.*

T. laxiusculum Boiss. & Bl. in Boiss., Diagn. ser. 2, 2 : 13 (1856).
T. desvauxii Boiss. & Bl. var. *laxiusculum* (Boiss. & Bl.) Boiss., Fl. 2 : 123 (1872).
T. purpureum Loisel. var. *desvauxii* (Boiss. & Bl.) Post, Fl. Syr. Pal. Sin. 236 (1883–96) p. p.
T. purpureum Loisel. var. *laxiusculum* (Boiss. & Bl.) Post, *loc. cit.* 235; Hossain, Not. Roy. Bot. Gard. Edinb. 23 : 420 (1961).
T. angustifolium L. subsp. *purpureum* (Loisel.) Gib. & Belli var. *desvauxii* (Boiss. & Bl.) Gib. & Belli, Mem. Accad. Sci. Torino ser. 2, 39 : 346 (1889) p. p.
T. angustifolium L. subsp. *purpureum* (Loisel.) Gib. & Belli var. *desvauxii* (Boiss. & Bl.) Gib. & Belli subvar. *laxiusculum* (Boiss. & Bl.) Gib. & Belli, *loc. cit.*
T. loiseleuri Rouy, Fl. Fr. 5 : 111 (1899).

Icon. : [Plate 173a]. Reichenb., Icon. 22 : t. 2144, f. 2, 9–14 (1903); Loisel., *loc. cit.* t. 14; Gib. & Belli, *loc. cit.* t. 6, f. 2.

Selected specimens : ROMANIA : Dobrogea, Caliacra, 1926, *Nayrady* 658 (W). FRANCE : Hérault, Montpellier, 1907, *Vichet* (W). YUGOSLAVIA : Serbia australis, in colle Hisar pr. Leskawatz, 1890, *Dörfler* (W). GREECE : Acarmania, pr. pagum Oropis, 1893, *Halácsy* (W). BULGARIA : Bulgaria australis : ad Sadovo, 1892, *Stribrny* (W). TURKEY : Prov. Mersin, distr. Anamur, Anamur-Gilindere, 20 m, 1956, *Davis* & *Polunin* D.25993 (K). LEBANON : Premières collines du Liban entre Saida et Beyrouth, 1853, *Blanche* 275a (G). SYRIA : S. of Kastel-Moauf, 800 m, 1938, *Dinsmore* 3879 (E). PALESTINE : Acco Plain, Acco sands near sea, 1964, *Zohary* et al. 601 (HUJ); Mt. Carmel, Wadi Shomriya, 1942, *Davis* 4382 (E); above Jabbok River, 500 m, 1911, *Meyers* & *Dinsmore* 613 (G, E). IRAQ : Jarmo, 1955, *Helbaek* 1538 (K); Kurdistan : supra pagum Schaqlawa, 1893, *Bornmüller* 1134 (W). IRAN : Prov. Kermanshah, Ghilane e Gharb, Vijinan, 1948, *Behbudi* 73 (E). EGYPT : Matruk, 250 km W. of Alexandria, *Balls* (K).

178b. T. purpureum Loisel. Var. **desvauxii** (Boiss. & Bl.) Post, *loc. cit.* 236 p. p. *Typus* : (Lebanon) Ras Beyrouth, 1850, *Blanche* 342 (G lecto).

T. desvauxii Boiss. & Bl. in Boiss., Diagn. ser. 2, 2 : 12 (1856); Boiss., Fl. 2 : 123 (1872) p. p.
T. angustifolium L. subsp. *purpureum* (Loisel.) Gib. & Belli var. *desvauxii* (Boiss. & Bl.) Gib. & Belli, *loc. cit.* p. p.

Selected specimens : TURKEY : Içel, 18 mi. from Silifke towards Mersin, near s.l., lime-

Plate 173a. *T. purpureum* Loisel. Var. *purpureum*
(Palestine : *Plitmann* 12/6, HUJ).
Plant in flower; flower.

stone, 1965, *Coode* & *Jones* 1079 (E); Kilyos, hillsides, 1936, *Post* (G). SYRIA: Djebel Druz, Kefr, 1932, *Eig* & *Zohary* (HUJ). PALESTINE: Jerusalem, 1925, *Zohary* 60 (HUJ).

178c. T. purpureum Loisel. Var. **pamphylicum** (Boiss. & Heldr.) Zoh. in Davis, Fl. Turk. 3 : 436 (1970). *Typus*: Turkey, in arenosis Pamphyliae ad fluvium Melas infra Monargat (Monavgat), 1846, *Heldreich* (G holo., iso. K).

T. pamphylicum Boiss. & Heldr. in Boiss., Diagn. ser. 1, 9 : 20 (1849) et Fl. 2 : 123 (1872).
T. angustifolium L. subsp. *purpureum* (Loisel.) Gib. & Belli var. *pamphylicum* (Boiss. & Heldr.) Gib. & Belli, *loc. cit.*
T. purpureum Loisel. subsp. *pamphylicum* (Boiss. & Heldr.) Holmboe, Stud. Veg. Cyprus, in Bergens Mus. Skr. n. r. 1, 2 : 105 (1914).
T. pamphylicum Boiss. & Heldr. var. *pamphylicum* Hossain; var. *dolichodontum* Hossain, Not. Roy. Bot. Gard. Edinb. 23 : 421 (1961).

Icon.: [Plate 173b].

Specimens seen: CRETE: Stalos, W. of Chania, sand dune about 10 m from shore, 1961, *Feinbrun* 510 (HUJ). TURKEY: Prov. Antalya, Karanlik Sokak, 150 m, 1936, *Tengwall* (W). CYPRUS: ad Matis litus pr. Daolu, 1880, *Sintenis* & *Rigo* 419 (P); Nicosia-Myrton, 500 ft, common as pioneer on eroding slopes, *Chapman* 579 (K).

Note: *T. purpureum* var. *pamphylicum* differs from *T. purpureum* var. *purpureum* only by the shorter flowers and the equal or almost equal calyx teeth (in var. *purpureum* and other varieties the lower tooth is much longer than the upper ones).
T. purpureum is one of the most common and handsome clovers in the Middle East, and also one of the most variable species. Only someone who has examined hundreds of specimens of this species can obtain a fairly comprehensive picture of its varietal make-up. From *T. angustifolium* L. it is well distinguished by the character of the calyx teeth and the size of the corolla, but it is closely related to and sometimes barely distinguishable from a whole series of binomials established chiefly by Boissier. These are: *T. blancheanum, T. desvauxii, T. laxiusculum, T. roussaeanum*, and *T. pamphylicum*. These binomials have been variously interpreted in the literature. In the present study, *T. blancheanum* is regarded as an independent species (see note on that species). Of the remaining four, *T. roussaeanum* is considered here as a species, but with certain reservations; it has patulous-hairy stems and opposite uppermost leaves; *T. desvauxii* can, at most, be regarded as a variety of *T. purpureum* and the same holds true for *T. pamphylicum*; *T. pamphylicum* is a dwarf, tiny plant which has almost equal calyx teeth and lanceolate-linear leaflets of the *T. arvense* type; *T. laxiusculum* is a maritime form of *T. purpureum*, hardly distinguishable from the other varieties. None of them, except *T. roussaeanum*, can be maintained at a specific level because of the rather quantitative nature of the characteristics. This is in marked disagreement with the procedure adopted by Hossain (*loc. cit.*). What has here been considered as var. *purpureum* is in itself a heterogeneous assemblage of forms which awaits further analytical and experimental study. *T. purpureum* var. *desvauxii* is also probably made up of various forms which have a few characteristics in common, namely: small, less compact flowering and fruiting heads, smaller flowers, etc. This variety grows under poor climatic and edaphic conditions; it is artificially delineated and intergrades with var. *purpureum*. This aspect, too, awaits further treatment as to composition and constancy.

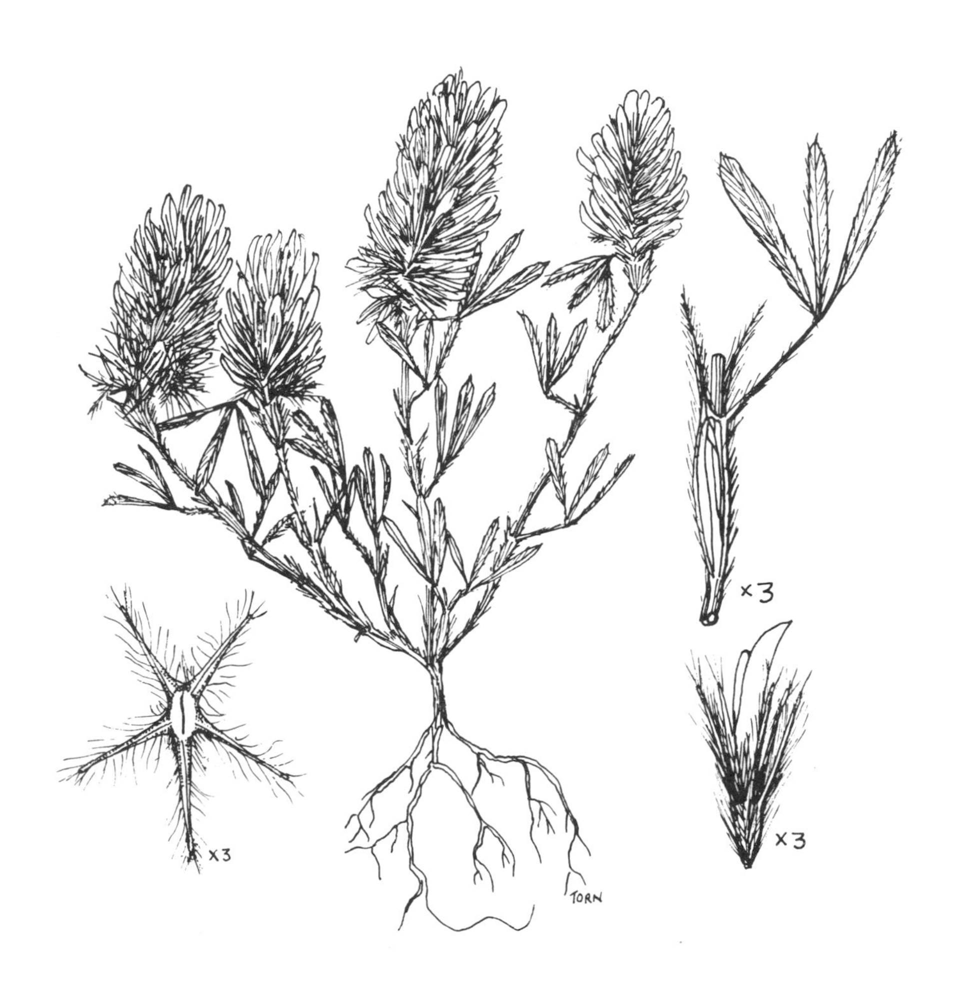

Plate 173b. *T. purpureum* Loisel. Var. *pamphylicum* (Boiss. & Heldr.) Zoh.
(Cyprus : *P. H. Davis* D. 3202, HUJ).
Plant in flower; fruiting calyx from above; leaf (enlarged); flower.

179. T. blancheanum Boiss., Diagn. ser. 2, 2 : 13 (1856). *Typus* : Lebanon, Beyrouth, bords des champs sablonneux, Avril 1850, *Blanche* (E lecto).

T. desvauxii Boiss. & Bl. var. *blancheanum* (Boiss.) Boiss., Fl. 2 : 124 (1872); Hossain, Not. Roy. Bot. Gard. Edinb. 23 : 420 (1961).
T. angustifolium L. subsp. *purpureum* (Loisel.) Gib. & Belli var. *desvauxii* (Boiss. & Bl.) Gib. & Belli subvar. *blancheanum* (Boiss.) Gib. & Belli, *loc. cit.*

Icon. : [Plate 174].

Annual, appressed- or subappressed-hairy, 10–20 cm. Stems many, procumbent, slender, terete, striate, leafy especially at base, scarcely branching. Leaves small, all short-petioled to subsessile; stipules oblong, almost scarious, prominently nerved, ciliate in upper part, terminating in lanceolate-cuspidate apex half as long as lower part; leaflets oblong-elliptical to elliptical, obtuse, usually 0.8–1 cm long, densely appressed-hairy, obliquely apiculate. Peduncles short in flower, elongating in fruit. Heads many-flowered, ovoid or globular, 1–1.5 cm long. Flowers 1–1.2 cm. Calyx tubular; tube prominently nerved, 3 mm long, densely beset with brown bristles; teeth unequal, setaceous, as long as or longer than tube, the lower one 1 mm longer than others. Corolla lilac or pink, about 1.5 times as long as calyx; standard ovate at base, abruptly attenuate, 1–2 mm longer than wings. Fruiting heads not increasing much in size, 1.3–2 cm long. Fruiting calyx somewhat indurated; throat closed by bilabiate callosity; teeth erect or divergent or stellately spreading. Pod membranous, disrupting, 1-seeded. Seed greenish-brown, 1.2 mm long. Mature heads hardly or not at all disarticulating at maturity. Fl. April–May.

Hab. : Sandy coast exposed to sea spray.

Gen. Distr. : **Lebanon, Palestine.**

Specimens seen : LEBANON : Saida, bords de la mer, 1854, *Gaillardot* 921 (G). PALESTINE : Acco Plain, Acco, light soil, *Eig* et al. 170 (HUJ); also collected by the senior author from Sharon Plain. [M. Z.]

Note : *T. blancheanum* has been misinterpreted by almost all those who have studied Middle Eastern flora, including the author of this species. It has been confused with *T. roussaeanum*, *T. laxiusculum* and others, and some authors included it within *T. purpureum*.

The senior author regards *T. blancheanum* as a species well distinguished from *T. purpureum* by a series of morphological characteristics and also by its habitat. It is a prostrate to procumbent plant which occurs in the littoral spray zone of Israel and Lebanon. This habitat has had a marked influence on speciation in various taxonomic groups. The striking features of this species are : small flowers, 1–1.2 cm; small, almost globular flowering heads; short fruiting heads; and small, elliptical, thick (somewhat fleshy) leaflets. We have collected this species both in the "locus classicus" and in habitats similar to the above in northern Israel.

There are, at first glance, some difficulties in distinguishing between *T. blancheanum* and a form of *T. purpureum* var. *desvauxii* which also grows in the coastal zone at some distance from the spray zone and has similarly small heads; however, the latter is readily recognizable by its linear-lanceolate leaflets.

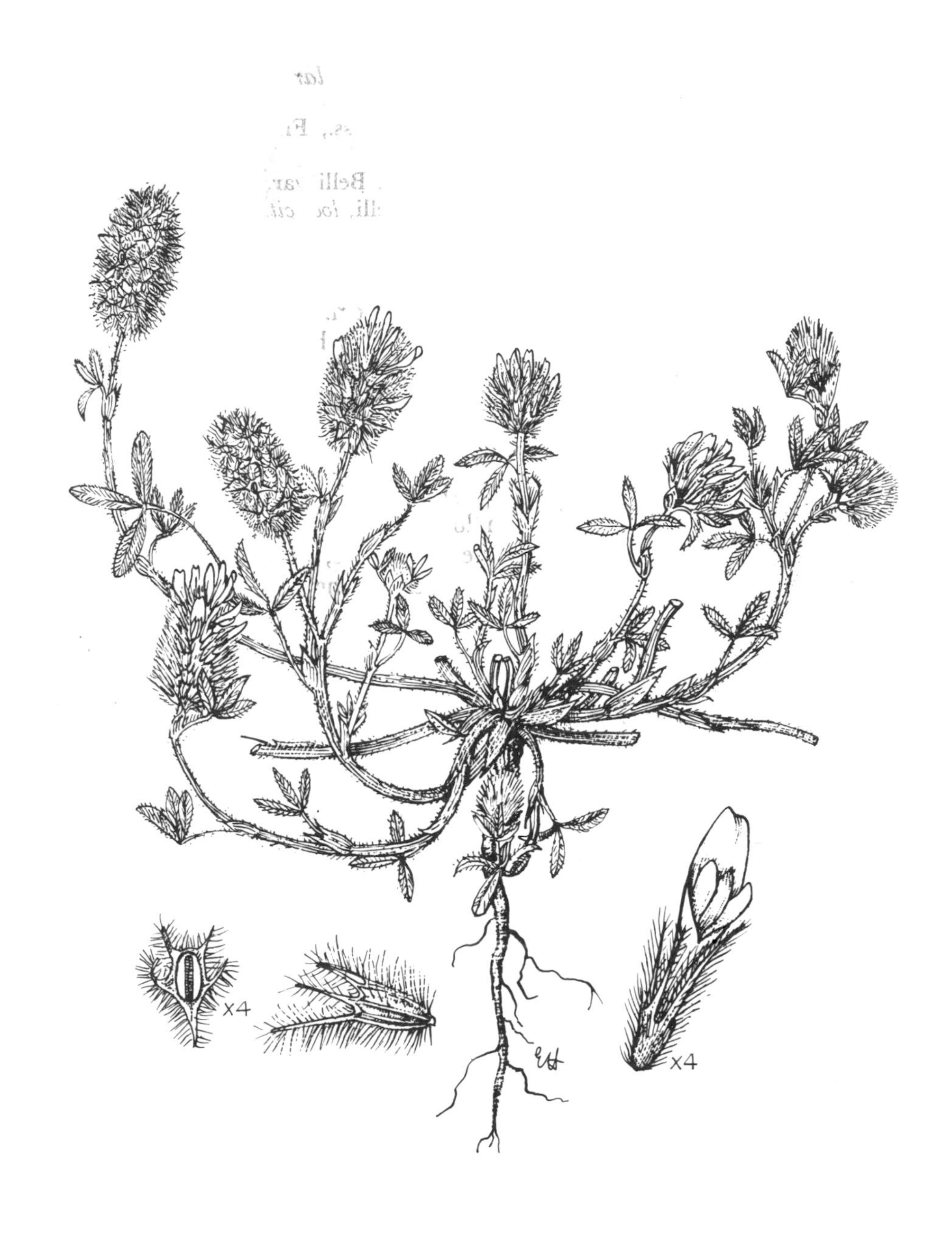

Plate 174. *T. blancheanum* Boiss.
(Palestine : *Zohary*, HUJ).
Plant in flower and fruit; fruiting calyces; flower.

180. T. roussaeanum Boiss., Diagn. ser. 1, 2 : 26 (1843); Hossain, Not. Roy. Bot. Gard. Edinb. 23 : 422 (1961). *Typus* : Hab. in Oriente, *Rousseau* (G lecto.); Anatolia in rupestribus Ciliciae, 1837, *Aucher* 1236 (G).

T. desvauxii Boiss. & Bl. var. *roussaeanum* (Boiss.) Boiss., Fl. 2 : 123 (1872).
T. angustifolium L. subsp. *purpureum* (Loisel.) Gib. & Belli var. *desvauxii* (Boiss. & Bl.) Gib. & Belli subvar. *roussaeanum* (Boiss.) Gib. & Belli, Mem. Accad. Sci. Torino ser. 2, 39 : 346 (1889).
T. purpureum Loisel. subsp. *roussaeanum* (Boiss.) Holmboe, Stud. Veg. Cyprus, in Bergens Mus. Skr. n. r. 1, 2 : 105 (1914).

Icon. : [Plate 175].

Annual, with patulous soft hairs, 5–15 cm. Stems few or single, poorly branching, erect or ascending. Leaves all, except for uppermost ones, alternate, small; petioles fairly short, almost at right angles to stipules; stipules linear-oblong, adnate to petioles for up to about half their length, patulous-hairy with shiny hairs, lower part membranous, green-nerved, free portion lanceolate-subulate, green; leaflets linear-oblong, 0.8–1.5 × 0.2–0.3 cm, sparingly covered with shiny hairs, tapering towards base, obtuse, denticulate at apex. Heads ovoid to oblong, 1.5–2 × 1–1.2 cm, short-peduncled, solitary or twin, often sessile, subtended by 2 uppermost leaves, one of which is sometimes reduced to a stipule. Flowers 1.3–1.5 cm, dense. Calyx tube cylindrical, covered with appressed, soft and shiny hairs, 10-nerved; throat closed at maturity by bilabiate callosity; teeth unequal, subulate-setaceous, blunt, the lower one longer than others, all considerably longer than tube. Corolla pink or flesh-coloured, about 1.5 times as long as calyx; standard oblong-acutish, much longer than wings and keel. Pod ovoid, membranous, with cartilaginous operculum. Seed 1, about 1.2–1.5 mm. Fl. March–May.
Hab. : Rocky ground.
Gen. Distr. : Turkey (endemic).

Specimens seen : TURKEY : Prov. Mersin, distr. Tarsus, gorge of Tarsus River, between Ulaş and Samlar, 150 m, rocky limestone slope, 1957, *Davis* & *Hedge* D. 26439 (E); Maraş, 800 m, *Balls* 908 (E).

181. T. dichroanthum Boiss., Diagn. ser. 1, 9 : 20 (1849) et Fl. 2 : 124 (1872).

Annual, patulous-, or subpatulous-, rarely appressed-hirsute, 10–50 cm. Stems few, striate, erect or ascending, rarely procumbent, often brownish. Leaves, at least uppermost ones, usually opposite, long-petioled; stipules membranous between nerves, oblong, the upper ones inflated, semi-ovate with long-subulate ends; leaflets 1–3 × 0.2–1 cm, oblong to elliptical and linear, with mucronate apex, broad, obsoletely toothed in upper part. Heads ovoid to cylindrical in flower, 2–6 × 1.5–2 cm, borne on thick peduncles, elongating in fruit. Flowers about 2 cm long, 2-coloured, young ones pale lilac, older ones dark lilac to purple, rarely altogether purple. Calyx 8–9 mm, densely subappressed-hirsute; tube almost cylindrical, much shorter than slightly unequal, subulate, blunt, plumose teeth. Corolla about 1.6–1.8 cm; limb of standard linear, one-third longer than wings; older flower heads becoming darker

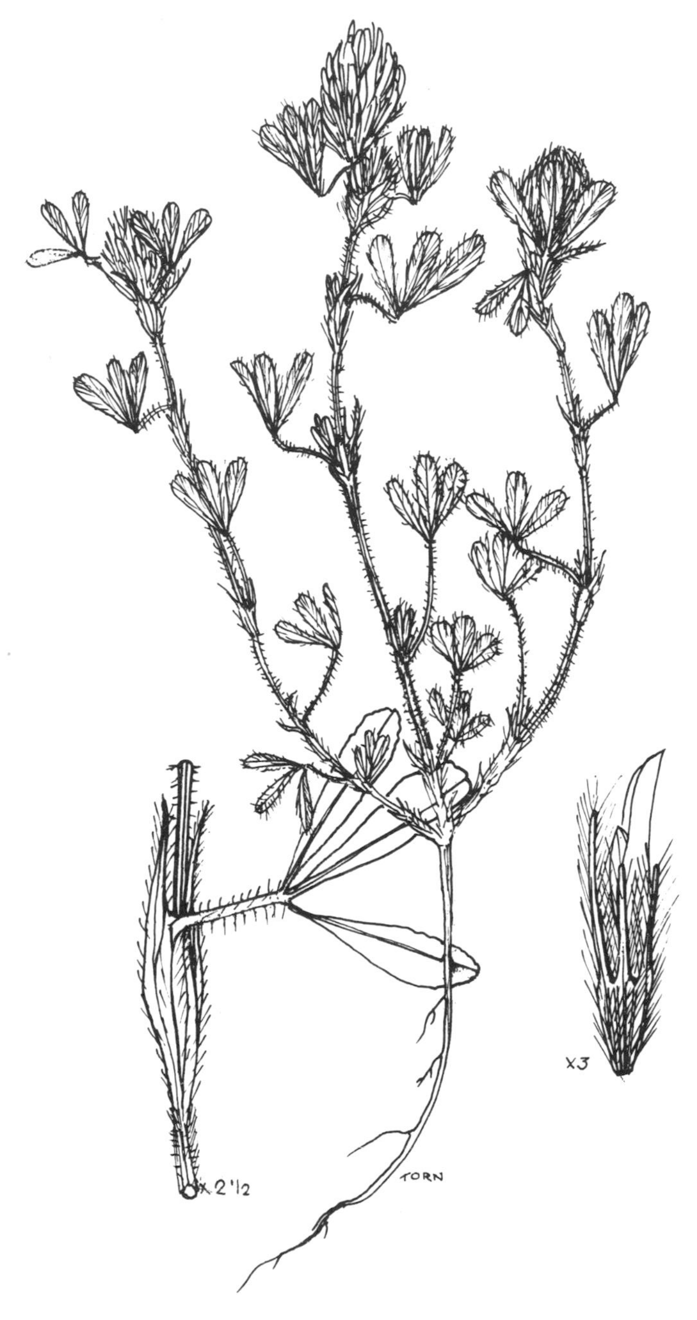

Plate 175. *T. roussaeanum* Boiss.
Plant in flower: leaf (enlarged); flower.

and uniform in colour. Fruiting calyx with closed, bilabiate throat, somewhat constricted below erect or divergent teeth. Pod membranous, ovoid, with cartilaginous apex, 1-seeded. Seed ovoid, brown, smooth. Fl. March–May. 2n = 16.

Hab. : Coastal belt, sandy soils.

Gen. Distr. : S. Turkey, Cyprus, Lebanon, Palestine, Egypt.

1. Plants appressed-hirsute. Leaflets elliptical, (0.6–)1–1.5 cm long. Heads up to 1.5 cm long. Stems not brownish. **181c.** var. **cyprium**
– Plants patulous- or subpatulous-hirsute. Leaflets and/or heads longer than above. Stems brownish 2
2. Petals pale lilac. **181a.** var. **dichroanthum**
– Petals (incl. young ones and at least wings and keel) intensely purple. **181b.** var. **ascalonicum**

181a. T. dichroanthum Boiss. Var. **dichroanthum**. *Typus* : Palestine, in collibus Judeae circa Gazam, 1846, *Boissier* (G lecto., iso. K).

T. angustifolium L. subsp. *dichroanthum* (Boiss.) Gib. & Belli, Mem. Accad. Sci. Torino ser. 2, 39 : 349 (1889).

Icon. : [Plate 176].

Selected specimens : LEBANON : Beyrouth, 1931, *Zohary* (HUJ). PALESTINE : Sharon Plain, S. of Magdiel, 1927, *Naftolsky* 16702 (HUJ), very common. EGYPT : Sinai : El Arish, 1926, *Eig* 160 (HUJ).

181b. T. dichroanthum Boiss. Var. **ascalonicum** Zoh., Candollea 27 : 155 (1972). *Typus* : Palestine, Philistean Plain, nr. Nitzanim, hills, sandy clay soil, 1962, *Zohary* 101 (HUJ holo.).

181c. T. dichroanthum Boiss. Var. **cyprium** Zoh., *loc. cit. Typus* : Cyprus, Hartcha, 10 ft, stony field near beach, 1950, *Chapman* 273 (HUJ holo.).

182. T. palaestinum Boiss., Diagn. ser. 1, 9 : 21 (1849) et Fl. 2 : 124 (1872). *Typus* : In arenosis circa Gazam Palestinae, Legi Aprili 1846, *Boissier* (G holo.).

Icon. : [Plate 177].

Annual, patulous-villous, 10–20(–40) cm. Stems many or few or solitary, thick, erect, ascending or decumbent, usually not branching. Lower leaves long-petioled, upper ones subsessile; stipules membranous between prominent nerves, oblong-linear, 1.5–2 cm, with long-cuspidate upper portion; leaflets 1.5–2.5 × 0.3–0.8 cm, oblong-linear to elliptical, obtuse or acute, apiculate, those of upper leaves toothed and often broader in upper part. Flowering heads often at end of dichotomous branches, ovoid, 1.5–2 cm long, elongating in fruit. Flower white, about

Plate 176. *T. dichroanthum* Boiss. Var. *dichroanthum* (Palestine : *Eig, Zohary* & *Feinbrun*, HUJ). Plant in flower and fruit; fruiting calyx; fruiting calyx from above; flower.

Plate 177. *T. palaestinum* Boiss.
(Palestine : *Zohary* & *Plitmann*, HUJ).
Plant in flower; flower; fruiting calyx from above.

1.5 cm. Calyx densely hispid; tube obconical, much shorter than slightly unequal, subulate, truncate, plumose teeth. Corolla usually not longer than calyx teeth; standard with oblong-linear limb, almost twice as long as wings and keel. Fruiting calyx about 1 cm long; throat closed by bilabiate callosity; teeth lanceolate, stellately spreading or slightly diverging. Pod membranous, oblong, obtuse, with 1 ovoid, brown, smooth, 2 mm long seed. Fl. March–May. 2n = 16.

Hab.: Coastal belt, loamy-sandy soils.

Gen. Distr.: Lebanon, Palestine.

Specimens seen: LEBANON: in Ras prope Beirut, 1855, *Kotschy* 1374 (W). PALESTINE: Coastal Sharon Plain, Benyamina-Caesarea, 1942, *Davis* 4410 (E).

Note: *T. palaestinum* is rather common. It displays a particular procumbent spray ecotype; varies slightly in size of flowers.

183. T. dichroanthoides Rech. fil., Ark. Bot. ser. 2/1, 5: 306 (1949). *Typus*: Syria, Jebel Ansarieh ad transitum Nebi Younes supra pagum Slenfe, 1933, *Samuelsson* 5812 (W holo.).

Icon.: [Plate 178].

Annual, with patulous or antrorse, shiny fine hairs, 8–15 cm. Stems few, poorly branching, erect or ascending, leafy. Leaves 1.3 cm, the lower ones petioled, the upper ones subsessile, all alternate, sometimes falsely opposite owing to approximation of two heads; stipules of mid-stem leaves lanceolate to oblong, white-membranous with green nerves, adnate to petioles for up to half their length, free portion lanceolate-acuminate, as long as lower part, hirsute; petioles longer than stipules; leaflets of lower leaves obovate-cuneate, 0.6–1 × 0.3–0.4 cm, with rounded denticulate apex, those of upper ones oblong-cuneate to oblong, 1–1.5 × 0.2–0.5 cm, acute or obtuse, indistinctly denticulate. Heads ovoid to oblong, subtended by uppermost leaves. Flowers 1.8–2 cm long. Calyx 0.5–0.7 cm, tubular-campanulate, densely covered with long, antrorse or subappressed, shiny, soft hairs; teeth erect, subulate, unequal, the shorter ones somewhat longer than tube. Fruiting calyx growing somewhat in size, with spreading teeth; throat closed by bilabiate callosity. Corolla about twice as long as calyx, often 2-coloured, pale purplish-pink when dry; standard oblong-linear, acute, much longer than wings and keel. Ovary short-stipitate. Fruit young, 1-seeded. Fl. April.

Hab.: Gravelly ground, mountains.

Gen. Distr.: Syria (endemic).

Specimens seen: SYRIA: Only the type specimen cited above.

Note: *T. dichroanthoides* is very close to *T. roussaeanum* Boiss., but differs from the latter in the following characters: larger flowers; calyx almost twice as long; calyx teeth unequal. The indumentum is, in both cases, soft, silky, shiny. The relations between these two species yet require study. It seems that *T. dichroanthoides* should be considered a variety of *T. roussaeanum*. The relations between *T. dichroanthoides* and *T. haussknechtii* Boiss. should

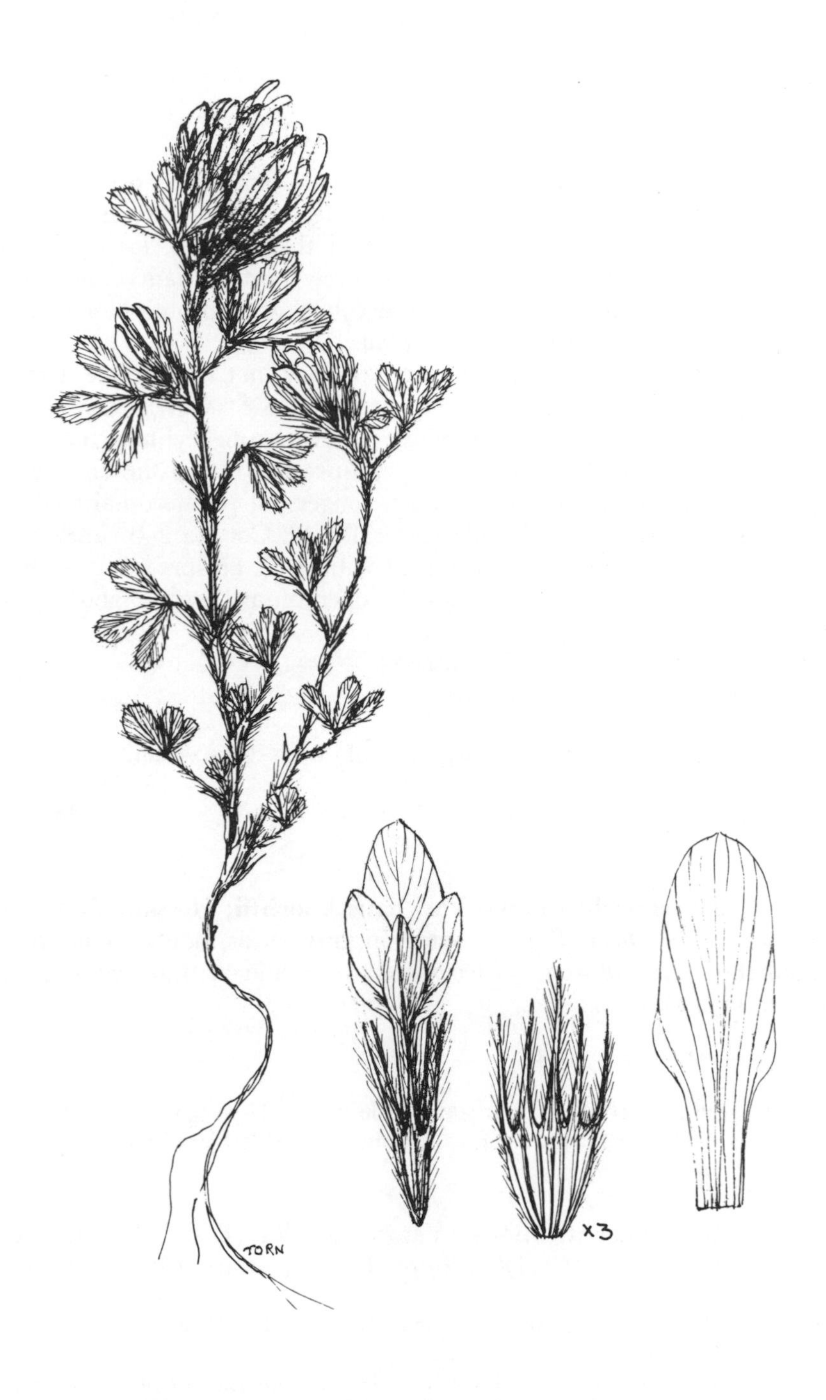

Plate 178. *T. dichroanthoides* Rech. f.
(Syria : *Samuelsson* 5812, W).
Plant in flower; flower; fruiting calyx; standard.

also be revised on the basis of more material (on *T. dichroanthoides*) than was available to the authors.

184. T. haussknechtii Boiss., Fl. 2 : 125 (1872).

Annual, antrorsely or appressed-villous, 10–20 cm. Stems ascending, striate, much branched, leafy. Leaves short-petioled; stipules of middle leaves about 1 cm or more in length, adnate to petioles for up to half their length, lower part membranous, oblong, with prominent green nerves, free portion lanceolate-subulate; leaflets 0.8–2.5 × 0.2–0.4 cm, narrowly oblanceolate or obovate-oblong, appressed-hairy, denticulate in upper part, obtuse. Heads 1.5–1.8 × 0.8–1 cm, ovoid to obovoid, slightly elongated in fruit, terminal and axillary, not involucrate, borne on capillary peduncles. Flowers about 1.5 cm, rather loose. Calyx densely covered with appressed, shiny, silky hairs concealing nerves; tube cylindrical, 10-nerved, 3–3.5 mm long; teeth subulate, 1-nerved, very unequal, blunt, the shorter ones about as long as tube, the lower one considerably longer, all plumose-hairy; throat of fruiting calyx entirely closed by thick, bilabiate callosity. Corolla 2–3 times as long as calyx, white to pink. Fruiting calyx obconical with erect or spreading teeth. Pod 1.5 mm, obovoid-membranous with cartilaginous operculum. Seed 1, about 2 mm, globular, brown, shiny. Fl. April–May. 2n = 16.

Hab. : Grassy places, steppes, scree.

Gen. Distr. : Turkey, Syria, Iraq.

1. Leaflets usually narrowly oblanceolate, 1.5–2.5 × 0.2–0.4 cm. **184a.** var. **haussknechtii**
– Leaflets obovate to oblong, 0.8–1 × 0.3–0.4 cm. **184b.** var. **candollei**

184a. T. haussknechtii Boiss. Var. **haussknechtii**; Hossain, Not. Roy. Bot. Gard. Edinb. 23 : 432 (1961). *Typus* : Syria, in graminosis montis Gebel Muhassan prope Aleppo, 1865, *Haussknecht* (G lecto., iso.: in K and W, labelled as *T. lasiocalycinum* Boiss. & Hausskn.).

Icon. : [Plate 179].

Specimens seen : TURKEY : Diyarbakir : 5 km S. of Diyarbakir town, 650 m, 1957, *Davis* & *Hedge* D. 28750 (E). SYRIA : Env. Hammath, 1913, *Haradjian* 1824 (K).

184b. T. haussknechtii Boiss. Var. **candollei** (Post) Hossain, Not. Roy. Bot. Gard. Edinb. 23 : 423 (1961). *Typus* : Turkey, prope Aintab (Herb. Post BEI).

T. candollei Post, Journ. Linn. Soc. Lond. Bot. 24 : 425 (1888).

Specimens seen : TURKEY : pr. Urfa, ca. 20 km from the turning to Sürüç towards Urfa, 600 m, 1957, *Davis* & *Hedge* D. 27986 (K). SYRIA : Env. of Muslemie, 1931, *Zohary* (HUJ). IRAQ : Jebel Sindjar, 330 m, rocks with red soil between them, 1933, *Eig* & *Zohary* (HUJ).

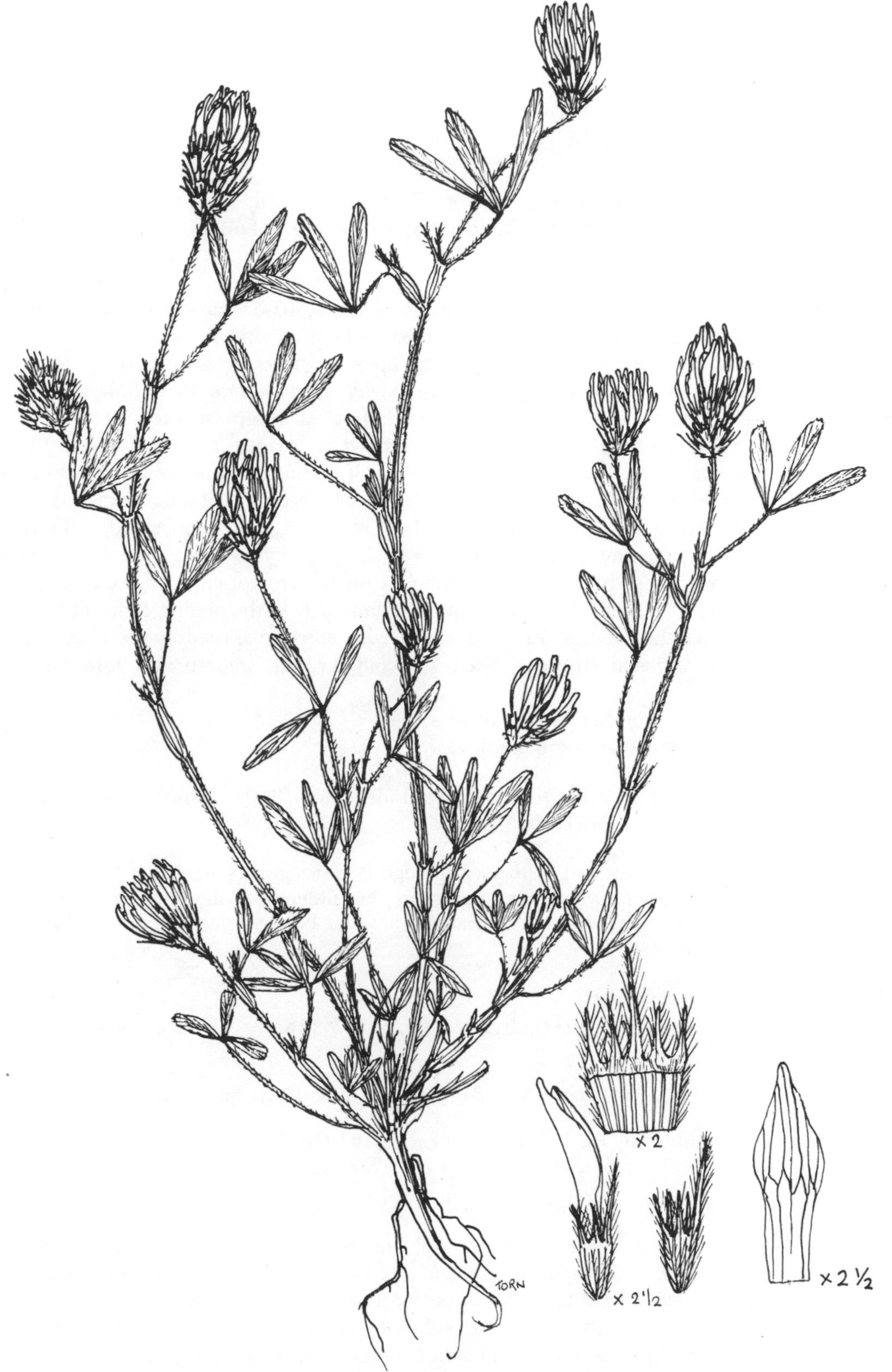

Plate 179. *T. haussknechtii* Boiss. Var. *haussknechtii*
(Turkey : *Davis* 42500, HUJ).
Plant in flower; dissected calyx; flower; fruiting calyx
(Turkey : *Zohary* & *Plitmann* 1960–8, HUJ); standard.

185. T. prophetarum Hossain, Not. Roy. Bot. Gard. Edinb. 23 : 421 (1961). *Typus* : Palestine, Mt. Gilboa, Beith Alfa, *Davis* 4142 (E holo., iso. K).

Icon. : [Plate 180].

Annual, with silky, appressed or subappressed hairs, 10–15 cm. Stems few or rarely solitary, delicate, slightly furrowed, erect or ascending, simple or rarely branching. Leaves with petioles shortening towards apex; stipules prominently nerved and membranous between nerves, oblong, the lower part adnate to petioles, 4–6 mm long, upper portion subulate; leaflets oblong-elliptical to almost linear, 1–2 cm long, 2–3 mm broad, terminating abruptly in somewhat oblique mucro. Heads often at end of false dichotomy (sometimes one branch of dichotomy undeveloped, and hence leaves opposite), many-flowered, obovoid in flower, becoming oblong or cylindrical in fruit, 1.5–3.5 cm long. Flowering calyx 6(–7) mm long, with appressed, silky hairs; tube 10-nerved, narrowly obconical, about 2 mm long; teeth equal, setaceous, plumose, 4 mm long, blunt. Corolla lilac to purple in upper part, exserting from calyx for about one-third of its length; standard with oblong-lanceolate limb, one-third longer than wings. Fruiting calyx with erect or spreading teeth and closed throat. Ovary and pod stipitate. Seed 1, ovoid, brown, smooth, 1.2 mm long. Fl. March–April.

Hab. : Batha, steppes and roadsides.
Gen. Distr. : Palestine (endemic).

Specimens seen : PALESTINE : Jerusalem, roadside, 800 m, 1808, *Dinsmore* 3162 (E); also found in Galilee and in the Jordan Valley.

Note : *T. prophetarum* is well delineated, though it varies greatly in size of flowers; its extreme forms resemble *T. arvense* in appearance, but differ from the latter in a number of diagnostic characteristics. Most of our material has long been considered as a small-flowered form of *T. dasyurum*.

186. T. dasyurum C. Presl, Symb. Bot. 1 : 53 (1831). *Typus* : Described from Crete.

T. formosum Urv., Mém. Soc. Linn. Paris 1 : 350 (1822) non Savi (1808–10); Boiss., Fl. 2 : 124 (1872).
T. velivolum Paine, Pal. Expl. Soc. Statement 3 : 103 (1875).
T. formosum Urv. var. *minus* Post, Fl. Syr. Pal. Sin. 236 (1883–96).

Icon. : [Plate 181]. C. Presl, *loc. cit.* t. 33.

Annual, appressed-pubescent, 10–30(–40) cm. Stems few or many, striate, diffuse, ascending, slightly branching above, often terminating in flower-bearing dichotomium. Leaves petiolate, the uppermost ones subsessile; stipules membranous, inflated, with arcuate ribs and long, subulate to cuspidate tip; leaflets elliptical to oblong, acute, 1–3.5 × 0.2–1 cm. Heads 1.5–3(–4) cm long, many-flowered, ovoid to rhomboid. Calyx appressed-hirsute; tube cylindrical to obconical; teeth equal, subulate, with lanceolate base, about twice as long as tube. Corolla purple at tip, whitish or pink below, as long as calyx or shorter; standard with oblong limb, some-

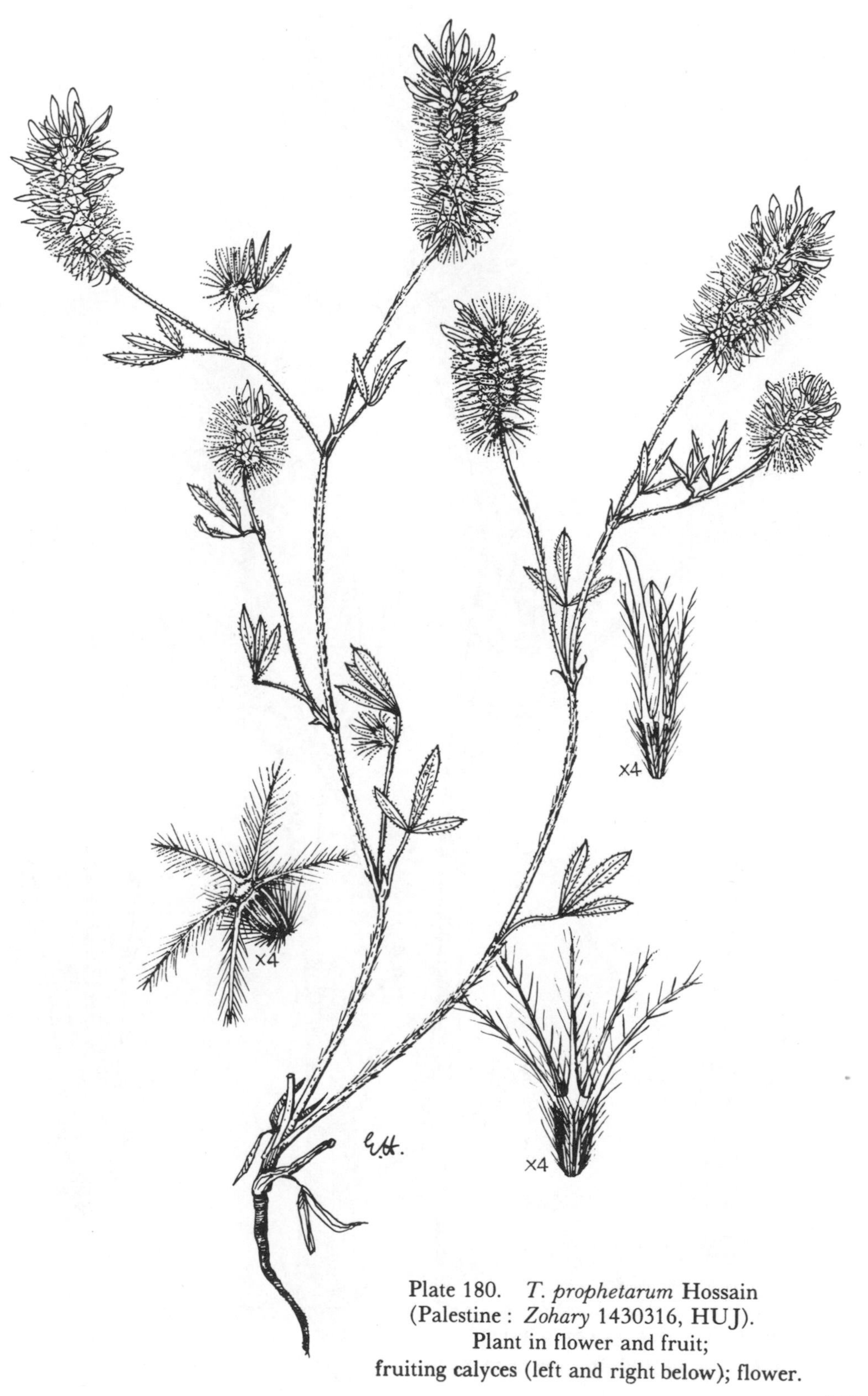

Plate 180. *T. prophetarum* Hossain
(Palestine : *Zohary* 1430316, HUJ).
Plant in flower and fruit;
fruiting calyces (left and right below); flower.

Plate 181. *T. dasyurum* C. Presl
(Palestine : *Zohary* 16778, HUJ).
Plant in flower and fruit; fruiting calyx; flower.

what longer than wings. Fruiting calyx with top-shaped tube, closed throat and somewhat divergent (not spreading) lobes. Pod stipitate, membranous, oblong, obliquely apiculate, with 1 globular to somewhat compressed, smooth, 1.8 mm long seed. Fl. March–May. 2n = 16.

Hab.: Batha and fields, stony ground, semi-arid areas.

Gen. Distr.: Greece, Aegean Islands, Turkey, Cyprus, Syria, Palestine, Iraq, Iran, Egypt, Libya.

Selected specimens: TURKEY: Prov. Urfa, Sürüç-Urfa, ca. 20 km from the turning to Sürüç, 600 m, 1957, *Davis* & *Hedge* D. 27983 (K). CYPRUS: near Agios Epiktitos (E. of Kyrenia), 60 m, 1956, *Casey* 1706 (K). SYRIA: 74 km S.W. of Damascus, basaltic stones and rocks, 950 m, 1933, *Eig* & *Zohary* 617 (HUJ). PALESTINE: Judean Mts., Deir Amar, garigue, 1954, *Jaffe* 17064 (HUJ). IRAQ: Jebel Bashiqah betw. Fadhihyed and Bashiqah (N.E. of Mosul), ca. 500 m, between rocks, 1933, *Eig* & *Zohary* (HUJ). IRAN: 16 mi. S. of Burugird, 1,350 m, 1929, *Cowan* & *Darlington* 1073 (K). EGYPT: Selum, sandy soil, 1952, *Shabetai* 39 (K). LIBYA: Cyrenaica: Benghasi, 1883, *Ruhmer* 100 (HUJ).

Subsection XIII. ALEXANDRINA Zoh.

Candollea 27 : 249 (1972)

Annuals. Calyx 10-nerved; tube cylindrical or campanulate, pubescent; throat open or slightly constricted, glabrous, hairy or ciliate; teeth of fruiting calyx erect or divaricate. Corolla white or cream-coloured, rarely pink, up to twice as long as calyx or longer, caducous.

187. T. salmoneum Mout., Fl. Djebel Druze 128 (1953). *Typus*: Syria, El Ayoun, 5.6.1943, *Mouterde* (phototype in Mouterde, *loc. cit.*)

Icon.: [Plate 182].

Annual, sparingly appressed-hairy, with dense indumentum on peduncles and sparse one in lower parts, (30–)50–80(–90) cm. Stems erect, usually solitary, striate, sometimes furrowed, fistulous, branching all along, sometimes becoming purple. Leaves all, except uppermost ones, alternate; adnate part of stipules oblong-linear, membranous, with dark, prominent longitudinal nerves, free portion shorter than lower part, subulate, herbaceous; leaflets (1.5–)2–3(–4) × (0.3–)0.5–1(–1.3) cm, linear-elliptical or oblong, cuneate at base, obtuse or acute at apex, appressed-pubescent with denticulate upper part. Heads terminal and axillary, pedunculate, the flowering ones 1.5–2.5 × 1.5–2 cm, obovoid to obconical. Flowers yellow to cream-coloured (rarely pinkish). Calyx tube cylindrical to obconical in flower, 10-nerved, appressed-hairy below and rather hispid at base of teeth; teeth unequal, triangular-subulate, mostly hispid or ciliate with hairs arising from tubercles, the lower one 1(–3)-nerved, about as long as tube, the others 1-nerved, shorter than tube. Corolla 1.5 times as long as wings and keel. Fruiting head 1.5–2(–2.5) × 1–1.5 cm, oblong-cylindrical to conical.

Plate 182. *T. salmoneum* Mout.
(Palestine : *Zohary*, HUJ).
Flowering branch; flower (right); fruiting head
(Palestine : *E. Ben-Shlomo*, HUJ)
and fruiting calyx from above.

Fruiting calyx tube obconical-campanulate, somewhat thickened; teeth erect; throat narrowed by annular callosity, glabrous or provided with few short hairs. Pod membranous, obovoid, leathery at apex, apiculate. Seed solitary, 1.5–2 mm long, obovoid or pyriform. Fl. April–May. 2n = 16.

Hab.: Damp places; near water. Rather rare.

Gen. Distr.: Syria, Palestine.

Selected specimens: PALESTINE: Hula Plain, env. of Gonen, slopes of basalt hills, swampy soil near spring, 1963, *Zohary* 11/6 (HUJ).

Note: *T. salmoneum* varies only slightly, chiefly in the colour of the petals (from white-cream to pinkish); the teeth of the calyx usually have a white tip, rarely a purple one; the throat is usually glabrous, very rarely with some scattered hairs.

188. T. apertum Bobrov in Komarov, Fl. URSS 11 : 391, 258, t. 16 (1945) et in Acta Inst. Bot. Acad. Sci. URSS ser. 1, 6 : 316 (1947).

Annual, appressed-hirtellous above, glabrescent below, (10–)20–60(–100) cm. Stems few or many, striate, slightly branching in upper part. Leaves all, except uppermost ones, alternate; stipules linear, membranous, with longitudinal, green nerves, free portion as long as or shorter than lower adnate part, subulate, herbaceous, ciliate; leaflets (0.6–)1–2(–2.3) × 0.3–0.9 cm, linear-lanceolate to broad-elliptical, tapering at both ends, rarely obovate-oblong, cuneate at base, apiculate with denticulate upper part. Heads terminal and axillary, pedunculate, in flower (1.2–)1.5–2.5(–4) × (1.2–)1.5–1.8 cm, rhomboid to conical when young, later becoming broad-linear. Flowers yellow to cream-coloured or pinkish. Calyx tube cylindrical to obconical in flower, appressed-hairy or glabrescent below and rather hispid above, 10-nerved; teeth lanceolate to subulate with long, triangular base, (3–)5-nerved at base, with purple-blotted tip (at least on lower tooth which is at least 1.5 times as long as others and tube). Corolla 1.5 times as long as calyx; standard with linear-lanceolate to linear-elliptical limb, longer than wings and keel. Fruiting head (1–)1.5–3.5 × (1–)1.2–1.5 cm, globular to oblong-conical. Fruiting calyces somewhat loose, thus individual tubes well discernible in head; tube obconical-campanulate, glabrescent below, somewhat membranous, with diverging nerves; teeth broad-triangular to triangular-lanceolate, divaricate or stellately spreading, the lower one sometimes much longer and broader than others; throat open with ring of hairs and lacking callosity. Pod submembranous with somewhat coriaceous upper part, obovoid, apiculate. Seed solitary, 1.8–2 × 1.1–1.4 mm. Fl. May–July. 2n = 16.

Hab.: Meadows, among bushes.

Gen. Distr.: Italy, Greece (probably introduced in these two countries), Turkey, Caucasus.

1. Plant 40 cm or more. Stems few, erect. Lower tooth of flowering calyx at least twice as long as other teeth and tube; teeth of fruiting calyx never reticulately nerved. Fruiting head 2–3.5 cm long. **188a**. var. **apertum**
– Plant 10–30 cm. Stems many, ascending. Lower tooth of flowering calyx 1.5 times as

Plate 183. *T. apertum* Bobrov Var. *apertum*
(Caucasus : *Bobrov* 3725, HUJ).
Flowering and fruiting branches; fruiting calyces in various positions; flower.
A–Var. *kilaeum* Zoh. & Lern.
(Turkey : *Zekerickmy* & *Scovnorovkmy*, HUJ).
Flowering and fruiting branch; fruiting calyces.

Plate 184. *T. berytheum* Boiss. & Bl.
(Palestine : *U. Plitmann*, HUJ).
Flowering and fruiting branches; flower; standard; wing; keel; fruiting calyces; dissected calyx.

long as other teeth and tube; teeth of fruiting calyx usually reticulately nerved at base. Fruiting head 1.2–1.5 cm long. **188b.** var. **kilaeum**

188a. T. apertum Bobrov Var.. **apertum**. *Typus*: Caucasus, circa Maikop, 1.7.1924, *Pastuchow* (LE holo., iso. HUJ).

Icon.: [Plate 183]. Fl. URSS 11 : 258, t. 16 (1945).

Selected specimens: ITALY: An campo Marzo, *Tommasini* 475 (probably introduced) (W). GREECE: Leucadia, in arvis Catunoe, 1951, *Mazyicri* 432 (W). TURKEY: Pr. Istanbul, bord de la route de Tache-Kichla à Dolmabaghtche, 19.6.1888, *Aznavour* 663 (G). CAUCASUS: Pr. Krasnodar, Majkop, in vicinitate stationis Experimentalis Majkopensis, in pratis inter frutices, 1949, *Bobrov* 3725 (G, HUJ).

188b. T. apertum Bobrov Var. **kilaeum** Zoh. & Lern., Not. Roy. Bot. Gard. Edinb. 29 : 322 (1969). *Typus*: Turkey, Pr. Istanbul, collines argileuses près de Kila (Kilyos), 29.6.1893, *Aznavour* (G holo.).

Icon.: [Plate 183].

Selected specimens: TURKEY: Kilyos, hillsides, 1936, *B. V. D. Post* (G).

189. T. berytheum Boiss. & Bl. in Boiss., Diagn. ser. 2, 2 : 15 (1856). *Typus*: Lebanon ad ostium riv. Sainib prope Sidonem, inter segetes, *Blanche* (G lecto, iso.: BM, E, K).

T. supinum Savi var. *tuberculatum* Boiss., Fl. 2 : 126 (1872).
T. alexandrinum L. var. *tuberculatum* (Boiss.) Gib. & Belli, Mem. Accad. Sci. Torino ser. 2, 39 : 290 (1889).
T. alexandrinum L. var. *berytheum* (Boiss. & Bl.) Trab., Bull. Div. Agr. Serv. Bot. 48 : 2 (1911); Opphr., Bull. Res. Counc. Israel D, 7 : 212 (1959).
T. constantinopolitanum Ser. var. *plumosum* Bornm., Bot. Centralbl. 31, 2 : 203 (1914).
T. echinatum M. B. var. *berythaeum* (Boiss. & Bl.) Dinsm., Repert. Sp. Nov. 30 : 126 (1931) et in Post, Fl. Syr. Pal. Sin. ed. 2, 1 : 338 (1932).

Icon.: [Plate 184]. Opphr., *loc. cit.* 219, f. 4.

Annual, appressed-hairy or glabrescent, (10–)20–50 cm. Stems erect, ascending or procumbent, few to many, slightly striate, branching from base and all along, appressed-hairy to glabrescent. Leaves all, except uppermost ones, alternate; adnate part of stipules oblong-linear, membranous, with dark, longitudinal nerves, free portion subulate, slightly denticulate at margin, ciliate; leaflets (1–)1.3–2.5(–3.5) × (0.3–)0.5–0.8(–1) cm, linear-elliptical to oblong-elliptical, obtuse, rarely acute, appressed-pubescent, denticulate in upper part. Heads terminal and axillary, pedunculate or sessile (involucrate by upper leaves), flowering ones (1–)1.5(–2) × 1–1.2 cm, obconical to obovoid. Flowers yellow to cream-coloured. Calyx tube cylindrical in flower, appressed-hairy below and rather hispid above (hairs later becoming tufted), 10-nerved; teeth unequal, lanceolate to triangular-subulate, mostly hispid or ciliate with long hairs usually arising from tubercles, generally 1-nerved, rarely base of lower one or 2 upper (posterior) ones 3-nerved, lower one often up to 1.5 times as long as tube, others usually as long as tube or shorter. Corolla 1.5–2 times as long as calyx; standard with oblong-linear limb, 1.5 times as long as wings, the latter being slightly longer than keel. Fruiting head (1.2–)1.5–2(–2.7) × 0.8–1.3 cm, ovoid to oblong-conical. Fruiting calyx tube obconical, somewhat thickened, with erect or partly stellately spreading teeth; throat of calyx provided with ring of long hairs, sometimes also with slight thickening. Pod membranous, obovoid, somewhat leathery at apex, apiculate. Seed solitary, 1.5–1.7 × 1–1.2 mm, obovoid, obliquely truncate at base. Fl. March–May. 2n = 16.

Hab.: Fields and damp places.

Gen. Distr.: Turkey, Lebanon, Palestine.

Selected specimens: TURKEY: Distr. Istanbul, collines entre Chichli et Kishathiani, 1893, *Aznavour* (G). LEBANON: Suq ul Charb, 1900, *Post* (HUJ). PALESTINE: Esdraelon Plain, Qishon River, E. of Mishmar Ha'emeq, field border, heavy soil, 1963, *Plitmann* 15/5 (HUJ, fairly common).

Note: *T. berytheum* Boiss. & Bl., which is morphologically well distinguished from all other species of Subsect. *Alexandrina*, has been wrongly reduced by Boissier to a variety of *T. supinum* (= *T. echinatum*), by Gibelli & Belli to a variety of *T. alexandrinum*, and by Bornmüller to a variety of *T. constantinopolitanum*. Inadequate knowledge of the three respective species could have led to such a procedure. It was Hossain (1961) who first restored *T. berytheum*, as conceived by Boissier in his Diagnoses, to specific rank. As a matter of fact, *T. echinatum* and *T. constantinopolitanum* belong to other subsections. The closest to *T. berytheum* are *T. alexandrinum*, *T. salmoneum* and *T. meironense*. *T. salmoneum* has a semi-closed calyx throat due to the presence of a protruding annular callosity. *T. meironense* has pink flowers. *T. berytheum* has an indurated fruiting calyx with coarse tufts of hairs, and a long-ciliate calyx throat, while in *T. alexandrinum* the calyx is white-membranous, the open throat very sparingly appressed-hairy, and the fruiting heads do not disarticulate into individual calyces.

190. T. meironense Zoh. & Lern. in Davis, Fl. Turkey 3: 438 (1970) et in Zoh., Fl. Palaest. 2: 183, t. 263 (1972). *Typus*: Palestine, Upper Galilee, Mt. Meiron, Adatir (grown in 1965 from seeds collected by *Katznelson* from the above locality), *Katznelson* 160 (HUJ holo.).

Icon.: [Plate 185].

Annual, appressed-pubescent, 15–30 cm. Stems many, ascending, slightly striate in lower part. Leaves alternate, the uppermost ones opposite; stipules linear-oblong, membranous with longitudinal purple nerves, free portion subulate, herbaceous; leaflets 0.8–1.5 × 0.2–0.4 cm, broad, linear-elliptical to oblong with cuneate base, mucronate, denticulate in upper part. Heads terminal and axillary, pedunculate, 1–1.5(–2) × ca. 1 cm, ovoid-oblong. Calyx tube cylindrical in flower, 10-nerved, with long, patulous hairs, especially at base of teeth; teeth unequal,

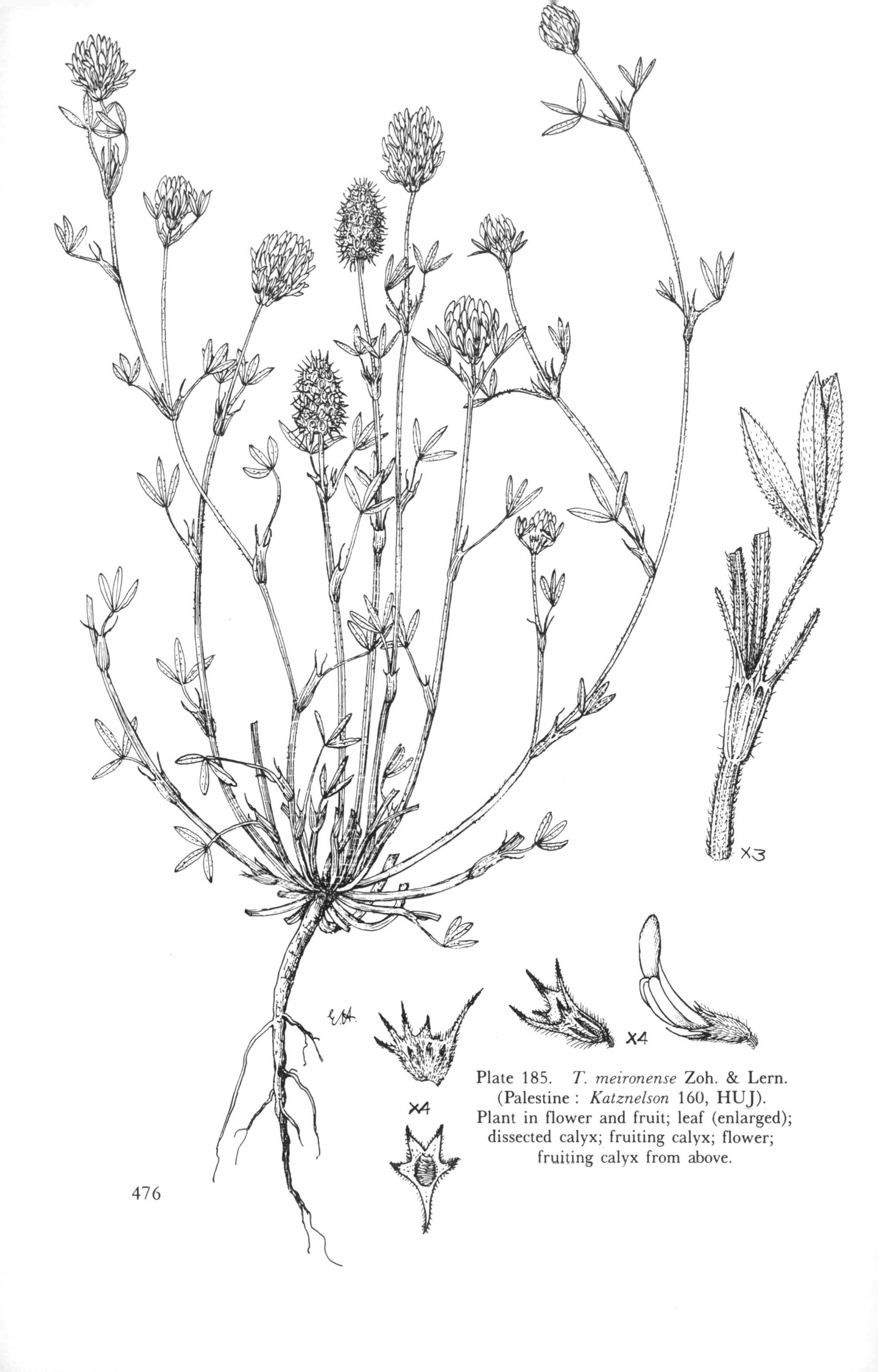

Plate 185. *T. meironense* Zoh. & Lern. (Palestine : *Katznelson* 160, HUJ). Plant in flower and fruit; leaf (enlarged); dissected calyx; fruiting calyx; flower; fruiting calyx from above.

triangular-subulate, long-haired, 1–3-nerved at base, all or at least lower one terminating in purple tip, lower one 1.5 times as long as tube, others usually as long as or shorter than tube. Corolla pale pink, 1.5–2 times as long as calyx; standard with linear limb; wings and keel three-quarters the length of standard. Fruiting head 0.8–1.8 × 0.6–1.2 cm, oblong-conical. Fruiting calyx tube obconical, thickened, with 10 costate nerves; teeth erect to somewhat spreading, triangular at base; throat open but provided with thin, callous and ciliate ring. Pod membranous, coriaceous above and apiculate. Seed solitary, ca. 1.5 × 1.1 mm, obovoid to pyriform, laterally very compressed. Dispersal by single calyces. Fl. April–May. 2n = 16.

Hab.: Batha and maquis.

Gen Distr.: Turkey, Palestine.

Specimens seen: TURKEY: in planitie inter Urfa et Siverek, 1841, *Kotschy* 86 (W).

191. T. vavilovii Eig, Bull. Appl. Bot. Leningrad ser. 7, 1 : 108 (1934); Opphr., Bull. Res. Counc. Israel D, 7 : 208 (1959). *Typus*: Palestine, Nahalal to Harosheth-Hagoim, 1927, *Eig* (HUJ lecto.).

Icon.: [Plate 186].

Annual, with patulous or appressed crisp hairs above and glabrescent below, (10–)20–35(–50) cm, erect or ascending. Stems few to many, slightly striate. Leaves all, except uppermost ones, alternate; connate part of stipules oblong, submembranous with dark longitudinal nerves, free portion as long as or longer than lower part, lanceolate to subulate, usually herbaceous; leaflets (1–)2–3(–3.5) × 0.5–1(–1.5) cm, all, or at least upper ones, lanceolate to elliptical, tapering at both ends, with denticulate upper part. Flowering head terminal, pedunculate, (1.5–)2–2.5(–3) × (1–)1.5(–2) cm, obconical to ovoid. Flowers 1–1.2 cm. Calyx tube cylindrical in flower, appressed-pubescent, 10-nerved; teeth lanceolate-subulate, distinctly purple-tipped, hairy (hairs usually not arising from tubercles), the lower one 1(–3)-nerved, thickened, 2–3 times as long as others which are 1-nerved and at most half the length of tube. Corolla white to cream-coloured, 1.5–2 times as long as calyx; standard with long-elliptical limb, up to 1.5 times as long as wings and keel. Fruiting head (1–)1.5–2(–3) × 1–1.5 cm, cylindrical to conical. Fruiting calyces not very compact, so that individual tubes are well discernible; tube campanulate, indurated, prominently nerved and rather glabrescent; lower tooth deflexed or recurved, elongated, thickened, others erect, never stellately spreading, triangular; throat open, naked, rarely provided with few short hairs. Pod membranous, hard, somewhat leathery above, obovoid, apiculate. Seed solitary, 1.5–1.8 × 0.8–1 mm, ovoid. Fl. March–May. 2n = 16.

Hab.: Open places, abandoned fields, roadsides and semi-arid batha.

Gen. Distr.: Syria, Palestine, Iraq.

Selected specimens: SYRIA: Sueda, Jebel Druz, 1931, *Zohary* (HUJ). PALESTINE: Upper Galilee, env. of Amiad, fallow fields and roadsides, basalt soil, 1963, *Zohary* 8/42 (HUJ). IRAQ: Kurdistan and Mosul, 1941, *Kotschy* (K).

Plate 186. *T. vavilovii* Eig
(Palestine : *U. Plitmann*, HUJ).
Flowering and fruiting branch; flower; fruiting calyces.

Note : *T. vavilovii* has long been overlooked by those who have studied the flora of the Near East. This was partly due to the inadequate description by Eig and to the inaccessibility of the Russian periodical in which this binomial was originally published. It is a very common species in Palestine, and it grows in semi-arid Mediterranean batha and other dry localities. It is readily distinguished from *T. alexandrinum* and *T. berytheum* by its smaller flowers, the length of the lower calyx tooth which is often twice that of the others, the feeble pubescence and leathery consistency of the calyx tube, the naked throat of the calyx, the purple tips of the calyx teeth, etc.
Hossain (*op. cit.* 427) synonymized *T. vavilovii* with *T. carmeli*, indicating that he was not sufficiently familiar with these two species. The latter differs considerably from the former by its foliage, its broadly lanceolate calyx teeth, its calyx throat which is closed by a callosity, and its non-disarticulating heads. Hossain was probably misled by the purple ("black") tips of the calyx teeth that occur both in *T. carmeli* and *T. vavilovii*, but this character is also encountered in *T. plebeium* and *T. constantinopolitanum*.
T. vavilovii varies slightly in the proportions of the calyx teeth, the pubescence of the calyx and also in dimensions of the leaflets. Future experimental studies will ascertain the taxonomic value of these forms.

192. T. alexandrinum L., Cent. Pl. 1 : 25 (1755) et Amoen. Acad. 4 : 286 (1759).

Annual, sparingly appressed-hairy or glabrescent, (10–)30–60 cm. Stems erect or ascending, few or many, striate, branching from base and above. Leaves all, except uppermost ones, alternate; adnate part of stipules oblong, membranous with green nerves, free portion as long as or shorter than the lower part, lanceolate-subulate, herbaceous, plumose-ciliate; leaflets 1.5–3.5(–5) × 0.6–1.5 cm, broad-elliptical to oblong-lanceolate, tapering at both ends, sparingly or densely appressed-pubescent, mucronate at apex, rarely retuse, denticulate in upper part. Heads terminal and axillary, pedunculate, flowering ones 1.5–2 × 1.5–2 cm, conical to ovoid, often provided with few bracts forming minute involucre at base of head. Flowers 0.8–1.3 cm. Calyx tube obconical to campanulate, with 10 prominent nerves and usually with appressed, silky hairs; teeth unequal, erect or somewhat spreading, triangular-subulate to triangular-lanceolate, plumose (with antrorse, silky hairs), terminating in white, glabrous tip, lower one usually as long as tube and slightly longer than others, often with oblique end or irregularly eroded at apex, 3-nerved at base, others 1–3-nerved. Corolla cream-coloured, 1.5–2 times as long as calyx; standard considerably longer than wings, with oblong, linear limb; wings somewhat longer than keel. Fruiting head 1.5–2.5 × 1–1.5 cm, oblong-conical to oblong-ovoid, not disarticulating at maturity. Fruiting calyx broad-campanulate; tube white-submembranous between nerves, somewhat oblique; throat open, provided with ring of short, sparingly appressed hairs. Pod 2.2–2.5 mm long, membranous in lower part, coriaceous and apiculate above. Seed solitary, 2–2.2 × 1.4–1.9 mm. Fl. April–May. 2n = 16.

Hab. : Cultivated in many countries, especially in S.W. Asia, as a forage plant. Wild races and progenitors not known.

Note : *T. alexandrinum* is subspontaneous in Palestine and some neighbouring countries. Of the many known cultivars of Egyptian clover, the most common are the "fahli" and "muscavi" The first is a spring form unable to regenerate after reaping and therefore grown

for seeds only. The second is an early summer crop and regenerates after harvest, producing 4–6 crops seasonally. Apart from this physiological property, the variety muscavi differs from the fahli in some other characters, namely the ability of the stem to branch profusely from the base, the configuration of the lower calyx tooth, etc.
After examining certain other morphological characters of these agriculturally very important strains, we have decided to rank them as varieties, and to name and describe them tentatively as follows:

192a. T. alexandrinum L. Var. **alexandrinum**; Boiss., Fl. 2 : 127 (1872) excl. syn. et var. β; Opphr., Bull. Res. Counc. Israel D, 7 : 204 (1959). *Typus* : Described from Egypt, Hb. Linn. 930/49.

Icon. : [Plate 187].

Plants not branched or only slightly branching at base. Stems slender, solid. Free portion of stipules on upper leaves mostly triangular. Involucre of bracts at base of head 0 or rudimentary. Teeth of calyx almost equal, lower one not oblique, with more or less regular apex. Fruiting calyx with lower tooth as long as or shorter than whitish, prominently nerved tube. Fruiting head ovoid, somewhat prickly to touch. Fl. March–April. Local name: "fahli".

Note : Widely cultivated as a single-harvest clover in the Mediterranean countries and elsewhere. Linné's specimen of this species in the Linnean Herbarium, London, represents the above variety.

192b. T. alexandrinum L. Var. **serotinum** Zoh. & Lern. in Zoh., Fl. Palaest. 2 : 185 (1972). *Typus* : Palestine, Lower Galilee, env. of Meskha, about 1.5 km west of the settlement, field border, alluvial basalt soil, 10.3.63, *Zohary* 2104/9 (HUJ holo.).

Plants profusely branching at base. Stems mostly thick and fistulous. Upper, free portion of stipules in uppermost leaves long-subulate. Base of heads mostly with distinct involucre of bracts, sometimes very short. Lower tooth of calyx conspicuously longer than others, and teeth of lower flowers usually irregular at apex (oblique, bifid or eroded with a unilateral lobe...), mostly 1.5 times as long as tube of fruiting calyx, which is herbaceous, greyish, with less prominent nerves; teeth less spreading, often erect. Fruiting head conical, more prickly to touch. Fl. May–July. Local name : "muscavi".

Note : Mouterde (1970) published some additional binomials related to *T. alexandrinum*. We have not seen these specimens but from the descriptions and the distribution of the published taxa it seems to us that all of them should be regarded as synonyms of some of the species recorded here within Subsect. *Alexandrina*.

Plate 187. *T. alexandrinum* L. Var. *alexandrinum* (Palestine : *Zohary*, HUJ). Plant in flower; flower; fruiting calyx.

Subsection XIV. SQUAMOSA Zoh.

Candollea 27 : 253 (1972)

Maritima Gib. & Belli, Mem. Accad. Sci. Torino ser. 2, 39 : 362 (1889) p. p. pro stirpe.

Annuals. Calyx 10-nerved; tube obconical or campanulate, whitish, glabrous or glabrescent; throat closed by bilabiate callosity; teeth triangular to lanceolate, shorter than tube, 3-nerved, divaricate in fruit. Corolla white or pink, caducous.

193. T. squamosum L., Amoen. Acad. 4 : 105 (1759). *Typus*: Described from England, Hb. Linn.

T. maritimum Huds., Fl. Angl. 284 (1762); Godr. in Gren. & Godr., Fl. Fr. 1 : 408 (1848–49); Boiss., Fl. 2 : 128 (1872); Gib. & Belli, Mem. Accad. Sci. Torino ser. 2, 39 : 384 (1889).
T. irregulare Pourr., Hist. Mem. Accad. Sci. Toulouse 3 : 311 (1788).
T. rigidum Savi, Fl. Pis. 2 : 154 (1798).
T. clypeatum sensu Lapyer., Hist. Abrég. Pyr. 436 (1813) non L. (1853).
T. glabellum C. Presl, Fl. Sic. 1 : XXI (1826).
T. albidum sensu Ten., Fl. Nap. App. 3 : 619 (1830) non Retz. (1786–87).
T. commutatum Ledeb., Fl. Ross. 1 : 543 (1843).
T. nigrocinctum Boiss. & Orph. in Boiss., Diagn. ser. 2, 6 : 47 (1859).
T. maritimum Huds. var. *nigro-cinctum* (Boiss. & Orph.) Boiss., Fl. 2 : 129 (1872); Gib. & Belli, *loc. cit.* 385 ("*nigrocinctum*").
T. maritimum Huds. var. *moriferum* Lojac., Tent. Monogr. Trif. Sic. 136 (1878).
T. xatardii DC., Fl. Fr. 5 : 558 (1815); Willk. & Lange, Prodr. Fl. Hisp. 3 : 369 (1877).
T. maritimum Huds. var. *irregulare* (Pourr.) Aschers. & Graebn., Syn. Mitteleur. Fl. 6/2 : 587 (1908).

Icon.: [Plate 188]. Reichenb., Icon. 22 : t. 2139, f. 2, 6–13 (1903); Gib. & Belli, *loc. cit.* t. 8, f. 8.

Annual, appressed-patulous-hairy to glabrescent, 10–50 cm. Stems mostly erect or ascending, rarely decumbent, rather branching. Leaves long-petioled with petioles decreasing in length towards apex of stem, all alternate except uppermost ones; stipules oblong, whitish, membranous, the lower part connate and adnate to petioles for up to about one-quarter of their length, green-nerved, free portion green, linear-acuminate, villous-ciliate; leaflets of lower leaves obovate, obtuse or retuse, those of middle and upper ones oblong to lanceolate, cuneate at base, denticulate in upper part, acute or mucronulate. Heads 1–2 × 1–1.5 cm, pedunculate, rarely sessile or subsessile and subtended by pair of opposite leaves, usually ovoid. Flowers many, dense, about 7–9 mm. Calyx tube obconical or campanulate, glabrous or glabrescent, with very prominent ribs or nerves, becoming effaced in fruit, leathery; throat of calyx closed by callosity; slit sometimes slightly broadening at maturity owing to growth of seeds; teeth sharp-pointed with purple tip, varying in length, triangular to lanceolate, 3-nerved, the lower one often slightly longer or all of the same length, stellately spreading in fruit, one-half to two-thirds the length of tube, often with

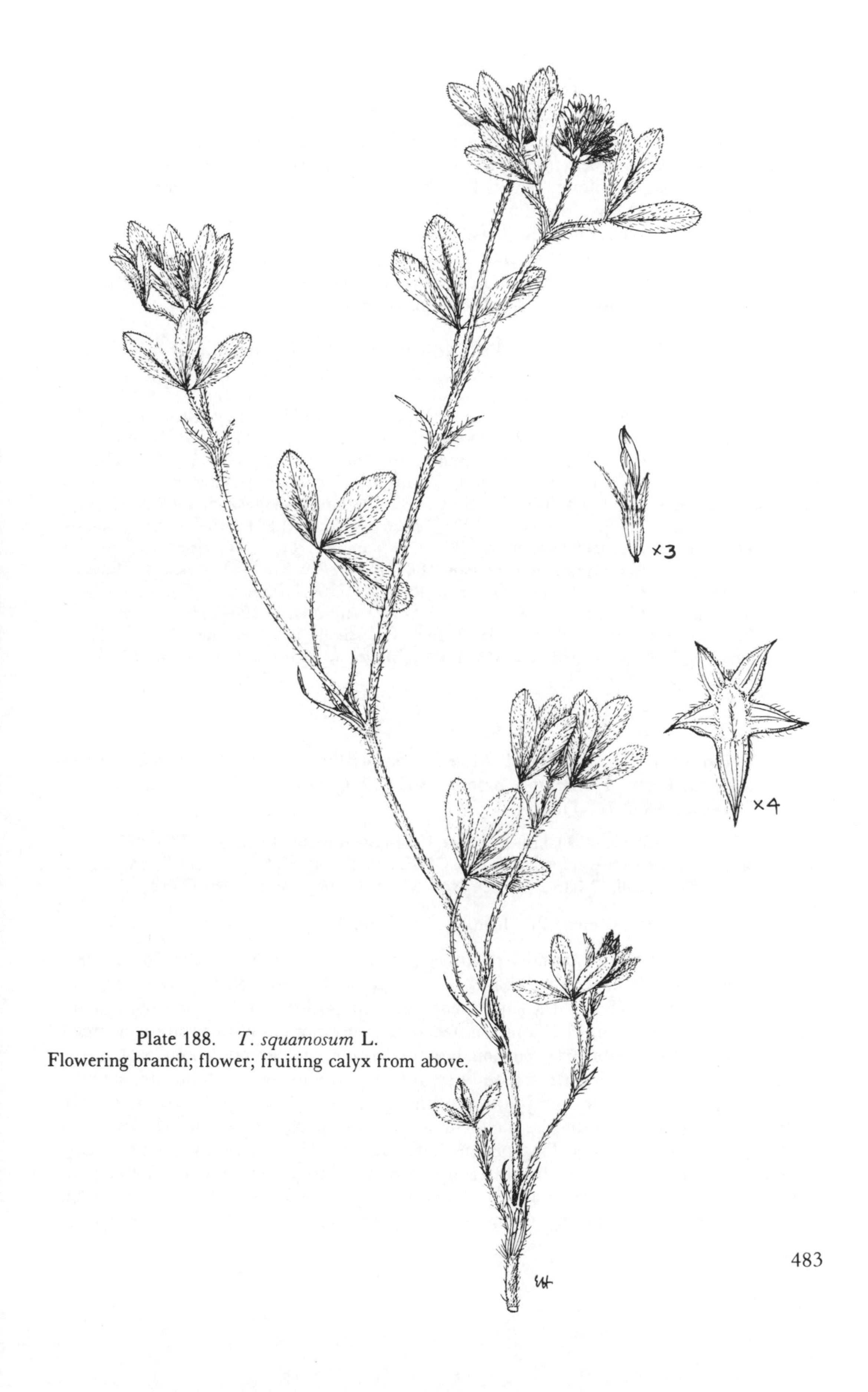

Plate 188. *T. squamosum* L.
Flowering branch; flower; fruiting calyx from above.

hairs scarcely tuberculate at base or with dark hairs forming black zone between tube and teeth (var. *nigrocinctum)*. Corolla about one-third longer than calyx, white or pale pink; standard oblong, retuse. Pod ovoid, dehiscent by 2 valves. Seed 1, ovoid, blackish, shiny. Fl. May–July. 2n = 16.

Hab.: Coastal areas, meadows, roadsides and fields.

Gen. Distr.: British Isles, Germany, Poland, Switzerland, Hungary, Romania, Portugal, Spain, France, Italy, Balkan Peninsula, Crimea; Turkey, Syria, Lebanon, Caucasus, Libya, Tunisia, Algeria, Morocco, Madeira.

Selected specimens: BRITISH ISLES: East Kent, Isle of Sheppey, rough grassland around salt marsh, 1953, *Cannon* et al. 2304 (W); prope Southend, 1884, *Fraser* (W). PORTUGAL: in herbosis agris Olinponensis prope Lumiar, 1891, *Welwitsch* (G). SPAIN: Champs humides à Lagos, Algarve, 1853, *Bourgeau* 1835 (G). FRANCE: Berges du Canal de la Vendée, Fontenay-le-Comte (Vendée), 1853, *Letourneux* 1163 (G); Montpellier, 1905, *Thellung* 5116 (HUJ); près humides au Havre, 1836, *Grenier* 52 (HUJ). SARDINIA: Thomas (G). ITALY: Calabria, Catanzaro in pascuis graminosis, 1898, *Rigo* 309 (W). YUGOSLAVIA: Istria, Wiesengräben in Val Baudori, bei Fasana, 1809 *Korb* (W); Dalmatia, Spalato oberhalb Salona, 1926, *Korb* (W). GREECE: in pratis humidis Peloponnesi pr. Tyrinthon Nauphas, 1857, *Orphanides* 613 (W, lectotype of *T. nigro-cinctum*). CRETE: Marais de Kissamos, 1884, *Reverchon* 239 (W). TURKEY: Env. of Istanbul, Pendik, 1892, *Aznavour* (G); San Stefano, 1900, *Aznavour* (G). LEBANON: Beyrouth, *Girandy* (G). LIBYA: Cirenaica, Al Gubba-Ain-Mara, 1934, *Pampanini* et al. 4022 (W). ALGERIA: Alger, champs, 1939, *Bové* (G); lit de la rivière de Chiens près de Constantine, 1857, *Cholette* 332 (G). MADEIRA: Madeira, 1838 (G); in pascuis Machico, 1865–66, *Mandon* 139 (W).

194. T. cinctum DC., Cat. Hort. Monsp. 152 (1813); Ser. in DC., Prodr. 2: 193 (1825); Hayek, Prodr. Fl. Penins. Balc. 1: 862 (1926). *Typus*: France, au Port Juvenal, *Saltzman* 1820 (G-DC holo).

T. succinctum Vis., Pl. Rar. Dalm., in Flora 12, Ergänzungsbl. 1: 21, no. 32 (1829).
T. maritimum Huds. subsp. *cinctum* (DC.) Gib. & Belli, Mem. Accad. Sci. Torino ser. 2, 39: 289 (1889); Thell., Mém. Soc. Sci. Nat. Math. Cherbourg 38: 320 (1912).

Icon.: [Plate 189]. Reichenb., Icon. 22: t. 2148, f. 1, 1–5 (1903).

Annual, patulous-hairy (especially in upper part), 25–50 cm. Stems erect or ascending, striate, sparingly branching above. Leaves sparse, long-petioled on lower part of stem, short-petioled above, the uppermost ones opposite, all hirsute, with hairs tuberculate at base; stipules oblong, united part membranous with prominent green nerves, adnate to petiole, free portion lanceolate-subulate, green, about as long as lower part; leaflets varying from base of stem to top, the lower ones obovate, about 1 cm long, the upper ones elliptical to oblong, 2–3(–4) × 0.6–1 cm, sparingly appressed-hairy on both surfaces, denticulate in upper part, mucronate. Heads 1.5–2 × 1–1.5 cm, ovate, rather long-pedunculate, subtended at base by membranous, prominently nerved, whitish involucre up to 8 mm long, cleft into 8–15 unequal, acute, dentate lobes. Flowers many, 1.1–1.3 cm long. Calyx tube 2–3 mm long, obconical or tubular, glabrous, white, with 10 prominent nerves; teeth very unequal,

Plate 189. *T. cinctum* DC.
Plant in flower; flower; standard;
fruiting calyx from above; dissected calyx.

sharp-pointed, often with purple tips, subulate with lanceolate base, long-hirsute at base and all along, hairs arising from bulbous base, the lower one 1.5–3 mm, 3-nerved, the others about half its length. Corolla white to cream-coloured, slightly longer than lower calyx tooth; standard oblong, limb 5–6 mm long, lingulate above, much longer than wings and keel. Fruiting calyx somewhat corky or leathery, with white, campanulate, glabrous, 5 mm long tube and prominent nerves only in lower part; teeth of fruiting calyx stellately spreading, markedly bulbous-hairy; throat closed by callosity. Fruiting head (vividly recalling that of *T. squamosum*) 1.5–2 × 1.5–2 cm (including spreading teeth). Pod 2–3 mm, membranous, indurated at top; style short. Seed 2–2.2 mm, ovoid, brown. Fl. April–June.

Hab.: Forests, meadows and probably elsewhere.

Gen. Distr.: France, Italy, Balkan Peninsula.

Selected specimens: YUGOSLAVIA: Dalmatia, *Velden* 824 (W) in fruit; *ibid.*: in pratis ad Salonam, *Pichler* 17 (W). ALBANIA: in herbidis silvaticis pr. Pristani, 1898, *Baldacci* 333 (G).

Note: *T. cinctum* should be kept apart from its nearest relative *T. squamosum*, not only on account of its involucre, but also because of the shape and indumentum of the calyx, and the length of the corolla. It is probably endemic in the Balkan Peninsula and has adventively spread into Italy and France.

Subsection XV. URCEOLATA Zoh.

Candollea 27 : 255 (1972)

Maritima Gib. & Belli, Mem. Accad. Sci. Torino ser. 2, 39 : 362 (1889) p. p. pro stirpe.

Calyx 10-nerved; tube urceolate, green or grey, pubescent or glabrous; throat closed by bilabiate callosity; calyx teeth triangular-lanceolate to lanceolate, 3–5-nerved, equal or lower one much longer. Corolla white or cream-coloured or pink, as long as or longer than calyx, caducous.

195. T. juliani Batt., Bull. Soc. Bot. France 34 : 387 (1887); Thell., Mém. Soc. Sci. Nat. Math. Cherbourg 38 : 318 (1912); Murb., Contr. Fl. N-W Afr. 1, in Lunds Univ. Arrskr. 33/12 : 64 (1897). *Typus*: Algeria, "in uliginosis montosis ca. Constantine, Djebel Ouach Meridj, ubi delectus fuit anno 1886", leg. *Julien* (P holo.)

Icon.: [Plate 190].

Annual, patulous-villous, 20–50 cm. Stems few, angular, hollow, erect or ascending, profusely branching. Leaves large, located all along stems and branches, alternate except for uppermost ones which are opposite; stipules of middle part of stem 2–3 cm, linear-oblong, brownish-green, glabrous or very sparingly hairy, united and

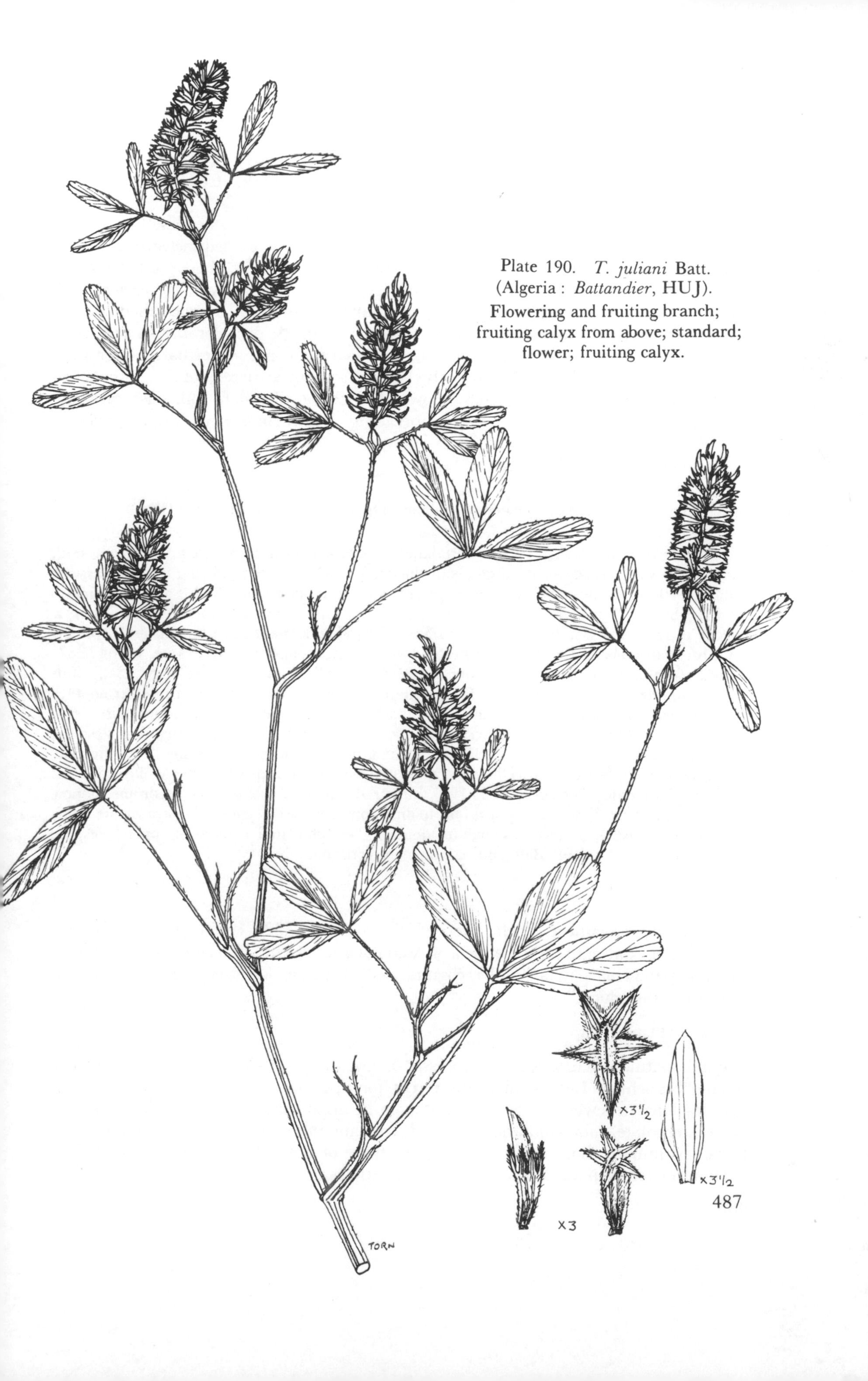

Plate 190. *T. juliani* Batt. (Algeria : *Battandier*, HUJ). Flowering and fruiting branch; fruiting calyx from above; standard; flower; fruiting calyx.

adnate to petioles for up to one-third of their length, free portion green, linear-lanceolate, up to 1.5 cm long, ciliate; leaflets 2–4 × 0.5–1.5 cm, appressed-hirsute (hairs with bulbous base), oblong-elliptical to linear-elliptical, obtuse or retuse with few teeth at apex. Heads terminal, short-peduncled, very loose, many-flowered, ovoid to cylindrical. Flowers 7–9 mm long, borne on furrowed rhachis. Flowering calyx about 3 mm; tube cylindrical to obconical, 10-nerved, with short patulous hairs; teeth triangular, acuminate, 3-nerved, ciliate, unequal, the lower one slightly longer than others, all sharp-pointed, considerably shorter than tube. Corolla twice as long as calyx tube; standard oblong, somewhat shorter than slightly auriculate wings. Fruiting calyx white, somewhat urceolate, i.e., broadened in middle and slightly constricted below teeth; throat closed by bilabiate callosity; teeth green, broadening and stellately spreading. Pod membranous. Seed 1, yellowish. Fl. May.

Gen. Distr. : Tunisia, Algeria.

Specimens seen : TUNISIA : in herbosis montis Dir el Kef, 1896, *Murbeck* (BM).

Note : The original description by Battandier was based on a specimen grown from seed. Later he probably changed his mind about this binominal, and the following is an interesting remark by him on the relations of this species to other related taxa (Bull. Soc. Dauphinoise ser.11, 1890) : "*Trifolium juliani* Batt. – Notre savant correspondant d'Algérie nous a envoyé cette plante sous le nom de *T. xatardii* Ser., abandonnant ainsi le nom de *T. juliani* qu'il lui avait d'abord donné dans le Bulletin de la Société botanique de France, vol.34 : 387, et tout récemment encore dans sa Flore d'Algérie 235. C'est sur une lettre de M. Belli, l'un des auteurs de la Monographie des Trèsfles italiens, que M. Battandier a cru devoir adopter cette nouvelle manière de voir. Mais la comparison minutieuse que nous avons faite de la plante de Constantine avec la description que De Candolle, véritable auteur de l'espèce, donne de son *T. xatardii*, en 1815, Fl. Fr. 5 : 558, nous laisse des doutes sur l'identification de la plante d'Algérie avec celle que Xatard avait récoltée à Prats de Mollo dans les Pyrénées-Orientales. Seringe (1825 : 193), auteur du genre *Trifolium* du Prodrome, donna une description du *T. xatardii* qui, il faut le dire, convient un peu mieux à notre plante, mais qui vu sa concision, ne lève pas tous les doutes. C'est pourquoi nous avons publié l'espèce sous le nom de *T. juliani* Batt., qui est un nom certain".

196. T. daveauanum Thell., Repert. Sp. Nov. 3 : 282 (1907). *Typus* : France, Montpellier (adventive), champ à l'Aiguelongue, 1898, *Daveau* (MPU holo.). "Planta peregrina patria haud recognita, sed affinitate Mediterranean in Galliam australem introducta".

Icon : [Plate 191].

Annual, patulous-hairy below, appressed-hairy above, 20–30 cm. Stems furrowed-striate, branching. Leaves alternate except for uppermost pair which are opposite; petioles long in lower leaves and very short in upper ones; stipules of lower leaves ovate-lanceolate, submembranous, nerved, connate and sheathing below, free portion linear-acuminate, shorter than adnate part, those of upper leaves narrower with free portion longer than adnate part, all pilose; petioles and leaflets white-haired; leaflets

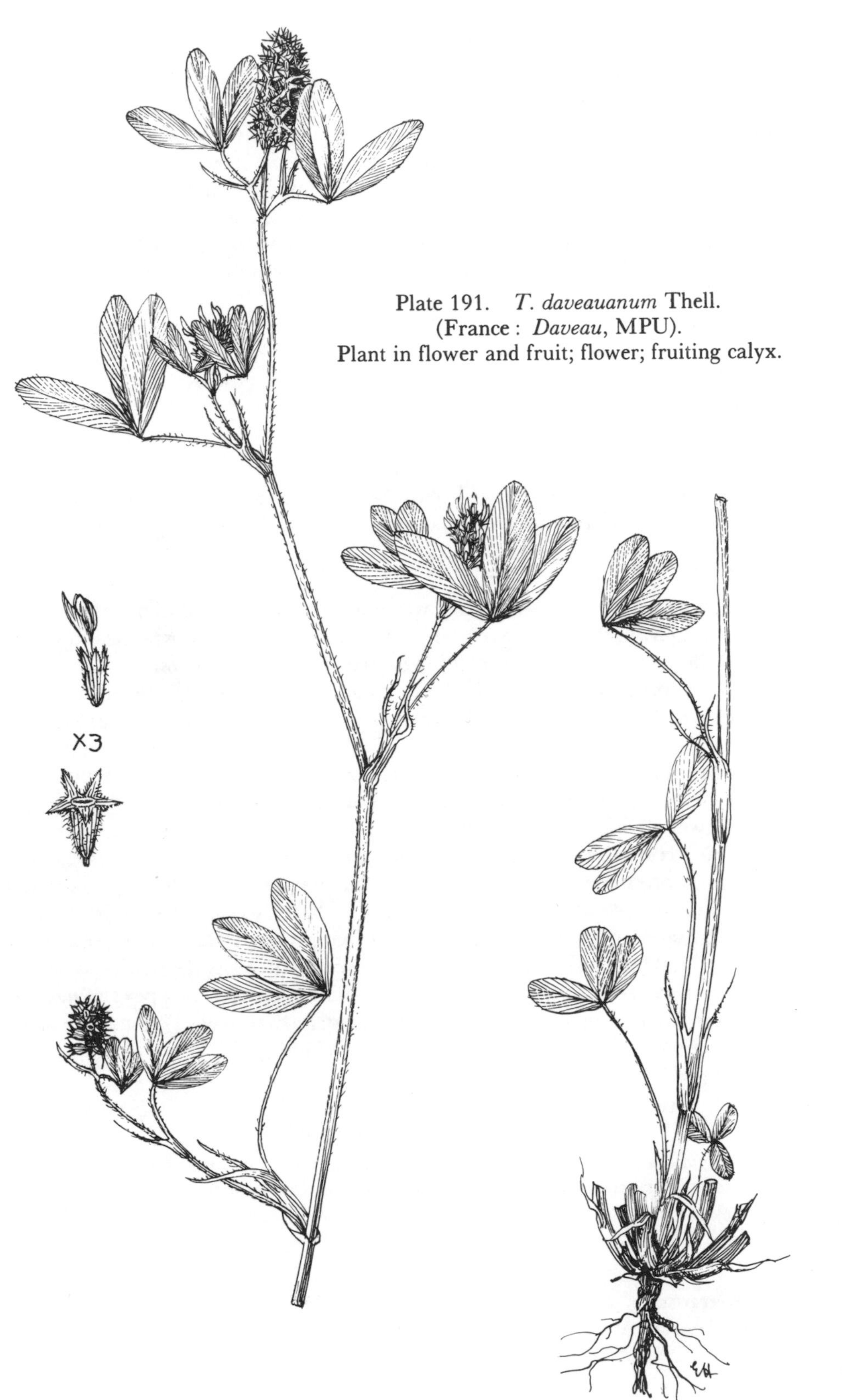

Plate 191. *T. daveauanum* Thell.
(France : *Daveau*, MPU).
Plant in flower and fruit; flower; fruiting calyx.

2–3 × 0.8–1 cm, sessile, oblong-elliptical, finely toothed in upper part, obtuse, mucronulate, those of lower leaves slightly notched. Fruiting heads 2.5–3 × 1 cm, short-peduncled, cylindrical. Flowers unknown. Fruiting calyx 4–6 mm, horizontally spreading, with callous base, sessile on pilose rhachis; tube urceolate, with 10 prominent nerves, subpatulous-hairy along nerves; throat closed by bilabiate callosity with only narrow slit between lips; teeth triangular-ovate, equal, much shorter than tube, 3-nerved, stellately spreading. Pod ellipsoidal, membranous, without operculum, indehiscent. Seeds solitary, yellowish-brown, compressed-ellipsoidal, smooth. Fl. May.

Hab.: Fields.

Gen.Distr.: Not certain.

Specimens seen: FRANCE: only the above-cited type specimen.

197. T. miegeanum Maire, Contr. Fl. Afr. N., in Bull. Soc. Hist. Nat. Afrique N. fasc. 19, 23 : 177 (1932). *Typus*: Morocco, env. Rabat, 1926, *Miège* (P holo).

Icon.: [Plate 192].

Annual, more or less appressed-villous above, later glabrous, 15–25 cm. Stems erect, simple, branching from base. Leaves all petiolate, the uppermost ones opposite; stipules of middle part of stem linear-oblong, sparingly villous, prominently nerved, adnate to petioles for up to about half their length, free portion green, narrow, linear, 1.2–2 cm; leaflets of lower leaves obovate, those of middle and upper ones ovate-oblong to lanceolate, 2–3(–4.5) × 0.8–1.5 cm, obtuse, slightly emarginate, with long hairs mostly tuberculate at base, obsoletely denticulate at margin. Heads 2–3.5 × 1–2 cm, short-pedunculate, terminal and axillary, ovoid-conoid, ovoid-oblong (in fruit), many-flowered. Flowers 7–8 mm long. Calyx tube almost tubular-cylindrical in flower, ovoid-urceolate in fruit, appressed-villous, with 10 prominent nerves; teeth of calyx with 3 prominent nerves, slightly unequal, sharp-pointed, triangular-lanceolate, acuminate-cuspidate at apex, broadest at base (compare *T. obscurum*), as long as or somewhat shorter than tube, ciliate at margin, stellately divergent in fruit; throat closed by bilabiate callosity, ciliate along central slit. Corolla white, as long as or shorter than calyx teeth; standard with oblong limb, rounded at apex; wings considerably shorter than standard and somewhat longer than keel. Pod membranous. Seed 1, ovoid, smooth, about 2 mm. Fl. April. $2n = 16$.

Hab.: Pastures.

Gen. Distr.: Morocco (endemic).

Note: *T. miegeanum* is very close to *T. obscurum*, but differs from it in a few characteristics, including the following one which is of great diagnostic importance: its lanceolate calyx teeth are not narrowed at the base.

198. T. squarrosum L., Sp. Pl. 768 (1753); DC., Fl. Fr. 5 : 531 (1815). *Typus*: Described from Spain, Hb. Linn. 930/31.

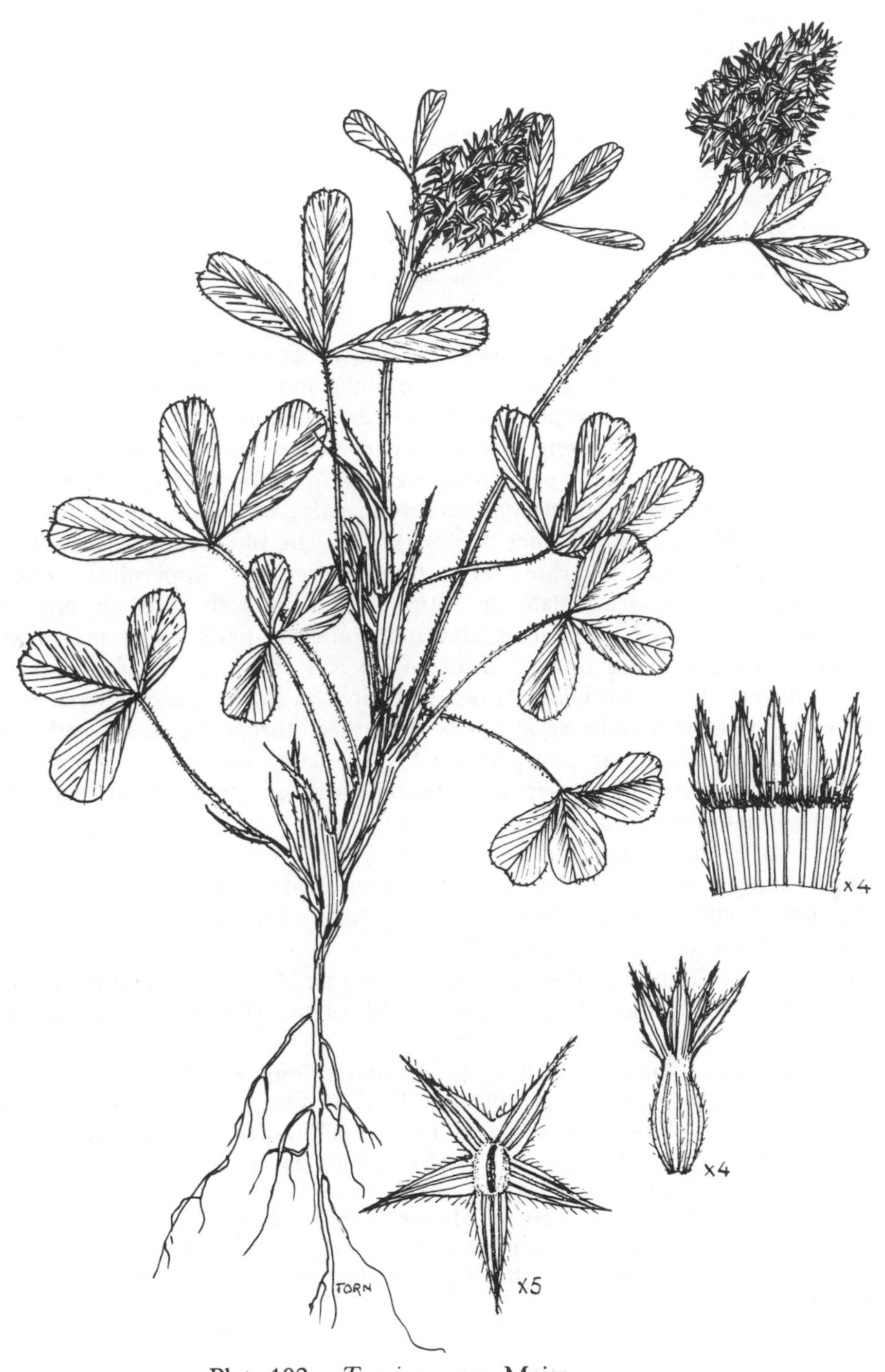

Plate 192. *T. miegeanum* Maire
(Morocco: *Miège*, P).
Plant in flower and fruit; dissected calyx; calyx; fruiting calyx from above.

T. dipsaceum Thuill., Fl. Env. Par. ed. 2 : 382 (1799); Gib. & Belli, Mem. Accad. Sci. Torino ser. 2, 39 : 362 (1889).
T. oblongifolium Ser. in DC., Prodr. 2 : 197 (1825).
T. panormitanum C. Presl., Fl. Sic. 1 : XXI (1826); Godr. in Gren. & Godr., Fl. Fr. 1 : 409 (1848–49); Boiss., Fl. 2 : 128 (1828).
T. longistipulatum Loisel., Fl. Gall. ed. 2, 2 : 122 (1828).
T. squarrosum L. var. *majus* et var. *minus* Rouy, Fl. Fr. 5 : 115 (1899).

Icon. : [Plate 193]. Reichenb., Icon. 22 : t. 2139, f. 1, 1–5 (1903); Gib. & Belli, *loc. cit.* t. 6, f. 5.

Annual, glabrescent or appressed-hairy, 25–60 cm. Stems erect or ascending, branching. Leaves long-petioled on lower part of stem and short-petioled to subsessile on upper part, all alternate except uppermost ones which are opposite; stipules (of middle leaves) about 3(–5) cm long, oblong-linear, connate and adnate to petioles for up to about one-third of their length, membranous, prominently nerved, glabrous, free portion green, subulate-caudate or lanceolate, hairy or ciliate; lower leaflets small, obcordate, middle and upper ones 3–5 × 0.5–1.5 cm, oblong-lanceolate or elliptical, obtuse, rarely ovate-truncate or slightly retuse, uppermost ones often acute-mucronate, all denticulate or entire, hairy on both sides or only beneath. Heads short-peduncled, sometimes almost involucrate by 2 upper leaves, terminal, solitary, ovoid or globular or conical, up to 2.5 × 1.5(–2) cm, many-flowered, rather dense. Flowers 0.8–1 cm. Calyx tube cylindrical in flower (urceolate in fruit), hirsute with hairs often bulbous at base, 10-nerved; throat closed by callosity, hairy along slit; teeth triangular, sharp-pointed, lanceolate, unequal, 3-nerved, the lower one longer than tube, the upper ones shorter, erect or spreading in fruit, the 2 uppermost connate for up to one-third of their length. Corolla whitish-pink, exceeding longer calyx teeth in length; wings much shorter than lanceolate standard. Fruiting calyx becoming constricted at end of tube. Pod membranous with hardened operculum. Seed 1, ovoid, brown, shiny. Fl. April–June. 2n = 14, 16.

Hab. : Meadows, pastures and fallow fields.

Gen. Distr. : Germany, Portugal, Spain, France, Corsica, Sardinia, Italy and islands, Balkan Peninsula, Crimea, Algeria, Morocco, Mauritania, Canary Islands.

Selected specimens : PORTUGAL : Prov. Estremadura, Sintra no Algueirao, nos terrenos em pousio, 180 m, 1950, *Bento Rainha* 1933A (HUJ). CORSICA : près de l'étang de Biguglia, 1867, *Mabille* 221 (W). SARDINIA : Santa Teresa Galhira par Tempio, prairies calcaires de Bancamino, 1881, *Reverchon* 193 (P). ITALY : pâturages et moissons de champs dans les terrains argileux près de Pise (Toscana), 1857, *Savi* 455 (W). SICILY : in cultis Palermo, *Todaro* (W). ALGERIA : Fossés près de Maison Blanche (dép. d'Alger), 1957, *Dubius* et al. 2969 (W). MOROCCO : Ripas fl. Lukas, 1930, *Font-Quer* 343 (BM). MAURITANIA : *Bové* 45 (W). CANARY ISLANDS : Teneriffa, Laguna, in collibus 500–600 m, 1900, *Bornmüller* 486 (W).

Note : *T. squarrosum* is a well-delineated species. It has the largest flowering and fruiting head in Subsect. *Urceolata*. The head, which may reach 2 cm in width, is never elongated to form a spike, but remains globular or ovoid in fruit and is most striking on account of the divergent, long calyx teeth; the lower tooth is much longer than the others, and the 2 upper

Plate 193. *T. squarrosum* L.
(Algeria : *A. Faure*, HUJ).
Plant in flower and fruit; dissected calyx;
fruiting calyx from above; flower.

ones are connate for about half their length. The basi-bulbous hairs which often cover the calyx are also very characteristic. Its corolla, unlike that of *T. obscurum*, is longer than the calyx. The attempt to follow Gibelli & Belli or Ascherson & Graebner in subdividing the species into varieties has failed.

199. T. obscurum Savi, Obs. Trif. 31 (1808–10); Caruel, Prodr. Fl. Tosc. 161 (1860) = *T. obscurum* Savi emend. Gib. & Belli, Mem. Accad. Sci. Torino ser. 2, 39 : 372 (1889). *Typus* : Italy, di Pisa dalla compaignon, *Michelli* 24 (FI ? holo.).

T. panormitanum C. Presl var. *aequidentatum* Perez-Lara, Fl. Gadit. 4 : 459 (1891).

T. dipsaceum Thuill. var. *aequidentatum* (Perez-Lara) Willk., Suppl. Prodr. 246 (1893).

T. isodon Murb., Contr. Fl. N-W Afr. 1, in Lunds Univ. Arrskr. 33/12 : 64, t. 3, f. 7, 8 (1897).

T. obscurum Savi subsp. *aequidentatum* (Perez-Lara) C. Vicioso, Rev. Gen. Trif., in Anal. Inst. Bot. Cavanilles 11 : 344 (1953).

Icon. : [Plate 194]. Savi, *loc. cit.* f. 1; Gib. & Belli, *loc. cit.* t. 7, f. 4.

Annual, sparingly appressed-hairy to glabrescent, 20–50 cm. Stems ascending or erect, branching from base. Lower leaves long-petioled, upper ones short-petioled, uppermost ones subsessile, opposite; stipules oblong-linear, membranous, green-nerved, connate and adnate to petioles for up to half their length, sparingly villous to glabrous, free portion lanceolate-subulate, green, ciliate; leaflets of lower leaves obovate-cuneate or obovate, truncate or retuse, those of middle and upper ones oblong, elliptical or lanceolate, denticulate in upper part, slightly emarginate. Heads 1.5–2 × 1.5 cm, solitary, terminal, pedunculate, ovate to globular, elongating in fruit. Flowers up to 1 cm long, loose. Calyx tube oblong-fusiform in flower, urceolate-inflated in fruit, 10-nerved, more or less villous, sometimes glabrescent; throat with callous thickening, closed in fruit; teeth lanceolate, more or less equal in length, sharp-pointed, 3-nerved, with base narrower than middle part, about as long as tube, loosely covered with long hairs. Corolla whitish to pink, somewhat shorter than or as long as calyx, caducous; standard oblong-linear, obtuse. Pod membranous, indehiscent, with leathery apex. Seed 1, ellipsoidal, yellowish. Fl. April–May. 2n = 16.

Hab. : Damp pastures.

Gen. Distr. : Italy, Turkey, Algeria, Morocco (and probably elsewhere).

Selected specimens : ITALY : San Casciano di Bagni, ad occid. urbis Chiasi, juxta confinum romanum, etc., 1881, *Levier* (P). TURKEY : Env. of Constantinopole (Istanbul), Halki ou près de Kichathanaköy, 1888, *Aznavour* (G). ALGERIA : Terny sur Tlemcen, 1890, *Battandier* & *Trabut* 526 (G). MOROCCO : in pascuis humidis Atlantis medii inter Azron et Ito, solo basaltico, 1,200–1,300 m, 1936, *Maire* (G, as *T. isodon* Murb.)

Note : *T. obscurum* is a well-delineated species, differing from all other allied species by its short corolla and especially by the particular shape of the equal calyx teeth, which are narrowed at the base. *T. isodon* is, without doubt, identical to *T. obscurum*. The few differences recorded by Murbeck do not appear to be constant.

Plate 194. *T. obscurum* Savi
(Italy : *Levier*, P).
Plant in flower and fruit; dissected calyx; young fruiting calyx; fruiting calyx from above; flower.

200. T. constantinopolitanum Ser. in DC., Prodr. 2 : 193 (1825); Boiss., Diagn. ser. 2, 2 : 14 (1856) non Fl. 2 : 127 (1872). *Typus* : Turkey ad Byzantium, *Castagne* (G-DC holo.).

T. alexandrinum L. var. *phleoides* (Boiss.) Boiss., Fl. 2 : 127 (1872); Hossain, Not. Roy. Bot. Gard. Edinb. 23 : 426 (1961).

Icon. : [Plate 195].

Annual, appressed- or patulous-hairy, (10–)15–35(–60) cm. Stems erect or ascending, few or many, moderately branching. Leaves all, except uppermost ones, alternate; adnate part of stipules linear, membranous, with longitudinal green or blackish nerves, free portion as long as or longer than lower part, lanceolate or subulate; leaflets (0.8–)1.5–2(–3) × 0.4–0.6(–0.8) cm, elliptical to obovate-oblong, tapering at both ends or cuneate at base, obtuse or mucronate at apex, appressed-hirsute, with denticulate upper part. Flowering heads 1.2–2.2 × 1–1.5(–2.2) cm, terminal, pedunculate, ovoid to obconical; flowers rather loose. Calyx tube cylindrical in flower, with 10 prominent nerves; teeth narrowly lanceolate to subulate, purple-tipped, 1-nerved, rarely with 3 nerves and then only middle one distinctly visible, densely hairy at base, lower tooth 1.5 times as long as others (sometimes shorter) and as long as or slightly longer than tube. Corolla 1.5–2 times as long as calyx; standard with broad-linear limb, up to 1.5 times as long as wings and keel; keel often with purple tip. Fruiting head 1.3–2(–2.5) × 1–1.5 cm, cylindrical-oblong to conical. Fruiting calyces rather loose so that individual tubes are easily discerned and somewhat two-ranked along rhachis; tube urceolate, appressed-hirtellous; teeth subulate, thickened, stellately spreading; throat entirely closed by bilabiate callosity, ciliate along slit. Pod membranous with coriaceous apex, oblong, apiculate. Seed solitary, about 1.5 mm, ovoid, yellowish-brown. Fl. April–June. 2n = 16.

Hab. : Damp fields, roadsides, river banks, etc.

Gen. Distr. : Switzerland, France, Italy (introduced), Turkey, Syria, Lebanon, Palestine, Algeria (probably adventive).

Selected specimens : SWITZERLAND : Solothurn, Bahnhof der Zollikofen, 1919, *Streun* (HUJ). FRANCE : Cherbourg, littoral, 1859, *Le Jolis* 35 (W). ITALY : Trieste, Campo Mayio (advena), 1882, *Marchesetti* (W, E). TURKEY : Antigone, 1893, *Aznavour* 663a (G); collines, bords des champs, Kathane-Chichli, 1893, *Aznavour* 663 (G); champs de l'île de Kilsali (golfe de Smyrna), 1854, *Balansa* 165 (G, W, E, P); Bafra, Orman Isletime, open field in grass, 20 m, 1963, *Tobey* 387 (E); Hatay, around marshes of Amouk, 1932, *Delbès* (HUJ). LEBANON : Marj Ayyoun, 1947, *Mouterde* (HUJ); env. de Saïda, 1854, *Blanche* (P). PALESTINE : Dan Valley, banks of Hatsbani River, near Ma'ayan Barukh, 1963, *Zohary* & *Plitmann* 21/3 (HUJ). ALGERIA : Oran, dans les forêts de l'est Algérie, 1914, *d'Alleizette* (P).

Note : In various herbaria one finds this species under different names, e.g., *T. scutatum, T., echinatum, T. alexandrinum,* etc.

201. T. leucanthum M. B., Fl. Taur.-Cauc. 2 : 214 (1808); Boiss., Fl. 2 : 128 (1872); Hayek, Prodr. Fl. Penins. Balc. 1 : 862 (1926); Post, Fl. Syr. Pal. Sin. 408

Plate 195. *T. constantinopolitanum* Ser.
(Palestine : *Zohary*, *Baum* & *Plitmann*, HUJ).
Flowering and fruiting branches; flower; fruiting calyces.

(1883–93) et ed. 2, 1 : 339 (1932). *Typus* : Crimea, in Tauriae meridionalis collibus siccis, *Bieberstein* (LE).

T. obscurum Guss., Cat. Pl. Boccadif. 65 (1821) non Savi (1808–10).
T. reclinatum sensu Griseb., Spicil. Fl. Rumel. 1 : 21 (1843) non Waldst. & Kit. (1810–11).
T. leucanthum M. B. var. *declinatum* Boiss., Fl. 2 : 128 (1872).
T. leucotrichum Petrovic, Fl. Agr. Nyss. 228 (1882).
T. dipsaceum Thuill. subsp. *leucanthum* (M. B.) Gib. & Belli, Mem. Accad. Sci. Torino ser. 2, 39 : 369, 422, t. 7 (1889).

Icon : [Plate 196]. Reichenb., Icon. 22 : t. 2148, f. 2, 6–17 (1903).

Annual, patulous- to appressed-pubescent, 10–20(–35) cm. Stems erect or ascending, few or many, striate, often dichotomously branching. Leaves all, except uppermost ones, alternate; adnate part of stipules linear, membranous, with longitudinal green nerves, free portion usually herbaceous, as long as lower part, triangular-caudate; leaflets of lower leaves obovate, cuneate, notched at apex, those of upper ones elliptical-oblong to obovate-oblong, obtuse or acute, usually denticulate at apex. Heads 1–1.5(–2) × 1–1.5 cm, terminal, with long peduncles, sometimes provided at base with long or short involucre, semi-globular to almost globular. Calyx tube obconical in flower, 10-nerved, densely covered with curved or patulous, long hairs; teeth slightly unequal, broad, lanceolate-subulate, usually 3-nerved, with or without purple tip, as long as or longer than tube. Corolla cream-coloured to pink and violet with purple blot at tip of keel, shorter, as long as, or slightly longer than calyx; standard slightly longer than wings and keel. Fruiting head semi-globular to globular. Fruiting calyces rather distinct from one another; tube urceolate; teeth stellately spreading, ciliate; throat closed by hairy callosity. Pod ovoid, membranous in lower part, coriaceous above, apiculate. Seed solitary, 1.2–2 × 1–1.6 mm, globular to obovoid. Fl. April–June. 2n = 14, 16.

Hab. : Mountain slopes, damp and grassy places, forest clearings.

Gen. Distr. : Germany, Spain, France, Corsica, Sardinia, Sicily, Balkan Peninsula, Crete, Crimea, Turkey, Cyprus, Syria, Palestine, Iraq, Iran, Algeria.

Specimens seen : SPAIN : inter Duruelo et Penaranda de Bracamonte (Ovila-Salamanca), in graminosis humidis, 1956, *Lainz* (W). FRANCE : Paris, 1843, *Spach* (P). SARDINIA : in herbidis prope Tortoli, *Moris* (P). SICILY : Palermo, *Todaro* (P); in pascuis submontosis Ficuzza, Portella del vento, 1856, *E.* & *A. Huet du Pavillon* 38 (P). GREECE : ad declivitates umbrosas montis Taktuli (G). BULGARIA : in graminosis ad Nova Mahala, 1894, *Stribrny* (K). CRIMEA : Jalta-Alupka, in declivitatibus siccis subaperis, 1901 (K). TURKEY : Turquie d'Europe, 1842, *Grisebach* (G); Malatya distr., 20–30 km W. of Malatya, *Quercus cerris* forest remnants, 1,600 m, 1963, *Zohary*, *Orshan* & *Plitmann* 27522 (HUJ); Sinus Smyrnaeus, in monte "Dyo-Adelphia" (Iki-Karadasch), 600–800 m, 1906, *Bornmüller* 9327 (G); village d'Alla-Dagh à 7 lieues au N.O. de Mersine (Cilicie), région montagneuse, 1855, *Balansa* 449 (G). CYPRUS : Platres, in pine forest by edge of overflow from a water tank, 1,219 m, 1941, *Davis* D. 3150 (HUJ). PALESTINE : Upper Galilee, Wadi Hish, *Naftolsky* 17003 (HUJ). IRAQ : Distr. Sulaimaniya, inter Sulaimaniya et Dokan, in jugo prope Surdash, in apertis quercetorum ca. 1,000 m, 1959, *Rechinger* 12470 (W).

Plate 196. *T. leucanthum* M. B.
(Greece : *F. Guiol* 146, HUJ).
Plant in flower and fruit;
flower; fruiting calyx from above.

Subsection XVI. ECHINATA Zoh.

Candollea 27 : 259 (1972)

Annuals. Calyx 10-nerved; nerves of fruiting calyx sometimes disappearing; throat closed by bilabiate callosity; teeth subequal, erect in fruit or divergent, not divaricate; calyx bases connate to one another and to rhachis of head; heads not disarticulate in fruit. Corolla longer than calyx.

202. T. latinum Seb., Rom. Pl. Fasc. 1, 7, t. 1, f. 2 (1813); Boiss., Fl. 2 : 126 (1872); Hayek, Prodr. Fl. Penins. Balc. 1 : 862 (1926). *Typus* : Italy, sylvula prope Romam versus mare vulgo dicta Macchi de Mattei, 1813, *Sebastiani*.

Icon. : [Plate 197].

Annual, appressed- to patulous-hirsute, 20–50 cm. Stems few, erect or ascending, slightly striate, branching above. Leaves all, except uppermost ones, alternate; adnate part of stipules linear, submembranous, with longitudinal green nerves, free portion much longer (at least in upper leaves), linear-subulate, herbaceous; leaflets 2–2.5(3.5) × 0.2–0.5(–1) cm, oblong-linear, cuneate at base, acuminate, appressed-pubescent, denticulate in upper part. Head terminal, pedunculate, 1.5–2(–2.5) × (1.3–)1.6–2 cm, obconical to ovoid. Calyx tube short-obconical, densely long-hirsute (hairs concealing 10 nerves); teeth subulate with triangular base, never purple-tipped, with 3 nerves at base and all along, rarely along lower tooth only, the latter with hairs sometimes arising from tubercles (hence margin appearing denticulate-ciliate), longer than others, at least 2.5–3 times as long as tube. Corolla (1.5–)2–2.5 times as long as calyx, white to pink, with keel purple-blotted at tip; standard with lanceolate to linear limb, somewhat longer than wings and keel. Fruiting head 1.4–1.8(–2) × 1–1.8(–2) cm, ovoid to globular, compact. Fruiting calyces dense and hardly separable from one another, united to each other at base and to rhachis and persisting on latter; tube increasing in size, obconical, villous; teeth erect, diverging, those of lower calyces reflexed, all 3-nerved at maturity; throat of calyx closed by bilabiate callosity with hairs along slit. Pod membranous in lower part, coriaceous above, ovoid, apiculate. Seed solitary, about 1.5 × 1.1. mm, ovoid, apiculate. Fl. April–June. 2n = 16, 28.

Hab. : Forests and maquis.

Gen. Distr. : France, Italy, Greece, Bulgaria, Turkey.

Selected specimens : GREECE : Nutades in Pinde tymphaeo, 1896, *Sintenis* 452 (P). BULGARIA : in collinis ad Suraw, 1910, *Stribrny* (G). TURKEY : Sinus Smyrnaeus, in monte "Dyo-Adelphia" (Iki-Karadasch), 600–800 m, 1906, *Bornmüller* 9321 (E).

203. T. echinatum M. B., Fl. Taur.-Cauc. 2 : 216 (1808).

Annual, appressed- or patulous-pubescent above, glabrescent below, (10–)20–50 cm. Stems few to many, erect, ascending or procumbent, striate, branching from base or

Plate 197. *T. latinum* Seb.
Plant in flower; flower;
fruiting calyx from above.

all along. Leaves all, except uppermost ones, alternate; adnate part of stipules linear-oblong, membranous, with dark longitudinal nerves, free portion as long as or longer than the lower part, triangular-subulate, herbaceous, long-villous; leaflets (0.8–)1.2–2(–2.5) × 0.4–1.5 cm, linear or broadly elliptical, tapering at both ends or obovate-cuneate, obtuse or acute, entire or retuse, appressed-pubescent, with denticulate upper part. Flowering heads pedunculate, 1–2.3 × 1–1.8(–2) cm, terminal and axillary, broad-ovoid to obconical or obovoid. Calyx tube cylindrical to obconical in flower, hirsute in upper part, with 10 nerves; teeth distinctly unequal, usually 1-nerved, subulate, with purple tip, the lower one usually with patulous hairs arising from tubercles, entire or serrulate at margin, 2–3 times (or more) as long as tube and considerably longer than lateral teeth which are 1.5 times as long as tube. Corolla cream-coloured (with or without purple blot at tip of keel) or pink (rarely violet); standard 1–1.5 cm, 1.5 times as long as wings and keel, oblong-spatulate. Fruiting head 1.4–2(–2.3) × 1.2–1.7(–2) cm, globular to ovoid, compact, echinate. Fruiting calyces adnate to each other at base and to rhachis, persistent; calyx tube campanulate, obsoletely nerved, becoming smooth and leathery, glabrescent; teeth erect, becoming spinulose with triangular, thickened base; throat closed by bilabiate, ciliate or glabrous callosity (throat of mature calyx opening slightly due to pressure of growing seed). Pod membranous, obovoid, with keel parallel to suture, apiculate. Seed solitary, about 1.3 mm long, ovoid. Fl. March–June. 2n = 16.

Hab. : Forest clearings, dwarf-shrub formations, waste places, river banks and roadsides.

Gen. Distr. : Germany, Switzerland, Hungary, Romania, Italy, Balkan Peninsula, Turkey, Cyprus, Syria, Lebanon, Palestine, Iraq, Transcaucasia, Libya.

1. Keel of corolla not purple at tip. Teeth of calyx entire, with sparse cilia. Fruiting head small, 1.2–1.6 × 1–1.4 cm. **203a.** var. **echinatum**
– Keel of corolla with purple tip. Teeth of calyx, especially lower one, serrulate, with long, patulous hairs. Fruiting head large, 1.5–2.3 × 1.5–2 cm. **203b.** var. **carmeli**

203a. T. echinatum M. B. Var. **echinatum**; Boiss., Fl. 2 : 126 (1872), pro syn.; Hayek, Prodr. Fl. Penins. Balc. 1 : 863 (1926). *Typus* : Ad Caucasum, *Bieberstein* (LE photo.).

T. supinum Savi, Obs. Trif. f. 2 : 46 (1808–10); Griseb., Spicil. Fl. Rumel. 1 : 22 (1843).
T. reclinatum Waldst. & Kit., Pl. Rar. Hung. 3 : 299, t. 269 (1810–11); Ser. in DC., Prodr. 2 : 197 (1825; err. "*reflexum*").
T. procerum Rochel, Pl. Banat. Rar. 50 (1828).
T. trichostomum Godr., Mem. Acad. Sci. Montp. sect. Med. 1, 4 : 427 (1853).
T. sefinense Freyn & Bornm. in Freyn, Mém. Herb. Boiss. 13 : 5 (1900).
T. echinatum M. B. var. *brevidens* Thell. in Zimmerm., Advent. Ruder. Fl. Mannheim 131 (1907).
T. echinatum M. B. subsp. *supinum* (Savi) Aschers. & Graebn., Syn. Mitteleur. Fl. 6/2 : 590 (1908).
T. echinatum M. B. subsp. *supinum* (Savi) Aschers. & Graebn. var. *trichostomum* (Godr.) Thell. in Aschers. & Graebn., *loc. cit.*

T. echinatum M. B. subsp. *supinum* (Savi) Aschers. & Graebn. var. *reclinatum* (Waldst. & Kit.) Aschers. & Graebn. et II *procerum* (Rochel) Aschers. & Graebn., *loc. cit.* 591.

Icon.: [Plate 198]. Reichenb., Icon. 22 : t. 2141, f. 2, 8–14 (1903).

Selected specimens: HUNGARY: in pratis fertilibus ad Danubium infra Alt-Moldova et ad Szakolovatz frequentissime; in loco posteriore cum *Trifolio resupinato* L. (locus classicus), *Borbàs* 1220 (P). ROMANIA: Reg. Constanta, raion Negru Voda, 1955, *Todar* & *Dinulescu* 58a (HUJ). ITALY: Toscane, lieux incultes et cultivés, bords des champs et des chemins moissons dans les terrains argileux, près de Pise, 1857, *Savi* 456 (HUJ). YUGOSLAVIA: Pratis siccis prope Belgrad, *Pancić* (W). GREECE: Thessalia: Larissa, in incultis et ad versuras planitiae 4 km oppido meridiem versus, 1961, *Rechinger* 22768 (W). BULGARIA: Gebüsche bei Obrasov-Irchifthik nächst Russe, 1930, *Zorny* (W). TURKEY: Thracia, 16 km N. of Tekirdag, shrub formation, 1963, *Orshan* (HUJ); prov. Adapazari, Sapanca Gölü, 50 m, eroded banks, 1962, *Davis* & *Coode* (E); S.E. Anatolia, E. of Diyarbakir, neglected field, heavy soil, 1964, *Zohary* & *Plitmann* 1962-9 (HUJ). CYPRUS: inter Heptakomi et Lionarissa, 1880, *Sintenis* & *Rigo* 414 (P). SYRIA: in humidis Aintab, 1865, *Haussknecht* (W). LEBANON: A'Eden, 1869, *Blanche* 816 (G). PALESTINE: Upper Galilee, Wadi Abu Ali, 1926, *Eig* & *Zohary* (HUJ). IRAQ: in collibus Dschebel Hamrin inter Baghdad et Kirkuk, 1892, *Bornmüller* 1138 (W). TRANSCAUCASIA: Azerbaidzhan, distr. Kjurdamir inter pag. Kjulalu et Kalagajly, ad vias, 1936, *Grossheim* 37 (HUJ); in graminosis Georgiae Caucasicae, 1834, *Hohenacker* (G). LIBYA: El Beda: Bir Tacar, 1934, *Pampanini* & *Pichi-Sermolli* 4023 (FI).

203b. T. echinatum M. B. Var. **carmeli** (Boiss.) Gib. & Belli, Mem. Accad. Sci. Torino ser. 2, 39 : 377 (1889). *Typus*: Palestine, Fauces Carmeli, April-May 1846, *Boissier* (G holo.).

T. carmeli Boiss., Diagn. ser. 2, 2 : 16 (1856) et Fl. 2 : 127 (1872); Post, Fl. Syr. Pal. Sin. 237 (1883–96) et ed. 2, 1 : 339 (1932); Opphr., Bull. Res. Counc. Israel D, 7 : 207 (1959).

T. constantinopolitanum Ser. in DC. var. *carmeli* (Boiss.) Thell., Vierteljahrschr. Nat. Ges. Zürich 52 : 454 (1907).

T. carmeli Boiss. var. *carmeli* Hossain, Not. Roy. Bot. Gard. Edinb. 23 : 427 (1961) p. p.

Icon.: [Plate 198].

Selected specimens: PALESTINE: Mt. Carmel, Shefeyia to Bat Shelomo, *Quercus* maquis, 1954, *Grizi* 16614 (HUJ). Probably also occurs in Lebanon and Syria.

Note: *T. echinatum* is exceedingly polymorphic in its vegetative parts, its habit and indumentum. There are also some variations in the parts of the flower. However, these are continuous and quantitative variations, many of which should be considered habitative deviations. Among others, stems range from dwarf to tall and from densely hairy to almost glabrous; leaves range from almost obovate to narrowly elliptical; flower colour varies from yellow to whitish and pink to violet. There is also marked variation in the proportion of the nervature and margin configuration of the teeth. All these are inadequate for dividing the species into varieties. Even var. *carmeli* is not sharply delineated from var. *echinatum* and intermediates between them occur in overlapping areas. It is, therefore, strange that Boissier (*loc. cit.*), Hossain (*loc. cit.*) and others considered *T. carmeli* as an independent species.

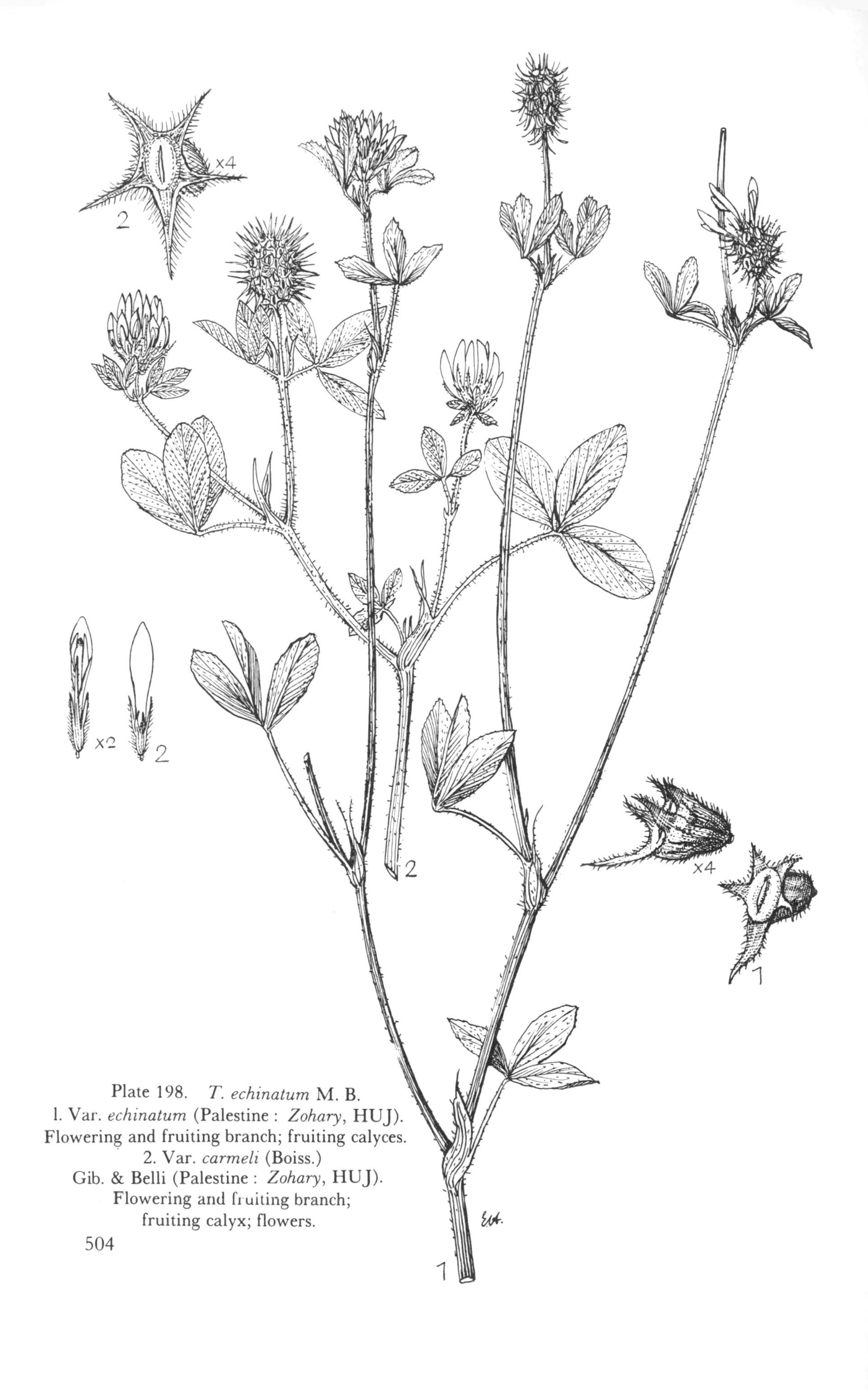

Plate 198. *T. echinatum* M. B.
1. Var. *echinatum* (Palestine : *Zohary*, HUJ).
Flowering and fruiting branch; fruiting calyces.
2. Var. *carmeli* (Boiss.)
Gib. & Belli (Palestine : *Zohary*, HUJ).
Flowering and fruiting branch;
fruiting calyx; flowers.

It is even more surprising that Ascherson & Graebner did not synonymize the many binomials of *T. echinatum*, an omission which has given rise to such confusion. Especially amazing is the fact that Hossain (*loc. cit.*) divided *T. carmeli* into 2 varieties according to the denticulation of the upper leaflet and the length of the lower calyx tooth.

Subsection XVII. CLYPEATA Gib. & Belli

Mem. Accad. Sci. Torino ser. 2, 39 : 393 (1889) pro stirpe

Annuals. Calyx 10-nerved; teeth unequal, reflexed in fruit, somewhat leaf-like, 3–5-nerved, the 2 posterior ones shorter than tube, long-connate; throat closed by bilabiate callosity. Corolla cream, white or pink, much longer than calyx. Heads not disarticulating (synaptospermic).

204. T. clypeatum L., Sp. Pl. 769 (1753); Boiss., Fl. 2 : 129 (1872); Gib. & Belli, Mem. Accad. Sci. Torino ser. 2, 39 : 394 (1889). *Typus* : Described from the Orient, Hb. Cliff. 373.

Icon. : [Plate 199]. Reichenb., Icon. 22 : t. 2152, f. 1, 1–8 (1903); Gib. & Belli, *loc. cit.* t. 8, f. 4.

Annual, patulous- to retrorsely-villous, 10–30 cm. Stems furrowed, erect or ascending, much branching, leafy. Leaves long-petioled; stipules almost membranous, semi-ovate, with delicate nerves branching above, free portion broad, ovate-triangular, acute, almost as long as or longer than united part; leaflets large, 1–3 × 0.5–2 cm, obovate to obovate-cuneate, sparingly appressed-hairy, dentate in upper part, the upper ones mucronate. Flowering heads long-peduncled, many-flowered, obovoid, 2–3 × 1–2.5 cm. Flowers pink to whitish, loose. Calyx tube green, tubular, prominently nerved, long-villous in upper part; teeth green, many-nerved, tuberculate-bristly, spiny-tipped, the lower one ovate-oblong, the others triangular to triangular-lanceolate, one-half to one-quarter the length of lower one. Corolla 2–3 times as long as calyx, pink to white; standard oblong-ovate, obtuse, slightly retuse, at least one-fourth longer than wings and keel. Fruiting head compact, 2–3 cm long. Fruiting calyx with campanulate, somewhat oblique, glabrescent tube and throat closed by broad, glabrous, callose cover; teeth stellately spreading, leaf-like, 3 mm or more broad, many-nerved, the lower one oblong-ovate, the lateral ones broad-ovate, one-third the length of the lower one, the upper ones very short, broad-triangular. Pod membranous-scarious, broad-ovoid, prominently nerved, 4 mm long, opening by an apiculate operculum. Seed cuneate-lenticular, brown, 3 mm long. Upper fruiting calyces detaching from rhachis, the others persisting on plant for a long time. Fl. January–May. 2n=16.

Hab. : Fields, roadsides and batha.

Gen. Distr. : Greece and Aegean Islands, Turkey, Cyprus, Syria, Lebanon, Palestine.

Plate 199. *T. clypeatum* L.
(Palestine : *Gabrielith*, HUJ).
Flowering branch; flower; fruiting head
(Palestine : *Zohary* & *Plitmann*, HUJ).

Selected specimens: GREECE: N. Syros in maquis on road up outside crater, 1963, *Gathorne-Hardy* 376 (E). TURKEY: Prov. Hatay, Antakya nr. St. Peter's Church, 300 m, rocky limestone slope, 1957, *Davis* & *Hedge* 27241 (E). CYPRUS: Platanisso, Karpas, 300 ft, edge of road, 1950, *Chapman* 207 (HUJ). SYRIA: Mts. Nussairy, Buhamra, 1909, *Haradjian* 2828 (G). PALESTINE: Upper Galilee, Kerem ben Zimra, Quercetum among basalt rocks, 1956, *Grizi* 16648 (HUJ). Also seen from Chios, Rhodes, Lebanon, Cyprus.

Note: *T. clypeatum* is one of the most common species. Although generally preferring habitats with a favourable moisture regime, it also occurs on stony hillsides and in open sunny areas. It ascends the mountains up to 1,200 m. According to habitat, it varies considerably in its dimensions, size of leaves, indumentum and flower heads. There is no room, however, for an infraspecific subdivision of this species.

205. T. scutatum Boiss., Diagn. ser. 1, 2 : 27 (1843) et Fl. 2 : 129 (1872); Post, Fl. Syr. Pal. Sin. 238 (1883–96) et ed. 2, 1 : 340 (1932). *Typus*: Turkey, in montosis Smyrnae supra Bournabat, V. 1842, *Boissier* (G holo.).

Icon.: [Plate 200].

Annual, densely or sparingly patulous-pubescent, 10–30 cm. Stems many, striate, ascending, branching all along. Leaves all, except uppermost ones, alternate; adnate part of stipules triangular to ovate, submembranous, with dark longitudinal nerves, sparingly villous, free portion as long as or longer than lower part, lanceolate-subulate, herbaceous, with long bristles; leaflets 1–1.4 × 0.5–0.8 cm, obtriangular to obovate with cuneate base, rounded or notched at apex, appressed-hirsute, with denticulate upper part. Head terminal, pedunculate, 1.8–2.3 × 1.3–1.8 cm in flower, ovoid to obovoid-globular in fruit. Flowers white to cream-coloured with purple blot at tip of keel. Calyx tube obconical in flower, glabrescent, with 10 prominent nerves; teeth unequal, never purple-tipped, ciliate, the lower one broad-lanceolate to broad-oblong, 3–5-nerved, somewhat broader than tube, 2–3 times as broad and at least 2 times as long as others which are triangular-acute to lanceolate. Corolla 2–3 times as long as calyx; standard with broad-linear to lanceolate limb, 1.5 times as long as wings and keel. Fruiting head 1.5–2.2 × 1.2–2.5 cm, globular-ovoid. Fruiting calyx tube obconical, almost glabrous, prominently furrowed; teeth stellately spreading to reflexed, ovate-triangular, leaf-like, reticulately nerved, the lower one ovate-lanceolate; tube compact, dense, concealed by spreading teeth; throat closed by bilabiate callosity with hairs along slit. Pod membranous below, somewhat coriaceous above, apiculate. Seed solitary, about 1.1 × 1 mm, obovoid. Dispersal by entire heads (synaptospermic). Fl. March–May. 2n = 16.

Hab.: Fields and among shrubs; calcareous and marly-calcareous soils.

Gen. Distr.: Turkey, Cyprus, Syria, Lebanon, Palestine, Iraq, Libya.

Selected specimens: TURKEY: prov. Adana, distr. Osmaniye, Toprakkale, basalt gulley, below the castle, 1957, *Davis* & *Hedge* D. 26906 (E); Lydia, Smyrna, Ilica, in collibus, 1906, *Bornmüller* 9338 (E). SYRIA: Tripoli, 1866, *Blanche* 137 (G). LEBANON: Ras Beyrouth, *Peyron* (G). PALESTINE: Upper Galilee, env. of Metula, roadsides and abandoned yards,

Plate 200. *T. scutatum* Boiss.
(Palestine : *E. Ben-Shlomo*, HUJ).
Plant in flower and fruit;
fruiting calyx; flower.

1963, *Zohary* & *Plitmann* 113/27 (HUJ). IRAQ: Mohammera, 1850, *Noë* 142 (G). LIBYA: Cirenaica, Ain Mara, 1931, *Pampanini* & *Sermolli* 4022 (K).

Note: *T. scutatum* is rather variable in the size of the leaves and heads and the number of flowers per head. Experiments have shown that these characters are adaptive. Transitions between it and *T. plebeium* occur in overlapping areas of the two species.

206. T. plebeium Boiss., Diagn. ser. 1, 9 : 23 (1849) et Fl. 2 : 129 (1872). *Typus*: Syria, Antilebanon inter Rasheya et Damascus, *Boissier* (G holo.).

T. alsadami Post, Journ. Linn. Soc. Lond. Bot. 29 : 425 (1888); Mout., Fl. Djebel Druze 129 (1953).

Icon.: [Plate 201].

Annual, densely or sparingly patulous-pubescent, 10–20(–30) cm. Stems usually many, slender, ascending and diffuse. Leaves all, except uppermost ones, alternate; adnate part of stipules short, linear or triangular, submembranous, with longitudinal dark nerves, free portion as long as or longer than the lower part, lanceolate, herbaceous; leaflets 0.6–1.2 × 0.3–1 cm, varying in shape in single plant from almost orbicular to obovate with often cuneate to oblong base, usually with round apex, rarely acute, obscurely denticulate in upper part. Head 1–2 × 1.2–1.8 cm, pedunculate, terminal, broad-ovoid to broad-obovoid. Flowering calyx 6–7 mm, cylindrical or somewhat campanulate, appressed-hairy, with 10 prominent nerves; teeth unequal, purple-tipped, long-hirsute, the lower one lanceolate-subulate, 1.5 times as long as others and as broad, the rest triangular-subulate, as long as or shorter than tube. Corolla white or cream-coloured, rarely pinkish or with purple tip on keel, 2–3 times as long as calyx; standard with lanceolate-oblong, slightly notched limb, 1.5 times as long as wings and keel. Fruiting head 1.5–2 × 1.5–1.7 cm, ovoid. Fruiting calyx tube ovoid, pubescent, prominently ribbed; lower calyx tooth lanceolate-subulate, somewhat prickle-like, others short-triangular, all stellately spreading, often recurved, with 3 prominent nerves; throat closed by bilabiate callosity with short hairs along slit. Pod membranous, dorsally carinate, about 2 mm long, obovate, apiculate. Seed solitary, about 1.5 × 1 mm, obovoid, obliquely truncate at base. Fl. March–May. 2n = 16.

Hab.: Hills, mountain slopes, batha, maquis, river banks.

Gen. Distr.: Turkey, Syria, Lebanon, Palestine.

Selected specimens: SYRIA: 74 km S.W. of Damascus, basalt rocks, 950 m, 1933, *Eig* & *Zohary* (HUJ). PALESTINE: Jerusalem, Beit Hakerem, 1952, *Orshan* 17297 (HUJ).

Note: *T. plebeium* has been well delineated by Boissier (*loc. cit.*), but Post (*loc. cit.*) was probably unaware of this and described it again as *T. alsadami* from Djebel Druze; his description, however, is short and the illustration of the calyx misleading. This became quite evident after Père Mouterde had kindly sent us a photograph and a drawing of *T. alsadami*.

Plate 201. *T. plebeium* Boiss.
(Palestine : *Lerner*, HUJ).
Flowering and fruiting branches; flower; fruiting calyces.

Hossain is incorrect in synonymizing *T. plebeium* with *T. scutatum* by stating that the former is only a depauperate form of the latter. The differences between the two species under discussion are tabulated here subsequent to the examination of scores of specimens (in HUJ):

	T. scutatum	*T. plebeium*
Tips of calyx teeth	Purple	White
Lower tooth of flowering calyx	Broadly lanceolate to broadly oblong, slightly longer than calyx tube, twice as long as other teeth and 2–3 times as broad	Lanceolate-subulate, 1.5 times as long as calyx tube and other teeth and as broad as latter
Upper teeth of flowering calyx	Triangular-lanceolate	Triangular-subulate to oblong-lanceolate
Lower tooth of fruiting calyx	Ovate-lanceolate, foliaceous	Oblong-lanceolate, often prickle-like
Fruiting calyx tube	Almost glabrous	Pubescent

Section Seven. TRICHOCEPHALUM Koch.

Syn. Fl. Germ. Helv. 171 (1835); Boiss., Fl. Or. 2 : 111 (1872)

Genus *Calycomorphum* C. Presl, Symb. Bot. 1 : 50 (1831).
Section *Calycomorphum* (C. Presl) Griseb., Spicil. Fl. Rumel. 1 : 31 (1843).
Section *Oliganthum* Bertol., Fl. Ital. 8 : 131 (1850) p. p.
Subgenus *Calycomorphum* (C. Presl) Hossain, Not. Roy. Bot. Gard. Edinb. 23 : 438 (1961).

Lectotypus : *T. subterraneum* L.

Annuals. Inflorescences capitate. Flowers mostly sessile, ebracteolate, the inner ones sterile without corollas, developing during or after flowering of fertile flowers. Pod 1-seeded. Calyces of sterile flowers consisting of solid tube and 5 long, bristly teeth in fruit concealing fertile ones, thus serving as means of dispersal either by cottony or silky cover that envelops fruiting head (anemochory) or by drilling abilities which facilitate penetration of fruiting head into ground (geocarpy), or by sticking to ground.

Synopsis of Species in Section Seven

Section Seven: TRICHOCEPHALUM Koch

207. T. pauciflorum D'Urv.
208. T. pilulare Boiss.
209. T. eriosphaerum Boiss.
210. T. meduseum Bl. ex Boiss.
211. T. globosum L.
212. T. batmanicum Katzn.
213. T. chlorotrichum Boiss. & Bal.
214. T. subterraneum L.
215. T. israëliticum D. Zoh. & Katzn.

Key to Species

1. Mature heads penetrating into the ground. Sterile flowers developing after maturation of fertile flowers — 2
– Mature heads not subterranean — 3
2. Leaflets mostly obovate, with rounded or truncate apex. Corolla red or violet, 1.4–1.8 cm, 3–4 times as long as calyx teeth. — **215. T. israëliticum**
– Leaflets mostly deeply notched. Corolla white, sometimes striped with pink, 0.8–1.4 cm, about as long as calyx teeth. — **214. T. subterraneum**
3 (1). Calyx teeth glabrous or glabrescent. — **213. T. chlorotrichum**
– Calyx teeth hairy, plumose or cottony — 4
4. Fertile flowers (1–)2–8, in one whorl — 5
– Fertile flowers 10–16, in two whorls — 8

5. Fertile flowers (4–)5–8 — 6
– Fertile flowers (1–)2–4 — 7
6. Fruiting heads 0.8–1 cm; hairs white-cottony not silky. Pod elongated. Calyx teeth plumose all over. — **209. T. eriosphaerum**
– Fruiting heads 1.2–1.5 cm; hairs not cottony but silky, greyish-white or pinkish. Pod ovoid. — **207. T. pauciflorum**
7 (5). Teeth of sterile calyces protruding from surface of head, plumose only in lower half. Corolla twice as long as calyx. Pod exserted from calyx for up to half its length. — **210. T. meduseum**
– Teeth of sterile calyces not protruding from surface of head, plumose all over. Corolla as long as calyx. — **208. T. pilulare**
8. Fruiting head hemispherical, adpressed to ground at maturity. Calyx tube ovate-cylindrical, twice as long as teeth. — **212. T. batmanicum**
– Fruiting head globular, not adpressed to ground. Calyx cylindrical; tube as long as or somewhat longer than teeth. — **211. T. globosum**

207. T. pauciflorum D'Urv., Mém. Soc. Linn. Paris 1 : 350 (1822) et Enum. Fl. 94 (1822). *Typus*: Constantinople, 1820, *d'Urville* (CN holo., iso. K).

T. globosum auct. non L. (1753).
T. oliverianum Ser. in DC., Prodr. 2 : 197 (1825).

Icon.: [Plate 202].

Annual, patulous-pubescent, 15–30 cm. Stems few, striate, erect or ascending, slightly branching. Lower leaves long-petioled, upper ones almost sessile; stipules membranous between nerves, semi-ovate, acuminate; leaflets 0.8–1.2 cm, cuneate at base, rhomboidal to obovate, denticulate at apex, mostly apiculate. Heads short-peduncled to subsessile, many-flowered, globular. Flowers sessile, corollate (fertile) ones (4–)5–7(–8), in one row, sterile ones numerous. Calyx densely hirsute; teeth equal in length, as long as or slightly longer than tube. Corolla purple, 0.8–1(–1.4) cm, one-fourth longer than calyx. Fruiting head 1.2–1.8 cm, compact, globular to ovoid, forming dense, greyish-pink, silky, opaque, hairy ball. Pod membranous, obovoid, included within calyx tube. Seed ovoid, compressed, shiny. Sterile flowers finally having flexuous, elongating, plumose-silky calyx lobes. Fruiting head with few seed-bearing calyces and many sterile ones separating as whole from stem. Fl. April–May. 2n = 16.

Hab.: Fallow fields, heavy soil, grassy steppes.

Gen. Distr.: Greece, Aegean Islands, Turkey, Lebanon, Syria, Israel, Iraq.

Selected specimens: TURKEY: Constantinople: in collibus prope "Flamour", haud procul ob "Bechiktache", 1896, *Aznavour* 3112 (E); Prov. Maraş, 10 km S. of Maraş, 550 m, 1957, *Davis* & *Hedge* D. 27345 (E). LEBANON: Baynu, 610 m, 1943, *Davis* 5963 (HUJ). SYRIA: 24 km S.W. of Damascus, 950 m, 1933, *Eig* & *Zohary* 7 (HUJ). ISRAEL: Upper Galilee, near Sa'sa, 1964, *Zohary* (HUJ); Hula Valley, Amir, at foot of Golan Mts., 1941, *F. Weissmann* (HUJ).

208. T. pilulare Boiss., Diagn. ser. 1, 2 : 29 (1843) et Fl. 2 : 135 (1872). *Typus*: In umbrosis montium Lydiae, V. 1842, *Boissier* (G lecto., iso. K).

Plate 202. *T. pauciflorum* D'Urv.
(Palestine : *Zohary*, HUJ).
Plant in flower and fruit; flower.

T. globosum L. subsp. *pilulare* (Boiss.) Gib. & Belli, Mem. Accad. Sci. Torino ser. 2, 43 : 211, t. 2 (1893).
T. pilulare Boiss. var. *longipedunculatum* Evenari in Opphr. & Evenari, Bull. Soc. Bot. Genève ser. 2, 31 : 295 (1941).

Icon. : [Plate 203].

Annual, patulous-hairy, 10–50 cm. Stems many, angulate-furrowed, erect, ascending or prostrate, branching throughout. Lower leaves long-petioled, upper ones short-petioled; stipules membranous between nerves, ovate, acuminate; leaflets 0.4–1.2 cm, obovate with cuneate base, rounded or truncate at apex, apiculate, obsoletely toothed in upper part. Peduncles capillary, lengthening in fruit and much longer than leaves, spreading or curved. Flowering heads 6–9 mm, ovoid or globular with 1–2 corollate fertile flowers and numerous sterile ones. Flowers sessile. Calyx of fertile flowers with densely villous tube and 5 equal, setaceous, loosely plumose teeth as long as or longer than tube. Corolla 6–7 mm, white; standard slightly longer than wings and keel. Fruiting head 0.8–1 cm, globular with greyish, silky hair-cover, translucent at margin. Pod membranous, ovoid-obtuse, included within calyx tube. Seeds 2 mm, ovoid-globular, dark brown, smooth, shiny. Sterile flowers many, with long bristly and plumose calyx teeth. Fruiting head with 1–2 seed-bearing calyces and many sterile ones, separating as whole from stem. Fl. February–May. 2n = 14.

Hab. : Batha, fields and roadsides, grassy and rocky slopes.

Gen. Distr. : Aegean Islands, Turkey, Cyprus, Syria, Lebanon, Israel, W. Iran, N. Iraq.

Selected specimens : AEGEAN ISLANDS : Mytileni, Montes Ordymnos, 1934, *Rechinger* It. 5964 (W). TURKEY : Prov. Çanakkale: Thymbra, 1883, *Sintenis* 54 (E); Mardin : 10–12 km W. of Savur, 900 m, 1966, *Davis* 42460 (E). CYPRUS : distr. Limasol, Agios Theodoras, 3,500–4,000 ft, *Davis* 3071 (K). LEBANON : Falougha, 1941, *T. Rayss* (HUJ); Hermon, Wadi Shelba, 1924, *Naftolsky* 17123 (HUJ). SYRIA : about Rina, 1931, *Zohary* (HUJ); about Aleppo, 1931, *Zohary* (HUJ). ISRAEL : Sharon Plain : Tantura, 1956, *Zohary* (HUJ); Upper Galilee, near Meiron, 1954, *Zohary* 17126 (HUJ). IRAN : Kechwar, 50 km S. of Khonanabad, 1937, *Köie* 1585 (W). IRAQ : Assyria Orientalis, Kuh Sefin, Erbil, 1893, *Bornmüller* 1141 (E).

209. T. eriosphaerum Boiss., Diagn. ser. 1, 9 : 25 (1849) et Fl. 2 : 134 (1872). *Typus* : In herbidis Palaestinae prope Hierosolymam, IV. 1846, *Boissier* (G holo., iso. LE).

T. globosum L. subsp. *eriosphaerum* (Boiss.) Gib. & Belli, Mem. Accad. Sci. Torino ser. 2, 43 : 208, t. 1, f. 3 (1893).
T. globosum L. var. *eriosphaerum* (Boiss.) Post, Fl. Syr. Pal. Sin. 239 (1896).

Icon. : [Plate 204].

Annual, appressed-hairy, especially in upper part, 10–40 cm. Stems few to many, striate, erect to ascending, sparingly branching. Lower leaves long-petioled, upper ones subsessile; stipules membranous between nerves, ovate-oblong, adnate to petioles, upper free portion short, acute; leaflets 0.6–1(–2) cm, obovate, cuneate at base, rounded at apex, the upper ones denticulate at tip. Heads 1.2–1.8 cm, axillary,

Plate 203. *T. pilulare* Boiss.
(Palestine : *Plitmann*, HUJ).
Plant in flower and fruit; flower.

Plate 204. *T. eriosphaerum* Boiss.
(Palestine : *N. Feinbrun* 16709, HUJ).
Plant in flower and fruit; flower.

ovoid to globular, with peduncles as long as or longer than leaves, elongating and spreading or recurving in fruit. Pedicels very short. Flowers many, corollate ones 4–6 in one row. Calyx white-wooly; teeth filiform with triangular base, plumose, as long as or somewhat longer than obscurely nerved tube. Corolla 1.2–1.8 cm, purple or flesh-coloured above. Fruiting heads about 1 cm or less, globular, with opaque, white, cottony hair-cover, composed of few seed-bearing calyces and many sterile ones, separating as whole from stem. Pod sessile, oblong, mucronate, much shorter than calyx. Seeds 1.5 mm, ovoid-oblong, brown, shiny, smooth. Dispersal unit: entire head, including many sterile calyces with their white cottony, setaceous, spreading, flexuous teeth. Fl. February–May. 2n = 14.

Hab.: Mainly batha and scrub.

Gen. Distr.: S. Syria, Lebanon, Palestine.

Selected specimens: SYRIA: Gebel Druz, Sueda, 1931, *Zohary* (HUJ). PALESTINE: Philistean Plain: Ben Shemen, 1923, *Naftolsky* 16721 (HUJ); E. Esdraelon Plain: Ein Harod, Mt. Gilboa, 1931, *Naftolsky* 16723 (HUJ); Amman: Amman to Salt, 1929, *Naftolsky* 16719 (HUJ).

Note: *T. eriosphaerum* is fairly homogeneous and common. It differs from all other species of this Section by its white, cottony heads.

210. T. meduseum Bl. ex Boiss., Fl. 2: 134 (1872); Hossain, *loc. cit.* 443. *Typus*: Lebanon, Djurd Hadet, 1866, *Blanche* 135 (G holo., iso. P).

T. globosum L. subsp. *meduseum* (Bl. ex Boiss.) Gib. & Belli, Mem. Accad. Sci. Torino ser. 2, 43: 210, t. 2 (1893).

T. globosum L. var. *meduseum* (Bl. ex Boiss.) Post, Fl. Syr. Pal. Sin. 239 (1896).

Icon.: [Plate 205].

Annual, patulous-hairy, 10–25 cm. Stems few, diffuse, rather thick, branching. Leaves 2–4 cm, the lower ones long-petioled, the upper ones almost sessile; stipules herbaceous, ovate, much longer than petioles; leaflets obovate-cuneate, rounded or retuse, toothed at apex. Peduncles rather longer than subtending leaves, erect, patulous in fruit. Heads cylindrical, later spherical with 2–4 fertile and many sterile flowers arising from white central body after anthesis. Calyx densely hirsute with setaceous, plumose teeth as long as tube. Corolla purple, almost twice as long as calyx. Sterile flowers with solid or partly hollow calyx tube and with long setaceous teeth plumose only in lower half. Fruiting head globular, 1.2–1.5 cm in diam., consisting of fertile calyces concealed by plumose cover of sterile ones and with teeth of latter protruding above surface of head. Pod ovoid, with reticulate upper part exserting from calyces. Seeds ovoid, black. 2n = 14.

Gen. Distr.: Syria, Lebanon.

Selected specimens: LEBANON: Sables ferrugineux, village de Harmoun, *Gaillardot* & *Hohenacker* 2645 (P, G, W); Sables et grès ferrugineux, Ain Hata, Antilebanon, entre Racheya et Hasbeya, 1879, *Gaillardot* 1733 (G). SYRIA: Hauran inter Damascus et Quneitra, 1957, *Rechinger* 13052 (W).

Plate 205. *T. meduseum* Bl. ex Boiss.
Plant in flower and fruit; flower.

211. T. globosum L., Sp. Pl. 767 (1753); Boiss., Fl. 2 : 134 (1872). *Typus* : Described from Montpellier, France, Herb. Cliff. 374, 12 (BM).

T. radiosum Wahlenb. in Berggren, Resor uti Europa och Oesterlanderne II Bih 43 (1827) et in Oken. Isis 21 : 992 (1828).
Calycomorphum globosum (L.) Presl, Symb. Bot. 1 : 50 (1831).
T. nidificum Griseb., Spicil. Fl. Rumel. 1 : 32 (1843).

Icon. : [Plate 206].

Annual, mostly patulous-hairy, rarely subappressed-hairy or glabrescent, 10–40 cm. Stems few to many, procumbent to ascending, branching. Leaves small, scattered, with petioles much longer than stipules; stipules herbaceous, with membranous margin, semi-ovate, acutish; leaflets 0.6–1.8 × 0.5–1.5 cm, obovate, cuneate and emarginate at apex to obcordate, appressed-hairy, denticulate mainly at apex. Peduncles 1–10 cm, longer than subtending leaves, densely appressed-hairy, later elongating and often deflexing. Heads ovoid to cylindrical, with 10–16 fertile flowers and numerous sterile ones developing after anthesis from feathery body in centre of head. Fertile flowers with calyx and white or pale pink corolla, 0.6–0.9(–1.1) cm, in 2 rows; calyx appressed-hirsute, two-thirds the length of corolla; teeth slightly unequal, as long as tube. Sterile flowers without corolla; calyx tube solid or part of it with hollow, long-plumose teeth much longer than tube; these flowers increasing in length and bending down, covering fertile ones so as to form together a depressed, globular,fleecy, 1.5–7.5 cm broad body, detaching from peduncles as dispersal unit. Pod membranous, enclosed within calyx. Seeds ovoid, compressed, brown-black. Fl. February–June. 2n = 10, 16.

Hab. : Open hillsides, dwarf-shrub formations.
Gen. Distr. : Greece, Macedonia, Bulgaria, Turkey, Cyprus.

Selected specimens : GREECE : Attica, Marathonae, 1875, *Hollzman* 73 (G). TURKEY : Dardanelles : ad versuras, 1883, *Sintenis* 55 (E); Istanbul : Collines près Bachiktache, 1908, *Aznavour* (G); Env. of Izmir, 1961, *Katznelson* (HUJ); Hatay : Kirikhan-Alexandrette, 1931, *M. Zohary* (HUJ). CYPRUS : near Dhrousha (distr. Paphos) on Vouni, 610 m, 1941, *Davis* 3213 (HUJ).

Note : *T. globosum* is readily distinguished from all other species by the number of fertile flowers per head (10–16), the highest in Sect. *Trichocephalum*. In the mature fruiting heads, the calyx teeth of the sterile flowers covering the head are thick (thicker than those of the related species) and wire-shaped.
T. globosum varies, as already pointed out by Katznelson (*loc. cit.*), in all its parts from almost glabrous to densely hairy, in the length of the petioles (2–15 cm), the length of the peduncles (2–20 cm), etc.

212. T. batmanicum Katzn., Israel J. Bot. 15 : 80 (1966). *Typus* : Turkey, Dyarbakir Province, Huseyn Kara, Batman, on shallow basalt soil, 800 m, *Cornelius* (grown in Albany, California, 1962) No. 137 (HUJ holo., iso. AHUC).

T. anatolicum Katzn., Israel J. Bot. 14 : 115 (1965) non Boiss. (1843).

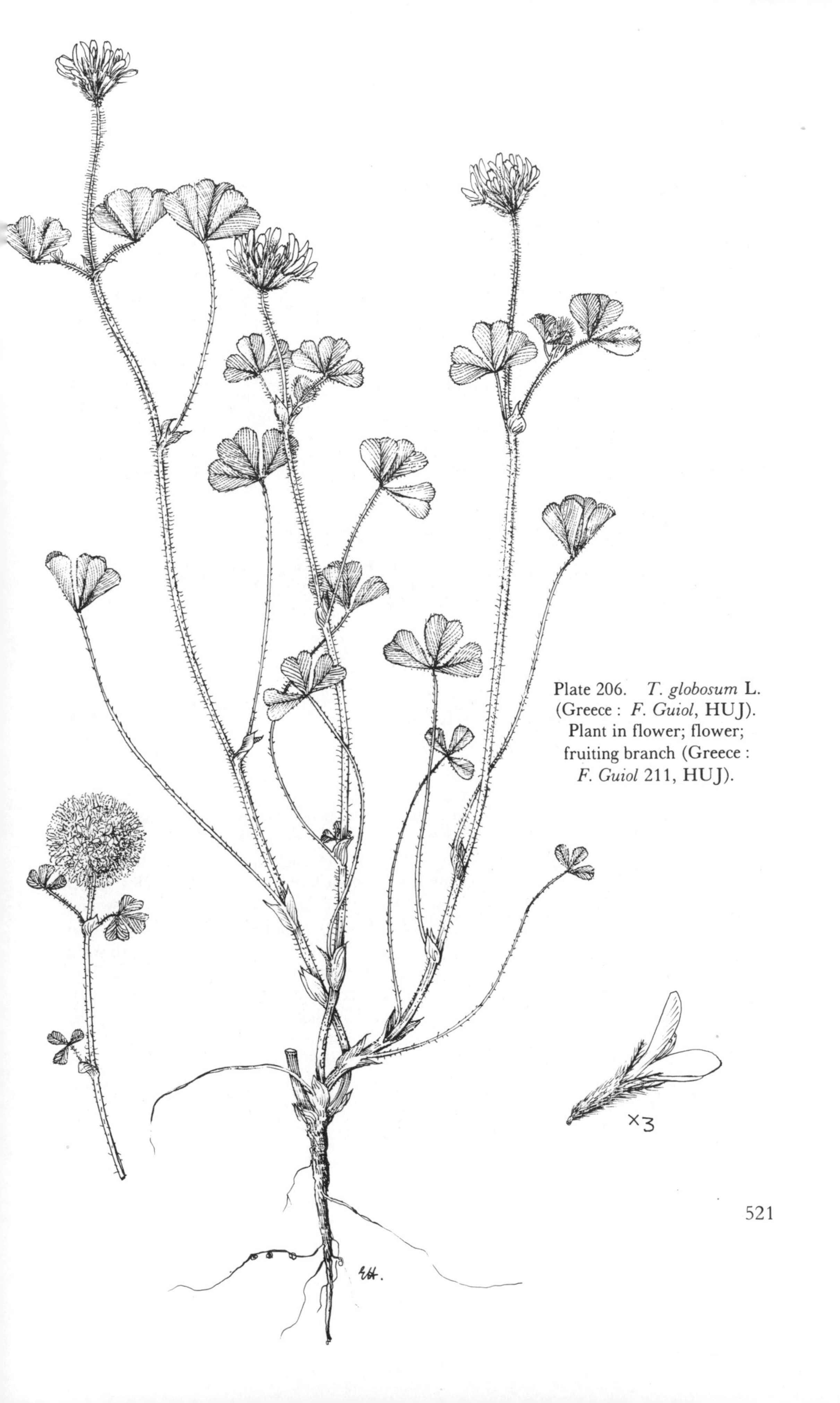

Plate 206. *T. globosum* L. (Greece : *F. Guiol*, HUJ). Plant in flower; flower; fruiting branch (Greece : *F. Guiol* 211, HUJ).

Icon.: [Plate 207]. Katzn., Israel J. Bot. 14 : ff. 1a, b, c, d (1965).

Annual, glabrous, 20–50 cm. Stems procumbent, thick, branching. Leaves petiolate; stipules herbaceous, ovate-triangular, acute with scarious margins; leaflets obovate, truncate or slightly notched, 1–1.5 × 1–1.5 cm. Peduncles becoming much longer than subtending leaves, bending towards ground after flowering. Heads semi-globular. Fertile flowers 10–14 in 2 whorls, deflexed in fruit; calyx tube ovate-cylindrical, twice as long as linear, slightly hairy, finally divaricate teeth; corolla cream-yellow, later turning rose-cream, 5–8 mm long, slightly longer than calyx teeth. Sterile flowers numerous, occurring before anthesis on rigid, glabrescent, conical body, 2–3 mm long, later developing into calyces with outer hollow and inner solid calyx tubes; teeth of lower sterile calyces 10–15 mm long, hairy; these flowers bending over fertile ones, together forming hemispherical or discoid, yellow-green fruiting head, 20–30 mm across. Pod membranous, included in calyx. Seed black, ovoid, oblong, smooth. 2n = 16.

Hab.: Basalt soils.

Gen. Distr.: Known only from type locality cited above.

Note: *T. batmanicum* differs from *T. globosum* L. in the following characters: plants (except calyces) glabrous; peduncle thick, pressing head to ground; seeds smooth; head not globular, etc.

213. T. chlorotrichum Boiss. & Bal. in Boiss., Diagn. ser. 2, 6 : 48 (1859) et Fl. 2 : 133 (1872). *Typus*: Turkey: Ad Karaguel Dere, à 2 lieues au SSO d'Ouchak (Phrygia), 29. V. 1857, *Balansa* 306 (G holo.)

Icon.: [Plate 208].

Annual, glabrous, 5–10 cm. Stems few to many, procumbent, branching. Leaves small, crowded at base, with petioles several times longer than leaflets; stipules semi-ovate, more or less obtuse, herbaceous with white, membranous margin; leaflets up to 8 mm, cuneate, truncate or obcordate, denticulate at apex. Peduncles much longer than subtending leaves, somewhat thickened, later deflexing and pressing heads to ground. Heads with one whorl of 6–8(–10) corollate flowers and several rows of sterile ones arising after anthesis from knob in centre of head. Fertile flowers with glabrous calyx; calyx tube cylindrical, longer than linear, patulous or deflexing, somewhat unequal teeth; corolla somewhat longer than calyx, white (when dry). Sterile flowers apetalous, glabrous, increasing in length and soon bending inward so as to cover fertile ones, thus together forming green globular body easily detached from plant as dispersal unit; calyx hollow in lower sterile flowers and solid in upper ones. Fruiting head green, 1.5–2 cm across. Pod ovoid, mucronate, entirely enclosed within calyx, membranous. Seeds ovoid, compressed, black. Fl. May.

Gen. Distr.: Known only from the type locality.

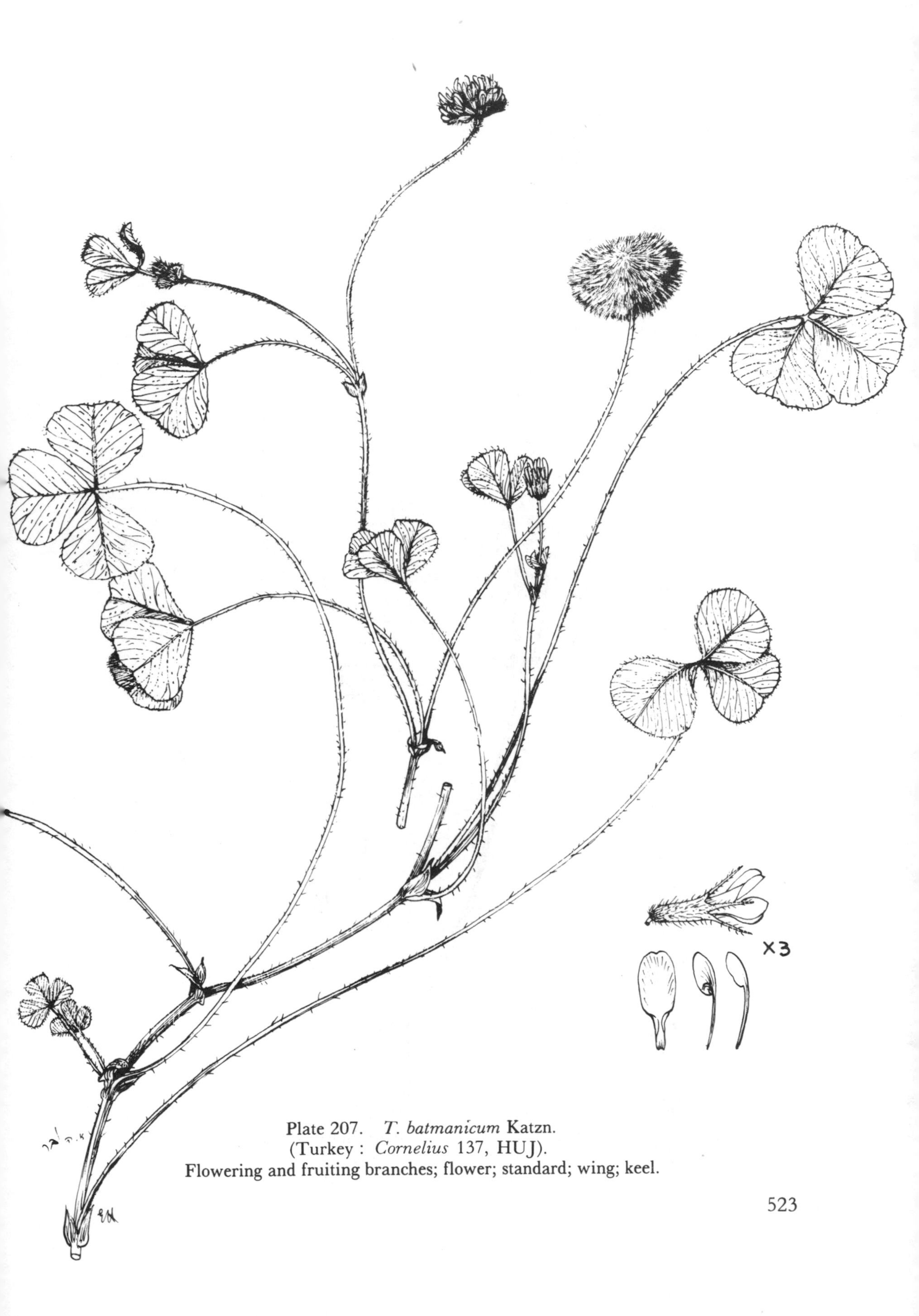

Plate 207. *T. batmanicum* Katzn.
(Turkey : *Cornelius* 137, HUJ).
Flowering and fruiting branches; flower; standard; wing; keel.

Plate 208. *T. chlorotrichum* Boiss. & Bal.
(Turkey : *Balansa* 306, G).
Plant in flower and fruit; corollate flower; sterile flower.

214. T. subterraneum L., Sp. Pl. 767 (1753).

Annual, glabrescent to patulous-hairy, 10–30 cm. Stems few, weak, prostrate or decumbent, sparingly branching or unbranched, slightly furrowed. Leaves long-petioled; stipules ovate, acute; leaflets 0.8–1.2(–2) cm, mostly obcordate with cuneate base, denticulate and notched at apex. Heads about 1 cm, long-peduncled, obovoid, loose, the fruiting ones globular. Corollate flowers few or many; sterile flowers developing after anthesis of fertile ones. Calyces of corollate flowers glabrous or sparingly pubescent; tube cylindrical, growing in fruit; teeth almost as long as tube, subequal, linear-subulate. Corolla about 1 cm, almost twice as long as calyx, white, pinkish-striped; standard elliptical. Calyces of sterile flowers with solid, stalk-like tube and linear teeth. Fruiting heads deflexed, ripening underground. Pod somewhat exserting from split calyx, membranous, obovoid or obtriangular. Seed 1, 2.5 mm, lenticular, black. Fruiting heads consisting of few seed-bearing calyces and many wire-like, sterile ones deflexed to cover fertile ones and pushing them into ground. Fl. March–April. 2n = 16.

*Key to Infraspecific Taxa of T. subterraneum**

1. Calyx transversely wrinkled. **214h.** var. **yanninicum**
– Calyx not transversely wrinkled 2
2. Calyx covering all or most of ripe pod 3
– Calyx covering only base to lower third (rarely half) of pod 5
3. Branches 10–45 cm long. Peduncles about as long as petioles. Heads 4–6-flowered 4
– Branches 5–15 cm long, densely hirsute. Heads 2–3(–4)-flowered. Calyx tube often red-purple. **214b.** var. **brachycladum**
4. Branches compact, 10–30 cm long. Petioles 5–10 cm long. **214a.** var. **subterraneum**
– Branches rather diffuse, 25–45 cm long. Petioles 15–25 cm long. **214c.** var. **majurculum**
5. Teeth of sterile calyces in fruiting heads about as long as tube 6
– Teeth of sterile calyces in fruiting heads about twice as long as tube 7
6. Heads 0.7–1 cm long. Delicate plants. Calyx covering less than one-fourth of pod. **214e.** var. **graecum**
– Heads 1–1.8 cm long. Robust plants. Calyx covering one-third to one-half of pod. **214g.** var. **oxaloides**
7. Stems thin. Stipules short, triangular. Seeds 2–5 mm long, "subprismatic-triangular". **214d.** var. **brachycalycinum**
– Stems rather thick. Stipules long-acuminate. Seeds ovoid-elongated. **214f.** var. **flagelliforme**

214a. T. subterraneum L. Subsp. **subterraneum** Var. **subterraneum.** *Typus*: Described from France and Italy, Herb. Cliff. 384, 11 (BM).

*This species has been critically revised by J. Katznelson & F. H. W. Morley (Israel J. Bot. 14 : 112–134, 1965). Its subdivision is largely based on the above revision, but the nomenclature has been partly changed.

Calycomorphum subterraneum (L.) Presl, Symb. Bot. 1 : 50 (1831).
T. subterraneum L. var. *vulgare* Guss., Enum. Pl. Ins. Iner. 50 (1854).
T. subterraneum L. var. *genuinum* Rouy, Fl. Fr. 5 : 98 (1899).
T. subterraneum L. var. *typicum* Aschers. & Graebn., Syn. 6, 2 : 596 (1908).

Icon. : [Plate 209].

Hab. : Fallow fields and grassy lands; batha.
Gen. Distr. : Tunisia, Algeria, Morocco, England, Portugal, Madeira, Spain, France, Corsica, Sardinia, Italy, Yugoslavia, Greece and Aegean Islands, Crete, Bulgaria, S. Russia, Turkey, Syria, Israel, Iran.

Selected specimens :* TUNISIA : Djebel Bir, Tabarka, 1930, *Eig* (HUJ); Kef en Nesour, El Beja, 1888, *Cosson* (P). ALGERIA : Oued Chiffa, *Neal Smith* CPI 1954; Environs d'Alger, Bouzarea, 1858, *Durand* (Al, P). MOROCCO : Grand Atlas, Timinkar, 1922, *Maire* (RAB). ENGLAND : Lyss, E. Hants, 1918, *Gamble* 30626 (K). PORTUGAL : 5 km S. of Alpendrinha, *Neal Smith* CPI 19470; Algarve, Serra de Moncique, 1945 (HUJ). MADEIRA : Cap Gisao, 1856, *Mandan* (P). SPAIN : La Sirena, *Donald* CPI 15262; Santander, Hinogedo, 1925, *Leroy* 5622 (W, G). FRANCE : Ville d'Avray, Env. de Paris (E, G, K, P, W). CORSICA : Ajaccio, Campo di Loro (HUJ, P). SARDINIA : près de St. Michel, *Pavillon* (G). ITALY : Firenze and environs, *Katznelson* Tsub 1108, Tsub 1110; Solfatare, near Napoli, *Katznelson* Tsub 1113, Tsub 1115; Sicilia, Messina, 1840 (FI, G, W). YUGOSLAVIA : Istria, Brioni, 1903, *Keller* (HUJ, W). GREECE AND AEGEAN ISLANDS : Mt. Pendelikon, *Katznelson* & *Critopoulos* Tsub 1107 (HUJ); Insula Melos, 1889, *Heldreich* (W); Insula Zante, 1893 (G). CRETE : Salinos, Levka Ori, 1942, *Rechinger* 12335, 13695 (G, K, W). BULGARIA : Graminis ad Tharkovi, 1899, *Stribny* (E). CAUCASUS : Abchazia, prope Suchum Kale, *Woronov* (P). TURKEY : Samsoun above Haci Ismail Koÿ, *Tobey* 1730 (E). SYRIA : Aleppo, 1841, *Kotschy* (W); 1843, *Hohenacker* 44 (W). IRAN : North Lahijan, Guilan, 1937, *Lindsay* 763 (K). ISRAEL : Magdiel, *Naftolsky* 17377 (HUJ); *Katznelson* 1144 (HUJ).

214b. T. subterraneum L. Var. **brachycladum** Gib. & Belli, Mem. Accad. Sci. Torino ser. 2, 4 : 183 (1893). *Lectotypus* : Constantinople, *Aucher* 1223 (FI, G, K).

Gen. Distr. : Algeria, Morocco, England, Portugal, Spain, France, Corsica, Sardinia, Italy, Sicily, Greece, Turkey.

Selected specimens : ALGERIA : Blida, 1883, *Debray* (AI). MOROCCO : in monte Beni Hosmar, prope Tetuan, 1871, *Ball* (G). ENGLAND : Farnham, Surrey, 1888 (G). PORTUGAL : Estramadura Lusitanica (P). SPAIN : Santander, Hinogedo, 1925 (HUJ). FRANCE : Charent. Inférieur, Ozillac, 1914, *Coste* 1245 (P). CORSICA : Ajaccio, Talus, 1914, *Coste* 1245 bis (P,G). SARDINIA : *Thomas* (G). ITALY : Tivoli, al Monte Ripoli, 1904, *Vaccari* (FI); Napoli, Pozzuoli, 1839 (G). SICILY : Messina, 1846, *Heldreich* (G, W). ISTRIA : Brioni, 1909, *Keller* (HUJ). INSULA ZANTE : 1839 (G). GREECE : near Salonica, 1917, *Wilson* (E). TURKEY : Antiphylles, Lycia, *Lambert* (K).

*Many of the geographical data recorded for this and other varieties of *T. subterraneum* are taken from the above-cited revision of Katznelson & Morley.

Plate 209. *T. subterraneum* L. Subsp. *subterraneum* Var. *subterraneum* (Palestine : *D. Zohary* & *Katznelson* 16970, HUJ). Plant in flower and fruit; fruiting head; fruiting calyx with projecting pod; flowering head.

214c. T. subterraneum L. Var. **majurculum** Adamović in scheddae emend. Katzn., Israel J. Bot. 14 : 124 (1965). *Typus* : Serbia, Nis. 8.5.1902, *Adamović* (G).

Gen Distr. : England, France, Italy, Turkey, Serbia.

Selected specimens : ENGLAND : Vice country, 1946, 5380 (K). FRANCE : Ager Parisiensis, 1829, *Enderfs* (E). ITALY : Piemonte Bastia, 1894 (FI). TURKEY : Prairie de Geuecksouyou, 1892, *Aznavour* 626 B (G); *ibid.* : Soultan Sou, bois, 1896, *Aznavour* 626 (G).

214d. T. subterraneum L. Subsp. **brachycalycinum** Katzn. & Morley, Israel J. Bot. 14 : 127 (1965) Var. **brachycalycinum**. *Typus* : Israel : Lower Galilee, Beit-Queshet, 200 m, *Katznelson* 917 (HUJ).

T. subterraneum L. var. *oxaloides* Eig, Bull. Inst. Agr. Nat. Hist. 6 : 25 (1927) non Bunge (1878).
T. brachycalycinum (Katzn. & Morley) Katzn., Israel J. Bot. 23 : 70 (1974).

Gen. Distr. : Turkey, Syria, Lebanon, Israel.

Selected specimens : TURKEY : Istanbul, in proximitatibus Constantinopolis, 1850, *Clementi* (E). ISRAEL : Upper Galilee : 1 km E. of Eilon, 1960, *Katznelson* (HUJ).

214e. T. subterraneum L. Var. **graecum** Katzn. & Morley, *loc. cit.* 129. *Typus* : Insula Kea, Menaria, 24.V.1898, *Heldreich* (W).

Gen. Distr. : Yugoslavia, Greece, Aegean Islands, Rhodes, S. Turkey.

Selected specimens : YUGOSLAVIA : Pola, Süd Istria, *Wetting* 7585 (W). GREECE : Thessalia, *Panos* 1102 (W). TURKEY : Bithynia : Brussa, Keshish Dagh, 1899, *Bornmüller* 4345 (G, K, W).

214f. T. subterraneum L. Var. **flagelliforme** Guss., Enum. Pl. Ins. Inarime 50 (1854). *Typus* : Insula Inarime (Ischia). In herbidis subarenosis elatis da Fonata a S. Nicola lungo la salita, a dritta sotto al telegrafo, *Gussone* (n. v.)

Gen. Distr. : Portugal, Spain, S. France, Italy, Corsica, Sardinia, Sicily, Greece, Turkey, Tunisia, Algeria, Morocco, Canary Islands.

Selected specimens : PORTUGAL : 8 km N.W. of Arronches, *Smith* CPI 19467 (CANB). SPAIN : Malaga, 1927, *Ellman* 175 (K). FRANCE : Freyuce, 1915, *Coste* 1945B (G). ITALY : Brindisi, 1913, *Vaccari* (FI). SICILY : Messina, 1840, *Heldreich* (G). GREECE : Andros Island, CPI 26221 (CANB). TURKEY : Istanbul, inter Zikizie Keuy et Scouenzoukeny, 1894, *Aznavour* (G). TUNISIA : Ein Draham, Kaldemelula, 1930, *Eig* (HUJ). ALGERIA : Djebel Midirana, 1879 (P). MOROCCO : 10 km S. of Rabat, 1926, *Emberger* (RAB). CANARY ISLANDS : in Campo Tafira, 1848, *Bourgeau* 585 (G).

214g. T. subterraneum L. Var. **oxaloides** Bge. in Nym., Consp. Fl. Eur. 177 (1878). *Typus*: Turkey: Constantinople, 1850, *Clementi* (G holo., iso.: E, W).

Gen. Distr.: Romania, Bulgaria, Greece, Turkey, Caucasus.

Selected specimens: ROMANIA: Insula Scrpilor, Ponto Euxinio, 1925, 561 (HUJ). BULGARIA: Sali Aga, Derbend, Macedonia, 1931, *Wisnievsky* (HUJ). GREECE: Monte Kralik Dagh, 1936, *Rechinger* 10452 (W). CAUCASUS: Kubensk, Temrink, 1892, *Lipsky* (HUJ).

214h. T. subterraneum L. Var. **yanninicum** (Katzn. & Morley) Zoh. (comb. nov.). *Typus*: Greece, N. of Yannina, Kranoula, low area nr. Lake Yannina, *Neal Smith* 1954, CPI 19476 W, Tsub 937 (HUJ).

T. subterraneum subsp. *yanninicum* Katzn. & Morley, Israel J. Bot. 14: 126 (1965).
T. yanninicum (Katzn. & Morley) Katzn., Israel J. Bot. 23: 69 (1974).

Gen. Distr.: Istria, Albania, Greece, Serbia.

Selected specimens: DALMATIA: Castel-Nuovo, 1876, *Studnizka* (G). SERBIA: Vranjam, *Adamović* 1903 (G).

215. T. israëliticum D. Zoh. & Katzn., Austral. Journ. Bot. 6: 179 (1958). *Typus*: Sharon Plain: between Pardes-Hanna and Karkur, on top of red sandy loam hill with open herbaceous vegetation and few bushes of *Calycotome villosa*, approx. 100 m, 1957, *D. Zohary* 16812 (HUJ holo.).

T. subterraneum L. var. *telavivense* ("*tel-avivensis*") Eig, Bull. Inst. Agr. Nat. Hist. 6: 25 (1927).

Icon.: [Plate 210].

Annual, sparingly patulous-villous, 5–40 cm. Stems few, weak, prostrate to decumbent, sparingly branching or unbranched, striate. Leaves long-petioled; stipules membranous, ovate to oblong, acuminate; leaflets 1–1.4 × 0.6–1.6 cm, obovate with cuneate base and rounded, rarely slightly emarginate apex, long-denticulate in upper part. Heads axillary, pedunculate, few-flowered, the lower peduncles thick and incurved. Corollate flowers 2–5, pedicellate; calyx tube tubular, white, glabrous, growing in fruit; teeth somewhat longer than tube, almost equal, setose, loosely long-villous; corolla 1.4–1.8 cm, 3–4 times as long as calyx, red or violet. Pod exserted from split calyx, much broader than long, leathery, transversely ovoid, dark violet to purplish, transversely wrinkled, upper margin prominently carinate. Seeds solitary, ovoid, usually yellowish. Fruiting heads ripening underground and consisting of few seed-bearing calyces and many indurated sterile ones bent downwards and surrounding fertile ones, thus serving as drilling tool for latter when penetrating ground. Fl. February–April. 2n = 12.
Hab.: Forests, batha, maquis and open fallow fields.
Gen. Distr.: Israel (endemic).

Plate 210. *T. israëliticum* D. Zoh. & Katzn.
(Palestine : *D. Zohary* 16877, HUJ).
Plant in flower and fruit; pods.

Section Seven: Trichocephalum

Selected specimens: ISRAEL: Upper Galilee, 1 km E. of Bar-Am, 1957, *D. Zohary* & *Katznelson* 16911 (HUJ); Lower Galilee: approx. 250 m W. of Beit Queshet, 170 m, 1957, *D. Zohary* & *Katznelson* 16834 (HUJ); Northern Samaria: hills approx. 2 km S. of Mishmar-Haemeq, 200m, 1957, *D. Zohary* & *Katznelson* 16905 (HUJ).

Section Eight. INVOLUCRARIUM Hook.

Fl. Bor. Amer. 1 : 132 (1833)

Genus *Lojaconoa* Bobr., Bot. Zhurn. 52 : 1598 (1967) p. max. p.

Type species : *T. wormskioldii* Lehm.

Annuals or perennials. Inflorescences pedunculate, umbellate; flowers whorled, many or sometimes few. Stipules deeply dentate-serrate, deeply laciniate or lobed, rarely entire. Involucre deeply divided into (2–)5–15 deeply incised or dentate, sometimes entire lobes; inner bracts of second whorls of flowers small. Pedicels very short or absent. Calyx 5–20-nerved, sometimes inflated in fruit; teeth unequal, simple, trifid or sometimes forked or branched. Standard sometimes inflated in fruit, becoming vesicular and including mature pod. Pod stipitate or sessile, 1–8-seeded. Twenty-two North and South American species.

Key to Subsections

1. Standard inflated in fruit, forming vesicle, constricted above into narrower appendage, including mature pod. Subsection II. **Physosemium**
– Standard not inflated and not forming vesiculate tube. Subsection I. **Involucrarium**

Synopsis of Species in Section Eight

Section Eight. INVOLUCRARIUM Hook.
Subsection I. INVOLUCRARIUM

216. T. antucoensis Heller
217. T. chilense Hook. & Arn.
218. T. microcephalum Pursh
219. T. microdon Hook. & Arn.
220. T. monanthum A. Gray
221. T. obtusiflorum Hook. f.
222. T. oliganthum Steud.
223. T. pinetorum Greene
224. T. polyodon Greene
225. T. triaristatum Bert. ex Colla
226. T. trichocalyx Heller
227. T. tridentatum Lindl.
228. T. variegatum Nutt.
229. T. vernum Phil.
230. T. wigginsii Gillett
231. T. wormskioldii Lehm.

Subsection II. PHYSOSEMIUM

232. T. barbigerum Torr.
233. T. cyathiferum Lindl.
234. T. depauperatum Desv.
235. T. fucatum Lindl.
236. T. minutissimum Heller & Zoh.
237. T. physanthum Hook. & Arn.

Key to Species

1. Entire calyx or only tube or teeth hairy or ciliate — 2
– Entire calyx glabrous — 11
2. Involucre much shorter than flowers, divided almost to base into lanceolate, entire lobes or lobules — 3
– Involucre longer, shallowly or deeply divided into several indistinct lobes or much toothed — 4
3. Caespitose perennials. Calyx about half the length of corolla. Standard and wings white or cream; keel purple. — **220. T. monanthum**
– Annuals. Calyx about as long as corolla or slightly shorter. Corolla pale purple. — **226. T. trichocalyx**
4. Teeth of calyx one-third the length of tube, triangular. — **219. T. microdon**
– Teeth of calyx as long as or (at least some of them) longer than tube, subulate or triangular at base, subulate above — 5
5. Plants glandular-pubescent. Stipules, leaflets and involucre deeply laciniate at margin with spinulose lobes or unequal teeth. — **221. T. obtusiflorum**
– Plants villous, not glandular. Stipules, leaflets and involucre entire or dentate-serrate at margin — 6
6. Calyx throat oblique; teeth unequal — 7
– Calyx throat horizontal; teeth more or less equal — 8
7. Corolla about as long as or shorter than calyx. — **232a. T. barbigerum** var. **barbigerum**
– Corolla much longer than calyx. — **232b. T. barbigerum** var. **andrewsii**
8. Teeth of calyx broadly triangular at base; margin broad-membranous, often serrate to ciliate. — **218. T. microcephalum**
– Teeth of calyx subulate from base; margin not membranous — 9
9. Plants 2–5 cm tall. Inflorescences 4–5 mm in diam., terminal. Peduncles longer than subtending leaves. — **236. T. minutissimum**
– Plants 5–25 cm tall. Inflorescences 0.7–1.5 cm in diam., axillary and terminal. Peduncles shorter than to as long as subtending leaves — 10
10. Stipules whitish, sheathing and covering internodes and overlapping in lower parts. Involucre shallowly lobed; lobes dentate-laciniate with unequal, subulate teeth. — **237. T. physanthum**
– Stipules green, not sheathing as above. Involucre deeply divided into triangular or lanceolate, simple or rarely bifid lobes. — **229. T. vernum**
11 (1). Involucre cleft to at least half its length into almost entire, free or distinct lobes — 12
– Involucre divided into laciniate or dentate, usually indistinct lobes — 16
12. Standard strongly inflated at maturity and becoming vesicular — 13
– Standard not inflated — 14
13. Leaflets broadly obovate to almost orbicular or rhombic-obovate, rounded at apex. Flowers 1–2.7 cm long. — **235. T. fucatum**

– Leaflets long-obovate to lanceolate or linear, cuneate, rounded to truncate or retuse at apex. Flowers (0.3–)0.5–1.1 cm long. **234. T. depauperatum**

14. Taprooted perennials, sometimes rooting at nodes. **230. T. wigginsii**

– Annuals or perennials, not rooting at nodes 15

15. Margin of stipules laciniate. Calyx teeth broadly triangular at base, abruptly setose above. Corolla 6–8 mm long. **222. T. oliganthum**

– Margin of stipules entire. Calyx teeth lanceolate-subulate. Corolla 1–1.5 cm long. **223. T. pinetorum**

16 (11). Calyx tube 13–20-nerved 17

– Calyx tube 10-nerved 20

17. Calyx teeth, at least some of them, 1–3 times trichotomously branched; branches of teeth setiform, divergent. Standard inflated in fruit. **233. T. cyathiferum**

– Calyx teeth simple or sometimes tridentate or 3–5-fid. Standard not inflated 18

18. Leaflets linear or lanceolate-oblong, mucronate. **227. T. tridentatum**

– Leaflets obovate-cuneate, oblanceolate to broad-elliptical 19

19. Calyx tube campanulate; teeth triangular-lanceolate, simple or sometimes one of them bifid or with small lateral tooth. **228. T. variegatum**

– Calyx tube tubular; teeth broad-triangular in outline, 3–5-fid, thus calyx sometimes appearing to have more than 5 teeth. **224. T. polyodon**

20 (16). Leaflets elliptical to linear or oblong-obovate 21

– Leaflets obovate, obcordate or oblanceolate 24

21. Calyx tube campanulate; teeth lanceolate-subulate, simple or sometimes one of them slightly trifid. Perennials. **231. T. wormskioldii**

– Calyx tube tubular; teeth broadly triangular, 3–5-fid, with terminal one longer than others. Annuals 22

22. Leaflets oblong-obovate to elliptical or rarely linear, cuneate at base, rounded at apex. **217. T. chilense**

– Leaflets linear-lanceolate to oblanceolate or lanceolate-oblong, acute, obtuse or truncate at apex 23

23. Involucre unequally laciniate or toothed; teeth considerably shorter than or as long as entire part. **227. T. tridentatum**

– Involucre divided into 10–16 unequally toothed lobes about as long as entire part. **225. T. triaristatum**

24 (20). Leaflets emarginate with broad apical sinus. Calyx tube tubular; teeth about one-third the length of tube, ovate-triangular, aristate, 1-nerved, with membranous and slightly denticulate margin. **216. T. antucoensis**

– Leaflets rounded, acute or retuse without sinus at apex. Calyx tube campanulate; teeth as long as or longer than tube, triangular-lanceolate or lanceolate-subulate, simple or sometimes one of them slightly bifid or trifid 25

25. Perennials. Stems fistulous, rhizomatous. Inflorescences 1.5–3 cm in diam. **231. T. wormskioldii**

– Annuals. Stems neither fistulous nor rhizomatous. Inflorescences 1–1.5 cm in diam. **228. T. variegatum**

Subsection I. INVOLUCRARIUM

Section *Micranthoidea* Lojac., Nuov. Giorn. Bot. Ital. 15 : 190 (1883).
Section *Tridentateae* McDerm., N. Am. Sp. Trif. 17 (1910).

Section *Variegateae* McDerm., *op. cit.* 35.
Section *Monantheae* McDerm., *op. cit.* 95.

Type species : *T. wormskioldii* Lehm.

Annuals or perennials. Calyx 10–20-nerved, not inflated in fruit. Standard not inflated and not enclosing mature pod. Pod sessile or stipitate, 1–4(–6)-seeded. Seventeen N. and S. American species.

216. T. antucoensis Heller (nom. nov.). *Typus* : Chile : Cresc. in prat. siccis Chile austr., ad Antuco, *Nov. lectum* 1868, *Pöpping* 181 (P !, iso. W !).

T. circumdatum Kunze in Linnaea 16 : 321 (1842) non Guss. ex Steud. (1841).
Lupinaster circumdatus (Kunze) Presl, Symb. Bot. 1 : 50 (1831).
T. inconspicuum Pöpping (inedit.) non (Fern.) Heller (1906).

Icon. : [Plate 211].

Annual, glabrous or sparingly hairy, 10–20 cm. Stems many, slender, flexuous, erect or ascending, branching. Leaves all more or less long-petioled; stipules 3–6 mm long, ovate, dentate, free portion triangular with long, lanceolate-subulate teeth; leaflets 0.4–2 × 0.4–0.8 cm, obcordate, cuneate at base, emarginate with broad apical sinus, short-mucronate, margin spinulose-dentate. Inflorescences 5 × 5–7 mm, semi-globular, few-flowered. Peduncles slender, 2–4 cm long, sparingly appressed-hairy. Involucre slightly shorter than flowers, 7–9-lobed; lobes about as long as entire part, triangular-lanceolate, cuspidate, laciniate. Calyx about 3 mm; tube tubular, membranous, 10-nerved; teeth about one-third the length of tube, ovate-triangular, aristate, 1-nerved, with membranous, slightly denticulate margin. Corolla 4–6 mm, whitish to flesh-coloured; standard ovate-elliptical, truncate. Pod oblong to narrowly elliptical, 1-seeded. Seed oblong, yellow. Fl. July–November.
Hab. : Dry meadows.
Gen. Distr. : Chile (endemic).

Specimen seen : CHILE : Valdivia, 1852, *Philip* 19 (W); *ibid.* : 1853, *Lechler* 302 (W); Colchagera, *Cummings* (W).

217. T. chilense Hook. & Arn., Bot. Beech. Voy. 16 (1832). *Typus* : Chile : Conception, *Cummings* 115 (K ! iso.).

Lupinaster ochreatus (Kunze) Presl, Symb. Bot. 1 : 47 (1831).
T. ochreatum Kunze ex Steud., Nom. ed. 2, 2 : 707 (1841).

Icon. : [Plate 212].

Annual, glabrous, 10–40 cm. Stems few to many, dichotomously branching in upper part, rigid, reddish, erect or ascending. Lower leaves long-petioled, upper ones very short-petioled; stipules 0.8–1.3 cm long, ovate, prominently nerved, coarsely dentate-lobate with spinulose-denticulate lobes, free portion as long as or longer than lower part; leaflets 1–3 × 0.2–0.6 cm, oblong-obovate to elliptical or rarely linear,

Plate 211. *T. antucoensis* Heller
(Chile: *Cummings*, W).
Plant in flower and fruit; involucre; flower.

Plate 212. *T. chilense* Hook. & Arn.
(Chile : *Lechler* & *Hohenacker* 297, P).
Plant in flower and fruit; flower; involucre.

spinulose-serrate, cuneate at base, rounded and mucronate at apex. Inflorescences 1–1.5 cm in diam., hemispherical, 10- to many-flowered. Peduncles rather long. Involucre somewhat shorter than flowers, very deeply lobed; lobes about as long as entire part, deeply incised into several dentate, lanceolate-aristate lobules; inner bract 0 to inconspicuous. Pedicels 0–0.5 mm. Calyx about 7 mm long; tube tubular, membranous, with 10 prominent nerves, rupturing lengthwise at maturity; teeth as long as or longer than tube, broadly triangular, spinulose, markedly tridentate with the middle tooth longer than 2 lateral ones. Corolla as long as or longer than calyx, pink. Pod about 4 mm long, broadly elliptical, stipitate, reticulately nerved, 1–2-seeded, unequallly inflated, sutures well developed. Seeds 1.5 mm, dark brown, ovoid. Fl. October–November. 2n = 16.

Hab. : In pastures.

Gen. Distr. : Chile (endemic).

Specimens seen : CHILE : Insula Valensaela, pr. Valdivia, 1850, *Lechler* & *Hohenacker* 297 (FI, P, W); Concon, 1827, *Pöpping* 221 (G, as *T. ochreatum* Kunze ex Steud.).

218. T. microcephalum Pursh, Fl. Am. Sept. 2 : 478 (1814). *Typus* : Idaho : Valley of Clark's River, 1 July, 1806, *M. Lewis* (PH).

T. microcephalum Pursh var. *lemmonii* Lojac., Nuov. Giorn. Bot. Ital. 15 : 193 (1883).
T. microcephalum Pursh f. *velutinum* McDerm., N. Am. Sp. Trif. 120 (1910).
Lojaconoa microcephala (Pursh) Bobr., Bot. Zhurn. 51 : 1598 (1967).

Icon. : [Plate 213]. McDerm., *loc. cit.* pls. 43–44.

Annual, more or less villous to glabrescent, 5–80 cm. Stems many, erect to ascending, branching from base and above. Leaves mostly long-petioled; stipules 0.5–1.5 cm, not membranous, ovate, serrate to nearly entire, free portion triangular, acuminate; leaflets 0.5–2 × 0.5–1 cm, obovate to obcordate or oblanceolate, rounded or retuse to deeply notched, margins dentate-serrate. Inflorescences 0.4–1.2 cm in diam., subglobular, 10–60-flowered, arranged in whorls. Peduncles 3–7 cm long, slender. Involucre shorter than or as long as flowers, villous, 6–12-lobed; lobes broadly triangular at base, long-subulate, entire or sometimes bifid. Calyx 3–5 mm long; tube tubular-campanulate, membranous, 10-nerved, hairy; throat horizontal; teeth about as long as or shorter than tube, broadly triangular at base with setaceous apex, margins broad, membranous, usually serrate to ciliate. Corolla 3–6 cm long, pale pink or white, as long as or longer than calyx; standard narrowly ovate-oblong, not inflated and not enclosing mature pod. Pod 1.5–2 mm long, sessile, broad-ellipsoidal, glabrous, membranous at base, cartilaginous at apex, 1-seeded. Seed 1.5 mm broad, ellipsoidal, brown. Fl. March–July. 2n = 16.

Hab. : Moist meadows, sandy river banks and drier hillsides.

Gen. Distr. : USA : California, Idaho, Nevada, Montana, Arizona, Oregon, Washington, Wyoming. British Columbia, Vancouver Island.

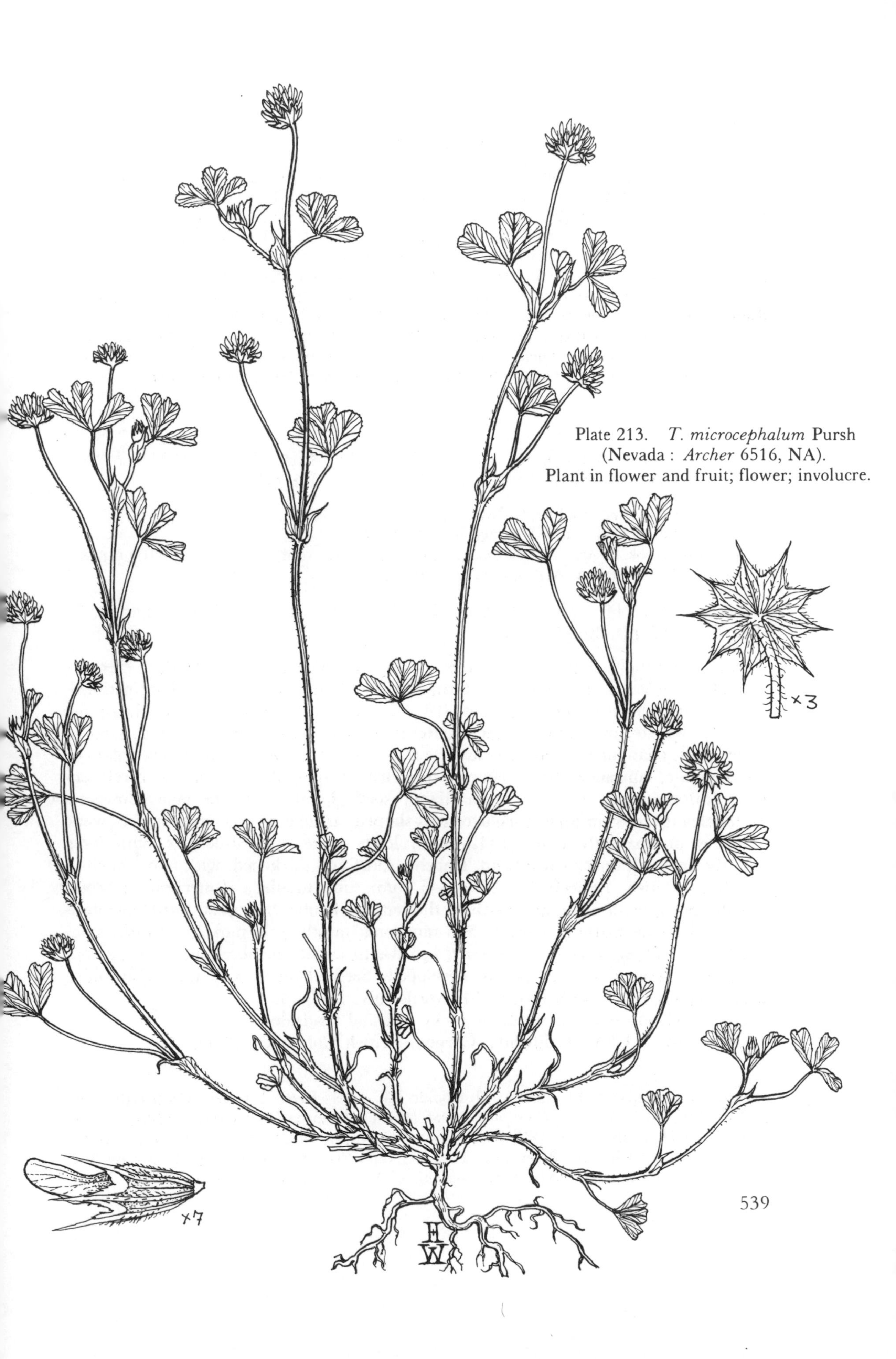

Plate 213. *T. microcephalum* Pursh (Nevada : *Archer* 6516, NA). Plant in flower and fruit; flower; involucre.

Selected specimens: CALIFORNIA: Monterey Co., Pacific Grove, 1903, *Heller* 6720 (HUJ); Placer Co., N.E. of Auburn, 1964, *Gillett* & *Moulds* 12657 (HUJ). IDAHO: E. of Rogerson, 1943, *Piemeisel* 43-1054 (UC); Valley Co., Payette River, 1940, *Ownbey* & *Meyer* 2105 (NA). NEVADA: Humboldt Co., Santa Rosa Natl. Forest, 1927, *Wooton* (NA); Jones Creek, N. of Galena Creek, Washoe Co., 6,500 ft, 1938, *Archer* 6516 (NA). ARIZONA: Pinaleno (Graham) Mts., 5,800 ft, 1935, *Maguire* & *Maguire* 11460 (NA). OREGON: Grant Co., Tex Creek, 4,700 ft, 1953, *Cronquist* 7624 (GH). WASHINGTON: Skamania Co., near Rands, 1938, *Hitchcock* & *Marsh* 3329 (NA); San Juan Islands, Friday Harbor, 1917, *Zeller* 754 (US). WYOMING: Dwyhee Co., Three Creek, 1912, *Nelson* & *Macbride* 1859 (E). BRITISH COLUMBIA: Thetis Lake, Victoria, 1941, *Eastham*, 8814 (NA). VANCOUVER ISLAND: Victoria, 1908, *Macoun* 78930 (E); *ibid.*: Sooke, 1959, *Holm* 342 (SP).

219. T. microdon Hook. & Arn. in Hook., Bot. Misc. 3: 180 (1833). *Typus*: Chile: Valparaiso, 1831, *Cummings* 747 (K!, iso.: E!, GL).

T. tricuspidatum Bert. ex Steud., Nom. ed. 2, 2: 704 (1841).
T. lechleri Phil., Anal. Univ. Santiago 84: 10 (1894).
T. microdon Hook. & Arn. var. *pilosum* Eastw., Proc. Cal. Acad. Sci. ser. 3, 1: 100 (1898).
T. chrysanthum Hook. & Arn. non Gaud. (1829).

Icon.: [Plate 214]. Hook. & Arn., Bot. Beech. Voy. t. 79 (1838); McDerm., N. Am. Sp. Trif. pls. 40–41 (1910).

Annual, sparingly hairy to glabrous, 5–60 cm. Stems many, slender, erect to ascending, grooved, dichotomously branched all along. Lower leaves long-petioled, upper ones short-petioled; stipules 0.8–1.2 cm long, ovate, slightly sheathing, coarsely dentate, membranous and ciliate at margin, free portion about as long as lower part, triangular acuminate; leaflets 0.5–2 × 0.4–1 cm, obovate to obcordate or oblanceolate, cuneate at base, truncate to retuse at apex, dentate-serrate, rarely entire. Inflorescences 0.5–1.5 cm in diam., semi-globular, 10- to many-flowered. Peduncles up to 8 cm long. Involucre cup-shaped, as long as or longer than flowers, glabrous or sparingly hairy, 8–12-lobed; lobes ovate, incised-dentate with long, slightly unequal, unbranched teeth. Rhachis short and thickened, with two whorls of cup-like cavities. Pedicels 0. Calyx 4–5 mm; tube tubular, 10-nerved, glabrous, membranous; teeth unequal, one-third the length of tube, triangular, with membranous and ciliate margins. Corolla 5–6 mm long, mostly pale pink to whitish; standard ovate-oblong. Pod 2–3 mm long, broad-ellipsoidal, membranous, cartilaginous at apex, 1-seeded, sutures poorly developed. Seed about 2 × 1 mm, ellipsoidal, darkly mottled, yellowish brown. Fl. April–July. 2n = 16.

Hab.: Meadows or on sandy or rocky soil and roadsides.

Gen. Distr.: USA: California, Oregon. British Columbia, Chile.

Selected specimens: CALIFORNIA: Humboldt Co., Eel River between Garberville and Phillipsville, 1966, *Gillett* & *Crompton* 12965 (HUJ); Alameda Co., Berkeley Hills, 120 m, 1907, Miss *H. A. Walker* 69 (HUJ). OREGON: Corvallis, 1922, *Epling* 5531 (UC). BRITISH COLUMBIA: Cluxewe River near Port McNeill, 20 ft, 1964, *Hett* & *Armstrong* 341 (HUJ); Vancouver Island, Mt. Finlayson, N. of Victoria, 1961, *Calder* & *McKay* 29421

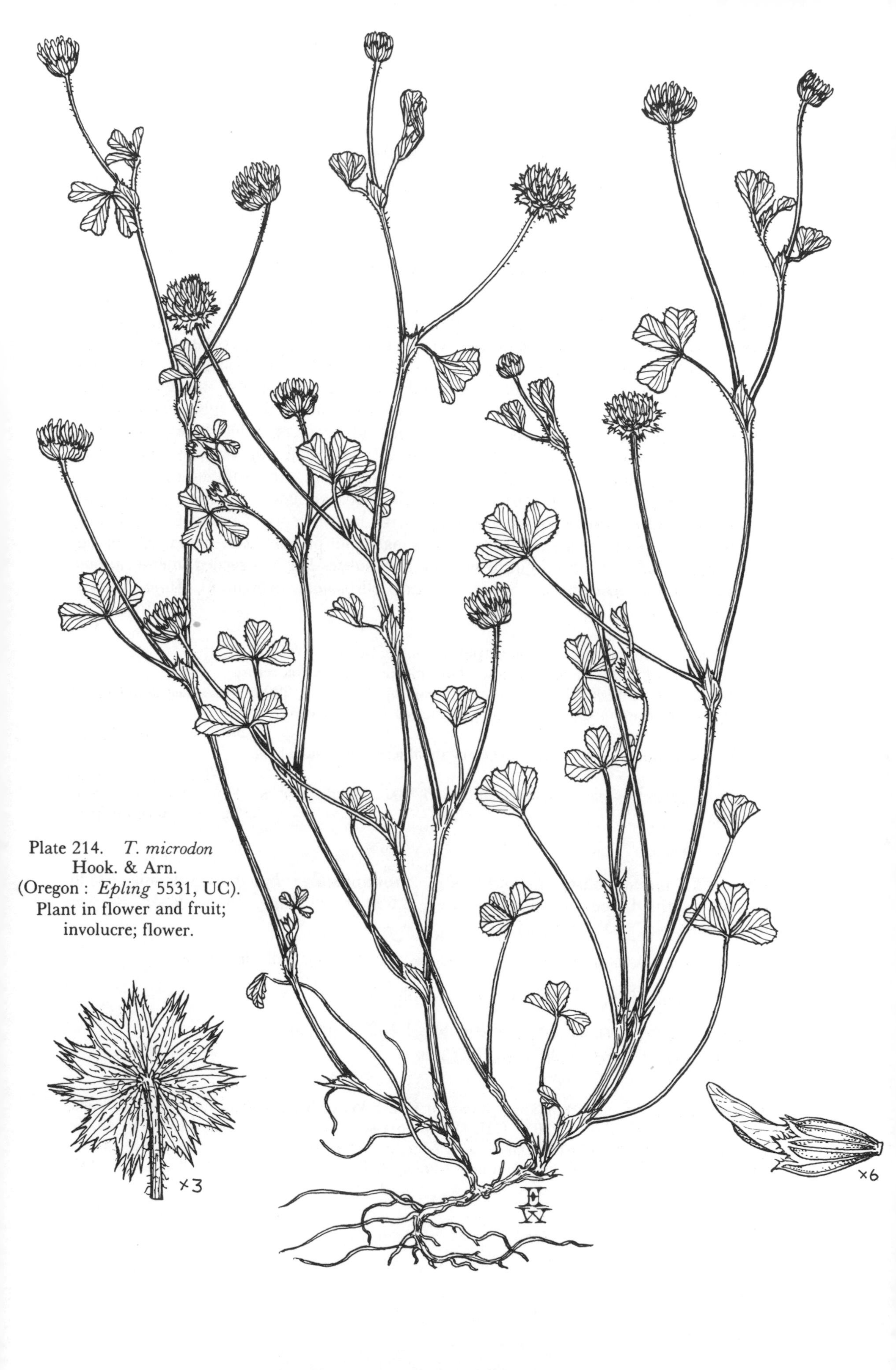

Plate 214. *T. microdon* Hook. & Arn. (Oregon : *Epling* 5531, UC). Plant in flower and fruit; involucre; flower.

(W). CHILE: Valdivia, 1852, *Philippe* & *Hohenacker* 19 (FI, G); *ibid.*, Lichte Waldestellen, 1898, *Buchtien* (E, G, W).

220. T. monanthum A. Gray in Proc. Am. Acad. 6 : 523 (1865).

Perennial, caespitose, villous to glabrous, (2–)5–12(–30) cm. Stems many, ascending to decumbent, slender, leafy, often rhizomatous at base. Leaves all long-petioled; stipules 0.3–1.2 cm, ovate to oblong, free portion triangular-lanceolate, herbaceous, few-nerved, mostly serrate; leaflets 0.3–2 × 0.2–0.5 cm, obovate, obcordate, oblanceolate to elliptical, acute, rounded to retuse at apex, margin serrate. Inflorescences 1–6(–10)-flowered. Peduncles often longer than subtending leaves. Involucre minute, 2–5 mm long, 2–5(–8)-lobed, glabrous or hairy; lobes divided almost to base into lanceolate lobules. Bracts 0 or small (in two-whorled head). Pedicels 1 mm long, erect. Calyx 2–5(–10) mm long; tube tubular-campanulate, 10-nerved, hairy; teeth triangular to lanceolate or subulate, as long as or slightly longer than tube, almost equal, sometimes with lateral tooth. Corolla 0.8–1.2(–1.6) cm, with white or cream standard and wings and purple keel; standard oblong, emarginate. Pod 2–3 mm, ovoid to subglobular, subsessile, 1–3-seeded, membranous, sparingly villous to glabrous. Seeds 1–2 mm, ellipsoidal, brown. Fl. June–August. 2n = 16.

1. Upper leaflets oblanceolate or elliptical, mostly acute — 2
– Upper leaflets obcordate to oblanceolate, mostly rounded or retuse — 3
2. Plants villous. Peduncles usually bent near top. — **220d.** var. **eastwoodianum**
– Plants glabrous to sparingly villous. Peduncles usually erect. — **220b.** var. **grantianum**
3. Plants villous, 5–30 cm tall. Lobes of involucre 0.5–2 mm long. — **220c.** var. **parvum**
– Plants glabrous to sparingly villous, 1–10 cm tall. Lobes of involucre 1–5 mm long. — **220a.** var. **monanthum**

220a. T. monanthum A. Gray Var. **monanthum**. *Typus*: California: Soda Springs in the Upper Tuolumne Valley, 8,600 ft, 26 June, 1863, *W. H. Brewer* 1704 (GH, iso.: DS, NY, UC).

T. monanthum A. Gray f. *spatiosum* McDerm., N. Am. Sp. Trif. 98 (1910).

Icon.: [Plate 215]. McDerm., *loc. cit.* pl. 35.

Hab.: Along springs and in meadows; 1,600–3,800 m.
Gen. Distr.: USA: California, Nevada.

Selected specimens: CALIFORNIA: Stanislau Co., S.W. of Sonora Pass, near Sardine Falls, 1966, *Gillett* & *Crompton* 13029 (HUJ); Mono Co., Cottonwood Creek, 11,000 ft, 1926, *Duran* 1629 (HUJ). NEVADA: Clark Co., McFarland springs, 2,425 m, 1938, *Clokey* 7984 (H, HUJ, UC, W); Nye Co., Toquima Range, 8,200 ft, 1964, *Holmgren* & *Reveal* 1445 (H).

Plate 215. *T. monanthum* A. Gray Var. *monanthum*
(Nevada : *Clokey* 7984, HUJ).
Plant in flower; involucre; flower.

220b. T. monanthum A. Gray Var. **grantianum** (Heller) Parish, Plant World 20 : 220 (1917). *Typus* : California : Mt. San Gorgonio, San Bernardino Mts., 23 July, 1904, *Geo. B. Grant* 6343 (POM, iso. : UC, WIS).

T. grantianum Heller, Muhlenbergia 1 : 136 (1906).
T. simulans House, Bot. Gaz. 41 : 341 (1906).
T. monanthum A. Gray var. *tenerum* Parish f. *grantianum* (Heller) McDerm., N. Am. Sp. Trif. 100 (1910).
T. monanthum A. Gray var. *tenerum* auct. non *T. tenerum* Eastw. (1902).

Icon. : McDerm., *loc.cit.* pls. 36–38.

Hab. : Stream banks and meadows.
Gen. Distr. : USA : California.

Specimens seen : CALIFORNIA : San Jacinto Mts., 1910, *Condit* (HUJ); San Bernardino Mts., 1908, *Heller* 8937 (K, UC); *ibid.* : White Water River, 8,000 ft, 1909, *Wilder* 1014 (HUJ).

220c. T. monanthum A. Gray Var. **parvum** (Kellogg) McDerm., N. Am. Sp. Trif. 105 (1910). *Typus* : California : Cisco, Sierra Nevada Mts., 6 July, 1870, *K. Kellogg* (CAS, iso. : ND-G, NY, UC).

T. pauciflorum Nutt. var. *parvum* Kellogg, Proc. Cal. Acad. Sci. 5 : 54 (1873).
T. multicaule Jones, Bull. Torrey Bot. Club 9 : 31 (1892).
T. parvum (Kellogg) Heller, Muhlenbergia 1 : 114 (1905).
T. monanthum A. Gray var. *parvum* (Kellogg) McDerm. f. *glabrifolium* McDerm., *op.cit.* 108.

Icon. : McDerm., *loc.cit.* pl. 39.

Hab. : Meadows, open areas in forest and along streams.
Gen. Distr. : USA : California, Nevada.

Specimens seen : CALIFORNIA : Placer Co., Yuba River at Cisco Flats, 1,740 m, 1910, *Hall* 8759 (E, POM, UC); between E. and W. Blue Lakes, 8,100 ft, 1933, *Wolf* 5274 (UC). NEVADA : Snow Valley, Ormsby Co., 2,460–2,615 m, 1902, *Baker* 1282 (W).

220d. T. monanthum A. Gray Var. **eastwoodianum** Martin, Madroño 8 : 233 (1946). *Typus* : California : Summit, trail to South Fork of King's River, 1–13 July, 1899, *A. Eastwood* (CAS, iso. UC).

T. tenerum Eastw., Bull. Torrey Bot. Club 29 : 81 (1902).

Hab. : Stream banks and meadows.
Gen. Distr. : USA : California.

Specimens seen : CALIFORNIA : Tulare Co., Coyote Creek, 1904, *Culberson* 4246 (HUJ, POM); Fresno Co., Huntington Lake, 7,000 ft, 1949. *Kappler* 1932 (UC).

221. T. obtusiflorum Hook. f., Bot. Beech. Voy. 331 (1838) et Ic. Pl. 3 : t. 281 (1840). *Typus* : Near Monterey, California, *D. Douglas* (K !, iso. W !).

T. tridentatum Lindl. var. *obtusiflorum* (Hook. f.) Wats., Proc. Am. Acad. 11 : 130 (1876).
T. roscidum Greene, Fl. Francisc. 31 (1891).
T. majus Greene, Pittonia 3 : 214 (1897).

Icon. : [Plate 216]. McDerm., N. Am. Sp. Trif. pls. 8–9 (1910).

Annual, glandular-pubescent, 30–100 cm. Stems few, erect to ascending, fistulous, profusely branching especially in upper part, grooved. Lower leaves long-petioled, upper ones short-petioled; stipules up to 1.5 cm long, broad-ovate, deeply laciniate with spinulose lobes, reflexed in mature plants, many-nerved, herbaceous; leaflets 1.5–4(–5) × 0.3–2.5 cm, elliptical to lanceolate or obovate to oblanceolate, acute to mucronate at apex, margins coarsely spinulose-serrate with unequal teeth. Inflorescences (1–)1.5–3 cm, globular, terminal and axillary, 10–50-flowered, whorled. Peduncles much longer than subtending leaves, stout and thick. Involucre flat, about one-third to two-thirds the length of flowers, glandular-hairy, deeply laciniate with spiny, lanceolate-subulate lobes. Pedicels about 1 mm. Calyx up to 1.3 mm long; tube tubular, about 20-nerved, slightly membranous below, inflated in fruit, glandular-hairy; teeth unequal, broadly triangular at base, subulate above, shorter than to as long as tube, the upper 4 shorter than lower one, sometimes one of them with 2 lateral teeth. Corolla 1–1.8 cm, much longer than calyx, white or pale-coloured with dark purple spot below; standard broad-elliptical, blunt at apex. Pod 4 × 2.5 mm, obovoid, membranous, with leathery margin, 1–2-seeded, finely reticulate. Seeds 2.5 mm, ellipsoidal. Fl. April–July. 2n = 16.

Hab. : Moist places below 1,600 m, many plant communities.

Gen. Distr. : USA : California, Oregon.

Selected specimens : CALIFORNIA : Trinity Co., Trinity River Ridge, 600 ft, 1964, *Gillett* & *Moulds* 12615 (HUJ); Yolo Co., N. of Vacaville, 1920, *Heller* 13374 (G); San Luis Obispo Co., N. of Santa Margarita, 1962, *Hurd* & *Stage* 62–30 (HUJ); Los Angeles Co., Glendale to Burbank, 1903, *Braunton* 886 (HUJ). OREGON : Rosenburg, Umpqua Valley, 1887, *Howell* (NA).

222. T. oliganthum Steud., Nom. ed. 2, 2 : 707 (1841). *Typus* : On the higher plains of the Oregon, near the outlet of the Wahlamet (=Willamette), *T. Nuttall* (NY, iso. : GH, K !).

T. pauciflorum Nutt. in Torr. & Gray, Fl. N. Am. 1 : 319 (1838) non D'Urv. (1822).
T. filipes Greene, Pittonia 1 : 66 (1887).
T. triflorum Greene, Pittonia 1 : 5 (1887).
T. oliganthum Steud. var. *sonomense* Greene, Man. Bot. Reg. San Franc. Bay 97 (1894).
T. oliganthum Steud. var. *triflorum* (Greene) Greene, Man. Bot. Reg. San Franc. Bay 97 (1894).
T. hexanthum Greene ex Heller, Muhlenbergia 2 : 215 (1906).
T. variegatum Nutt. var. *pauciflorum* (Nutt.) McDerm., N. Am. Sp. Trif. 67 (1910).
T. sonomense Greene (inedit.).

Icon. : [Plate 217]. McDerm., *loc. cit.* pls. 28–30.

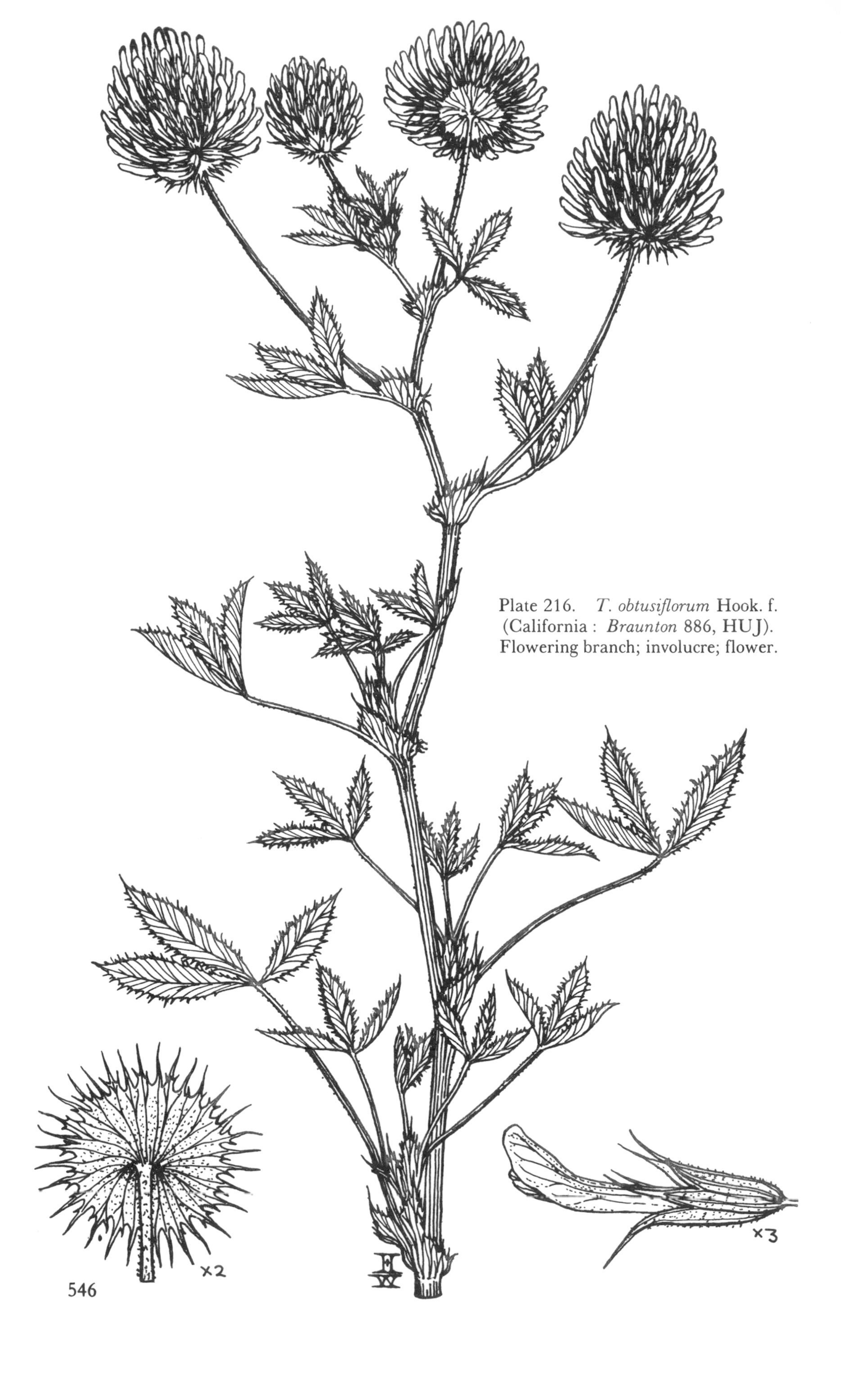

Plate 216. *T. obtusiflorum* Hook. f. (California : *Braunton* 886, HUJ). Flowering branch; involucre; flower.

Plate 217. *T. oliganthum* Steud.
(California: *Baker* 549, B).
Plant in flower and fruit; flower; involucre.

Annual, glabrous, rarely glandular-hairy, 10–50 cm. Stems few, erect, slender, branching from base and above. Leaves rather long-petioled; stipules 0.4–1.2 cm, lanceolate to oblong, few-nerved, free portion subulate-cuspidate, margin laciniate; leaflets 1–4 × 0.1–0.5 cm, linear, oblong, oblanceolate to elliptical, rounded, truncate and mucronate at apex, margin spinulose-dentate to entire. Inflorescences 0.3–1.2 × 0.3–0.8 cm, obovate to obconical, 3–15-flowered. Peduncles much longer than subtending leaves, filiform. Involucre one-third to one-fourth the length of (rarely almost as long as) flowers, divided almost to base into 3–14 lanceolate-subulate lobes. Pedicels 0.5–2 mm long, erect. Calyx 4–6 mm long; tube campanulate, 10-nerved, glabrous; teeth shorter than to as long as tube, broadly triangular at base, abruptly setose above. Corolla 6–8 mm, lavender with white tips and purple keel; standard narrowly obovate-oblong. Pod 2–3.5 mm, membranous, sessile, 1–2(–3)-seeded, somewhat constricted between seeds, ellipsoidal. Seeds 1.2 mm, lentiform to reniform, dark brown. Fl. March–June.

Hab.: Meadowland to dry rocky soil.

Gen. Distr.: Vancouver Island (British Columbia), USA: California, Washington, New Mexico, Oregon.

Selected specimens: VANCOUVER ISLAND: Victoria, 1908, *Macoun* 98, 932 (E). CALIFORNIA: Sonoma Co., near Healdsburg, 1897, *Alice King* (HUJ); Santa Clara Co., near Stanford Univ., 1902, *Baker* 549 (B, UC, W). WASHINGTON: Yakima Co., Satus Pass, 1958, *Hitchcock* & *Muhlick* 21590 (HUJ); San Juan Islands, Friday Harbor, 1917, *Zeller* 791 (UC). NEW MEXICO: Mogollan Mts., W. Fork of the Gila River, Socorra Co., 8,000 ft, 1903, *Metcalfe* 582 (E).

223. T. pinetorum Greene, Erythea 2: 182 (1894). *Typus*: New Mexico: Pinos Altos Mts., 15 September, 1880, *E. L. Greene* 372 (lecto. GH, isolecto. NY).

T. triaristatum Bert. ex Colla f. *pinetorum* (Greene) McDerm., N. Am. Sp. Trif. 60 (1910).
T. longicaule Woot. & Standl., Contrib. U. S. Nat. Herb. 16: 141 (1913).
T. wormskioldii Lehm. var. *longicaule* (Woot. & Standl.) Benson, Am. J. Bot. 28: 363 (1941).

Icon.: [Plate 218]. McDerm., *loc. cit.* pls. 19–20.

Perennial, glabrous, 10–30 cm. Stems few, erect to decumbent, leafy, slightly branching below and above. Leaves long-petioled; stipules 1–1.5 cm, narrowly ovate-oblong, almost white with green nerves, free portion triangular-lanceolate, entire; leaflets 0.4–2.5 × 0.3–1 cm, obovate to narrowly obovate or oblong, rounded to truncate or retuse, mucronulate, margin spinulose-dentate. Inflorescences 1.2–2 cm in diam., obovoid to semi-globular, 3–15-flowered, whorled. Peduncles slender, longer than subtending leaves. Involucre at most half the length of flowers, divided almost to base into 8–12 lanceolate, mostly entire lobes; inner bracts of upper whorl inconspicuous, broadly ovate, membranous. Pedicels up to 2 mm long in fruit, partly deflexed. Calyx 6–10 mm long; tube campanulate, 10-nerved, white-membranous, with green nerves; teeth up to twice as long as tube, lanceolate-subulate, almost equal. Corolla 1–1.5 cm, whitish or cream-coloured with dark purplish wings and keel;

Plate 218. *T. pinetorum* Greene
(Mexico : *Pringle* 1581, W).
Plant in flower; involucre; flower.

standard oblong, retuse at apex, not inflated. Pod 4–5 mm, membranous, stipitate, 2-seeded, elliptical. Seeds 1.5 mm across, reniform, brown, darkly mottled. Fl. July–September.

Hab. : Moist ground and near springs.

Gen. Distr. : USA : Utah, Arizona, New Mexico. Mexico.

Selected specimens : ARIZONA : Cochise Co., head of Carr Canyon, Huachuca Mts., 7,000 ft, 1944, *Darrow* et al. 1453 (NY, UC); Grand Canyon, 8,175 ft, *Cooper* 1791 (UC). NEW MEXICO : Grant Co., 15 mi. N. of Pinos Altos, Gila National Forest, 1969, *Isley* 10931 (NY, US). MEXICO : Coahuila, Villa Acuna Coa, 1936, *Piedra Blanco* 3690 (MEXU); State of Chihuahua, Sierra Madre, 9,700 ft, 1888, *Pringle* 1581 (W, UC).

224. T. polyodon Greene, Pittonia 3 : 215 (1897). *Typus* : California : Pacific Grove, Monterey Co., 27 May, 1895, *E. L. Greene* 31905 (ND-G).

T. tridentatum Lindl. var. *polyodon* (Greene) Jeps., Fl. Calif. 2 : 292 (1936).
T. variegatum Nutt. var. *polyodon* (Greene) Martin, Ph. D. Thesis 26 (1943).

Icon. : [Plate 219]. McDerm., N. Am. Sp. Trif. pl. 34 (1910).

Annual, glabrous, 30–60 cm. Stems many, branching all along, leafy, ascending to decumbent, grooved. Lower leaves long-petioled, upper ones short-petioled; stipules up to 1 cm, broad-ovate, deeply and unequally laciniate at margin, many-nerved, sometimes reflexed; leaflets 0.5–2 × 0.4–1.2 cm, obovate to broad-elliptical, cuneate at base, rounded or truncate, retuse at apex, dentate-serrulate at margin. Inflorescences 1–1.8 cm across, many-flowered, whorled, semi-globular to globular. Peduncles slender, much longer than subtending leaves. Involucre shorter than flowers, divided up to half its length into 8–15 unequally dentate-laciniate, spinulose lobes. Pedicels up to 1 mm. Calyx 4–7 mm long; tube tubular, 20-nerved; teeth shorter than or about as long as tube, broad-triangular in outline, each 3–5-fid with main tooth longer, subulate-spinulose and lateral ones shorter. Corolla 8–10 mm long, pink with purple-tipped keel; standard elliptical, tapering at both ends, slightly retuse. Pod 3.5 cm, membranous, ellipsoidal, hairy at apex, 2-seeded. Seeds 1.5 mm, reniform, brown. Fl. May.

Hab. : Along streams and in moist meadows.

Gen. Distr. : USA : California (endemic).

Specimens seen : CALIFORNIA : Monterey Co., Pacific Grove, 1903, *Heller* 6707 (E, K); *ibid.*, 1904, *Kennedy* (UC).

225. T. triaristatum Bert. ex Colla in Mem. Acad. Torino 37 : 54 (1834). *Typus* : Chile : in monte la Leona, anno 1829, *Berterro* (TO n. v.).

Icon. : [Plate 220]. Colla, *loc. cit.* t. 8.

Annual, sparingly hairy to glabrous, 10–30 cm. Stems few, erect, grooved, sparingly

Plate 219. *T. polyodon* Greene (California : *Heller* 6707, E). Flowering branch; flower; involucre.

Plate 220. *T. triaristatum* Bert. ex Colla
(Chile : *Germain*, W).
Plant in flower and fruit; flower; involucre.

branching, leafy. Leaves long-petioled; stipules 1–2 cm, oblong-lanceolate, green, the upper ones with deeply laciniate margin and free portion 3 times as long as adnate lower part; leaflets 1–3.5 × 0.3–0.8 cm, linear-lanceolate to oblanceolate, obtuse, mucronate, margins sharply serrulate-dentate. Inflorescences 1.5–.8 cm broad, hemispherical, 6- to many-flowered, whorled. Peduncles terminal and axillary, longer than or as long as subtending leaves. Involucre somewhat shorter than calyces, divided into 10–16 lobes; lobes about as long as entire part, unequally toothed, spinulose. Calyx 7–9 mm long, indurated; tube broadly tubular, with 10 prominent nerves; teeth somewhat longer than tube, broadly triangular in outline, 3–5 (or more)-toothed, the terminal tooth longer, all spiny-tipped. Corolla slightly longer or sometimes shorter than calyx, reddish to dark reddish when dry; standard obovate, slightly emarginate. Pod ovoid, 2–4-seeded.

Hab. : Near springs.

Gen. Distr. : Chile, Mexico(?), USA : New Mexico(?).

Specimens seen : CHILE : Monte la Leona, 1829, *Berterro* 709 (G, K probably isotypes); S. Juan, 1852, *Philippi* 301 (W); Valparaiso, *Philippi* (W); Env. de Conception, 1855, *Germain* (W).

226. T. trichocalyx Heller, Muhlenbergia 1 : 55 (1904). *Typus* : California : Pacific Grove, Monterey Co., 13 May, 1903, *A. A. Heller* 6721 (ND-G, iso. : DS, GH, MO, NY !, PH, POM !).

T. oliganthum Steud. var. *trichocalyx* (Heller) McDerm., N. Am. Sp. Trif. 84 (1910).

Icon. : [Plate 221]. McDerm., *loc. cit.* pl. 31.

Annual, sparingly villous to glabrescent, 10–45 cm. Stems few to many, decumbent to erect, leafy, sparingly to profusely branched. Leaves mostly long-petioled; stipules 0.4–1.2 cm long, ovate, acute, serrate to laciniate, terminating in long cusp; leaflets 0.5–1.5 × 0.4–1 cm, obovate, cuneate at base, truncate, retuse and shortly mucronate at apex, remotely dentate. Inflorescences 0.3–1.5 cm in diam., semi-globular, 5–20-flowered. Peduncles slender, longer than subtending leaves. Involucre much shorter than flowers, deeply divided into 5–15 entire, unequal, lanceolate-acuminate lobes. Pedicels up to 1.5 mm long. Calyx 6–9 mm long; tube 10-nerved, tubular; teeth longer than tube, unequal, lanceolate-subulate, plumose at base. Corolla 7–10 mm, pale purple; standard obovate, retuse, not inflated. Pod sessile, membranous, 4–6-seeded. Seeds 1.5 mm. Fl. May.

Hab. : Sandy soil of open pine woods.

Gen. Distr. : USA : California (endemic).

Specimens seen : CALIFORNIA : Monterey Co., Pacific Grove, 1907, *Kennedy* (H, HUJ); *ibid.*, 1907, *Heller* 8499 (NY).

227. T. tridentatum Lindl., Bot. Reg. 13 : t. 1070 (1827). *Typus* : Oregon : Near the Columbia River, *D. Douglas* (CGE n. v., iso. K !).

Plate 221. *T. trichocalyx* Heller (California : *Kennedy*, HUJ). Plant in flower; flower; involucre.

T. involucratum Sm. in Rees, Cycl. 36 : 28 (1817) non Ortega (1797).
T. aciculare Nutt. in Torr. & Gray, Fl. N. Am. 1 : 319 (1838).
T. polyphyllum Nutt. in Torr. & Gray, *loc. cit.*
T. nuttallii Steud., Nom. ed. 2, 2 : 707 (1841).
T. tridentatum Lindl. var. *melananthum* Wats., Proc. Am. Acad. 11 : 130 (1876).
T. watsonii Lojac., Nuov. Giorn. Bot. Ital. 15 : 186 (1883).
T. scabrellum Greene, Pittonia 1 : 159 (1888).
T. segetum Greene, Pittonia 3 : 221 (1897).
T. trimorphum Greene, *op. cit.* 220.
T. tridentatum Lindl. var. *aciculare* (Nutt.) McDerm., N. Am. Sp. Trif. 26 (1910).
T. tridentatum Lindl. var. *aciculare* (Nutt.) McDerm. f. *watsonii* (Lojac.) McDerm., *op. cit.* 28.
T. tridentatum Lindl. var. *segetum* (Greene) McDerm., *op. cit.* 24.
T. tridentatum Lindl. f. *trimorphum* (Greene) McDerm., *op. cit.* 20.
T. tridentatum Lindl. var. *watsonii* (Lojac.) Jeps., Fl. Calif. 2 : 292 (1936).

Icon. : [Plate 222]. McDerm., *loc. cit.* pls. 3, 4, 6, 7.

Annual, glabrous, 10–100 cm. Stems few, erect to ascending, more or less leafy, branching especially above, grooved. Leaves long-petioled; stipules 1–2 cm, membranous, many-nerved, ovate to lanceolate, margin coarsely dentate-serrate (at least in upper ones) with subulate teeth; leaflets 1–5 × 0.2–1.5 cm, linear, lanceolate-oblong to elliptical, finely to coarsely spinulose-dentate, acute or truncate and retuse, mucronate. Inflorescences 0.6–3 cm in diam., ovate to semi-globular, 6- to many-flowered, whorled. Peduncles slender, much longer than subtending leaves. Involucre shorter than to as long as lower calyces, flat, disk-like, unequally laciniate or toothed; teeth considerably shorter than or as long as entire part, spiny. Pedicels very short or up to 1 mm long. Calyx 6–9 mm long; tube tubular, 10–15-nerved, membranous; teeth as long as or shorter than tube, broadly triangular, simple or with 2 or more small lateral teeth, apex with abrupt, short or long cusp. Corolla 1.2–2 cm, pale purple or whitish often with darker keel turning brown in fruit; standard narrowly oblong, not inflated in fruit. Pod about 3 mm, broad-ellipsoidal, 1–2-seeded, membranous, reticulately nerved, sessile. Seeds 1.5–2.5 mm, reniform to ellipsoidal, dark brown. Fl. March–July. 2n = 16.

Hab. : Grassy hillsides and meadowland.

Gen. Distr. : Vancouver Island, USA : California, Oregon, Washington, Arizona.

Selected specimens : VANCOUVER ISLAND : Nanaimo, 1908, *Macoun* 78, 936 (E); Gordon Head, 15 ft, 1965, *Davies* 407 (H). CALIFORNIA : Santa Clara Co., Mt. Hamilton, 1907, *Heller* 8628 (HUJ); Kern Co., near Caliente, 1905, *Heller* 7621 (E); vicinity of Berkeley, 1907, *Walker* 592 (HUJ); Baja, 18 mi. E. of Rosario, 1952, *Gentry* & *Fox* 11697 (MEXU). OREGON : Josephine Co., Grant's Pass, 1910, *Heller* 10029 (E). WASHINGTON : Bingen, Klickitate Co., 1906, *Suksdorf* 5590 (HUJ); Brown's Island, 1917, *Zeller* 793 (UC). ARIZONA : Devil's Canyon, near Superior, 1935, *Nelson* 1855 (G); Sierra Estrella, 1928, *Peebles* & *Harrison* 5279 (UC).

Note : *T. tridentatum* is one of the most variable clover species of N. America. Many attempts were made to divide it into varieties or other taxa, but so far the division was based on insufficiently distinct characters. Only experimental investigations will make possible clarification of the taxonomy of this species.

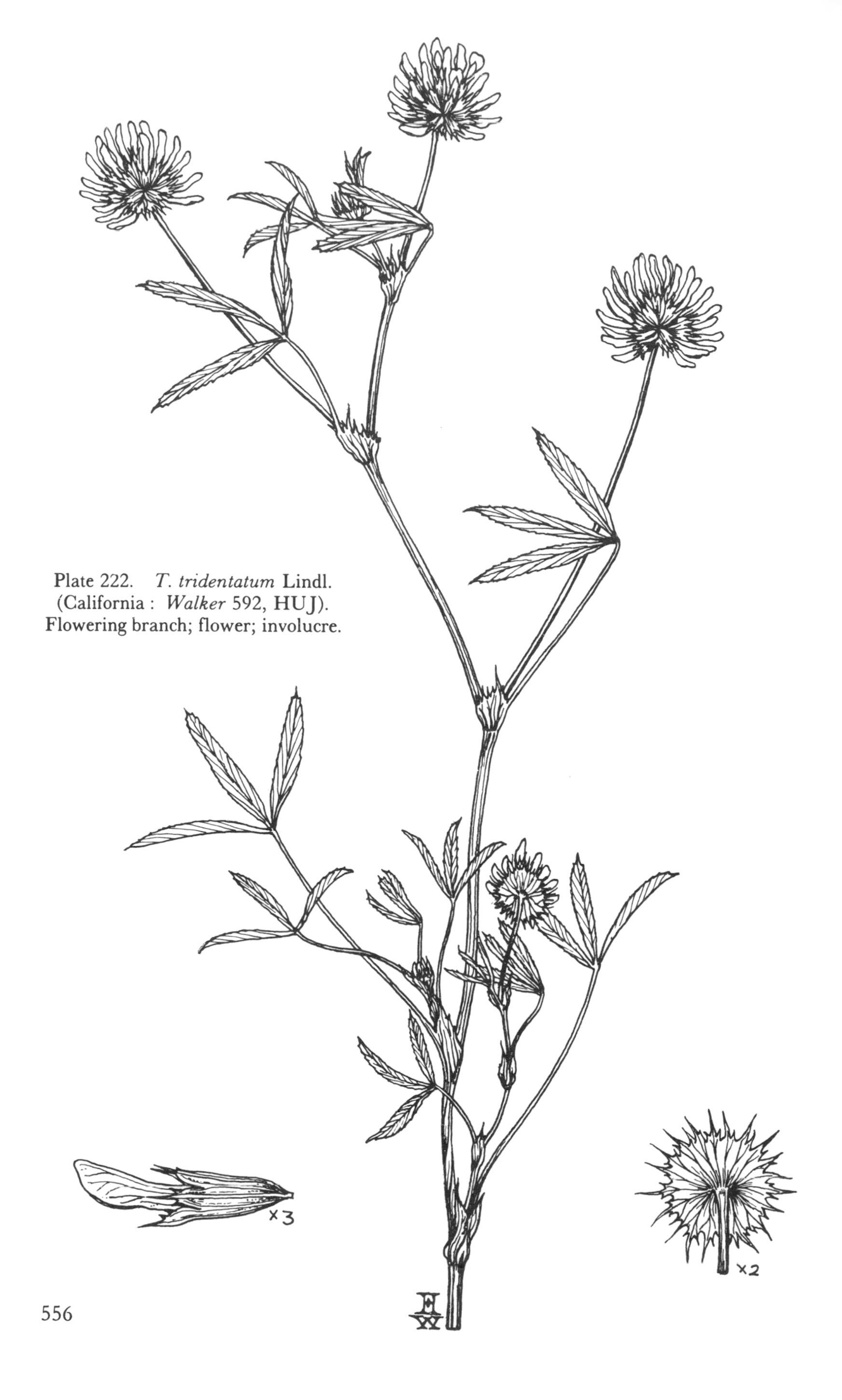

Plate 222. *T. tridentatum* Lindl. (California : *Walker* 592, HUJ). Flowering branch; flower; involucre.

228. T. variegatum Nutt. in Torr. & Gray, Fl. N. Am. 1 : 317 (1838).

Annual, glabrous to glabrescent, 5–100 cm. Stems few to many, erect, decumbent or ascending, stout or filiform, much branching. Leaves rather short-petioled; stipules 0.5–1.3 cm long, broad-ovate, green, prominently nerved, margin unequally laciniate; leaflets 0.3–5 × 0.1–1.5 cm, oblanceolate to obovate, rounded, acute or retuse at apex, cuneate at base, margins dentate-serrate. Inflorescences 1–1.5 cm in diam., semi-globular, few- to many-flowered. Peduncles terminal and axillary, as long as or somewhat longer than subtending leaves, stout. Involucre mostly shorter than calyces, deeply divided into 4–12 unequally laciniate lobes; teeth subulate-spinulose. Calyx 5–8 mm; tube campanulate, 10–20-nerved; teeth slightly unequal, the longest one much longer than tube, all triangular-lanceolate, simple or sometimes one of them bifid or with small lateral tooth. Corolla 0.6–1 cm, purple, white-tipped, becoming dark brown in fruit; standard obovate, retuse or sometimes apiculate. Pod about 3 mm, broad-elliptical, short-stipitate, membranous, 1–2-seeded. Seeds 1.2 mm, dark brown, glossy. Fl. March–August. 2n = 16.

228a. T. variegatum Nutt. Var. **variegatum**. *Typus* : Oregon : Springy places near the mouth of Wahlamet (= Willamette), *Thomas Nuttall* (NY, iso. : GH, K !).

T. melananthum Hook. & Arn., Bot. Beech. Voy. 331 (1838).
T. pauciflorum Lojac., Nuov. Giorn. Bot. Ital. 15 : 183 (1883).
T. trilobatum Jeps., Bull. Torrey Bot. Club 18 : 322 (1891).
T. morleyanum Greene, Erythea 3 : 47 (1895).
T. calophyllum Greene, Pittonia 3 : 213 (1897).
T. dianthum Greene, *op. cit.* 217 (1897).
T. geminiflorum Greene, *op. cit.* 216.
T. phaeocephalum Greene, *op. cit.* 216.
T. pusillum Greene, *op. cit.* 217.
T. subsalinum Greene, *op. cit.* 219.
T. ultramontanum Greene, *op. cit.* 218.
T. variegatum Nutt. var. *pauciflorum* (Nutt.) McDerm., N. Am. Sp. Trif. 67 (1910).
T. variegatum Nutt. var. *pauciflorum* (Nutt.) McDerm. f. *phaeocephalum* (Greene) McDerm., *op. cit.* 78.
T. variegatum Nutt. var. *trilobatum* (Jeps.) McDerm., *op. cit.* 73.
T. variegatum Nutt. var. *melananthum* (Hook. & Arn.) McDerm., *op. cit.* 74.
T. variegatum Nutt. var. *melananthum* (Hook. & Arn.) McDerm. f. *morleyanum* (Greene) McDerm. et f. *major* (Lojac.) McDerm., *op. cit.* 76.

Icon. : [Plate 223]. McDerm., *loc. cit.* pls. 21–27.

Keel beakless or with a broad beak not over 0.3 mm long.

Hab. : Dry sandy soil to moist meadows.

Gen. Distr. : Vancouver Island, USA : California, Nevada, Idaho, Montana, Washington, Oregon, Utah, Arizona, Wyoming. Mexico.

Selected specimens : VANCOUVER ISLAND : Victoria, 1908, *Macoun* 78, 939 (E). CALIFORNIA : San Mateo Co., Menlo Heights, 1914, *Abrams* 5077 (NA); Mendocino Co., road to Dos Rios, 1966, *Gillett* & *Crompton* 12951 (HUJ); Placer Co., Squan Valley, 1916,

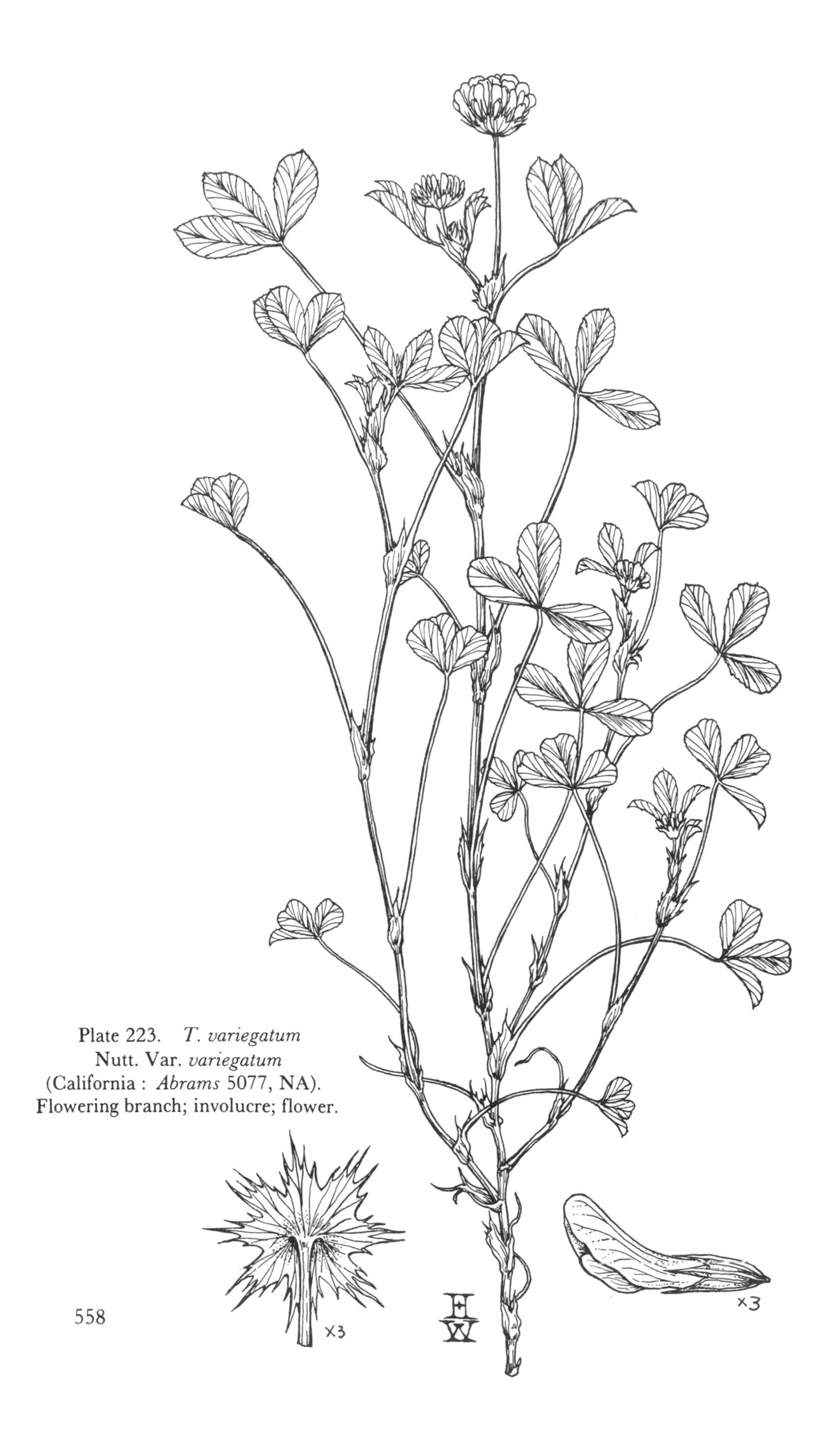

Plate 223. *T. variegatum*
Nutt. Var. *variegatum*
(California : *Abrams* 5077, NA).
Flowering branch; involucre; flower.

Smith 406 (HUJ). NEVADA: Ormsby Co., King's Canyon, 1,700–2,000 m, 1902, *Baker* 1449 (W); Deeth, Elko Co., 5,340 ft, 1908, *Heller* 9121 .(HUJ). IDAHO: Boise, meadow lands, 2,880 ft, 1911, *Clark* 255 (E). WASHINGTON: San Juan Islands, Friday Harbor, 1917, *Zeller* 792 (UC); Yakima Co., Satus Pass, 1958, *Hitchcock* & *Muhlick* 21590 (HUJ). OREGON: Corvallis, 1922, *Epling* 5931 (UC); Jackson Co., 7 mi. S.W. of Prospect., 2,100 ft, 1939, *Hitchcock* & *Martin* 4995 (UC). ARIZONA: near Prescott, 1927, *Peebles* & *Harrison* 4194 (UC). WYOMING: Evanston, 1897, *Williams* 2400 (NA). MEXICO: Valley of Palms, 1882, *M. E. Jones* 3677 (NY).

228b. T. variegatum Nutt. Var. **rostratum** (Greene) Martin emend. Heller. *Typus*: California: Lake Merritt, Oakland, April 1889, *V. K. Chestnut* (ND-G iso.)

T. appendiculatum Lojac., Nuov. Giorn. Bot. Ital. 15 : 181 (1883).
T. rostratum Greene, Proc. Acad. Nat. Sci. Philad. 47 : 547 (1896).
T. splendens Heller, Muhlenbergia 1 : 115 (1905).
T. appendiculatum Lojac. f. *rostratum* (Greene) McDerm., N. Am. Sp. Trif. 92 (1910).
T. appendiculatum Lojac. var. *rostratum* (Greene) Jeps., Man. Fl. Pl. Calif. 539 (1925).

Icon.: McDerm., *loc. cit.* pls. 32–33.

Keel slender with a 0.5 mm long beak.
Hab.: Low, wet fields and grassy places, Coastal Prairie.
Gen. Distr.: USA: California.

Specimens seen: CALIFORNIA: Marin Co., Ignacio, 20 ft, 1946, *Rose* 46037 (B, H, HUJ, W); Pacific Grove, 1907, *Kennedy* (HUJ); Mendocino Co., 4 mi. S. of Laytonville, 1929, *Kennedy* 10309 (H, R).

229. T. vernum Phil. in Anal. Univ. Chile 687 (1872). *Typus*: Chile: San Fernando, Sept. 1864, *F. Philippi* 404 (SGO !).

Icon.: [Plate 224].

Annual, sparingly to densely villous, 5–25 cm. Stems few, erect to ascending, leafy, branching all along. Lower leaves long-petioled, upper ones short-petioled; stipules 2–8 mm, broadly ovate, deeply laciniate to entire, green, upper free portion lanceolate-subulate; leaflets 0.3–1.3 × 0.2–0.8 cm, obovate, obcordate to oblong, cuneate at base, remotely dentate, retuse to emarginate. Inflorescences small, 0.7–0.9 mm in diam., globular to semi-globular, few- to 15-flowered. Peduncles mostly shorter than subtending leaves, densely villous to almost glabrous. Involucre almost covering heads, deeply divided into 6–10 triangular-cuspidate or lanceolate, simple, rarely bifid lobes, about as long as entire part; inner bracts up to 1 mm, hyaline, lanceolate. Calyx 4–7 mm long; tube narrowly campanulate, densely villous-wooly or with only few hairs near teeth, 10-nerved; teeth somewhat longer than tube, subulate, more or less equal, simple. Corolla 4–5 mm long, white to cream; standard elliptical. Pod about 2 mm broad, ellipsoidal, 1-seeded. Seed up to 2 mm, ellipsoidal, brown. Fl. May–June, December. 2n = 16.

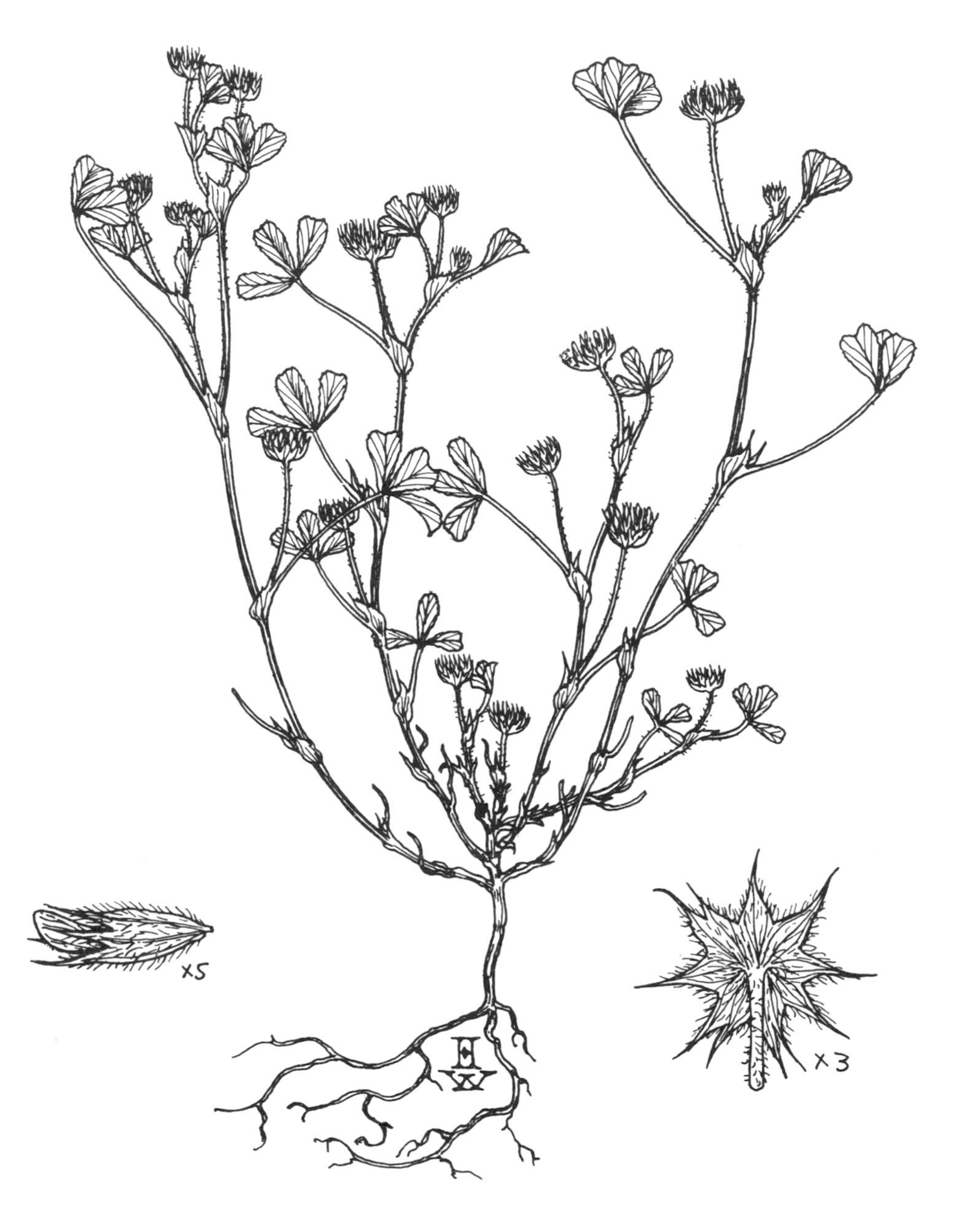

Plate 224. *T. vernum* Phil.
(Chile : *Morrison* 17044, G).
Plant in flower and fruit; flower; involucre.

Hab. : Wet soil along streams, rocky crevices.
Gen. Distr. : Chile.

Specimens seen : CHILE : Prov. Aconcagua, Depto. Los Andes, Cerro Chache, 1,900 m, 1938, *Morrison* 17044 (G, NA); San Fernando, 1876, *Philippi* 2027 (B, G, NY, SGO, W).

Note : *T. vernum* is very close to *T. microcephalum* Pursh, but differs from the latter by having calyx teeth which are subulate and longer than the tube.

230. T. wigginsii Gillett, Madroño 23 : 335, f.1 (1976). *Typus* : Mexico : Baja California, Sierra San Pedro Martir, W. of Vallecitos, 2,450 m, 23 Aug., 1968, *Reid Moran* 15400 (RSA, iso. SD).

Perennial, glabrous, 50–60 cm. Stems spreading and prostrate, branching, with well-developed taproot, rhizomatous and often rooting at nodes. Leaves on 0.3–2 cm long petioles; stipules 4–7 mm long, united for up to one-third of their length, ovate with attenuate tips, lacerate at margin; leaflets 2–7 × 1.5–4 mm, cuneate to obovate, truncate to rounded or slightly retuse, minutely apiculate, with dentate to serrate margin. Inflorescences hemispherical, 10–15-flowered, in 1–3 whorls. Peduncles 1–6 cm long. Involucre about half the length of calyces, deeply divided into 7–9 oblong lobes; lobes with 3 acute to acuminate teeth; inner bracts small, fused, hyaline, blunt. Pedicels very short to sessile. Calyx about 1 cm long, slightly oblique; tube 10-nerved; teeth about as long as tube, almost equal, subulate. Corolla 1.2 cm long, white or lavender towards tips or wings reddish-violet and keel purplish-violet; standard elliptical, parallely veined, dichotomising apically, flared upwards in upper third. Immature pod about 2 mm long, oblong, 2-seeded. Fl. June–August.
Hab. : Meadows, dry gravel and hillsides in open pine forest; 2,200–2,500 m.
Gen. Distr. : Mexico (endemic).

Note : Description follows Gillett, *loc. cit.*

231. T. wormskioldii Lehm., Ind. Sem. Hort. Bot. Hamb. 17 (1824). *Typus* : Grown from seed collected by *Wormskjold* in California (S).

T. involucratum Orteg., Nov. Rar. Pl. Hort. Bot. Matrir. Desc. 33 (1797) non Lam. (1778).
T. fimbriatum Lindl., Bot. Reg. 13 : t. 1070 (1827).
Lupinaster wormskioldii (Lehm.) Presl, Symb. Bot. 3 : 47 (1831).
T. spinulosum Dougl. in Hook., Fl. Bor. Am. 1 : 133 (1831).
T. calocephalum Nutt. ex Torr. & Gray, Fl. N. Am. 1 : 318 (1838).
T. heterodon Torr. & Gray, *loc. cit.*
T. spinulosum Dougl. var. *triste* Torr. & Gray, *loc. cit.*
T. fendleri Greene, Pittonia 3 : 221 (1897).
T. involucratum Orteg. var. *fendleri* (Greene) McDerm., N. Am. Sp. Trif. 47 (1910).
T. involucratum Orteg. var. *fimbriatum* (Lindl.) McDerm., *op. cit.* 52.
T. involucratum Orteg. var. *kennedianum* McDerm., *op. cit.* 56.
T. kennedianum (McDerm.) Nels. & Macbride, Bot. Gaz. 61 : 31 (1916).

Icon. : [Plate 225]. McDerm., *loc. cit.* pls. 10–18.

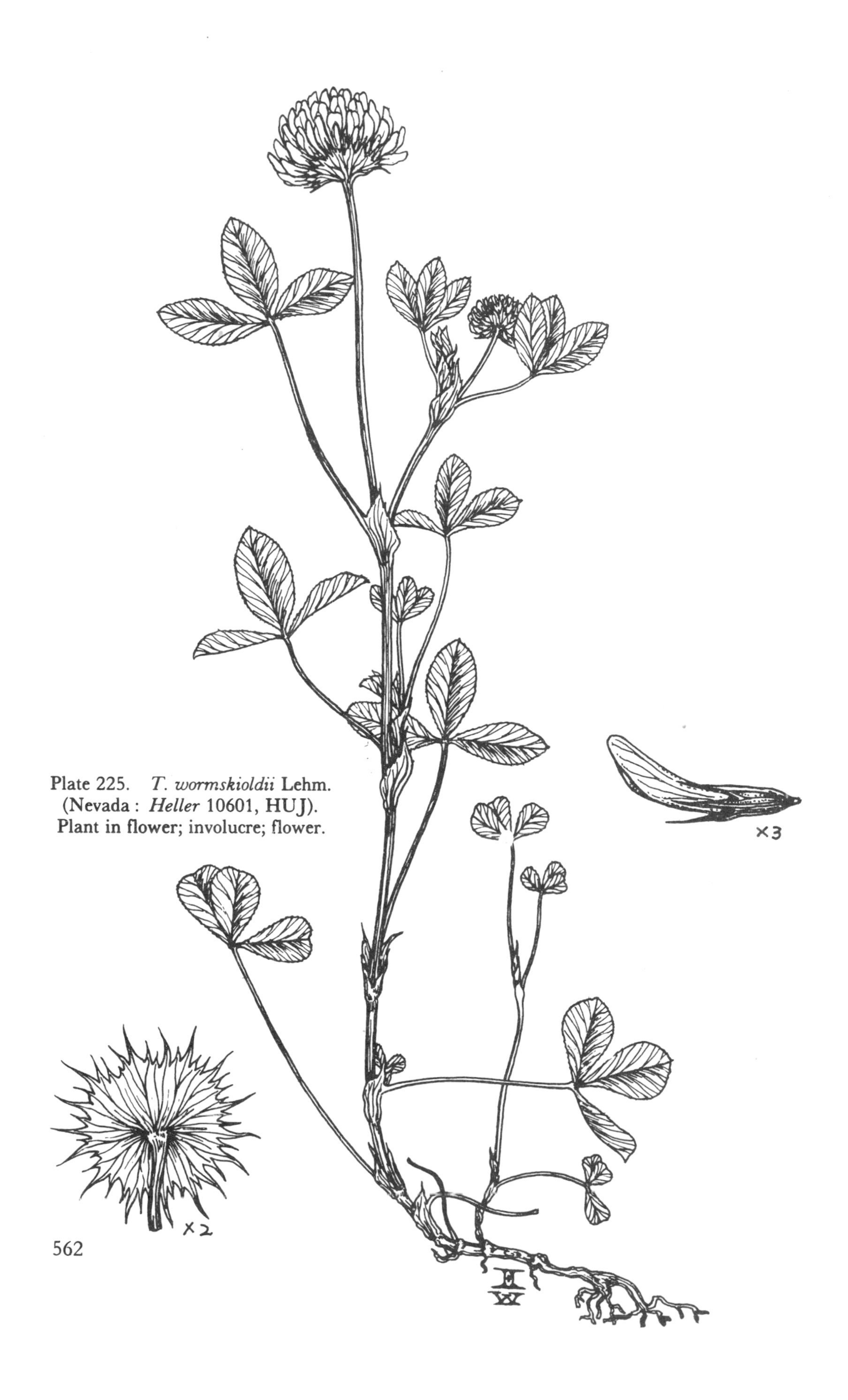

Plate 225. *T. wormskioldii* Lehm.
(Nevada : *Heller* 10601, HUJ).
Plant in flower; involucre; flower.

Perennial, glabrous to sparingly hairy, 10–100 cm. Stems single or few, ascending to decumbent, striate, mostly fistulous, rhizomatous, leafy. Leaves long-petioled to subsessile; stipules 1–2 cm, ovate to lanceolate, deeply dentate-serrate, free portion triangular to lanceolate; leaflets 0.7–4.5 × 0.2–1.5 cm, obovate, obcordate, oblong to lanceolate or linear, acute or obtuse, rounded or truncate and retuse, with dentate-serrate margin. Inflorescences 1.5–3 cm in diam., hemispherical, many-flowered, whorled. Peduncles terminal and axillary, rather stout, longer than subtending leaves. Involucre about half the length of head, with 5–15 deep to shallow lobes; lobes broad and toothed or narrow and entire; inner bracts minute to 1 mm or forming involucre. Pedicels short, up to 2 mm long. Calyx 6–9 mm; tube campanulate, 10-nerved; teeth lanceolate-subulate, longer than tube, simple or sometimes one of them slightly trifid. Corolla 0.9–1.8 cm, purple, lighter in upper part; standard obovate-oblong, emarginate, not inflated in fruit. Pod about 4 mm, ellipsoidal, membranous, 1–4-seeded, sessile to short-stipitate. Seeds about 1.5 mm long, brown. Fl. April–September. 2n = 16, 32, ca. 48.

Hab.: Meadows, stream banks and edges of coastal sand dunes.

Gen. Distr.: Vancouver Island, USA: California, Idaho, Utah, Colorado, Nevada, Oregon, Washington, New Mexico, Arizona. Mexico.

Selected specimens: VANCOUVER ISLAND: Jordan River, 1959, *Holm* 367 (SP); edge of sandy beach, 1901, *Rosendal* & *Brand* 13 (E). CALIFORNIA: San Jacinto Mts., Saunders Meadows, 5,500 ft, *Meyer* 591 (HUJ); Humboldt Co., salt flat near Arcata, 1923, *Heller* 13781 (HUJ, NY, US). UTAH: Cliff above beach, Brookings, 1946, *Ricker* 6226 (NA). COLORADO: Gunnison Watershed, 7,680 ft, 1901, *Baker* 826 (E). NEVADA: between Washoe and Franktown, 5,100 ft, 1912, *Heller* 10601 (HUJ); Ormsby Co., 6 mi. from head of Clear Creek, 5,000 ft, 1937, *Nichols* & *Lund* 667 (NA). OREGON: Tillamook Co., Sand Is. near Sand Lake, 1962, *Gillett* & *Taylor* 11156 (HUJ); Josephine Co., 1 mi. S. of Waldo, 1925, *Kennedy* (HUJ). WASHINGTON: near Spokane, 1892, *Sandberg* et al. 1028 (E); Port Crescent, Clallam Co., 1904, *Lawrence* 233 (NA). NEW MEXICO: White Mts., Lincoln Co., 6,300 ft, 1897, *Wocton* 235 (E); Catron Co., Gilila Camp, 20 mi. N.E. of Mogollon, 8,000 ft, 1938, *Hitchcock* et al. 4408 (NA, UC). ARIZONA: Willow Spring, 1890, *Palmer* 507 (E); White Mts., 8 mi. N. of Hanmgan Ranger Station, 7,000 ft, 1927, *Goddard* 674 (UC). MEXICO: Tamaulipas: between Marcella and Hermosa, 1949, *Taylor* 2669 (HUJ); San Luis Potosi, 1879, *Schaffner* 602 (R).

Subsection II. PHYSOSEMIUM (Lojac.) Heller & Zoh. (comb. nov.)

Section *Physosemium* Lojac., Nuov. Giorn. Bot. Ital. 15: 230 (1883).
Section *Cyathiferum* Lojac., *op. cit.* 187.
Section *Physantha* Nutt., Proc. Acad. Phil. 4: 8 (1848).
Section *Cyathiferae* McDerm., N. Am. Sp. Trif. 110 (1910) p. p.
Section *Vesiculeae* McDerm., *op. cit.* 127.

Type species: *T. fucatum* Lindl.

Annuals. Calyx 5–20-nerved, sometimes oblique, inflated in fruit. Standard inflated

in fruit, forming vesicule, constricted above into narrower appendage, including mature pod. Pod stipitate to sessile, 1–8-seeded.
Six N. and S. American species.

232. T. barbigerum Torr., Pac. Rail. Reports 4 : 79 (1857).

Annual, villous to glabrescent, 5–30 cm. Stems many, erect, ascending to decumbent, branched. Leaves all long-petioled; stipules 0.8–2 cm long, mostly crowded on lower part of stems, white, membranous, broadly ovate, partly sheathing, deeply incised-dentate with acute to acuminate lobes, free portion longer than lower part; leaflets 0.4–2.5 × 0.3–1.3 cm, obovate to elliptical or oblanceolate, cuneate, rounded to acute or truncate and retuse at apex, with coarsely dentate to serrate margins. Inflorescences 0.5–3 cm in diam., 7–30-flowered, slightly whorled, semi-globular to globular. Peduncles very long, erect. Involucre 1–1.5 cm wide, hairy to glabrous, from one-third the length of to almost as long as flowers, with 5–15 shallow lobes at most about one-third the length of entire part of involucre; lobes finely to coarsely dentate; teeth triangular-lanceolate, acute to acuminate or sharp-pointed. Rhachis inflated, as long as broad. Pedicels almost 0. Calyx 5–8 mm long; tube membranous, sparingly villous, campanulate, 5-nerved; throat slightly oblique; teeth almost as long as tube, subequal, plumose, subulate-setaceous, one of the lower ones bifid or trifid. Corolla 0.5–1.2 mm long, purple; standard broad-ovate, tapering at apex, inflated in fruit, forming tube with constricted throat above calyx. Pod 3 × 2 mm, stipitate, broad, ovoid, membranous, irregularly rupturing, 1–2-seeded. Seeds 1.5 mm, reniform, light brown, somewhat mottled. Fl. April–June.

1. Standard about as long as or shorter than calyx. **232a.** var. **barbigerum**
– Standard about 7 mm, longer than calyx. Plants somewhat larger. **232b.** var. **andrewsii**

232a. T. barbigerum Torr. Var. **barbigerum**. *Typus* : California : San Francisco, 1853–4, *J. M. Bigelow* (GH, iso. NY !).

Icon. : [Plate 226]. McDerm., N. Am. Sp. Trif. pl. 45 (1910).

Hab. : Open, low, moist places. Coastal Prairie. Mixed evergreen forests.
Gen. Distr. : USA : California (endemic).

Selected specimens : CALIFORNIA : near Santa Cruz, *Anderson* (HUJ); Solano Co., S. of Dixon at Dozier Station, 1956, *Beecher Crampton* 3290 (HUJ); Monterey Co., Point Pinos, 1907, *Heller* 8498 (E).

232b. T. barbigerum Torr. Var. **andrewsii** Gray, Proc. Am. Acad. Arts Sci. 7 : 335 (1868). *Typus* : *Andrews*, in 1856 (locality not stated) (G, H); better material cited in type description is *H. N. Bolander* 4781 in 1866 from California (G !, GH !, MO, UC !, W !, iso. : G !, K !, MO, UC !, W !).

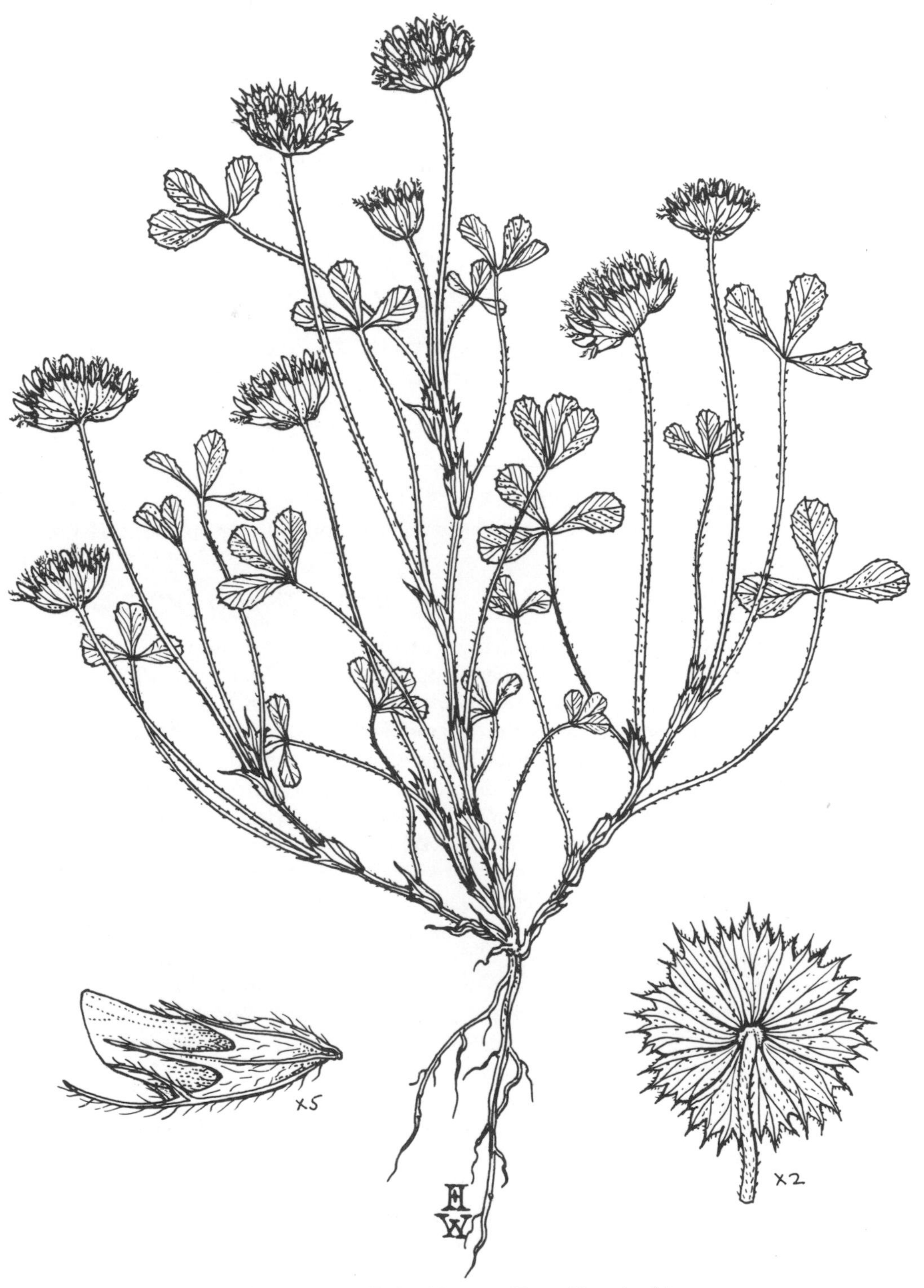

Plate 226. *T. barbigerum* Torr. Var. *barbigerum*
(California : *Anderson*, HUJ).
Plant in flower and fruit; flower; involucre.

T. grayi Lojac., Nuov. Giorn. Bot. Ital. 15 : 189 (1883).
T. lilacinum Greene, Proc. Acad. Philad. 1895 : 447 (1896).
T. andrewsii (Gray) Heller, Muhlenbergia 1 : 114 (1905).
T. barbigerum Torr. var. *lilacinum* (Greene) Jeps., Fl. Calif. 2 : 189 (1936).

Icon. : McDerm., N. Am. Sp. Trif. pl. 46 (1910).

Hab. : Wet meadows.
Gen. Distr. : USA : California (endemic).

Selected specimens : CALIFORNIA : Santa Rosa Valley, Olivet School, 1929, *Sinclair* (HUJ); Santa Cruz, *Anderson* 107 (HUJ); Santa Cruz Co., Glenwood, 3,000, ft, 1947, *Rose* 47120 (G).

233. T. cyathiferum Lindl., Bot. Reg. 12 : t. 1070 (1827). *Typus* : From seeds collected by *D. Douglas* in neighbourhood of Columbia River, in 1826 (CGE, iso. K !).

Icon. : [Plate 227]. Hitchcock et al., Vasc. Pl. Pac. N. W. 3 : 359 (1961).

Annual, glabrous, 5–50 cm. Stems many, erect to ascending, simple or branching all along, grooved, leafy. Leaves long-petioled, petioles of lower ones 10 cm or more; stipules 0.6–1.2 cm, membranous, many-nerved, ovate to lanceolate, with entire, serrate or laciniate margin, free portion triangular, apical tooth long-aristate at tip; leaflets 0.5–2.5 × 0.3–1.5 cm, obovate, elliptical to oblanceolate, rounded to retuse at apex, margins usually serrulate to denticulate. Inflorescences 0.5–2 cm in diam., semi-globular to ovate, few- to many-flowered. Peduncles 1–8 cm long, slender. Involucre nearly as long as to longer than flowers, bowl-shaped, membranous, with prominent green nerves, margin with 3–14 shallow lobes; lobes many-toothed; teeth triangular, terminating in long cusp. Calyx 0.6–1.3 cm long; tube campanulate, membranous, 13–20-nerved, inflated on lower side; throat markedly oblique; lower teeth 1–2 times as long as tube, with 1–3 trichomotomous, setiform, divergent branches; upper teeth as long as tube, simple or forked. Corolla 0.6–1.3 mm, cream, with rose or pink tips, becoming brown in fruit; standard obovate-elliptical, inflated in fruit. Pod about 3 mm, stipitate, leathery, ellipsoidal, 1–2-seeded. Seeds about 0.5 mm, yellowish-brown. Fl. May–August. 2n = 16.
Hab. : Wet meadows to fairly dry sandy soils.
Gen. Distr. : USA : Nevada, Oregon, Idaho, California, Washington. Vancouver Island, Mexico.

Selected specimens : NEVADA : Ormsby Co., King's Canyon, 1,700–2,000 m, 1902, *Baker* 1127 (UC, W). OREGON : near Mt. Keamath, 1,470 m, 1894, *Leiberg* 648 (E); Union Co., Blue Mts., Meadow Creek, 5,000 ft, 1963, *Hermann* 18832 (W). IDAHO : Cave Lake, Blaine Co., 6,000 ft, 1916, *Macbride* & *Payson* 3030 (E). CALIFORNIA : Humboldt Co., E. of Willow Creek near county line, 1966, *Gillett* & *Crompton* 12986 (HUJ); Placir Co., Canon above Coldstream, 1884, *Sonne* (HUJ). WASHINGTON : Yakima Co., Satus pass., 1958, *Hitchcock* & *Muhlick* 21589 (HUJ); Olympic Mts., Lake Crescent, 1910, *Webster* 26

Plate 227. *T. cyathiferum* Lindl.
(Washington : *Webster* 26, UC).
Plant in flower and fruit; involucre; flower

(UC). VANCOUVER ISLAND: Mount Benson, 1908, *Mucoun* 78.916 (E). MEXICO: Mounterry, 1910, *Nicolus* (G).

234. T. depauperatum Desv., Journ. Bot. 4 : 69 (1814).

Annual, glabrous or very sparingly glandular-hairy, 5–50 cm. Stems usually many, erect to decumbent, leafy, branching all along. Leaves long- to short-petioled; stipules 0.5–1.3 cm long, ovate-oblong to lanceolate, sheathing at base only, free portion as long as or longer than lower part, short-triangular, cuspidate, with entire to dentate margin; leaflets 0.5–3 × 0.3–1.4 cm, long-obovate to lanceolate or linear, cuneate, margin entire or dentate-serrate to deeply laciniate, apex rounded to truncate or retuse, also acute and short- to long-mucronate. Inflorescences 0.5–1(–1.5) cm in diam., 3–15-flowered, semi-globular to globular in outline. Peduncles axillary and terminal, much longer than subtending leaves. Outer bracts 1–6 mm long, linear to elliptical, connate at base, 5–7-lobed, sometimes reduced to ring or cup-like structure; inner bracts inconspicuous, whorled. Pedicels 0–0.5 mm. Calyx 2–5 mm long, glabrous; tube broadly campanulate, membranous and scarious, 5–6-nerved, with oblique throat; teeth very unequal, the longer one as long as or longer than tube, all subulate or triangular-lanceolate. Corolla (3–)5–11 mm, white to pink, turning brown with age; standard broadly ovate, strongly inflated at maturity and becoming vesicular. Pod about 4 mm long, stipitate to sessile, ovoid, membranous, reticulately nerved, 1–6-seeded. Seeds about 2 mm, reniform, dark brown. Fl. March–June. 2n = 16.

1. Involucre reduced to non-lobed ring or cup-like structure. **234a.** var. **depauperatum**
– Involucre bracts evident 2
2. Involucre bracts broad, irregular, with wide, hyaline margins. **234b.** var. **amplectens**
– Involucre bracts oblong, with very narrow, hayaline margins 3
3. Involucre bracts as long as calyx. **234c.** var. **stenophyllum**
– Involucre bracts half the length of calyx. **234d.** var. **diversifolium**

234a. T. depauperatum Desv. Var. depauperatum. *Typus*: "Western coasts of both North and South America" (n. v.).

Lupinaster depauperatus (Desv.) Presl, Symb. Bot. 1 : 50 (1831).
T. laciniatum Greene, Pittonia 1 : 7 (1887).
T. laciniatum Greene var. *angustatum* Greene, Man. Bot. Reg. San Franc. Bay 101 (1894).
T. depauperatum Desv. var. *laciniatum* (Greene) McDerm., N. Am. Sp. Trif. 133 (1910).
T. depauperatum Desv. var. *angustatum* (Greene) Jeps., Man. Fl. Pl. Calif. 538 (1925).

Icon.: McDerm., *loc. cit.* pls. 47–48.

Hab.: Open grassy slopes and clay meadows; 0–3,800 ft.
Gen. Distr.: USA: Oregon, California, Washington. Peru, Chile.

Selected specimens: CALIFORNIA: ElDorado Co., Webber Creek, North Fork, E. of Camino, 3,200 ft, 1943, *Robbins* 999 (HUJ); San Joaquin Co., Corral Hollow, 400 ft,

1938, *Rose* 38119 (B). CHILE : Concepcion, 1952, *Kunkel* 2226 (SI); Concon, *Pöppig* 184 (W).

Note : The deep laciniation of the leaflets, which led to the establishing of *T. laciniatum* Greene, is a character which also occurs in species of other genera, e.g., *Medicago, Trigonella,* and *Melilotus*, and seems to lack value as a diagnostic character.

234b. T. depauperatum Desv. Var. **amplectens** (Torr. & Gray) Wats., Bot. Calif. 2 : 441 (1880). *Typus* : California, *D. Douglas* (GH, iso. : K !, NY !).

T. amplectens Torr. & Gray, Fl. N. Am. 1 : 319 (1838).
T. quercetorum Greene, Pittonia 1 : 172 (1888).
T. stenophyllum Greene, Fl. Francisc. 34 (1891).

Icon. : [Plate 228]. McDerm., *loc. cit.* pl. 52; Hook. & Arn., Bot. Beech. Voy. t. 78 (1838).

Hab. : As above.
Gen. Distr. : California.

Selected specimens : CALIFORNIA : Almeda Co., Livermore, 1904, *Heller* 7319 (HUJ); San Mateo Co., Coloma, 400 ft, 1907, *Pendleton* 628a (HUJ); Sonoma Co., Bennett Val. road near Vernal Pool, 1934, *Baker* 7583 (UC).

234c. T. depauperatum Desv. Var. **stenophyllum** (Nutt.) McDerm., N. Am. Sp. Trif. 137 (1910). *Typus* : California : Santa Catalina, San Pedro, *Gambel* (GH, iso. K !).

T. stenophyllum Nutt., Journ. Acad. Philad. N. S. 1 : 151 (1848) non Boiss. (1849).
T. franciscanum Greene, Man. Bot. Reg. San Franc. Bay 100 (1894).
T. franciscanum Greene var. *truncatum* Greene, *loc. cit.*
T. truncatum (Greene) Greene, Proc. Acad. Philad. 1895 : 564 (1896).
T. minutiflorum Greene, Pittonia 3 : 215 (1897).
T. anodon Greene, Pittonia 5 : 107 (1903).
T. brachyodon Greene, *loc. cit.* non Celak. (1888).
T. decodon Greene, *op. cit.* 108.
T. depauperatum Desv. var. *amplectens* (Torr. & Gray) Wats. f. *truncatum* (Greene) McDerm., *op. cit.* 144.
T. depauperatum Desv. var. *stenophyllum* (Nutt.) McDerm. f. *franciscanum* (Greene) McDerm., *op. cit.* 140.
T. amplectens Torr. & Gray var. *truncatum* (Greene) Jeps., Man. Fl. Pl. Calif. 537 (1925).
T. amplectens Torr. & Gray var. *stenophyllum* (Nutt.) Jeps., *loc. cit.*
T. depauperatum Desv. var. *truncatum* (Greene) Martin ex Isley, Brittonia 32 : 56 (1980).

Icon. : McDerm., *loc. cit.* pls. 50–51.

Hab. : Along coastal ranges and in interior valleys.
Gen. Distr. : USA : California.

Selected specimens : CALIFORNIA : San Diego, 1903, *Brandegee* 3371 (B, W, isotypes of

Plate 228. *T. depauperatum* Desv. Var. *amplectens* (Torr. & Gray) Wats. (California : *Baker* 7583, UC). Plant in flower and fruit; mature flower; involucre.

T. decodon Greene); San Mateo Co., Colma, 400 ft, 1907, *Pendleton* 628 (HUJ); San Diego Co., Witch Creek, 1893, *Alderson* (HUJ).

234d. T. depauperatum Desv. Var. **diversifolium** (Nutt.) McDerm., N. Am. Sp. Trif. 135 (1910). *Typus*: California: St. Simeon, *Gambel* (n. v., iso. K !).

T. diversifolium Nutt., Journ. Acad. Philad. N. S. 1 : 152 (1848).
T. hydrophilum Greene, Man. Bot. Reg. San Franc. Bay 100 (1894).
T. amplectens Torr. & Gray var. *hydrophilum* (Greene) Jeps., Fl. W. Mid. Calif. 311 (1901).
T. amplectens Torr. & Gray var. *diversifolium* (Nutt.) Jeps., Man. Fl. Pl. Calif. 537 (1925).
T. depauperatum Desv. var. *hydrophilum* (Greene) Martin ex Isley, Brittonia 32 : 55 (1980).

Icon.: McDerm., *loc. cit.* pl. 49.

Hab.: Wet adobe soil to open, dry pine woodland.
Gen. Distr.: USA: California.

Specimens seen: CALIFORNIA: Bennett Valley Road, E. of Santa Rosa, 1935, *Baker* 8077 (UC); grown from seeds collected at San Jose, 1908, *Kennedy* (HUJ, UC).

235. T. fucatum Lindl., Bot. Reg. 22 : t. 1883 (1836). *Typus*: California: raised from seeds collected by *D. Douglas* (seen photo HUJ, iso. K !).

T. physopetalum Fisch. & Mey., Ind. Sem. Petrop. 3 : 47 (1837).
T. furcatum Lindl. ex Hook. & Arn., Bot. Beech. Voy. 332 (1838, typogr. error).
T. gambelii Nutt., Jour. Acad. Nat. Sci. N. S. 1 : 151 (1848).
T. flavulum Greene, Pittonia 2 : 223 (1892).
T. virescens Greene, *loc. cit.*
T. fucatum Lindl. var. *flavulum* (Greene) Jeps., Fl. W. Mid. Calif. 310 (1901).
T. fucatum Lindl. var. *gambelii* (Nutt.) Jeps., *op. cit.* 311.
T. fucatum Lindl. var. *virescens* (Greene) Jeps., *op. cit.* 311.
T. fucatum Lindl. var. *gambelii* (Nutt.) Jeps. f. *flavulum* (Greene) McDerm., N. Am. Sp. Trif. 158 (1910).
T. fucatum Lindl. f. *virescens* (Greene) McDerm., *op.cit.* 152.

Icon.: [Plate 229]. Lindl., *loc. cit.*; McDerm., *loc. cit.* pls. 55–60.

Annual, glabrous to glabrescent, 10–80 cm. Stems many, dichotomously branched, erect to ascending, fistulose, grooved. Leaves long-petioled; stipules 0.8–2 cm broad, ovate, membranous, free portion as long as lower part, triangular, acuminate-bifid; leaflets 0.8–4 × 0.7–3 cm, broadly obovate to almost orbicular or rhombic-obovate, rounded, with margins remotely dentate to densely serrulate-dentate. Inflorescences (1–)1.5–3(–4) cm, hemispherical to spherical, few- to many-flowered. Peduncles very long, much longer than petioles. Outer bracts 3–9, united at base for up to half of their length, forming involucre, one-half to three-quarters the length of flowers, membranous-margined, lanceolate to ovate, entire or bifid, acuminate; inner bracts

Plate 229. *T. fucatum* Lindl.
(California : *Baker* 1981, HUJ).
Plant in flower and fruit; flower; involucre.

inconspicuous, whorled. Pedicels about 1 mm. Calyx 3–8 mm long; tube campanulate; throat oblique, 10-nerved, membranous; teeth very unequal, the lower one 1–8 mm long, the upper ones 0–3 mm long, simple to trifid. Corolla 1–2.7 cm long, yellowish or cream, becoming pink with age; keel sometimes dark purple; standard scarious, persistent, very broad, ovate, becoming strongly inflated and vesicular in fruit. Pod 7–8 mm long, linear, stipitate, 3–8-seeded, reticulately nerved. Seeds 1–4 mm, globular, finely tuberculate. Fl. April–June. 2n = 16.

Hab. : Moist places, in many plant communities below 1,000 m.

Gen. Distr. : USA : California, Oregon.

Selected specimens : CALIFORNIA : Humboldt Co., Trinidad, 1946, *Gillett* & *Moulds* 12604 (HUJ); Santa Clara Co., 450 ft, 1907, *Pendleton* 721 (HUJ); Solano Co., N. of Vacaville, 1920, *Heller* 13369 (G, HUJ); Napa Co., Calistoga, 1903, *Baker* 1981 (HUJ). OREGON : Siskiyou, Eden Ridge, 2,000 ft, 1926, *Ingram* 1991 (HUJ).

Note : *T. fucatum* is very polymorphic with regard to the size of the calyx and the forked or simple teeth, as well as the proportions of the lower and upper teeth. It is not possible to delineate the 3 recorded varieties (var. *fucatum*, var. *gambelii* and var. *virescens*) as clear-cut taxa, even of the rank of variety, because of the nature of the characteristics.

236. T. minutissimum Heller & Zoh. (sp. nov.).* *Typus* : California : Monterey in Clemens Co., 1921, *Miss Gibbs* (BM).

Icon. : [Plate 230].

Annual, villous, 2–5 cm. Stems few, almost naked, scape-like. Leaves long-petioled, almost crowded at base; stipules up to 8 mm, membranous, ovate-oblong, many-nerved, free portion triangular, with dentate margin; leaflets 4–6 × 2–4 mm, obtriangular, densely hairy, deeply retuse at apex, dentate all along margin. Inflorescences 4–5 × 3–4 mm, terminal, ovoid to hemispherical, 4–10-flowered. Peduncles very long, up to 4 cm in length. Involucre about 4 mm long, hairy, 4–6-lobed; lobes about half the length of entire part, unequally dentate; teeth triangular at base, subulate-setose. Calyx up to 5 mm long; tube tubular, with 10 obscure nerves; teeth unequal, subulate-setose, the lower one longer than tube, sometimes bifid. Corolla 5–6 mm, slightly longer than calyx teeth, dark purple when dry; standard oblong-elliptical, tapering at apex, somewhat inflated in fruit. Pod unknown.

Gen. Distr. : California (endemic).

Note : Known only from the type collection.

237. T. physanthum Hook. & Arn. in Hook., Bot. Misc. 3 : 180 (1833). *Typus* : Chile : Valparaiso, 1831, *H. Cumming* 748 (GL, iso. : E !, FI !, G !, K !, W !).

* See Appendix at the end of this volume.

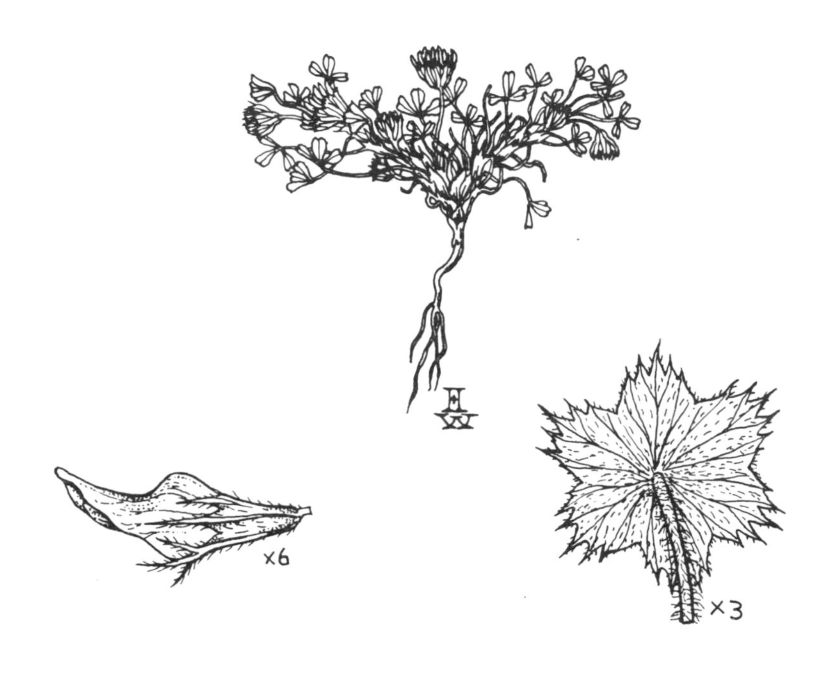

Plate 230. *T. minutissimum* Heller & Zoh.
(California : *Miss Gibbs*, BM).
Plant in flower; flower; involucre.

T. valdivianum Phil. (inedit.)

Icon. : [Plate 231].

Annual, glabrous or sparsely patulous-hairy, 10–15 cm. Stems many, poorly branched, ascending to decumbent, lower part covered by crowded sheathing stipules. Leaves long-petioled; stipules 0.5–2 cm long, ovate, whitish, membranous, with darker nerves, laciniate and spinulose in upper part, sheathing and covering internodes and overlapping in lower part; leaflets 0.6–1.6 × 0.3–0.8 cm, narrowly obovate, rounded to truncate, retuse at apex, cuneate at base, spinulose-dentate. Inflorescences 0.8–1.5 cm in diam., semi-globular, many-flowered, terminal and axillary. Peduncles about as long as subtending leaves, rather thick. Involucre about as long as flowers, glabrous, reticulately nerved, with 6–8 shallow lobes; lobes dentate-laciniate with unequal, subulate teeth. Calyx 7–8 mm; tube campanulate, membranous, hairy, inflated in fruit, 5-nerved; teeth considerably longer than tube, subulate-setaceous, one or two of them bifid or trifid. Corolla 6–8 mm, purple; standard elliptical, tapering towards apex, strongly inflated and including pod, constricted above calyx tube. Pod 2–3 mm, sessile, 2-seeded, ellipsoidal. Seeds about 1 mm, brown. Fl. October–November.

Gen. Distr. : Chile (endemic).

Specimens seen : CHILE : Corral, *Philippi* & *Hohenacker* 709 (G); Prov. Valdivia, 1862, *Philippi* (G type of *T. valdivianum* Phil.); Bancagua, *Philippi* (W)

Note : *T. physanthum* is very close to the Californian *T. barbigerum* and perhaps should not be specifically distinguished from it. One of the obvious differences between them is the size and dentation of the involucre lobes and the calyx teeth.

Plate 231. *T. physanthum* Hook. & Arn.
(Chile : *Philippi*, G).
Plant in flower and fruit; involucre; flower.

Addenda

By the time the manuscript was submitted to press, the following perennial species had been treated by J. M. Gillett, Can. Jour. Bot., 58 : 1425–1448 (1980), and should be added to Section *Involucrarium*, Subsection *Involucrarium* :

T. mucronatum Willd. ex Spreng., Syst. Veg. 3 : 208 (1826) Subsp. **mucronatum**. *Typus* : Mexico : Valladido, *Humboldt* (B, Herb. Willd. No. 14182).

T. involucratum Ortega, Hort. Reg. Bot. Matr. 33 (1797) non Lam. (1778).
T. ortegae Greene, Pittonia 3 : 186 (1897).
T. fistulosum A. Vaughan, Am. Midl. Nat. 22 : 575 (1939).

Hab. : Wet meadows, springy places, open wet places in pink-oak forests and on stream banks; 1,900–3,100 m. 2n = 16.
Gen. Distr. : Mexico.

T. mucronatum Willd. ex Spreng. Subsp. **lacerum** (Greene) J. M. Gillett, Can. Jour. Bot. 58 : 1444 (1980). *Typus* : New Mexico : Hidalgo Co., Animas Valley, 1851, *C. Wright* 997 (US, iso. : GH, NY).

T. lacerum Greene, Erythea 2 : 182 (1894).
T. arizonicum Greene, Erythea 3 : 18 (1895).
T. oxyodon Greene ex Rydb., Fl. Colorado 201 (1906).
T. involucratum Ortega var. *arizonicum* (Greene) McDerm., Ill. Key N. Am. Sp. Trif. 150 (1910).

Hab. : Gravelly damp pastures along streams or wet places in pine forest. 2n = 16.
Gen. Distr. : Mexico, USA : Arizona, California, Colorado, New Mexico, Texas, Utah.

T. mucronatum Willd. ex Spreng. Subsp. **vaughanae** J. M. Gillett, Can. Jour. Bot. 58 : 1445 (1980). *Typus* : Mexico : San Luis Potosi, 6,000 ft., 1878, *Palmer* & *Parry* 135 (GH, iso : ND-G).

T. ortegae Greene f. *pumilum* Vaughan, Am. Midl. Nat. 22 : 578 (1939).

Hab. : Sandy stream banks in pasture land.
Gen. Distr. : Mexico.

T. siskiyouense J. M. Gillett, Can. Jour. Bot. 58 : 1441 (1980). *Typus* : Oregon : Josephine Co., Grants Pass, 10 June, 1904, *C. V. Pipper* 5048 (GH, iso. : US).

Hab. : Wet meadows
Gen. Distr. : USA : California, Oregon.

Chromosome number and karyotype of 60 species of the genus *Trifolium* from Bulgaria were recently studied in : A.V. Petrova & S.I. Kozuharov, 'Cytotaxonomic Study of the Genus *Trifolium* L. in Bulgaria. I.', *Phytology,* IXX (1982), pp. 3–23 (to be continued in Vol. XX).

APPENDIX

Latin Diagnoses of New Taxa

Series *Grandiflora* Heller & Zoh. (ser. nov.) [p. 108]
Perennes, dense caespitosae. Inflorescentiae vulgo pauciflorae, scapis vel scapiformibus ramis sustentae. Flores conspicui, 1.5–3 cm longi.

Series *Phyllodon* Heller & Zoh. (ser. nov.) [p.119]
Annuae, sparse pubescentes vel glabrae. Pedunculi prope basim capituli geniculati. Calyx bilabiatus, labium superius inferiore longius; dentes inaequales, foliacei, nervis reticulatis vel ramosis percursi. Vexillum apice paulo dentatum vel erosum.

Series *Brachyantha* Heller & Zoh. (ser. nov.) [p. 121]
Annuae, glabrae. Caules numerosi, erecti vel ascendentes. Inflorescentia pedunculata, hemisphaerica, multiflora, densa. Calyx corolla longior, aliquantum coriaceus; duo dentes superiores paululum ceteris longiores, dentes omnes post anthesem demum valde recurvati, deinde capitula fructifera spinosa.

Series *Pectinata* Heller & Zoh. (ser. nov.) [p.184]
Annuae. Caules erecti. Bracteae minutae, saepe connatae flores verticillatos sustinentes. Pedicelli conspicui, post anthesem deflexi. Rachis nonnunquam ultra flores producta. Dentes calycini inaequales, triangulari-lanceolati, hyalini, valde ciliati; margine dentati usque ad pectinati.

Series *Curvicalyx* Heller & Zoh. (ser. nov.) [p. 186]
Perennes, villosae vel rarius glabrae. Stipulae superiores 2–5 cm longae, foliaceae, late lanceolatae. Flores spiratim dispositi. Pedunculi infra capitula curvati – deinde capitula inversa vel horizontalia. Tubus calycinus post anthesem curvatus; dentes calycini inaequales, nonnunquam curvati et contorti, inferiores superioribus longiores, subulati.

Series *Producta* Heller & Zoh. (ser. nov.) [p. 190]
Perennes, glabrae vel sparse villosae. Caules erecti, ramosi vel caespitosi. Inflorescentiae pauci- ad multiflorae, in verticillis dispositae; rachis ultra flores producta, simplex vel furcata, saepe ad verticillium superiorem fasciculo florum sterilium gemmiformium instructa.

Trifolium vestitum Heller & Zoh. (sp. nov.) [p. 206]
Perennnis, glabra vel glabrescens, 7–10 cm. Caules numerosi, procumbentes, ascendentes, caespitosi, stipulis persistentibus usque ad 2–3 cm eorum longitudinis dense obtecta, e corona lignosa ramosi. Folia petiolis longis tenuibus; stipulae oblongae, membranaceae, parte superiore libera triangulari, acuta ad acuminata; foliola 3–7 × 3–4 mm, obtriangularia, apice biloba, mucronata, supra profunde et acute dentata infra integra, conspicue nervosa, glabra. Inflorescentia umbellata, 2–5-flora, pedunculo tenui, villoso, foliis longiore. Bracteae usque ad 2 mm, oblongo-

lipticae, hyalinae, prominenter nervosae, acutae vel acuminatae, 2–6-bracteae involucrum ad capituli basin formantes. Pedicelli 3–6 mm longi, villosi, post anthesem deflexi. Calyx 4–5 mm longus, villosus; tubus calycinus 10-nervius, campanulatus; dentes paululum inaequales, dens longissimus tubo longior, triangulari-lanceolatus, acuminatus, basi membranaceo-marginatus, 1-nervius. Corolla 7–9 mm longa, carnae (in sicco); vexillum late, ellipticum, apice rotundatum, usque ad aliquantulum retusum, vexillo et alis longius; alae auriculatae, unguis 7 mm longus. Ovarium 2–3 mm, sessile, lineare, 3-ovulatum; stylus 3 mm longus. Legumen et semina ignota. Floret Januario.

Subsection *Oxalioidea* (Gillett) Heller [p. 206]
Perennes, sparse pubescentes usque ad dense villosae. Caules prostrati, stoloniferi, repentes ad nodos radicantes. Foliola late obcordata vel obovato-cuneata, retusa usque ad profunde emarginata. Inflorescentiae et flores dimorphi: inflorescentia terminalis longe pedunculata multiflora; inflorescentia basalis, axillaris brevipedunculata, caule stolonifero portata, cum 1–8 floribus cleistogamis et subterraneis.

Series *Macrochlamis* Heller & Zoh. (ser. nov.) [p. 238]
Perennes, glabrae vel pubescentes ad villosae. Caules semper caespitosi vel repentes, ad nodos radicantas, e radice magna ramosi; stipulis persistentibus obtecti. Inflorescentia bracteis conspicuis magnis suffulta; bracteae involucrum formantes; nonnunquam bracteae basi parum connatae; bracteae superiores minores.

Series *Phyllocephala* Heller & Zoh. (ser. nov.) [p. 246]
Perennes, adpresse pubescentes. Caules dense caespitosi, stipulis persistentibus obtecti. Inflorescentia axillaris, plerumque capitulis duobus sessilibus composita, stipulis membranaceis, supremis foliisque pseudoinvolucrum formantibus suffulta.

Series *Stipitata* Zoh. (ser. nov.) [p. 314]
Annuae vel biennes. Inflorescentiae pauciflorae, laxae. Flores 0.8–1 cm longi, rosei ad purpurei. Vexillum conduplicatum, in fructei naviculiforme. Legumen longissime stipitatum.

Trifolium minutissimum Heller & Zoh. (sp. nov.) [p. 573]
Annuum, villosum, 2–5 cm altum. Caules pauci, subnudi, scapiformes. Folia petiolis longis, basi fere conferta; stipulae ad 8 mm, membranaceae, ovato-oblongae, multinerves, parte libera triangulari, margine dentata; foliola 4–6 × 2–4 mm, obtriangularia, dense hirsuta, apice profunde retusa, tota margine dentata. Inflorescentia 4–5 × 3–4 mm, terminalis, ovoidea ad hemisphaerica, 4–10 florae. Pedunculi longissimi, usque ad 4 cm longi. Involucrum fere 4 mm longum, villosum, lobis 4–6; lobi fere dimidio partis integrae, inaequaliter dentati; dentes basi triangulares, subulato-setosi. Calyx usque ad 5 mm longi; tubus calycinus tubularis, nervis 10 obscuris; dentes inaequales, subulato-setosi, inferior tubo longior, aliquando bifidus. Corolla 5–6 mm, dentibus calycinis paulo longior, in sicco atropurpurea; vexillum oblongo-ellipticum, apice angustatum, in fructi subinflatum. Legumen ignotum.

We acknowledge with gratitude the assistance of Dr. E. Kollmann in the preparation of the Latin version.

BIBLIOGRAPHY

Ahuja M. R. (1955) 'Chromosome Numbers of Some Plants', *Ind. Jour. Gen. Pl. Br.*, 15 : 142–143.

Almeida J. L. F. de & P. de Carvalho (1964) 'Trevos autotetraploides, III, Ocaso de *T. alexandrinum* L.', *Agron. Lusit*, 24 : 45-76

Anderson M. K. , N. L. Taylor & G. B. Collins (1972) 'Somatic Chromosome Numbers in Certain *Trifolium* Species', *Can. J. Genet. Cytol.*, 14(1) : 139–145.

Angulo M. D. & A. M. Sánchez de Rivera (1975) 'Studies of *Trifolium subterraneum* ecotypes', *Cytologia,* 40 : 415–423.

Angulo M. D., A. M. Sánchez de Rivera & F. González-Bernáldez (1969) 'Estudios cromosomicos en el genero *Trifolium*, II', *Anal. Est. Exper. Aula Dei*, 9 : 97–110.

–(1970) 'Estudios cromosomicos en el genero *Trifolium*, III', *Bol. Soc. Brot.*, ser. 2, 44 : 13–26.

–(1971) 'Estudios cromosomicos en el genero *Trifolium*, V', *Bol. Soc. Brot.*, ser. 2, 45 : 253–267.

–(1972a) 'Estudios cromosomicos en el genero *Trifolium*, VI', *Lagascalia*, 2 : 3–13.

–(1972b) 'Estudios cromosomicos en el genero *Trifolium*, VII : Revision cariologica sobre especies de la subsection *Probostoma*', *Genet. Iber.*, 24 : 305–324.

Ascherson P. F. A. & K. O. P. Graebner (1907–8) '*Trifolium* L.', in : *Synopsis der Mitteleuropäischen Flora*, Leipzig, Vol. 6(2 Abt.) : 472–617.

Bandel G. (1974) 'Chromosome Numbers and Evolution in the Leguminosae', *Caryologia*, 27 : 17–32.

Belli S. (1892) 'Sui rapporti sistematico-biologici del *Trifolium subterraneum* L. cogli affini del gruppo *Calycomorphum* Presl', *Malpighia* 6 : 1–41.

–(1894) 'Revista critica della species de "*Trifolium*" Italiane della sezione *Lupinaster*', *Mem. Reale Accad. Sci. Torino*, ser. 2, 44 : 233–292.

–(1896) 'Endoderma e periciclo nel G. *Trifolium* in rapporto colla teoria della Stelia di V. Thieghem e Douliot', *Mem. Accad. Sci. Torino*, ser. 2, 46 : 353–443, tt. 1–6.

Bentham G. & J. D. Hooker (1865) *Genera Plantarum*, Vol. 1, London.

Bleier H. (1925) 'Chromosomenstudien bei der Gattung *Trifolium*', *Jahrb. Wissensch. Bot.*, 64, 4 : 604–636.

Bobrov E. G. (1947) 'Vidi Kleverov SSSR (The clovers of USSR)', *Acta Inst. Bot. Acad. Sci. URSS*, Ser. 1, 6 : 164–336 (in Russian with English summary).

–(1967) 'Ob ob"eme roda *Trifolium* s.l. (On the span of the genus *Trifolium* s.l.)', *Bot. Zhurn.*, 52(11) : 1593–1599.

Bogdan A. V (1956) 'Indigenous Clovers of Kenya', *E. Afr. Agric. J.*, 22 : 40–45.

–(1966) 'Pollination and Breeding Behaviour in the *Trifolium rueppellianum* Complex in Kenya', *New Phytologist*, 65 : 417–422.

Boissier E. (1843, 1849, 1856) '*Trifolium* L.' in : *Diagnosis Plantarum Orientalium Novarum*, ser. 1, 2 : 25–35, 9 : 20–31; ser. 2, 2 : 12–19, Genevae, Parisiis, Lipsiae.

–(1872) '*Trifolium* L.', in : *Flora Orientalis*, Vol. 2 : 110–156, Genevae et Basileae.

Bolkhovskikh Z., V. Grif, T. Matvejeva & O. Zakjaryeva (1969) *Chromosome Numbers of Flowering Plants*, Academy of Sciences of the USSR.

Britten E. J. (1963) 'Chromosome Numbers in the Genus *Trifolium*', *Cytologia*, 28 : 428–449.

Bujoreau G. & S. Grigore (1969) 'Clover Species in Banat and Their Distribution', *Comun. Bot.*, 9 : 115–119.

Calabuig E. L. & J. M. G. Gutierrez (1976) 'Nueva cita de *Trifolium michelianum* Savi para la peninsula Iberia', *Lagascalia* 6 : 3–6.

Čelakovsky L. (1874) 'Ueber den Aufbau der Gattung *Trifolium*', *Oesterr. Bot. Zeitschr.*, 24 : 37–45, 75–82.

Chen Chi-Chang & P. B. Gibson (1970a) 'Meiosis in Two Species of *Trifolium* and Their Hybrids', *Crop. Sci.*, 10(2) : 188–189.

–(1970b) 'Chromosome Pairing in Two Interspecific Hybrids in *Trifolium*', *Can. J. Genet. Cytol.*, 12 : 790–794.

–(1971a) 'Karyotypes of Fifteen *Trifolium* Species in Section *Amoria*', *Crop. Sci.*, 11 : 441–445.

–(1971b) 'Seed Development Following the Mating of *T. repens* × *T. uniflorum*', *Crop Sci.*, 11 : 667–672.

–(1972a) 'Barriers to Hybridization of *Trifolium repens* with Related Species', *Can. J. Genet. Cytol.*, 14 : 381–389.

–(1972b) 'Chromosome Relationship of *Trifolium uniflorum* to *T. repens* and *T. occidentale*', *Can. J. Genet. Cytol.*, 14 : 591–595.

–(1973) 'Effect of Temperature on Pollen-Tube Growth in *T. repens* after Cross- and Self-Pollination', *Crop Sci.*, 13 : 563–566.

Coombe D. E. (1961) '*Trifolium occidentale*, a New Species Related to *T. repens* L.', *Watsonia*, 5 : 68–87.

–(1968) '*Trifolium* L.', in :Tutin T. G. et al., *Flora Europaea*, Vol. 2 : 157–172, Cambridge.

Cufodontis G. (1958) '*Trifolium* in systematischer Bearbeitung der in Süd-Aethiopien gesammelten Pflanzen', *Senck. biol.*, 39 : 296.

Endlicher S. L. (1841) *Genera Plantarum Secundum Ordines Naturales Disposita*, Wien.

Erdtman G. , J. Praglowski & S. Nilsson (1963) *An Introduction to a Scandinavian Pollen Flora*, Vol. 2, Almquist & Wiksell, Stockholm.

Evans A. M. (1962) 'Species Hybridization in *Trifolium*, I : Methods of Overcoming Species Incompatibility', *Euphytica*, 11 : 164–176.

–(1976) 'Clovers', in : Simonds N.W., *Evolution of Crop Plants*, Longman, pp. 175–179.

Fahn A. & M. Zohary (1955) 'On the Pericarpal Structure of the Legumen, its Evolution and Relation to Dehiscence', *Phytomorphology*, 5 : 99–111.

Fernandes A. & M. F. Santos (1971) 'Contribution à la connaissance cytotaxonomique des Spermatophyta du Portugal, IV : Leguminosae', *Bol. Soc. Brot.*, ser. 2, 45 : 185–201, Coimbra.

Fiori A. (1948) 'Chiave analitica ed illustrazione delle species di *Trifolium* dell'Abyssinia', *Nuov. Gior. Bot. Ital.*, n.s. 55 : 335–346.

Free J. B. (1970) *Insect Pollination of Crops*, Academic Press, pp. 215–241.

Gadella T. W. J. & E. Kliphuis (1963) 'Chromosome Numbers of Flowering Plants in the Netherlands', *Acta Bot. Neerl.*, 12 : 195–230.

Gibelli G. & S. Belli (1887) 'Intorno alla morfologia differenziale esterna ed alla nomenclatura delle specie di *Trifolium* della sezione *Amoria* Presl, crescenti spontanee in Italia', *Atti Accad. Fis. Mat. et Nat.*, 22 : 628–672.

–(1889a) 'Rivista critica delle specie di *Trifolium* italiani della sezione *Chronosemium* Ser.', *Malpighia*, 3 : 193–233, 305–319.

–(1889b) 'Rivista critica e descrittiva delle specie di *Trifolium* italiane e affini comprese nella sezione *Lagopus* Koch', *Mem. Reale Accad. Sci. Torino*, ser. 2, 39 : 245–428.

–(1890–1893) 'Rivista critica delle specie di *Trifolium* italiane comparata con quelle del resto d'Europa e delle regioni circummediterranee ...', *Mem. Reale Accad. Sci. Torino*, ser.

2, 41 : 149–222; 42 : 3–46; 43 : 176–232.
Gibson P. B. & G. Beinhart (1969) 'Hybridization of *T. occidentale* with Two Other Species of Clover', *J. Hered.*, 60(2) : 93–96.
Gibson P. B. , Chi-Chang, Chen, J. T. Gillmgham & O. W. Barnett (1971) 'Interspecific Hybridization of *T. uniflorum* L.', *Crop. Sci.*, 11(6) : 895–899.
Gillett J. B. (1952) 'The Genus *Trifolium* in Southern Arabia and in Africa South of the Sahara', *Kew Bull.*, 7 : 367–404.
–(1970) 'Further Notes on *Trifolium* L. in Tropical Africa', *Kew Bull.*, 24 : 217–220.
Gillett J. M. (1965) 'Taxonomy of *Trifolium* : Five American Species of Section *Lupinaster* (Leguminosae)', *Brittonia*, 17 : 121–136.
–(1966) 'Type Collection of *Trifolium* in the Greene Herbarium at Notre Dame', *The American Midland Naturalist*, 76, 2 : 468–474.
–(1969) 'Taxonomy of *Trifolium*, II : The *T. longipes* Complex in North America', *Can. J. Bot.*, 47 : 93–113.
–(1970) 'On the Taxonomy of the Genus *Trifolium* L.', Report of the *Trifolium* Research Work Conference, Clemson, S. Carolina, pp. 1–46.
–(1971) 'Taxonomy of *Trifolium* (Leguminosae), III : *T. eriocephalum*', *Can. J. Bot.*, 49(3) : 395–405.
–(1972a) 'Two New Species of *Trifolium* from California and Nevada', *Madroño*, 21 : 451–455.
–(1972b) 'Taxonomy of *Trifolium* (Leguminosae), IV : The American Species of Section *Lupinaster*', (Adan.) Ser., *Can. J. Bot.*, 50 : 1975–2007.
–(1976) A New Species of *Trifolium* from Baja California, Mexico', *Madrono*, 23 : 334–337.
–(1980) 'Taxonomy of *Trifolium*, V : The Perennial Species of Section *Involucrarium*', *Can. J. Bot.*, 58 : 1425–1488.
Gillett J. M. , I. T. Bassett & C. W. Crompton (1973) 'Pollen Morphology and its Relationship to the North American *Trifolium* Species', *Pollen and Spores*, 15(1) : 91–108.
Gillett J. M. & T. Mosquin (1967) 'IOPB Chromosome Number Reports', *Taxon*, 16 : 149–156.
González-Bernáldez F. , A. M. Sánchez de Rivera & M. D. Angulo (1970) 'Estudios cromosomicos en el genero *Trifolium*, IV', Prenetado en el "XXIX congreso Luso-Español para el Progreso de las Ciencias", Lisboa.
–(1973) 'Estudios cromosomicos en el genero *Trifolium*, IV', *Lagascalia*, 3(2) : 195–205.
Graham A. & A. S. Tomb (1977) 'Palynology of *Erythrina* (Leguminosae, Papilionoideae) : The Subgenera, Section and Genera Relationships', *Lloydia*, 40(5) : 413–435.
Gruenberg-Fertig I. & W. T. Stearn (1972) 'Nomenclatural Remarks on *Trifolium*', *Israel J. Bot.*, 21 : 1–8.
Gunn C. R. & D. E. Barnes (1977) 'Seed Morphology of *Erythrina* (Fabaceae)', *Lloydia*, 40(5) : 454–470.
Gwendolyn E. A. (1924) *White Clover* (T. repens *L.*) – A monograph, Duckworth and Co., London, 150 pp.
Hedberg O. (1973) 'Adaptive Evolution in a Tropical Alpine Environment', in : Heywood : *Taxonomy and Ecology*, The Systematics Ass. Spec. Vol. 5 : 71–92, London & New York.
Heinisch O. (1955) *Samenatlas der wichtigsten Futterpflanzen und ihrer Unkräuter*, Deutsche Akademie der Landwirtschaftswissenschaften zu Berlin.
Heller D. (1978) 'Systematic Studies in Section *Lotoidea* of the Genus *Trifolium*', Ph.D. thesis, Hebrew Univ. Jerusalem (in Hebrew and English abstract).
Hermann F. (1935–36) 'Übersicht über die europäischen Rotten und Arten und einige andere Arten der Gattung *Trifolium*', *Feddes Repert.*, 39 : 332–351.
–(1937) 'Berichtigung zu meiner Arbeit : Übersicht über die europäischen Rotten und Arten der Gattung *Trifolium*', *Feddes Repert.*, 42 : 111.

–(1938) '*Trifolium*. Conspectus Sectionum et Subsectionum Europaearum Genesis', *Feddes Repert.*, 43 : 316–319.

Hermann F. J. (1953) 'A Botanical Synopsis of the Cultivated Clovers', U. S. Dept. Agr. Monograph 22, 45 pp.

Heyn C. C. (1968) 'An Evolutionary Study of Fruit Morphology in the Tribe Trigonelleae (*Medicago, Trigonella* and *Melilotus*)', *Phytomorphology*, 18(1) : 54–59.

Heyn C. C. & I. Herrnstadt (1977) 'Seed Coat Structure of Old World *Lupinus* Species', *Bot. Not.*, 130 : 427–435.

Hossain M. (1961) 'A Revision of *Trifolium* in Nearer East', *Not. Roy. Bot. Gard. Edinb.*, 23 : 387–481.

House D. H. (1906) 'New and Noteworthy North American Species of *Trifolium*', *Bot. Gaz.*, 41 : 334–347.

Hutchinson J. (1964) *The Genera of Flowering Plants*, Oxford.

Igoshina K.N. (1966) 'Flora of the Mountain and Plain Tundras and Open Forests of the Urals', in: *Vascular Plant of the Siberian North and Northern Far East*, Edit. A. Tolmachev.

Isely D. (1948) 'Seed Characters of Common Clovers', *Iowa State Coll. J. Sci.*, 23 : 125–136.

Karpechenko G. D. (1925) 'Karyologische Studien über die Gattung *Trifolium*', *Bull. Appl. Bot. Genet. Pl. Breed.*, 14 : 143–148, 271–279 (Russian, with German summary).

Katznelson J. (1965) 'A Taxonomic Revision of Sect. *Calycomorphum* of the Genus *Trifolium*, II : The Anemochoric Species', *Israel J. Bot.*, 14 : 171–183.

–(1967) 'Interspecific Hybridization in *Trifolium*', *Crop Sci.*, 7 : 307–310.

–(1974) 'Biological Flora of Israel, 5 : The Subterranean Clovers of *Trifolium* Subsect. *Calycomorphum* Katzn. *T. subterraneum* L. (*sensu latu*)', *Israel J. Bot.*, 23 : 69–108.

Katznelson J. & F. H. Morley (1965) 'A Taxonomic Revision of *Trifolium* Sect. *Calycomorphum* of the Genus *Trifolium*, I : The Geotropic Species', *Israel J. Bot.*, 14 : 112–134.

Katznelson J. & E. Putiyevsky (1974) 'Cytogenetic Studies in *Trifolium* spp. Related to Berseem, II : Relationship within the Echinata Group', *Theor. Appl. Genet.*, 43 : 87–94.

Katznelson J. & D. Zohary (1970) 'Seed Dispersal in *Trifolium* Sect. *Calycomorphum*', *Israel J. Bot* 19 : 114–120.

Kazimierski T. & E. M. Kazimierska (1972) 'Hybrids in *Hiantia* Bobr. Section *Trifolium* L. Genus. Morpnological Characters and Fertility', *Gen. Pol.*, 13 : 67–90.

–(1973) 'Investigation on Hybrids in the Genus *Trifolium* L., V : Fertility and Cytogenetics of the Hybrid *T. nigrescens* Viv. × *T. isthmocarpum* Brot.', *Acta Soc. Bot. Pol.*, 42 : 567–589.

Kennedy P. B. (1909–1913) 'Studies in *Trifolium*, I–VII', *Muhlenbergia*, 5 : 1–13, 37–46, 58–61, 100–104, 157–161; 7 : 97–100; 9 : 1–29.

Kliphuis E. (1962) 'Chromosome Numbers of Some Annual *Trifolium* Species Occurring in the Netherlands', *Acta Bot. Neerl.*, 11 : 90–92.

Koch C. (1835) '*Trifolium*', in : *Synopsis Florae Germanicae et Helveticae*, Vol. 1 : 167–176, Wilmans, Francofurtis ad Moenum.

Koller D. (1964) 'The Survival Value of Germination-Regulating Mechanisms in the Field', *Herbage Abst.*, 34 : 1–7.

Kozuharov S. J. , A. V. Petrova & T. A. Markova (1975) in : Löve, A., IOPB chromosome number reports XLVII. *Taxon*, 24(1) : 145–146.

Kuzamanov B. A. & G. Stancev (1972) in : Löve, A., IOPB chromosome number reports XXXVIII. *Taxon*, 21 : 681.

Larsen K. (1960) 'Cytological and Experimental Studies on the Flowering Plants of the Canary Islands', *K. Danske Vedenskab. Selskab. Biol. Skr.*, 11, 3 : 1–60.

Lindley J. (1846) *The Vegetable Kingdom*, London.

Lojacono M. (1878a) *Monografia dei Trifogli di Sicilia*, Palermo, 172 pp.

–(1878b) *Tentamen monographiai Trifoliorum sive generis species Panarini*, Palermo.
–(1883a) 'Revisione dei Trifogli dell'America settentrionale', *Nuov. Giorn. Bot. Ital.*, 15 : 113–198.
–(1883b) 'Clavis Specierum Trifoliorum', *Nuov. Giorn. Bot. Ital.*, 15 : 225–278.
Martin J. S. (1943) 'A Revision of the Native Clovers of the United States', Ph. D. thesis, Univ. Washington, Seattle (unpubl.).
McDermott L. F. (1910) *An Illustrated Key to the North American Species of* Trifolium, San Francisco, 325 pp.
Melchior H. (1964) *Engler Syllabus der Pflanzenfamilien*, ed. 12, 2, Berlin.
Moench C. (1794) *Methodus Plantas horti botanici*, Merburg.
Morley F. H. & J. Katznelson (1965) 'Colonization in Australia by *Trifolium subterraneum* L.', *The Genetics of Colonizing Species*, New York, pp. 269–282.
Mosquin T. & J. M. Gillett (1965) 'Chromosome Numbers in American *Trifolium* (Leguminosae)', *Brittonia*, 17 : 136–143.
Noda K. (1946) 'Chromosome Studies with Clover', *Jap. Jour. Genet.*, 21 : 93–96.
Norris D. O. (1956) 'Legumes and *Rhizobium* Symbiosis', *Emp. J. Exp. Agric.*, 24 : 247–270.
Norris D. O. & L. t'Mannetje (1964) 'The Symbiotic Specialization of African *Trifolium* spp. in Relation to Their Taxonomy and to Their Agronomic Use', *E. African Arg. and Forestry Journ.*, 29, 3 : 214–235.
Oppenheimer H. R. (1959) 'The Origin of the Egyptian Clover with Critical Revision of Some Closely Related Species', *Bull. Res. Counc. Israel*, D7 : 202–221.
–(1961) 'Essai d'une révision des Trèfles de la Palestine', *Bull. Soc. Bot. Fr.*, 108 : 47–61.
Peinado Lucena E., M. Medina Blanco & A. G. Gómez Castro (1971) 'Vegetal Taxonomy, IV : A Biometric Study of Seeds of Some *Trifolium* spp.', *Archivos de Zootecnia*, 20 : 67–86.
Piergentili D. (1970) 'Pollination of Cleistogamic Flowers of *Trifolium argentinense* Speg.', *Rivista de la Faculdad de Agronomia Universidad Nacional de la Plata*, 46(2) : 233–235.
Presl C. B. (1830–32) '*Trifolium*', in : *Symbolae Botanicae. Sumpt. Auctoris*, Vol. 1 : 44–50, Pragae.
Pritchard A. J. (1962) 'Number and Morphology of Chromosomes in African Species in the Genus *Trifolium* L.', *Austr J. Agric. Res.*, 13 : 1023–1029.
–(1967) 'The Somatic Chromosomes of *Trifolium cherleri* L., *T. hirtum* All., *T. ligusticum* Balb. and *T. scabrum* L.', *Caryologia*, 20 : 323–331.
–(1969) 'Chromosome Numbers in Some Species of *Trifolium*', *Austr. J. Agric. Res.*, 20 : 883–887.
Pritchard A. J. & K. F. Gould (1964) 'Chromosome Numbers in Some Introduced and Indigenous Legumes and Grasses', Div. Trop. Pastures Tech. Pap. 2 (CSIRO, Australia), 18 pp.
Pritchard A. J. & L. t'Mannetje (1967) 'The Breeding System and Some Interspecific Relations of a Number of African *Trifolium* spp.', *Euphytica*, 16(3) : 324–329.
Putiyevsky E. & J. Katznelson (1970) 'Chromosome Number and Genetic System in Some Clovers Related to *T. alexandrinum* L.', *Chromosoma*, 30 : 476–482.
–(1972) 'Cytology and Crossability of Several Mediterranean *Trifolium* Species', *Israel J. Bot.*, 21 : 179–181.
–(1973) 'Cytogenetic Studies in *Trifolium* spp. Related to Berseem, I : Intra- and Interspecific Hybrid Seed Formation', *Theor. Appl. Genet.*, 43 : 351–358.
Saubert S. & J. G. Scheffer (1967) 'Strain Variation and Host Specificity of *Rhizobium*, 3 : Host Specificity of *Rhizobium trifolii* on African Clovers', *S. Afr. Jour. Agric. Sci.*, 10(2) : 357–364.

Savi G. (1808–1810) *Observationes in varias* Trifolium *species*, Piatti Florentinae, 118 pp., 1 pl.
Schulz O. E. (1901) 'Monographie der Gattung *Melilotus*', *Bot. Jahrb.*, 29 : 660–735.
Seringe N. (1825) '*Trifolium*', in : De Candolle A. P., *Prodromus Systematis Naturalis Regni Vegetabilis*, Vol. 2 : 189–207, Parisiis.
Sokolovskaja A. P. & O. S. Strelkova (1948) 'Geograficheskoye raspredelenie poliploidov, III : Issledovanie flory al'pujskoj oblasti tsentralnovo Kavkazskovo chrebta', *Uchenye Zapiski LGU*, 66 : 195–216.
Stebbins G. L. (1950) *Variation and Evolution in Plants*, New York.
–(1971) *Chromosomal Evolution of Higher Plants*, Edward Arnold Ltd., London, 216 pp.
–(1974) *Flowering Plants, Evolution above the Species Level*, Cambridge, Mass.
Tatuno S. & A. Kodama (1965) 'Cytological Studies on Root Nodules of Some Species in Leguminosae, I', *Bot. Mag. Tokyo*, 78 : 503–509.
Taubert P. (1896) '*Trifolium*', in : Engler und Prantl, *Die natürlichen Pflanzenfamilien*, W. Engelmann, Leipzig, Vol. III, 3 : 249–254.
Taylor N. L. , K. H. Quesenberry & M. K. Anderson (1979) 'Genetic System Relationships in *Trifolium*', *Econ. Bot.*, 33 : 431–441.
Thulin M. (1976) 'Two New Species of *Trifolium* from Ethiopia', *Bot. Notiser*, 129 : 167–171.
Tolmachev A. (Israel Program for Scientific Translation (1969) pp. 285–286).
True G. H. (1974) 'Lemmon's Clover Rediscover (*T. lemmonii*)', *Four Seasons*, 4(4) : 22–23.
Tschechov W. (1930) 'Karyologisch-systematische Untersuchung des Tribus *Galegeae*. Fam. Leguminosae', *Planta*, 9 : 673–680.
Turner B. L. & O. S. Fearing (1960) 'Chromosome Numbers in the Leguminosae, III : Species of the Southwestern United States and Mexico', *Am. Jour. Bot.*, 47 : 603–608.
Ulbrich E. (1927) *Biologie der Früchte und Samen (Karpobiologie)*, Springer, Berlin.
Uphof J. C. Th. (1968) *Dictionary of Economic Plants*, Cramer.
Van der Pijl L. (1972) *Principles of Dispersal in Higher Plants*, Springer-Verlag, Berlin.
Vicioso C. (1952–53) 'Triboles Espanoles. Revision del genero *Trifolium*', *Anal. Inst. Bot. Cavanilles*, 10 : 347–398 (1952); *ibid.*, 11 : 289–383 (1953).
Weaver N. (1965) 'Foraging Behaviour of Honeybees on White Clover', *Insects Soc.*, 12 : 231–240.
Weaver N. & R. M. Weihing (1960) 'Pollination of Several Clovers by Honey-Bees', *Agron. Jour.*, 52 : 183–185.
Wettstein F. (1935) *Handbuch der systematischen Botanik*, Vol. 2, ed. 2, Leipzig und Wien.
Wexelson H. (1928) 'Chromosome Numbers and Morphology in *Trifolium*', *Univ. Calif. Publ. Agr. Sci.*, 2 : 355–376.
Williams W. (1945) 'Varieties and Strains of Red and White Clover – British and Foreign', *Bull. Welsh Pl. Breed. Stn. Series H*, No. 16 : 1–26.
Wipf L. (1939) 'Chromosome Numbers in Root Nodules and Root-Tips of Certain Leguminosae', *Bot. Gaz.*, 101, 1 : 51–67.
Zohary M. (1937) 'Die verbreitungsökologischen Verhältnisse der Pflanzen Palästinas, I : Die antitelechorischen Erscheinungen', *Beih. Bot. Centralbl.*, 56 : 1–155.
Zohary M. (1962) *Plant Life of Palestine*, Ronald Press.
–(1970) '*Trifolium*', in : Davis P.H., *Flora of Turkey*, 3 : 384–448, Edinburgh.
–(1971, 1972a) 'A Revision of the Species of *Trifolium* Sect. *Trifolium*, I : Introduction, II, III : Taxonomic Treatment.' I. *Candollea*, 26 : 297–308; II. *Candollea*, 27 : 99–158; III. *Candollea*, 27 : 249–264.
–(1972b) '*Trifolium*', in : *Flora Palaestina*, 2 : 157–193, Jerusalem.
–(1972c) 'Origins and Evolution in the Genus *Trifolium*', *Bot. Not.*, 125(4) : 501–511.

Zohary M. & D. Heller (1970) 'The *Trifolium* Species of Sect. *Vesicaria* Crantz', *Israel J. Bot.*, 19 : 314–335.

Zorin H. , F. W. Hely & B. S. Dear (1976) 'Host Strain Relationships in Symbiosis between Hexaploid *Trifolium ambiguum* Bieb. (Caucasian Clover) and Strains of *Rhizobium trifolii* , *Austr. CSIRO Div. Pl. Ind. Field Stn Rec.*, 15 : 63–71.

INDEX

Synonyms and page references to synonyms are printed in *italics;* the page references to main entries are in **bold** type; page references to illustrations and maps are followed by an asterisk (*)

כתבי האקדמיה הלאומית הישראלית למדעים
החטיבה למדעי-הטבע

הסוג תלתן

מאת

מ׳ זהרי וד׳ הלר

ירושלים תשמ״ד